Lecture Notes in Computer Science   3375

Commenced Publication in 1973
Founding and Former Series Editors:
Gerhard Goos, Juris Hartmanis, and Jan van Leeuwen

Marco Ajmone Marsan   Giuseppe Bianchi
Marco Listanti   Michela Meo (Eds.)

Quality of Service in Multiservice IP Networks

Third International Workshop, QoS-IP 2005
Catania, Italy, February 2-4, 2005
Proceedings

 Springer

Volume Editors

Marco Ajmone Marsan
Politecnico di Torino, Dipartimento di Elettronica e IEIIT-CNR
Corso Duca degli Abruzzi 24, 10129 Torino, Italy
E-mail: ajmone@polito.it

Giuseppe Bianchi
Università di Roma Tor Vergata, Dipartimento di Ingegneria Elettrica
Via del Politecnico 1, 00133 Roma, Italy
E-mail: giuseppe.bianchi@uniroma2.it

Marco Listanti
Università di Roma "La Sapienza", Dipartimento di INFOCOM
Via Eudossiana, 18, 001864 Roma, Italy
E-mail: marco@infocom.uniroma1.it

Michela Meo
Politecnico di Torino, Dipartimento di Elettronica
Corso Duca degli Abruzzi 24, 10129 Torino, Italy
E-mail: michela.meo@polito.it

Library of Congress Control Number: 2004118330

CR Subject Classification (1998): C.2, D.2, H.4.3, K.6

ISSN 0302-9743
ISBN 3-540-24557-X Springer Berlin Heidelberg New York

Springer is a part of Springer Science+Business Media

springeronline.com

© Springer-Verlag Berlin Heidelberg 2005
Printed in Germany

Typesetting: Camera-ready by author, data conversion by Olgun Computergrafik
Printed on acid-free paper   SPIN: 11377108   06/3142   5 4 3 2 1 0

Message from the QoS-IP Organizing Committee Chairs

This volume contains the proceedings of the *3rd International Workshop on QoS in Multiservice IP Networks* (QoS-IP 2005), held in Catania, Italy, during February 2–4, 2005. In this third edition, QoS-IP was particularly successful. The 50 high-quality papers that you can enjoy in this book were selected from an open call, out of about 100 submissions. The papers cover most of the hot topics in the field of QoS provisioning in multiservice IP networks, from network architectures to routing and scheduling algorithms, from analytical models for performance evaluation to traffic characterization.

Much like previous editions, QoS-IP 2005 was organized at the end of a research program funded by the Italian Ministry of Education, University and Research; the program was named TANGO (Traffic Models and Algorithms for Next Generation IP Networks Optimization) and contributed to the implementation of methodologies for QoS design, dimensioning, configuration and management of next-generation multiservice IP networks.

Behind a successful workshop there is the tireless effort of many people who participate with enthusiasm in its organization. First of all, we are in debt to all the researchers involved in the TANGO project, who contribute to making our Italian networking community strong and lively, to the point where we could organize a relevant international event such as QoS-IP 2005. We would like to sincerely thank Maurizio Munafò, our Information System Chair, for taking care of paper handling through the information system, for contributing to publicizing the event, and for his help in the preparation of the technical program. A debt of gratitude is owed to our Finance Chair, Alfio Lombardo, and to our Local Arrangements Chair, Giovanni Schembra. They helped us with their invaluable talent and time to make this workshop a reality. Sincere thanks are due to Raffaele Bolla, our Publication Chair, and to Alfred Hofmann and all the LNCS staff: the high quality of this edition is due to their invaluable effort. Finally, many thanks are due to the members of the Technical Program Committee and to the reviewers who thoroughly helped us select the best-quality papers among many submitted manuscripts.

November 2004

Marco Ajmone Marsan
Giuseppe Bianchi
Marco Listanti
Michela Meo

Organization

General Chair

Marco Ajmone Marsan (Politecnico di Torino and IEIIT-CNR)

Program Committee

Program Committee Co-chairs

Giuseppe Bianchi (Università di Roma Tor Vergata)
Marco Listanti (Università di Roma La Sapienza)
Michela Meo (Politecnico di Torino)

Program Committee Members

Mohammed Atiquzzaman
Tulin Atmaca
Andrea Baiocchi
Mario Baldi
Roberto Battiti
Ernst Biersack
Nicola Blefari-Melazzi
Chris Blondia
Franco Callegati
Pietro Camarda
Andrew T. Campbell
Augusto Casaca
Claudio Casetti
Giovanni Colombo
Giorgio Corazza
Eugenio Costamagna
Rene Cruz
Franco Davoli
Edmundo A. de Souza e Silva
Juan Carlos De Martin
Christophe Diot
Lars Dittmann
Tien Van Do
Josè Enrìquez-Gabeiras
Serge Fdida

Erina Ferro
Enrica Filippi
Viktoria Fodor
Luigi Fratta
Andrea Fumagalli
Rossano Gaeta
Richard Gail
Ayalvadi Ganesh
Stefano Giordano
Weibo Gong
Annie Gravey
Enrico Gregori
Mounir Hamdi
Boudewijn Haverkort
Paola Iovanna
László Jereb
Daniel Kofman
Anurag Kumar
T.V. Lakshman
Luciano Lenzini
Emilio Leonardi
Christoph Lindemann
Renato Lo Cigno
Saverio Mascolo
Francesco Masetti-Placci

Marco Mellia
Sándor Mólnar
Masayuki Murata
Ilkka Norros
Antonio Nucci
Sara Oueslati
Elena Pagani
Michele Pagano
Sergio Palazzo
Achille Pattavina
Krzysztof Pawlikowski
Xavier Pérez-Costa
Ramón Puigjaner
Guy Pujolle
Peter Rabinovitch
Carla Raffaelli
Gian Paolo Rossi
Aldo Roveri
Harry Rudin
Roberto Sabella

Stefano Salsano
Fausto Saluta
M.Y. Sanadidi
Andreas Schrader
Matteo Sereno
Dimitris Serpanos
Aleksandra Smiljanic
Josep Solè-Pareta
Heinrich Stuettgen
Tatsuya Suda
Guillaume Urvoy-Keller
Nina Taft
Yutaka Takahashi
Tetsuya Takine
Asser Tantawi
Phuoc Tran-Gia
Kurt Tutschku
Adam Wolisz
Bulent Yener

Organizing Committee

Alfio Lombardo (Università di Catania)
Giovanni Schembra (Università di Catania)

Sponsoring Institutions

Alcatel
Ericsson
Fastweb
IEIIT-CNR
Marconi Selenia Communications
STMicroelectronics

Table of Contents

MPLS Failure and Restoration

Network Planning and Dimensioning

DiffServ and IntServ

Routing I

Software Routers

Network Architectures for QoS Provisioning

Routing II

Multiservice Wireless Networks

TCP in Special Environments

Scheduling

Author Index . . . .

Of Mice and Models

Marco Ajmone Marsan, Giovanna Carofiglio, Michele Garetto, Paolo Giaccone,
Emilio Leonardi, Enrico Schiattarella, and Alessandro Tarello

Dipartimento di Elettronica, Politecnico di Torino, C.so Duca degli Abruzzi 24, Torino, Italy*
firstname.lastname@polito.it

Abstract. Modeling mice in an effective and scalable manner is one of the main challenges in the performance evaluation of IP networks. Mice is the name that has become customary to identify short-lived TCP connections, that form the vast majority of packet flows over the Internet. On the contrary, long-lived TCP flows, that are far less numerous, but comprise many more packets, are often called elephants. Fluid models were recently proved to be a promising effective and scalable approach to investigate the dynamics of IP networks loaded by elephants. In this paper we extend fluid models in such a way that IP networks loaded by traffic mixes comprising both mice and elephants can be studied. We then show that the newly proposed class of fluid models is quite effective in the analysis of networks loaded by mice only, since this traffic is much more critical than a mix of mice and elephants.

1 Introduction

The traffic on the Internet can be described either at the packet level, modeling the dynamics of the packet generation and transmission processes, or at the flow level, modeling the start times and durations of sequences of packet transfers that correspond to (portions of) a service requested by an end-user. Examples of flows can be either the train of packets corresponding to an Internet telephone call, or the one corresponding to the download of a web page. In the latter case, the flow can be mapped onto a TCP connection. This is the most common case today in the Internet, accounting for the vast majority of traffic. The number of packets in TCP connections is known to exhibit a heavy-tailed distribution, with a large number of very small instances (called TCP mice) and few very large ones (called elephants).

Models for the performance analysis of the Internet have been traditionally based on a packet-level approach for the description of Internet traffic and of queuing dynamics at router buffers. Packet-level models allow a very precise description of the Internet operations, but suffer severe scalability problems, such that only small portions of real networks can be studied.

Fluid models have been recently proposed as a *scalable* approach to describe the behavior of the Internet. Scalability is achieved by describing the network and traffic dynamics at a higher level of abstraction with respect to traditional discrete packet-level models. This implies that the short-term random effects typical of the packet-level network behavior are neglected, focusing instead on the longer-term deterministic flow-level traffic dynamics.

* This work was supported by the Italian Ministry for Education, University and Research, through the FIRB project TANGO.

M. Ajmone Marsan et al. (Eds.): QoS-IP 2005, LNCS 3375, pp. 15–32, 2005.

In fluid flow models, during their activity period, traffic sources emit a continuous information stream, the fluid flow, which is transferred through a network of fluid queues toward its destination (also called sink).

The dynamics of a fluid model, which are continuous in both space and time, are naturally described by a set of ordinary differential equations, because of their intrinsic deterministic nature.

Fluid models were first proposed in [1–4] to study the interaction between TCP elephants and a RED buffer in a packet network consisting of just one bottleneck link, either ignoring TCP mice [1–3], or modeling them as unresponsive flows [4] introducing a stochastic disturbance. In this case, fluid models offer a viable alternative to packet-based simulators, since their complexity (i.e., the number of differential equations to be solved) is independent of both the number of TCP flows and the link capacity. In Section 2 we briefly summarize the fluid models proposed in [1–3], which constitute the starting point for our work.

Structural properties of the fluid model solution were analyzed in [5], while important asymptotic properties were proved in [6, 7]. In the latter works, it was shown that fluid models correctly describe the limiting behavior of the network when both the number of TCP elephants and the bottleneck link capacity jointly tend to infinity.

The single bottleneck model was then extended to consider general multi-bottleneck topologies comprising RED routers in [3, 8].

In all cases, the set of ordinary differential equations of the fluid model are solved numerically, using standard discretization techniques.

An alternative fluid model was proposed in [9, 10] to describe the dynamics of the average window for TCP elephants traversing a network of drop-tail routers. The behavior of such a network is pulsing: congestion epochs, in which some buffers are overloaded (and overflow), are interleaved to periods of time in which no buffer is overloaded and no loss is experienced, due to the fact that previous losses forced TCP sources to reduce their sending rate. In such a setup, a careful analysis of the average TCP window dynamics at congestion epochs is necessary, whereas sources can be simply assumed to increase their rate at constant speed between congestion epochs. This behavior allows the development of differential equations and an efficient methodology to solve them. Ingenious queueing theory arguments are exploited to evaluate the loss probability during congestion epochs, and to study the synchronization effect among sources sharing the same bottleneck link. Also in this case, the complexity of the fluid model analysis is independent of both the link capacities and the number of TCP flows.

Extensions that allow TCP mice to be considered are outlined in [9, 10] and in [8]. In this case, since the dynamics of TCP mice with different size and/or different start times are different, each mouse must be described with two differential equations; one representing the average window evolution, and one describing the workload evolution. As a consequence, one of the nicest properties of fluid models, the insensitivity of complexity with respect to the number of TCP flows, is lost.

In [11] a different description of the dynamics of traffic sources is proposed, that exploits *partial* differential equations to analyze the asymptotic behavior of a large number of TCP elephants through a single bottleneck link fed by a RED buffer.

In [12] we built on the approach in [11], showing that the partial differential equation description of the source dynamics allows the natural representation of mice as well as elephants, with no sacrifice in the scalability of the model.

The limiting case of an infinite number of TCP mice is considered in [13], where it is proved that, even in the case of loads lower than 1, deterministic synchronization effects may lead to congestion and to packet losses.

When the network workload is composed of a finite number of TCP mice, normally the link loads (given by the product of the mice arrival rate times the average mice length) are well below link capacities. In this case, the deterministic nature of fluid models leads to predict that buffers are always empty, and this fact contradicts the observations made on real packet networks. This discrepancy is due to the fact that, in underload conditions, the stochastic nature of the input traffic plays a fundamental role in the network dynamics, which cannot be captured by the determinism of fluid models.

In [12] we first discussed the possibility of integrating stochastic aspects within fluid models. We proposed a preliminary solution to the problem, exploiting a hybrid fluid-Montecarlo approach. In [14] we further investigate the possibilities for the integration of randomness in fluid models, proposing two additional approaches, which rely on second-order Gaussian approximations of the stochastic processes driving the network behavior.

In this paper, we consider the hybrid fluid-Montecarlo approach proposed in [12], further investigating the impact that different modeling choices can have in different dynamic scenarios. We present numerical results to show that the hybrid fluid-Montecarlo approach is capable of producing reliable performance predictions for networks that operate far from saturation. In addition, we prove the accuracy and the flexibility of the modeling approach by considering both static traffic patterns, from which equilibrium behaviors can be studied, and dynamic traffic conditions, that allow the investigation of transient dynamics.

2 Fluid Models of IP Networks

In this section we briefly summarize the fluid model presented in [1–3] and the extension presented in [12].

Consider a network comprising K router output interfaces, equipped with FIFO buffers, feeding links at rate C (the extension to non-homogeneous data rates is straightforward). The network is fed by I classes of TCP elephants; all the elephants within the same class follow the same route through the network, thus experiencing the same round-trip time (RTT), and the same average loss probability. At time $t = 0$ all buffers are assumed to be empty. Buffers drop packets according to their instant occupancy, as in drop tail buffers, or their average occupancy, as in RED (Random Early Detection [15]) active queue management (AQM) schemes.

2.1 Elephant Evolution Equations

In [1–3], simple differential equations were developed to describe the behavior of TCP elephants over networks of IP routers adopting a RED AQM scheme. We refer to this original model with the name MGT.

Consider the ith class of elephants; the temporal evolution of the average window of TCP sources in the class, $W_i(t)$, is described by the following differential equation:

$$\frac{dW_i(t)}{dt} = \frac{1}{R_i(t)} - \frac{W_i(t)}{2}\lambda_i(t) \tag{1}$$

where $R_i(t)$ is the average RTT for class i, and $\lambda_i(t)$ is the loss indicator rate experienced by TCP flows of class i. The differential equation is obtained by considering the fact that elephants can be assumed to be always in congestion avoidance (CA) mode, so that the window dynamics are close to AIMD (Additive Increase, Multiplicative Decrease). The window increase rate in CA mode is linear, and corresponds to one packet per RTT. The window decrease rate is proportional to the rate with which congestion indications are received by the source, and each congestion indication implies a reduction of the window by a factor two.

In [12] we extended the fluid model presented in [1–3]. In our approach, that will be named PDFM, rather than just describing the average TCP connection behavior, we try to statistically model the dynamics of the entire population of TCP flows sharing the same path. This approach leads to systems of partial derivatives differential equations, and produces more flexible models, which scale independently from the number of TCP flows.

To begin, consider a fixed number of TCP elephants. We use $P_i(w, t)$ to indicate the number of elephants of class i whose window is $\leq w$ at time t. For the sake of simplicity, we consider just one class of flows, and omit the index i from all variables. The source dynamics are described by the following equation, for $w \geq 1$:

$$\frac{\partial P(w,t)}{\partial t} = \int_w^{2w} \lambda(\alpha, t)\frac{\partial P(\alpha,t)}{\partial \alpha}\, d\alpha - \frac{1}{R(t)}\frac{\partial P(w,t)}{\partial w} \tag{2}$$

where $\lambda(w, t)$ is the loss indication rate. The intuitive explanation of the formula is the following. The time evolution of the population described by $P(w, t)$ is governed by two terms: i) the integral accounts for the growth rate of $P(w, t)$ due to the sources with window between w and $2w$ that experience losses; ii) the second term describes the decrease rate of $P(w, t)$ due to sources increasing their window with rate $1/R(t)$.

2.2 Network Evolution Equations

In both models, $Q_k(t)$ denotes the (fluid) level of the packet queue in the kth buffer at time t; the temporal evolution of the queue level is described by:

$$\frac{dQ_k(t)}{dt} = A_k(t)\left[1 - p_k(t)\right] - D_k(t) \tag{3}$$

where $A_k(t)$ represents the fluid arrival rate at the buffer, $D_k(t)$ the departure rate from the buffer (which equals C_k, provided that $Q_k(t) > 0$), and the function $p_k(t)$ represents the instantaneous loss probability at the buffer, which depends on the packet discard policy at the buffer. An explicit expression for $p_k(t)$ is given in [2] for RED buffers, while for drop-tail buffers:

$$p_k(t) = \frac{\max(0, A_k(t) - C)}{A_k(t)} \mathbb{1}_{\{Q_k(t)=B_k\}} \tag{4}$$

If $T_k(t)$ denotes the instantaneous delay of buffer k at time t, we can write:

$$T_k(t) = Q_k(t)/C_k$$

If $\mathcal{F}_k$ indicates the set of elephants traversing buffer k, $A_k^i(t)$ and $D_k^i(t)$ are respectively the arrival and departure rates at buffer k referred to elephants in class i, so that:

$$A_k(t) = \sum_{i \in \mathcal{F}_k} A_k^i(t)$$

$$\int_0^{t+T_k(t)} D_k(a)\,\mathrm{d}a = \int_0^t A_k(a)\left[1 - p_k(t)\right]\mathrm{d}a$$

$$\int_0^{t+T_k(t)} D_k^i(a)\,\mathrm{d}a = \int_0^t A_k^i(a)\left[1 - p_k(t)\right]\mathrm{d}a$$

which means that the total amount of fluid arrived up to time t at the buffer leaves the buffer by time $t + T_k(t)$, since the buffer is FIFO. By differentiating the last equation:

$$D_k^i(t + T_k(t))\left(1 + \frac{\mathrm{d}T_k(t)}{\mathrm{d}t}\right) = A_k^i(t)\left[1 - p_k(t)\right]$$

2.3   Source-Network Interactions

Consider the elephants of class i. Let $k(h, i)$ be the h-th buffer traversed by them along their path P_i of length H^i. The RTT $R_i(t)$ perceived by elephants of class i satisfies the following expression:

$$R_i\left(t + g_i + \sum_{h=1}^{H^i} T_{k(h,i)}(t_{k(h,i)})\right) = g_i + \sum_{h=1}^{H^i} T_{k(h,i)}(t_{k(h,i)}) \tag{5}$$

where g_i is the total propagation delay[1] experienced by elephants in class i, and $t_{k(h,i)}$ is the time when the fluid injected at time t by the TCP source reaches the h-th buffer along its path P_i. We have:

$$t_{k(h,i)} = t_{k(h-1,i)} + T_{k(h-1,i)}(t_{k(h-1,i)}) \tag{6}$$

Now the instantaneous loss probability experienced by elephants in class i, $p_i^F(t)$, is given by:

$$p_i^F(t) = 1 - \prod_{h=1}^{H^i} \left[1 - p_{k(h,i)}(t_{k(h,i)})\right]$$

---

[1] Equation (5) comprises the propagation delay g_i in a single term, as if it were concentrated only at the last hop. This is just for the sake of easier reading, since the inclusion of the propagation delay of each hop would introduce just a formal modification in the recursive equation of $t_{k(h,i)}$.

By omitting the class index from the notation, the loss rate indicator $\lambda(w,t)$ in our PDFM model can be computed as follows:

$$\lambda\left(w, t+R(t)\right) = \frac{w\, p^F(t)}{R(t)} \tag{7}$$

where $w/R(t)$ is the instantaneous emission rate of TCP sources, and the source window at time $t + R(t)$ is used to approximate the window at time t. Intuitively, this loss model distributes the lost fluid over the entire population, proportionally to the window size.

Finally, in both models:

$$A_k(t) = \sum_i \sum_q r^i_{qk} D^i_q(t) + \sum_i e^i_k \frac{W_i(t)}{R_i(t)} N_i \tag{8}$$

where $e^i_k = 1$ if buffer k is the first buffer traversed by elephants of class i, and 0 otherwise; r^i_{qk} is derived by the routing matrix, being $r^i_{qk} = 1$ if buffer k immediately follows buffer q along P_i.

2.4   Mice

The extension of fluid models to a finite population of mice was provided only in [12], thus, here we refer only to the PDFM model.

The model of TCP mice discussed in this section extends the PDFM model reported in [12], and takes into account the effects of the sources maximum window size W_{max}, of the TCP fast recovery mechanism which prevents from halving the window more than once in each round-trip time, of the initial slow-start phase up to the first loss, and of time-outs.

We assume flow lengths to be exponentially distributed, with average L. Thanks to the memoryless property of the exponential distribution, we can write:

$$\frac{\partial P_{nl}(w,t)}{\partial t} = -\frac{1}{R(t)} \frac{\partial P_{nl}(w,t)}{\partial w} - \frac{1}{R(t)L} \int_1^w \alpha \frac{\partial P_{nl}(\alpha,t)}{\partial \alpha}\, d\alpha$$
$$- \int_1^w \lambda(\alpha,t) \frac{\partial P_{nl}(\alpha,t)}{\partial \alpha}\, d\alpha + \frac{1}{R(t)} P_l(w,t) \tag{9}$$

$$\frac{\partial P_l(w,t)}{\partial t} = \int_1^{\min(2w, W_{max})} \lambda(\alpha,t) p_{NTO}(\alpha) \frac{\partial P_{nl}(\alpha,t)}{\partial \alpha}\, d\alpha$$
$$+ \int_1^{\min(2w, W_{max})} \lambda(\alpha,t) p_{NTO}(\alpha) \frac{\partial P_s(\alpha,t)}{\partial \alpha}\, d\alpha$$
$$+ \int_1^{W_{max}} \lambda(\alpha,t) p_{TO}(\alpha) \frac{\partial P_{nl}(\alpha,t)}{\partial \alpha}\, d\alpha$$
$$+ \int_1^{W_{max}} \lambda(\alpha,t) p_{TO}(\alpha) \frac{\partial P_s(\alpha,t)}{\partial \alpha}\, d\alpha - \frac{1}{R(t)} P_l(w,t) \tag{10}$$

$$\frac{\partial P_s(w,t)}{\partial t} = -\frac{w}{R(t)}\frac{\partial P_s(w,t)}{\partial w} - \frac{1}{R(t)L}\int_1^w \alpha\frac{\partial P_s(\alpha,t)}{\partial \alpha}\,d\alpha$$

$$-\int_1^w \lambda(\alpha,t)\frac{\partial P_s(\alpha,t)}{\partial \alpha}\,d\alpha + \gamma(t) \quad (11)$$

where $P_{nl}(w,t)$ represents the window distribution of TCP sources in congestion avoidance mode which have not experienced losses in the last round-trip time; $P_l(w,t)$ represents the windows distribution of TCP sources in congestion avoidance mode which have experienced losses in the last round-trip time; $P_s(w,t)$ represents the window distribution of TCP sources in slow-start mode; and W_{max} represents the maximum window size. Finally, $p_{TO}(w) = 1 - p_{NTO}(w) = \min(1, 3/w)$ represents the probability that packet losses induce a time-out of TCP transmitters having window size w, as proposed in [16].

The left hand side of (9) represents the variation of the number of TCP sources in congestion avoidance which at time t have window less than w and have not experienced losses in the last round-trip time. The negative terms on the right hand side of the same equation indicate that this number decreases because of the window growth (first term), of connection completions (second term), and of losses (third term). The fourth positive term indicates an increase due to the elapsing of a round trip time after the last loss.

Similarly, the left hand side of (10) represents the variation of the number of TCP sources in congestion avoidance which at time t have window less than w and have experienced losses in the last round-trip time. The positive terms on the right hand side indicate that this number increases because of losses that either do not induce or do induce a timeout (respectively, first and second, and third and fourth terms) for connections that either were in slow-start mode (second and fourth terms) or had not experienced losses in the last round-trip time (first and third terms). The fifth negative term indicates a decrease due to the elapsing of a round trip time after the last loss.

Finally, the left hand side of (11) represents the variation of the number of TCP sources that, at time t, have window less than w and are in slow-start mode. The negative terms on the right hand side indicate that this number decreases because of the window growth (first term), of connection completions (second term), and of losses (third term). The fourth positive term indicates an increase due to the generation of new connections, that always open in slow-start mode with window size equal to 1.

We wish to stress the fact that (9)-(11) provide quite a powerful tool for an efficient representation of TCP mice, since a wide range of distributions (including those incorporating long range dependence) can be approximated with a good degree of accuracy by a mixture of exponential distributions [17].

3 The Role of Randomness

The fluid models presented so far provide a deterministic description of the network dynamics at the flow level, thus departing from the traditional approach of attempting a probabilistic description of the network at the packet level by means of stochastic models, such as continuous-time or discrete-time Markov chains and queueing models.

Table 1. Correspondence of input parameters and dynamic variables in the original fluid model and in the transformed model.

	Unchanged		Multiplied by η
Input	g, L		N, B, C
parameters	p_max		$\gamma(t,l), \gamma(t), min\_th, max\_th$
Dynamic	$R(t), \lambda(w,t), p(t), W(t)$		$Q(t), D(t), A(t), P(w,t), P_{max}(t)$
variables	$p^F(t), T(t), \bar{p}_L, \bar{w}(t)$	$P_O(w,t), P_L(w,t), P(w,t,l), P_s(w,t,l), P_s(w,t)$

Deterministic fluid models were proven to correctly represent the asymptotic behavior of IP networks when the number of active TCP elephants tends to infinity [7]. Indeed, when considering scenarios with only elephants, randomness, which is completely lost in fluid models, plays only a minor role, because links tend to be heavily congested, and the packet loss rate is determined by the load that TCP connections offer in excess of the link capacity.

Deterministic fluid models also exhibit nice invariant properties, as proven in [5, 18]. In particular, the whole set of equations describing our PDFM model is invariant under the linear transformation of parameters and variables summarized in Table 1. The top rows of Table 1 report the transformations which map the original network parameters into those of the transformed network, being $\eta \in I\!\!R^+$ the multiplicative factor applied to the model parameters. Basically, the transformed network is obtained from the original network by multiplying by a factor η the number of elephants and the arrival rate of mice γ, as well as all transmission capacities and buffer dimensions. Table 1 in the bottom rows also reports the transformations which relate the modeled behavior of the original network to that of the transformed network. These invariance properties are extremely interesting, since they suggest that the behavior of very large systems can be predicted by scaling down network parameters and studying small-scale systems. This result was confirmed by simulation experiments reported in [5, 18], in case of heavily congested links.

However, deterministic fluid models are not suitable for the study of network scenarios where link capacities are not saturated. In particular, fluid models fail to correctly predict the behavior of networks loaded by a finite number of TCP mice only. This fact can be explained by looking at a network with just a single bottleneck link of capacity C, loaded by TCP mice that open according to a stationary process with rate γ, and consist of a unidirectional transfer of data with arbitrary distribution and mean size L. The average utilization ρ of the bottleneck link can be expressed as

$$\rho = \frac{\gamma L}{C} \tag{12}$$

Clearly, ρ must be smaller than 1 in order to obtain a stable system behavior. Neglecting pathological behaviors which may appear for particular initial conditions, as shown in [13], for all values $\rho < 1$ a deterministic fluid model predicts that the buffer feeding the link is always empty, so that queueing delays and packet loss probabilities are equal to zero. This is due to the fact that deterministic models consider only the first moment of the packet arrival process at the buffer, $A(t)$, and of the opening process of TCP mice, $\gamma(t)$. By so doing, $A(t)$ remains below the link capacity C, and the buffer remains empty. Since the loss probability is zero, no retransmissions are needed, and the actual traffic intensity on the link converges to the nominal link utilization ρ computed by (12). This prediction is very far from what is observed in either *ns-2* simulation experiments or measurement setups, that show how queuing delays and packet losses are non-negligible for a wide range of values $\rho < 1$. This prediction error is essentially due to the fact that, in underload conditions, randomness plays a fundamental role that cannot be neglected in the description of the network dynamics. Indeed, randomness impacts the system behavior at two different levels:

– **Randomness at Flow Level** is due to the stochastic nature of the arrival and completion processes of TCP mice. The number of active TCP mice in the network varies over time, and the offered load changes accordingly.
– **Randomness at Packet Level** is due to the stochastic nature of the arrival and departure processes of packets at buffers. In particular, the burstiness of TCP traffic is responsible for high queueing delays and sporadic buffer overloads, even when the average link utilization is much smaller than 1.

3.1 The Hybrid Fluid-Montecarlo Approach

The approach we propose to account for randomness consists in transforming the deterministic differential equations of the fluid model into stochastic differential equations, which are then solved using a Montecarlo approach.

More in detail, we consider two levels of randomness:

– **Randomness at Flow Level.** The deterministic mice arrival rate $\gamma(t)$ in (10) is replaced by a Poisson counter with average $\gamma(t)$. The deterministic mice completion process can be replaced by an inhomogeneous Poisson process whose average at time t is represented by the sum of the second terms on the right hand side of (9) and (11). As opposed to the mice arrival process, which is assumed to be exogenous, the mice completion process depends on the network congestion: when the packet loss probability increases, the rate at which mice leave the system is reduced.
– **Randomness at Packet Level.** The workload emitted by TCP sources, rather than being a continuous deterministic fluid process with rate $W_i(t)N_i/R_i(t)$, can be taken to be a stochastic point process with the same rate. Previous work in TCP modeling [19, 20] showed that an effective description of TCP traffic in networks with high bandwidth-delay product is obtained by means of a batched Poisson process in which the batch size is distributed according to the window size of TCP sources. The intuition behind this result is that, if the transmission time for all packets in a window is much smaller than the round trip time, packets transmissions are clustered at the beginning of each RTT. This introduces a high correlation in the inter-arrival times of packets at routers, that cannot be neglected, since it heavily impacts the buffer behavior. By grouping together all packets sent in a RTT as a single batch, we are able to capture most of the bursty nature of TCP traffic. This approach is very well suited to our model, which describes the window size distribution of TCP sources.

An important observation is that the addition of randomness in the model destroys the invariance properties described in Section 3. This can be explained very simply by considering that the loss probability predicted by any analytical model of a finite queue with random arrivals (e.g., an M/M/1/B queue) depends on the buffer size, usually in a non linear way. Instead, according to the transformations of Table 1, the loss probability should remain the same after scaling the buffer size. In underload conditions, it is clearly wrong to assume that the packet loss probability does not change with variations in the buffer size.

4   Results for Mice

In this section we discuss results for network scenarios comprising TCP mice. First, we investigate the impact of the source emission model in a scenario where only mice are active. Second, we study the impact of the flow size distribution. Third, we investigate the invariance properties of the network when mice are present. Finally we study a dynamic (non-stationary) scenario in which a link is temporarily overloaded by a flash crowd of new TCP connection requests.

4.1   Impact of the Source Emission Model

Consider a single bottleneck link fed by a drop-tail buffer, with capacity equal to 1000 packets. The link data rate C is 1.0 Gbps, while the propagation delay between TCP sources and buffer is 30 ms. In order to reproduce a TCP traffic load close to what has been observed on the Internet, flow sizes are distributed according to a Pareto distribution with shape parameter equal to 1.2 and scale parameter equal to 4.

Using the algorithm proposed in [17], we approximated the Pareto distribution with a hyper-exponential distribution of the 9-th order, whose parameters are reported in Table 2. The resulting average flow length is 20.32 packets. Correspondingly, 9 classes of TCP mice are considered in our model. The maximum window size is set to 64 packets for all TCP sources. Experiments with loads equal to 0.6, 0.8 and 0.9 were run; however, for the sake of brevity, we report here only the results for load equal to 0.9.

Fig. 1 compares the queue lengths distributions obtained with *ns-2*, and with the stochastic fluid model. While in the model the flow arrival and completion processes have been randomized according to a non-homogeneous Poisson process (see Section 3.1), different approaches have been tried to model the traffic emitted by sources:

- *Poisson:* the emitted traffic is a Poisson process with time-varying rate;
- *Det-B:* the emitted traffic is a batch Poisson process with time-varying rate and constant batch size, equal to the instantaneous average TCP mice window size;
- *Exp-B:* the emitted traffic is a batch Poisson process with time-varying rate and exponential batch size, whose mean is equal to the instantaneous average TCP mice window size;
- *Win-B:* the emitted traffic is a batch Poisson process with time-varying rate, in which the batch size distribution is equal to the instantaneous TCP mice window size distribution.

The last three approaches were suggested by recent results about the close relationship existing between the burstiness of the traffic generated by mice and their window size [19].

Table 2. Parameters of the hyper-exponential distribution approximating the Pareto distribution.

Prob.	mean length
$7.88 \; 10^{-1}$	6.48
$1.65 \; 10^{-1}$	23.26
$3.70 \; 10^{-2}$	80.65
$8.34 \; 10^{-3}$	279.7
$1.87 \; 10^{-3}$	970.2
$4.22 \; 10^{-4}$	3376
$9.46 \; 10^{-5}$	11862
$2.10 \; 10^{-5}$	43086
$4.52 \; 10^{-6}$	176198

If we use a Poisson process to model the instants in which packets (or, more precisely, units of fluid) are emitted by TCP sources, the results generated by the fluid model cannot match the results obtained with the *ns-2* simulator, as can be observed in Fig. 1. Instead, the performance predictions obtained with the fluid model become quite accurate when the workload emitted by TCP sources is taken to be a Poisson process with batch arrivals. The best fitting (confirmed also by several other experiments, not reported here for lack of space) is obtained for batch size distribution equal to the instantaneous TCP mice window size distribution (case *Win-B*). Note that our proposed class of fluid models naturally provides the information about the window size distribution, whereas the MGT model provides only the average window size.

Table 3 reports the average loss probability, the average queue length, and the average completion time for each class of TCP mice, obtained with *ns-2*, with the *Poisson* and the *Win-B* models. The *Poisson* model significantly underestimates the average queue length and loss probability, thus producing an optimistic prediction of completion times. The *Win-B* model moderately overestimates the average queue length and loss probability, as pointed out in [19]. However, for very short flows, completion time predictions obtained with the *Win-B* model are slightly optimistic; this is mainly due to the fact that an idealized TCP behavior (in particular, without timeouts) is considered in the model.

4.2   Impact of the Flow Size

We now discuss the ability of our model to capture the impact on the network behavior of the flow size variance.

We consider three different scenarios, in which flow lengths are distributed according to either an exponential distribution ("Distr.1"), or hyper-exponentials of the second order ("Distr.2" and "Distr.3"). For all three scenarios, we keep the average flow size equal to 20.32 (this is the average flow size used in the previous subsection), and we vary the standard deviation σ. Detailed parameters of our experiments are reported in Table 4.

Table 5 shows a comparison between the results obtained with the *Win-B* model and with *ns-2*. As in previous experiments, the model moderately overestimates both the average loss probability and the average queue length. The discrepancies in the average completion times between model and *ns-2* remain within 10%.

Fig. 2, which reports the queue length distributions obtained by the model in the three scenarios, emphasizes the significant dependency of the queue behavior on the flow size variance. This dependency is mainly due to the complex interactions between the packet level and the flow level dynamics which are due to the TCP protocol.

4.3   Impact of the Link Capacity

Finally, we discuss the effect on performance of the link capacity. The objective of this last study of networks loaded with TCP mice only is to verify whether the performance of networks which differ for a multiplicative factor in capacities show some type of invariance, like in the case of elephants.

More precisely, we wish to determine whether the queue length distribution exhibits any insensitivity with respect to the bottleneck link capacity, for the same value of the traffic intensity. This curiosity is motivated by the fact that in many classical queuing models (e.g. the M/M/1 queue, possibly with batch arrivals) the queue length distribution depends only on the average load, not on the server speed.

We consider the third scenario ("Distr.3") of the previous experiment, we fix the traffic load at 0.9, and we study four different networks, in which the bottleneck capacity is equal to 10 Mbps, 100 Mbps, 1 Gbps, 10 Gbps, respectively.

The results of the fluid model, depicted in Fig. 3, show that, in general, the queue length distribution exhibits a dependency on the link capacity. The packet level behavior, indeed, strongly depends on flow level dynamics, which cause a slowly varying modulation of the arrival rate at the packet level. The flow level dynamics, however, do not scale up with the capacity of the system, since the random variable which represent the number of active flows has a coefficient of variation which decreases as we increase the system capacity (consider, for example, the Poisson distribution of the number of active flows proposed in [17]).

Nevertheless, when the capacity of the system becomes very large (in the considered example, greater than 1 Gbps) the dependence of the queue distribution on capacity tends to vanish, and the queueing behavior becomes indeed independent from the link capacity. This phenomenon was confirmed by *ns-2* simulations.

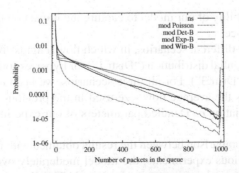

Fig. 1. Queue length distribution for single drop tail bottleneck, varying the random process modeling the workload emitted by the TCP sources; comparison with *ns-2* simulator.

Table 3. Average loss probability (ALP), average queue length (AQL) and average completion times (ACT) in seconds of the nine classes of mice for the setup of Section 4.1.

	ALP	AQL	ACT[s]
Poisson	$1.23\ 10^{-6}$	17.61	0.0932, 0.129, 0.169 0.274, 0.613, 1.68 5.48, 19.2, 73.7
win Poisson	$1.22\ 10^{-4}$	143.12	0.0991, 0.138, 0.187 0.297, 0.658, 1.92 6.29, 22.4, 95.1
ns-2	$5.34\ 10^{-5}$	101.63	0.104, 0.160, 0.219 0.327, 0.661, 1.83 6.10, 21.3, 87.0

Table 4. Parameters of the three flow length distributions.

	σ	mean length 1	mean length 2
Distr. 1	20.32	20.32	-
Distr. 2	28.89	6.48	80.65
Distr. 3	215.51	6.48	3376.24

This behavior is mainly due to the fact that when the capacity becomes very large, the coefficient of variation of the number of active flows becomes small. As a consequence, the effects of the flow level dynamics on the network performance tend to become negligible, and the packet-level behavior resembles that of a single server queue loaded by a stationary Poisson (or batched Poisson) process, for which the queue length distribution is independent of the server capacity.

To confirm this intuition, we solved the fluid model by eliminating the randomness at the flow level (i.e., in the flow arrival and departure processes), and we observed that the dependency on the capacity disappears.

We would like to remark, however, that the flow length distribution plays a major role in determining the system capacity above which the queue length distribution no longer depends on the system capacity – the invariance phenomenon appears at higher data rates when the variance of the flow length distribution increases.

4.4   Flash Crowd Scenario

We now study a non-stationary traffic scenario, in which a network section is temporarily overloaded by a flash crowd of new TCP connection requests.

Consider a series of two links at 100 Mb/s, as shown in Figure 4. The first link is fed by a drop-tail buffer, while the second link is fed by a RED buffer. Both buffers can store up to 320 packets. As shown in Figure 4, three classes of TCP mice traverse this network section. Mice of class 1 traverse only the first link, mice of class 2 traverse only the second link, and mice of class 3 traverse both links. The length of mice in all three classes is geometrically distributed with mean 50 packets. Mice of classes 1 and 3 offer a stationary load equal to 45 and 50 Mb/s, respectively. Mice of class 2 offer a

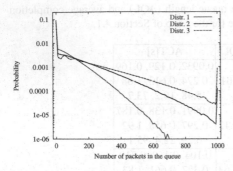
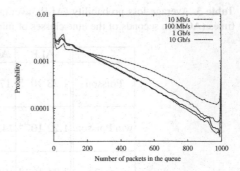

Fig. 2. Queue size distribution for single drop tail bottleneck, varying the flow length distribution.

Fig. 3. Queue size distribution for single drop tail bottleneck, varying the bottleneck capacity.

time-varying load that follows the profile shown in Figure 5, and determine a temporary overload of the second queue between $t_1 = 50$ s and $t_2 = 95$ s.

Figures 6 and 8 report, respectively, the queue lengths and the source window sizes versus time, averaging values obtained during intervals of duration 1 s for a better representation. For the sake of comparison, Figures 7 and 9 report the same performance indices obtained with *ns-2*. First of all, it is necessary to emphasize the fact that the reported curves refer to sample paths of the stochastic processes corresponding to the network dynamics, not to averages or moments of higher order; thus, the curves are the result of particular values taken by the random variables at play, and the comparison between the *ns-2* and the model results requires a great amount of care. In spite of this fact, we can observe a fairly good agreement between the *ns-2* and the model predictions. For example, in both cases it is possible to see that congestion arises at the second queue as soon as the offered load approaches 1, i.e., around $t = 50$ s. After a while, congestion propagates back to the first queue, due to retransmissions of mice of class 3, which cross both queues. At time $t = 100$ s, when the offered load at the second buffer decreases below 1, congestion ends at the second queue, but persists until $t = 160$ s at the first queue, due to the large number of active mice of class 3.

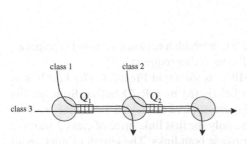
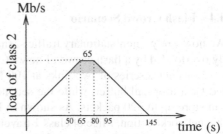

Fig. 4. Network topology of the flash crowd scenario.

Fig. 5. Temporal evolution of the load produced by the flash crowd.

Table 5. Average loss probability (ALP), average queue length (AQL) and average completion times (ACT) in seconds of the different classes of mice for the setup of Section 4.2, having introduced random elements.

	ALP	AQL	ACT[s]
Distr. 1 (model)	0.0	63.9	0.131
Distr. 1 (ns)	0.0	40.9	0.128
Distr. 2 (model)	$4.01 \ 10^{-5}$	123	0.0985, 0.185
Distr. 2 (ns)	$9.00 \ 10^{-6}$	98.0	0.0878, 0.191
Distr. 3 (model)	$3.29 \ 10^{-4}$	167	0.0999, 2.01
Distr.3 (ns)	$1.23 \ 10^{-4}$	142	0.0915, 1.85

For the sake of comparison, in this case we also show in Figure 10 the predictions generated by the fully deterministic fluid model. The fact that in this case the network becomes overloaded allows the deterministic fluid model to correctly predict congestion. However, the deterministic fluid model predicts a behavior at the first queue after the end of the overload period which is different from *ns-2* observations. Indeed, the fluid model predicts a persistent congestion for the first queue, which gets trapped into a "spurious" equilibrium point. This phenomenon is not completely surprising, and confirms the results presented in [13], where it was recently proved that asymptotic mean

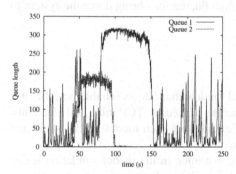

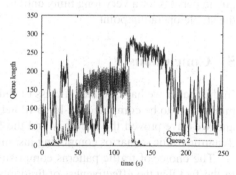

Fig. 6. Queue size evolution obtained with the Fluid-Montecarlo approach.

Fig. 7. Queue size evolution obtained with *ns-2*.

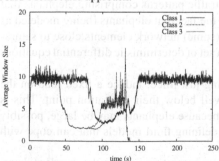

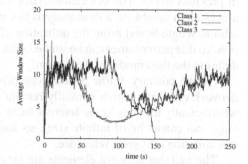

Fig. 8. Window size evolution obtained with the Fluid-Montecarlo approach.

Fig. 9. Window size evolution obtained with *ns*.

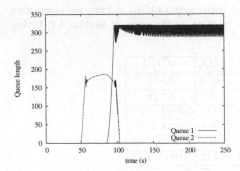
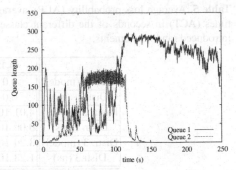

Fig. 10. Queue size evolution obtained with the deterministic fluid model.

Fig. 11. *ns-2* sample path showing persistent congestion at queue 1.

field models of networks loaded by a large number of short-lived TCP flows may exhibit permanent congestion behaviors also when the load is less than 1. The existence of two equilibrium points at loads close to 1, one stable and the other unstable, was also described in [21]. This pseudo-chaotic dynamics of deterministic fluid models reflects a behavior that can be observed in *ns-2* simulations. See for example the *ns-2* sample path reported in Figure 11, for the same scenario. In this case congestion at the first queue persists for a very long time, until random fluctuations bring down the system to the stable operating point.

5   Conclusions

In this paper we have defined a class of fluid models that allows reliable performance predictions to be computed for large IP networks loaded by TCP mice and elephants, and we have proved the accuracy and the flexibility of such models under static and dynamic traffic scenarios comprising just mice.

The choice of traffic patterns comprising just mice in the model validation is due to the fact that the effectiveness of fluid models in the performance analysis of IP networks loaded by just elephants or by mixes of elephants and mice was already proved in previous works. The key characteristic of traffic patterns comprising elephants that makes them suitable to a fluid analysis lies in the fact that elephants (being modeled as infinite-length flows) bring the utilization of (some) network elements close to saturation, so that performance can be studied with a set of deterministic differential equations defining the fluid model of the system.

On the contrary, traffic patterns comprising just mice, induce a fixed load on the network elements, which normally remain well below their saturation point. This is what actually happens in the Internet today (because elephants may be large, possibly huge, but cannot be of infinite size), so that defining fluid models that can cope with this situation is of great relevance.

The fact that network elements are far from saturation, makes the standard deterministic fluid models useless for performance analysis, because they predict all buffers to be deterministically empty, and all queuing delays to be deterministically zero. This is not what we experience and measure from the Internet. The reason for this discrepancy lies in the stochastic nature of traffic, which is present on the Internet, but is not reflected by the standard deterministic fluid models.

The fluid model paradigm that we defined in this paper is capable of accounting for randomness in traffic, at both the flow and the packet levels, and can thus produce reliable performance predictions for networks that operate far from saturation.

The accuracy and the flexibility of the modeling paradigm was proved by considering both static traffic patterns, from which equilibrium behaviors can be studied, and dynamic traffic conditions, that allow the investigation of transient dynamics.

References

1. V.Misra, W.Gong, D.Towsley, "Stochastic Differential Equation Modeling and Analysis of TCP Window Size Behavior", *Performance'99*, Istanbul, Turkey, October 1999.
2. V.Misra, W.B.Gong, D. Towsley, "Fluid-Based Analysis of a Network of AQM Routers Supporting TCP Flows with an Application to RED", *ACM SIGCOMM 2000,* Stockholm, Sweden, August 2000.
3. Y.Liu, F.Lo Presti, V.Misra, D.Towsley, "Fluid Models and Solutions for Large-Scale IP Networks", *ACM SIGMETRICS 2003*, San Diego, CA, USA, June 2003.
4. C.V.Hollot, Y.Liu, V.Misra and D.Towsley, "Unresponsive Flows and AQM Performance," *IEEE Infocom 2003*, San Francisco, CA, USA, March 2003.
5. R.Pan, B.Prabhakar, K.Psounis, D.Wischik, "SHRiNK: A Method for Scalable Performance Prediction and Efficient Network Simulation", *IEEE Infocom 2003*, San Francisco, CA, USA, March 2003.
6. S.Deb, S.Shakkottai, R.Srikant, "Stability and Convergence of TCP-like Congestion Controllers in a Many-Flows Regime", *IEEE Infocom 2003*, San Francisco, CA, USA, March 2003.
7. P.Tinnakornsrisuphap, A.Makowski, "Limit Behavior of ECN/RED Gateways Under a Large Number of TCP Flows", *IEEE Infocom 2003*, San Francisco, CA, USA, March 2003.
8. M.Barbera, A.Lombardo, G.Schembra, "A Fluid-Model of Time-Limited TCP flows", to appear on *Computer Networks*.
9. F.Baccelli, D.Hong, "Interaction of TCP Flows as Billiards", *IEEE Infocom 2003*, San Francisco, CA, USA, March 2003.
10. F.Baccelli, D.Hong, "Flow Level Simulation of Large IP Networks", *IEEE Infocom 2003*, San Francisco, CA, USA, March 2003.
11. F.Baccelli, D.R.McDonald, J.Reynier, "A Mean-Field Model for Multiple TCP Connections through a Buffer Implementing RED", *Performance Evaluation,* vol. 49 n. 1/4, pp. 77-97, 2002.
12. M.Ajmone Marsan, M.Garetto, P.Giaccone, E.Leonardi, E.Schiattarella, A.Tarello, "Using Partial Differential Equations to Model TCP Mice and Elephants in Large IP Networks", *IEEE Infocom 2004,* Hong Kong, March 2004.
13. F.Baccelli, A.Chaintreau, D.Mc Donald, D.De Vleeschauwer, "A Mean Field Analysis of Interacting HTTP Flows", *ACM SIGMETRICS 2004*, New York, NY, June 2004.
14. G.Carofiglio, E.Leonardi, M.Garetto, M.Ajmone Marsan, A.Tarello, "Beyond Fluid Models: Modelling TCP Mice in IP Networks with Non-Stationary Random Traffic", submitted for publication.
15. S.Floyd, V.Jacobson, "Random Early Detection Gateways for Congestion Avoidance", *IEEE/ACM Transactions on Networking*, vol. 1, n. 4, pp. 397-413, August 1993.

16. J.Padhye, V.Firoiu, D.Towsley, and J.Kurose, "Modeling TCP Throughput: A Simple Model and its Empirical Validation," *ACM SIGCOMM'98 - ACM Computer Communication Review*, 28(4):303–314, September 1998.
17. A.Feldmann, W.Whitt, "Fitting Mixtures of Exponentials to Long-Tail Distributions to Analyze Network Performance Models", *IEEE Infocom 97*, Kobe, Japan, April 1997.
18. M.Ajmone Marsan, M.Garetto, P.Giaccone, E.Leonardi, E.Schiattarella, A.Tarello, "Using Partial Differential Equations to Model TCP Mice and Elephants in Large IP Networks", Technical report available at http://www.telematics.polito.it/garetto/papers/TON-2004-leonardi.ps
19. M.Garetto, D.Towsley, "Modeling, Simulation and Measurements of Queuing Delay Under Long-Tail Internet Traffic", *ACM SIGMETRICS 2003,* San Diego, CA, USA, June 2003.
20. M.Garetto, R.Lo Cigno, M.Meo, M.Ajmone Marsan, "Modeling Short-Lived TCP Connections with Open Multiclass Queuing Networks", *Computer Networks Journal*, vol. 44, n. 2, pp. 153–176, February 2004.
21. M.Meo, M.Garetto, M.Ajmone Marsan, R.Lo Cigno, "On the Use of Fixed Point Approximations to Study Reliable Protocols over Congested Links", *IEEE Globecom 2003*, San Francisco, CA, December 2003.

Service Curve Estimation by Measurement:
An Input Output Analysis of a Softswitch Model

Luigi Alcuri, Giuseppe Barbera, and Giuseppe D'Acquisto

Dipartimento di Ingegneria Elettrica
Viale delle Scienze-Parco D'Orleans 90128 Palermo
{luigi.alcuri,giuseppe.barbera,giuseppe.dacquisto}@tti.unipa.it

Abstract. In this paper, we describe a new method to solve a methodological problem for the analysis of telecommunications systems in packet networks. The basic building block of the framework is the concept of service curve, which defines a general QoS model. In terms of min-plus theory, a system is characterized by a service curve, which is determined by the latency and the allocated rate. Unfortunately, there are no analytical expressions that compute the parameters of a service curve for a non-FIFO and work-conserving systems. Our simulation based approach, instead, can be applied to all these systems and gives an estimate of the service curve in a very simple way, with minimum information about system architecture. We applied this method to a very simple model of a Softswitch node in a NGN telephone service environment. The main result of the work is the selection of a proper rate-latency service curve.

Keywords: Network calculus, service curve, softswitch model, performance bounds, delay bounds.

1 Introduction

Catching the characteristics of a telecommunication system with a partial knowledge the system itself, that is regarding it as a blackbox, may require a complex and long analytical work. The method we provide allows to determine an estimate of the typical parameters of a system (rate and delay), with minimum information about the system and without input-output synchronization. The approach we followed is based on the concepts of the min-plus theory, also called Network Calculus, which is a theory of mainly deterministic queuing systems for computer networks. In particular, the concept of service curve, which abstracts the details of packet scheduling, is the most relevant for our study. This concept defines a general QoS model taking into account both bandwidth and delay requirements.

The paper deals with the problem of characterizing the service curve of a non-FIFO system and presents a calculation of worst-case bounds of offered service, by introducing service curve estimation and simulation. Basically, the estimation of the service curve requires only the knowledge of the input and output curves.

Our method has been tested by the simulation program Extend, a commercial tool from ImagineThat Inc [8], which has revealed helpful in reproducing the case study under observation: the performance evaluation of a softswitch node loaded by various streams of voice signaling. Signaling traffic streams and the dynamic of the system have been reproduced using Extend. In order to take into consideration the variety of

M. Ajmone Marsan et al. (Eds.): QoS-IP 2005, LNCS 3375, pp. 49–60, 2005.

services and applications with different QoS requirements which a softswitch is able to manage, a non-FIFO and work-conserving representation of this node is proposed.

Numerical results we show refer to the superposition of two types of services: the basic telephone call service and the 800 Free Number which requires access to a Database (DB) for address resolution. Furthermore, considering recent studies on the statistical characteristics of telecommunication traffic [2], in the simulations we used both self-similar and poissonian input traffic streams.

The remaining of the paper is organized as follows. In Section 2 we give some insight on the min-plus theory and its results, we define the service curve concept and derive some applications in the context of computer networks. In section 3, we present our method and describe the simulation based estimation technique for the service curve. In Section 4, we illustrate the model for a Softswitch, finally in Section 5 and 6, we show our experimental results and make a point on further developments.

2   Background

2.1   Network Calculus

Network Calculus is set of rules and results that can be used for computing tight bounds on delay, backlogs, and arrival envelopes in a lossless setting applicable to packet networks. With Network Calculus, we are able to understand some fundamental properties of integrated services networks, of window flow control, scheduling and buffer or delay dimensioning [4].

The main difference between Network Calculus, which can be regarded as the system theory that applies to computer networks, and traditional system theory, which was so successfully applied to design electronic circuits, is the algebra under which operations take place. The mathematics involved in Network Calculus uses min-plus algebra, as described in [4].

Applications of the min-plus algebra have been proposed for the analysis of communication networks. For example, the min-plus algebra is a powerful tool for analyzing discrete event systems: these can be described using linear min-plus equations, even though they are nonlinear in the sense of the conventional system theory. This is one of the most significant advantage of the Network Calculus, as the majority of communication network systems are non-linear and so their analysis is rather complex because of this inherent non-linearity.

Now, we show how Network Calculus results can be applied to derive general concepts for packet networks with guaranteed service.

2.2   Arrival and Departure Processes

We define a process to be a function of the time, which is a mapping form the real numbers into the extended non-negative real numbers. A process could count the amount of data arriving or departing to/from some network element, and in this case we may call the process an arrival process R(t) or departure process R*(t), respectively. All processes are assumed to be non-decreasing and right continuous. We shall often consider what we call causal processes, which are simply processes which are identically zero for all negative times [5].

In a system based on the fluid model, both R and R^* are continuous functions of t. In the packet-by-packet model, we assume that R increases only when the last bit of a packet is received by the server; likewise, R^* is increased only when the last bit of the packet in service leaves the server. Thus, the fluid model may be viewed as a special case of the packet-by-packet model with infinitesimally small packets.

In this sense, these processes are cumulative functions. It is convenient to describe data flows by means of the cumulative function $R(t)$, defined as the number of bits seen on the flow in time interval $[0,t]$. Considering a system S, which we view as a blackbox, that receives input data and delivers it after a variable delay; we call $R(t)$ the cumulative function that describes the input data, and $R^*(t)$ the output function, namely, the cumulative function at the output of system S.

From the input and output functions, we derive the following quantity of interest.

Definition 1 (Backlog). For a lossless system, the backlog at time t is:

$$B(t) = R(t) - R^*(t) . \tag{1}$$

The backlog is the amount of bits that are held inside the system, namely, the number of bits "in transit", assuming that we can observe input and output simultaneously. A Backlogged Period (BP) for session is any period of time during which traffic belonging to that session is continuously queued in the system [4].

If we call t_i the beginning of BP and t_a the end of BP: $R(t_i) = R^*(t_i), R(t_a) = R^*(t_a)$ and $B(t) = R(t) - R^*(t) \neq 0$ for $t \in (t_i, t_a)$.

2.3 Service Curve

We introduce the concept of service curve as a common model for a variety of network nodes. A network element can be modeled in terms of the transformation effects the element has on the stream of packets belonging to a session. The service curve is a mathematical model which specifies how the arriving stream of packets (the arrival process) is converted into departing stream (the departure process). Furthermore, the concept of service curve abstracts the details of packet scheduling.

We first introduce an operation called "min-plus convolution", which is the basis of min-plus algebra.

Definition 2 (Convolution). Given two processes f and g, the min-plus convolution of f and g is defined to be the function $f \otimes g : \Re \to \Re_+ \cup \{+\infty\}$ such that

$$(f \otimes g)(t) = \inf_{0 \leq s \leq t} \{f(t-s) + g(s)\} . \tag{2}$$

If $t < 0$, $(f \otimes g)(t) = 0$.

Definition 3 (Service Curve). Consider a system S and a flow through S with input and output functions R and R^*. We say that S offers to the flow a service curve SC if and only if SC is wide sense increasing, $SC(0) = 0$ and

$$R^*(t) \geq R(t) \otimes SC . \tag{3}$$

The most classical service curve is the "rate-latency" service curve, which has recently been considered by the IETF for traffic and quality of service specifications [5].

Definition 4 (Rate-Latency Function $\beta_{r,T}$). $\beta_{r,T}(t) = r[t-T]^+$, for some $r \geq 0$ and $T \geq 0$ and where $[x]^+ = max\{x, 0\}$. The parameter T is called "latency", the parameter r is called "rate".

With non-linear service curve, both priority (delay) and bandwidth allocation are taken into account in a unique representation.

The main result used in the concatenation of network elements is the following theorem: its advantage is that a network may be collapsed into a single element.

Theorem (Concatenation of Nodes). Assume a flow through systems S_1 and S_2 in sequence. Assume that S_i offers a service curve β_i ($i=1,2$) to the flow. Then the concatenation of the two systems offers a service curve $SC_1 \otimes SC_2$ to the flow.

Example. Consider two nodes offering each a rate-latency service curve β_{R_i, T_i}, $i=1,2$. After simple computation we have

$$\beta_{R_1, T_1} \otimes \beta_{R_2, T_2} = \beta_{min(R_1, R_2), T_1 + T_2} .\qquad(4)$$

Thus, concatenating rate-latency nodes amounts to adding the latency components and taking the minimum of the rates.

Besides the concept of service curve, the concept of envelope curve is necessary to derive performance results in the deterministic network calculus.

Definition 5 (Envelope Curve). Let $R(t, t+\tau)$ be the number of bits from a source in the time interval $[t, t+\tau]$. A function $E(t)$ is an envelope curve if [10]

$$R(t+\tau) - R(t) \leq E(t), t, \tau \geq 0 .\qquad(5)$$

The most accurate envelope curve that can be obtained is called empirical envelope, defined as

$$E(t) = sup\{R(t+\tau) - R(t)\}, t, \tau \geq 0 .\qquad(6)$$

3   Method to Estimate a Service Curve

The analysis of a telecommunications system, developed in terms of the min-plus theory, requires that the system is characterized by a service curve. We have chosen a rate-latency curve, which is determined by two parameters: the maximum delay and a constant service rate. The main result of the work is the selection of a proper rate-latency service curve, which has been proposed by IETF for the characterization of network elements in both IntServ and DiffServ scenarios.

In [6] it has been developed a general model for the analysis of traffic scheduling algorithms for FIFO system in packet networks. FIFO scheduling, is perhaps the simplest to implement.

Unfortunately, the approach in [6] cannot be applied to non-FIFO models, to our knowledge, no analytical expressions of rate-latency service curve is available for non-FIFO and work-conserving systems. The analysis has been carried out combining an analytical approach with simulations. This approach has allowed to find a closed form solution for both parameters of the service curve with minimum information about system internal architecture and no input-output synchronization.

3.1 Procedure

The procedure we adopt for the estimate of the service curve or, more precisely, its parameters, is inspired by the deductions of [6] and [7].

A session is said to be guaranteed a service curve SC, if, for any time t, there exists a time $t_i < t$, which is the beginning one of session's BP (not necessarily including t), such that the following holds

$$SC(t-t_i) \leq R^*(t_i,t) = R^*(t) - R^*(t_i) .$$

(7)

where $R^*(t_i,t)$ is the amount of service received by session during the time interval $[t_i,t]$. For packet systems, we restrict t to be packet departure times [7].

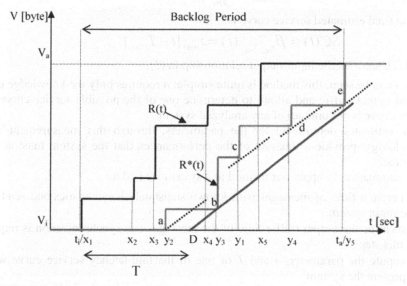

Fig. 1. The first BP. t_i = beginning of BP, t_a = end of BP, $V_i = R^*(t_i)$ and $V_a = R^*(t_a)$; x_i is i-th arrival and y_i is i-th departure

A guaranteed service curve can be derived knowing the input and the output traffic functions of our system in advance and proceeding with a backlog analysis. The method describes a simple technique to estimate rate and latency by looking at input and output traces simultaneously. Our method proceeds in the following steps, starting with the first BP (Fig. 1):

n-th backlogged period
1. Determine: t_i, t_a, $V_i = R^*(t_i)$ and $V_a = R^*(t_a)$.
2. Compute the throughput:

$$r = \frac{V(t_a) - V(t_i)}{t_a - t_i} .$$

(8)

3. Estimate the maximum value of the r of the n BPs with technique in section 3.2:

$$r_{max} = \max_{BPs}\{r\} .$$

(9)

4. If with this estimate the desired precision of estimate P is achieved, go to step 5; else, return at step 1 with (n+1)-th BP.
5. Trace the bundle of parallel straight lines with r_{max} slope, crossing the "outside" points (a, b, c, d, e in figure 1) of every BP.
6. For each BP, take the more external straight line (the blue one in the figure), compute its intersection with the horizontal axis (D in the figure) and then compute the respective delay:

$$T = D - t_i .$$ (10)

7. Compute the maximum of the delays of every available BP:

$$T_{max} = \max_{BPs}\{T\} .$$ (11)

8. The final estimated service curve is:

$$SC(t) = \beta_{r_{max}, T_{max}}(t) = r_{max}[t - T_{max}]^+ .$$ (12)

which satisfies the input-output relationship in (3).

As it can be seen, this method is quite simple; it requires only the knowledge of input and output traffic and allows to determine one of the possible service curve (the service curve is not unique) of any analyzed system.

The estimates determined for the parameters, through this measurement-based methodology, provide a measure of the performances that the system must at least guarantee.

To summarize, to apply our method to a system we had to:

− generate a flow of messages with known statistical characteristics and send it to the input system;
− determine the output traffic flow of the messages (no synchronization is required at this step);
− compute the parameters r and T of one of the rate-latency service curve which represent the system.

3.2 Maximum Estimate Technique

The procedure in section 3.1 shows a maximum estimate problem concerning the rate r. In order to solve this problem, we use a technique based on Chebyshev inequality. The scope of this technique is to obtain reliable estimates using few numbers of samples.

Let $x_{(1)}; x_{(2)}; ...; x_{(n)}$ be a random sample, that represent the rates until the n-th BP, with continuous distribution function $F(x)$ and density $f(x)$. Denote the ordered random variables $x_1; x_2; ...; x_n$, where $x_1 \le x_2 \le ... \le x_n$ and where x_n is the maximum. These ordered random variables are called *order statistics*. The density of the maximum x_n is given by [3]:

$$f_n(x) = n \cdot f(x) \cdot F^{n-1}(x) .$$ (13)

where $f(x)$ denotes the probability density function of the x_i samples and n is the number of observed samples. Further results suggest that the distribution of x_n is asymptotically normal [3].

Chebyshev inequality, which is the essence of the estimate technique, states that:

$$\Pr\{|x - m| \geq k\} \leq \frac{\sigma^2}{k^2} .$$

(14)

where m is the mean, σ^2 is the variance and k is a variable.

In our case, Chebyshev inequality has to be applied to the x_n variable, then:

$$m = \int_{x_1}^{x_n} x f_n(x)dx .$$

(15)

$$\sigma^2 = m_2 - (m)^2$$

where m2 is the mean square:

$$m_2 = \int_{x_1}^{x_n} x^2 f_n(x)dx .$$

(16)

Chosen k, an estimate of x_n at a desired precision P is obtained if $\dfrac{\sigma^2}{k^2} \leq P$, that is

if the percentage of samples falling out of the interval $[m-k, m+k]$ is less or equal to a fraction P of the n samples.

From formulae (15) and (16), it follows that the computation of m and m_2 requires $f(x)$ and $F(x)$. These two functions can be determined numerically, by partitioning the segment $[x_1, x_n]$ into a given number M of sub-interval of the same width $\Delta = \dfrac{x_n - x_1}{M}$, and observing that in the i-th sub-interval $[x_1+(i-1)\cdot\Delta, x_i+i\cdot\Delta]$ we have:

$$f^{(i)}(x) = \frac{n_i}{n\Delta} .$$

(17)

$$F^{(i)}(x) = f^{(i)}(x) \cdot (x - x_i) + \sum_{j=1}^{i-1} \Delta \cdot f^{(j)}(x) .$$

(18)

where n_i is the number of samples falling in the i-th sub-interval and i = 1...M.

In this way, we have all the data necessary to make an estimate. Once reached the desired precision, at *n-th* step, we set x_n as the estimate of the maximum of that distribution.

4 Softswitch Model

In order to apply and test our method on a Softswitch node, we carried out some discrete-event simulations in the EXTEND environment. In our simulated model, we used packets (i.e., messages) as clients.

The model chosen to represent the Softswitch (fig. 2) is based on a feedback queue system, which allows to describe the three main effects of the Softswitch:

- System capability: the CPU extracts signaling messages from the input queue with infinite buffer length, serving the packets of the flow in first in first out (FIFO) order.
- Feedback: it takes into consideration the interactions, that are due to the providing of intelligent services, between the Softswitch and the DB, mainly for address resolution.
- Delayer: the DB is represented just by a delayer, to take into account the delay caused by every interactions with the DB.

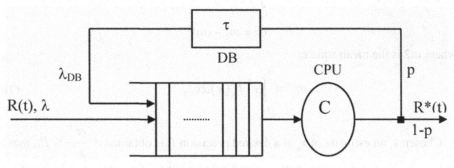

Fig. 2. Softswitch model

In this way, the Softswitch is modeled as a non-FIFO and work-conserving system: because of the presence of the feedback line, the order of the output messages differs from the order of the input messages; for this reason, the blackbox representing the Softswitch shows some "mixing phenomena", which are the effect of the interactions with the DB.

The input and output traffics are expressed in terms of min-plus theory by the input $R(t)$ and output $R^*(t)$ cumulative functions respectively. They are a superposition of flows of signaling messages from basic and enhanced telephony services.

The meaning of the parameters of the model is the following:

- λ [pkt/s] is the mean input rate of signaling messages.
- p is the percentage of messages that need an interaction with the DB.
- λ_{DB} [pkt/s] is the frequency of the messages returning in forward line after a DB-query:

$$\lambda_{DB} = \frac{p}{1-p}\lambda .$$ (19)

- C [bps] is the processing capability of the CPU:

$$C = \frac{(\lambda + \lambda_{DB})\overline{L}}{\rho} .$$ (20)

where ρ is the utilization level ($0 \le \rho \le 1$) and $\overline{L}$ the average packet size.
- τ [ms] is the DB query-response delay.

5   Results

Simulations have been carried out to assess the correctness of the estimation techniques, with a realistic traffic scenario developed only for modeling purposes. The values of system parameters λ, ρ, τ, as reported in table 1, with a fixed number N=1000 of total users of the Softswitch which can be assumed as a reasonable load for this kind of nodes.

To simplify the testing of the method, we chose only two types of service: the basic phone call service and the 800 Free Number which requires access to DB information. The values of the parameters λ, λ_{DB}, C and p are determined by direct inspection of a typical call pattern and signaling message length.

The latencies of servers computed in this section are based on the assumption that a packet is considered serviced only when its last bit has left the server.

First of all, we verified the sampling technique for the throughput r, evaluating the number of BP (N_{BP}) necessary to make an estimate of r at several values of the precision P. To this aim, we chose $k = 0.1 \cdot m$, in (14), so to guarantee a bounded relative error. From figure 3, it can be noted that N_{BP} gets higher the lower P is, at the same rate both for Poisson and self-similar traffic, and also that to achieve an excellent estimate of r ($P=10^{-6}$) some thousands of BP are sufficient.

Table 1. Values of simulation parameters

Type of users	λ [pkt/s]	$\overline{L}$ [byte]	p	λ_{DB} [pkt/s]	ρ / C [byte/s] $^{-1}$	τ [ms]
100% 800	61.11	180.18	9.1%	6.12	0.7 / 17304.73	a)50 b) 100
					0.9 / 13459.23	
50% 800 50%basic call phone	58.33	176.86	4.77%	2.92	0.7 / 15475.69	
					0.9 / 12036.65	

The delay T_{max} can be determined trace by trace, and several parameters may affect its value. Really, the delay T in our softswitch model depends at least on the following quantities: the constant DB query τ, the number of loop of feedback in a BP (N_{fb}) which occur in our traffic mix due to the percentage of packets p requiring a DB interaction, the maximum packet size of the BP (L_m) and the maximum packet size in the feedback line for each BP ($L_m^{(fb)}$). Thus, the resulting delay T_{max} depends on the particular input trace and the delay is larger the longer the trace is. We are in search of an empirical functional relation between the observed delay T, and N_{fb}, L_m, $L_m^{(fb)}$ of the BP and DB-delay τ.

Simulation shows that the law that puts these quantities together is of the form:

$$T = \frac{L_m}{C} + \alpha \cdot N_{fb} \cdot \tau + \beta \cdot \frac{L_m^{(fb)}}{C} . \qquad (21)$$

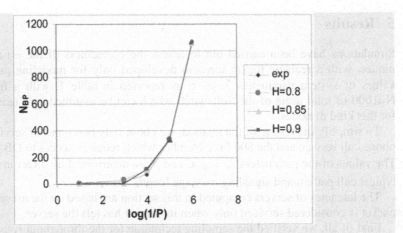

Fig. 3. N_{BP} necessary to estimate the throughput r with precision P for 4 input traces - with the same mean input rate - to the same system

where $0 \leq \alpha, \beta \leq 1$. The coefficients α and β depend on how the packets are mixed inside the BP. This dependence on the phenomena of reordering taking place in the BP can be formalized by saying that α and β are functions of a "mixing factor" $\gamma \in [0,1]$, which represent in mathematical way the mixing phenomena. Such parameter reaches its minimum when no mixing occurs, and its maximum value in correspondence with the maximum mixing, γ being a function of τ and p. Thus, the mixing phenomena – that are evident in figure 1 – in the non-FIFO systems cause a dependence that is not perfectly linear in τ, N_{fb} and $L_m^{(fb)}$, whereas T is a linear function of L_m for fixed values of the other quantities involved.

Our conjecture is confirmed by the graphics of figure 4. In graphic 4a, the approximation in (21) is represented for $N_{fb} = 0$ (and so $L_m^{(fb)} = 0$). In this case no mixing takes place ($\gamma = 0$), the system becomes FIFO, and (21) becomes $T = L_m/C$, as many analytical approaches confirm [6]. Graphic 4b represents the BPs with $N_{fb}=1$ for the same simulation before mentioned: it shows that the couples (T, L_M), where $L_M = L_m + L_m^{(fb)}$ lies within a region which for very small values of L_M has a width equal to τ and then gets wider. The region is delimited in the lower part by the straight line $T = L_m/C$ and in the upper part by the straight line $T = L_m/C + \tau + L_m^{(fb)}/C$.

Experimental results show a similar trend also at higher number of feedbacks within a busy period, we can then write the following series of inequalities:

$$\frac{L_m}{C} \leq T \leq \frac{L_m}{C} + N_{fb} \cdot \tau + \frac{L_m^{(fb)}}{C} \leq \frac{2L_m}{C} + N_{fb} \cdot \tau. \tag{22}$$

An easy upper bound on the parameter T_{max} of the service curve representing the system is:

$$T_{\max} \leq \frac{2 \cdot (L_m)_{\max}}{C} + (N_{fb})_{\max} \cdot \tau . \tag{23}$$

where $(L_m)_{\max} = \max_{BPs}\{L_m\}$ and $(N_{fb})_{\max} = \max_{BPs}\{N_{fb}\}$.

The same behavior has been observed for both Poisson and self similar input traffic at the same mean input rate. The estimates of the rate r with self-similar traffics with several values of Hurst parameter H are the same as the estimate made with exponential traffic.

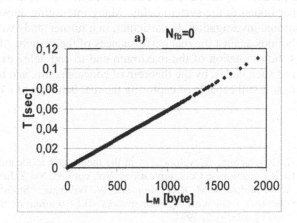

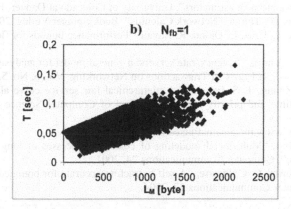

Fig. 4. Graphics *a)* and *b)* represent the BPs with N_{fb}=0 and with N_{fb}=1 respectively of the simulation with only 800 Free Number users, C=17304.73 byte/s, τ=50 msec

This simulation outcome confirms a basic property of the service curve concept, namely traffic independence. Essentially any service curve expresses through rate parameter the amount of work that a system can at least offer in a time interval, and through delay parameter the server availability, or the time that must elapse before information is processed due to priority or other system impairments (as database query or feedback loops). Numerical results show that this property holds in both long range dependent and short range dependent traffic scenarios.

6 Conclusions and Future Works

In this paper, we presented a simulation based method for studying the worst-case behavior of individual sessions in a network of non-FIFO schedulers; this method gives performance and delay bounds for a flow traversing a non-FIFO and work-conserving system. We showed that the worst-case offered service is independent of the input traffic characterization, being based only on the allocated rate and on the internal parameters of the scheduler. In particular, our worst-case analysis of the system indicates the parameters that have to be controlled in order to achieve low delays, and gives an upper bound of the system delay. Many interesting questions arise, and they require additional investigation. In particular, in a further study we will check the tightness of our bounds against the real performances of the system. Also, an interesting open issue is the estimation of the maximum end-to-end delay experienced by a packet in a chain of such nodes, by the theorem of concatenation, and to what extent a worst case approach is still capable of capturing the real behavior of a network.

References

1. P. M. Hahn, M. C. Jeruchim. "Developments in the theory and application of importance sampling", IEEE Transaction on Communications, Vol. com-35, No. 7, July 1987.
2. D. E. Duffy, A. A. McIntosh, M. Rosenstein and W. Willinger. "Statistical Analysis of CCSN/SS7 Traffic Data from Working Subnetworks" IEEE Journal on Selected Areas in Communications, Vol. 12, No. 3, 1994.
3. S. R. Sain. "Properties of estimators", University of Colorado at Denver, February 2004.
4. J. Y. Le Boudec, P. Thiran. "Network Calculus", Book Spinger Verlag, 2004.
5. R. Agrawal, R. L. Cruz, C. Okino, R. Rayan. "Performance bounds for flow control protocol", 1998.
6. D. Stiliadis, A. Varma. "Latency-rate servers: a general model for analysis of traffic scheduling algorithms", IEEE/ACM Transactions on Networking, Vol. 6, No. 5, October 1998.
7. I. Stoica, H. Zhang, T. S. Eugene. "A hierarchical fair service curve algorithm for link-sharing, real-time and priority services", School of Computer Science Carnegie Mellon University.
8. EXTEND V.6, www.imaginethatinc.com.
9. J. Gao, I. Rubin. "Multifractal modeling of counting processes of long-range dependent network traffic", Computer Communications 24, 2001.
10. M. D. de Amorim, O. C. Duarte. "A self-extracting accurate for bounded-delay video services", Computer Communications 27, 2004.

Utility Proportional Fair Bandwidth Allocation: An Optimization Oriented Approach

Tobias Harks*

Konrad-Zuse-Zentrum für Informationstechnik Berlin (ZIB),
Department Optimization, Takustr. 7, 14195 Berlin, Germany
harks@zib.de

Abstract. In this paper, we present a novel approach to the congestion control and resource allocation problem of elastic and real-time traffic in telecommunication networks. With the concept of utility functions, where each source uses a utility function to evaluate the benefit from achieving a transmission rate, we interpret the resource allocation problem as a global optimization problem. The solution to this problem is characterized by a new fairness criterion, *utility proportional fairness*. We argue that it is an application level performance measure, i.e. the utility that should be shared fairly among users. As a result of our analysis, we obtain congestion control laws at links and sources that are globally stable and provide a utility proportional fair resource allocation in equilibrium. We show that a utility proportional fair resource allocation also ensures utility max-min fairness for all users sharing a single path in the network. As a special case of our framework, we incorporate utility max-min fairness for the entire network. To implement our approach, neither per-flow state at the routers nor explicit feedback beside ECN (Explicit Congestion Notification) from the routers to the end-systems is required.

1 Introduction

In this paper, we present a network architecture that considers an application-layer performance measure, called *utility*, in the context of bandwidth allocation schemes. In the last years, there have been several papers [1–7] that interpreted congestion control of communication networks as a distributed algorithm at sources and links in order to solve a global optimization problem. Even though considerable progress has been made in this direction, the existing work focuses on elastic traffic, such as file transfer (FTP, HTTP) or electronic mail (SMTP). In [8], elastic applications are characterized by their ability to adapt the sending rates in presence of congestion and to tolerate packet delays and losses rather gracefully. From a user perspective, common to all elastic applications is the request to transfer data in a short time. To model these characteristics, we resort

---

* This work has been supported by the German research funding agency 'Deutsche Forschungsgemeinschaft' under the graduate program 'Graduiertenkolleg 621 (MAGSI/Berlin)'.

M. Ajmone Marsan et al. (Eds.): QoS-IP 2005, LNCS 3375, pp. 61–74, 2005.

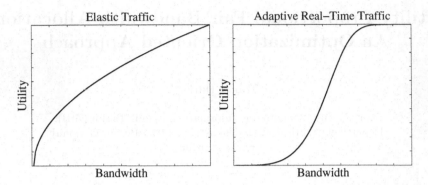

Fig. 1. Utilities for elastic traffic and adaptive real-time traffic.

to the concept of utility functions. Following [8] and [2], traffic that leads to an increasing, strictly concave (decreasing marginal improvement) utility function is called *elastic* traffic. We call such a utility function *bandwidth utility* since the utility function evaluates the benefit from achieving a certain transmission rate. The proposed source and link algorithms are designed to maximize the aggregate bandwidth utility (sum over all bandwidth utilities) subject to capacity constraints at the links. Kelly introduced in [2] the so called *bandwidth proportional fair* allocation, where bandwidth utilities are logarithmic. The algorithms at the links are based on Lagrange multiplier methods coming from optimization theory, so the concavity assumption seems to be essential. As shown in [8], some applications, especially real-time applications have non-concave bandwidth utility functions. A voice-over-IP flow, for instance, receives no bandwidth utility, if the rate is below the minimum encoding rate. Its bandwidth utility is at maximum, if the rate is above its maximum encoding rate. Hence, its bandwidth utility can be approximated by a step function. According to Shenker [8], the bandwidth utility of adaptive real-time applications can be modeled as an S-shaped utility function (a convex part at low rates followed by a concave part at higher rates) as shown in Figure 1. The paradigm of the work dealing with bandwidth utility functions of elastic applications in the context of congestion control is to maximize the bandwidth utilization of the network (bandwidth system optimum) under specific bandwidth fairness aspects (bandwidth max-min, bandwidth proportional fair).

The central part of this work is to turn the focus on fairness of user-received utility of different applications including non-elastic applications with non-concave bandwidth utility functions. A user running an application does not care about any fair bandwidth shares, as long as his application performs satisfactory. Hence, we argue that it is an application performance measure, i.e. the utility that should be shared fairly among users. To motivate this new paradigm, we refer to the concept of *utility max-min fairness* introduced by Cao and Zegura in [9]. Let us consider a network consisting of a single link of capacity one shared by two users. One user transfers data according to an elastic application with strictly increasing and concave bandwidth utility $U_1(\cdot)$. The other user trans-

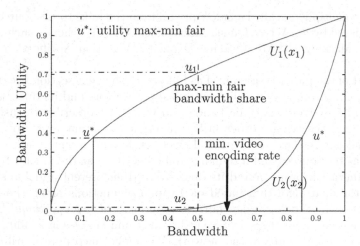

Fig. 2. Utility max-min and bandwidth max-min fairness.

fers real-time video data with a non-concave bandwidth utility function $U_2(\cdot)$. Figure 2 shows, how different bandwidth allocations affect the received utility. If the bandwidth is shared equally, what is referred to as *max-min bandwidth* allocation in this example, user 1 receives a much larger utility than user 2. Conversely, user 2 would not be satisfied since he does not receive the minimum video encoding bandwidth. If we want to share utility equally, instead of bandwidth, we would like to have a resource allocation, where the received utilities are equal or *utility max-min fair*, i.e. $U_1(x_1) = U_2(x_2) = u^*$.

In [9], Cao and Zegura present a link algorithm that achieves a utility max-min fair bandwidth allocation, where for each link the utility functions of all flows sharing that link is maintained. In [10], Cho and Song present a utility max-min architecture, where each link communicates a supported utility value to sources using that link. Then sources adapt their sending rates according to the minimum of these utility values.

In this paper, we extend the utility max-min architecture and propose a new fairness criterion, *utility proportional fairness*, which includes the utility max-min fair resource allocation as a special case. A utility proportional fair bandwidth allocation is characterized by the solution of an associated optimization problem. The benefit a user s gains when sending at rate x_s is evaluated by a new *second order utility* $F_s(x_s)$ and the objective is to maximize aggregate second order utility subject to capacity constraints. The second order utilities are assumed to be strictly concave, whereas the bandwidth utilities can be chosen arbitrarily. We only assume that the bandwidth utilities are monotonic increasing in a given interval. This is a natural assumption since any application will profit from receiving more bandwidth in a certain bandwidth interval. We emphasize, that our distributed algorithm does not need any per-flow information at the links. The feedback from links to sources does not include overhead, such as explicit utility values as done in [10]. It merely relies on the communication of Lagrange multipliers, called shadow prices, from the links to the sources. This can be achieved by an Active Queue Management (AQM) scheme, such as Random Early Marking (REM) [6] using Explicit Congestion Notification (ECN) [11].

The rest of the paper is organized as follows. In the next section, we describe our model, the second order utility optimization problem and its dual based on ideas of [1, 2, 5]. Given a specific bandwidth utility, we describe a constructive method to find the second order utility function $F_s(\cdot)$ that leads to a utility proportional fair resource allocation. In Section 3, we present a static primal algorithm at the sources and a dynamic dual algorithm at the links solving the global optimization problem and its dual. We further present a global stability result for the dual algorithm based on Lyapunov functions along the lines of [12]. In Section 4, we define a new fairness criterion, *utility proportional fairness*, and show that our algorithms achieve utility max-min fairness in equilibrium for users sharing a single path in the network. We further incorporate utility max-min fairness for the entire network as a special case of our framework. Finally, we conclude in Section 5 with remarks on open issues.

2   Analytical Model

Considerable progress has recently been made in bringing analytical models into congestion control and resource allocation problems [1–5]. Key to these works has been to explicitly model the *congestion measure* that is communicated implicitly or explicitly back to the sources by the routers. It is assumed that each link maintains a variable, called *price*, and the sources have information about the aggregate price of links in their path.

In this section, we describe a fluid-flow model, similar to that in [1, 2, 5]. We interpret an equilibrium point as the unique solution of an associated optimization problem. The resulting resource allocation is aimed to provide a fair share of an application layer performance measure, i.e. the utility to users. In contrast to [1–7, 12] we do not pose any restrictions on the bandwidth utility functions, except for monotonicity.

2.1   Model

We model a packet switched network by a set of nodes (router) connected by a set L of unidirectional links (output ports) with finite capacities $c = (c_l, l \in L)$. The set of links are shared by a set S of sources indexed by s. A source s represents an end-to-end connection and its route involves a subset $L(s) \subset L$ of links. Equivalently, each link is used by a subset $S(l) \subset S$ of sources. The sets $L(s)$ or $S(l)$ define a routing matrix

$$R_{ls} = \begin{cases} 1 & \text{if } l \in L(s), \\ 0 & \text{else.} \end{cases}$$

A transmission rate x_s in *packets per second* is associated with each source s. We assume, that the rates $x_s, s \in S$ lie in the interval $X_s = [0, x_s^{max}]$, where x_s^{max} is the maximum sending rate of source s. This upper bound may differ substantially for different applications. A subset of sources $S_r \subset S$ transferring real-time data, for instance, may have a maximum encoding rate $x_s^{max}, s \in S_r$, which can be much lower than the upper bound $x_s^{max}, s \in S \setminus S_r$ of elastic applications, which are greedy for any available bandwidth in the network. Thus, sending rates of elastic applications are constrained by bottleneck links in the network.

Definition 1. *A rate vector $x = (x_s, s \in S)$ is said to be* feasible *if it satisfies the conditions:*

$$x_s \in X_s \; \forall s \in S \text{ and } Rx \leq c.$$

With each link l, a scalar positive congestion-measure p_l, called *price*, is associated. Let

$$y_l = \sum_{s \in S} R_{ls} x_s$$

be the aggregate transmission rate of link l, i.e. the sum over all rates using that link, and let

$$q_s = \sum_{l \in L} R_{ls} p_l$$

be the end-to-end congestion measure of source s. Note that taking the sum of congestion measures of a used path is essential to maintain the interpretation of p_l as dual variables [1]. Source s can observe its own rate x_s and the end-to-end congestion measure q_s of its path. Link l can observe its local congestion measure p_l and the aggregate transmission rate y_l. When the transmission rate of user s is x_s, user s receives a benefit measured by the bandwidth utility $U_s(x_s)$, which is a scalar function and has the following form:

$$U_s \; : \; X_s \to Y_s$$
$$x_s \mapsto U_s(x_s),$$

where $Y_s = [U_s(0), U_s(x_s^{max})] = [u_s^{min}, u_s^{max}], U_s(0) = u_s^{min}, U_s(x_s^{max}) = u_s^{max}$.

Assumption 1. *The bandwidth utility functions $U_s(\cdot)$ are continuous, differentiable, and strictly increasing, i.e. $U_s'(x_s) > 0$ for all $x_s \in X_s$, $s \in S$.*

This assumption ensures the existence of the inverse function $U_s^{-1}(\cdot)$ over the range $[u_s^{min}, u_s^{max}]$. Before we present a constructive method to generate second order utility functions, we briefly restate the overall paradigm. An optimal operation point or equilibrium should result in almost equal utility values for different applications. The exact definition of the proposed resource allocation, i.e. *utility proportional fair* resource allocation, will be given below. If we want to follow this paradigm, we must translate a given congestion level of a path, represented by q_s, into an appropriate utility value the network can offer to source s. We model this utility value, the *available utility*, as the transformation of the congestion measure q_s by a *transformation function* $f_s(q_s)$. This function is assumed to be strictly decreasing.

Assumption 2. *The transformation function $f_s(\cdot)$ describing the available utility of a path used by sender s is assumed to be a continuous, differentiable, and strictly decreasing function of the aggregate congestion measure q_s, i.e. $f'_s(q_s) < 0$ for all $q_s \geq 0$ and $s \in S$.*

This assumption is reasonable, since the more congested a path is, the smaller will be the available utility of an application. The main idea is, that each user s should send at data rates x_s in order to match its own bandwidth utility with the available utility of its path. This leads to the following equation:

$$U_s(x_s) = [f_s(q_s)]_{u_s^{min}}^{u_s^{max}}, s \in S, \tag{1}$$

where $[w]_a^b := min\{max\{w,a\}, b\} = \begin{cases} w, & \text{if } a \leq w \leq b \\ a, & \text{if } w < a \\ b, & \text{if } w > b. \end{cases}$

Note that the utility a source can receive is bounded by the minimum and maximum utility values u_s^{min} and u_s^{max}. Hence, the source rates x_s are adjusted according to the available utility $f_s(q_s)$ of their used path as follows:

$$x_s = U_s^{-1}([f_s(q_s)]_{u_s^{min}}^{u_s^{max}}), \quad s \in S. \tag{2}$$

A source $s \in S$ reacts to the congestion measure q_s in the following manner: if the congestion measure q_s is below a threshold $q_s < q_s^{min} := f_s^{-1}(u_s^{max})$, then the source transmits data at maximum rate $x_s^{max} := U_s^{-1}(u_s^{max})$; if q_s is above a threshold $q_s > q_s^{max} := f_s^{-1}(u_s^{min})$, the source sends at minimum rate $x_s^{min} := U_s^{-1}(u_s^{min})$; if q_s is in between these two thresholds $q_s \in Q_s := [q_s^{min}, q_s^{max}]$, the sending rate is adapted according to $x_s = U_s^{-1}(f_s(q_s))$.

Lemma 1. *The function $G_s(q_s) = U_s^{-1}([f_s(q_s)]_{u_s^{min}}^{u_s^{max}})$ is positive, differentiable, and strictly monotone decreasing, i.e. $G'_s(q_s) < 0$ on the range $q_s \in Q_s$, and its inverse $G_s^{-1}(\cdot)$ is well defined on X_s.*

Proof. Since $U_s(\cdot)$ is defined on X_s, $U_s^{-1}(\cdot)$ is always nonnegative. Since $f_s(\cdot)$ is differentiable over Q_s, and $U_s^{-1}(\cdot)$ is differentiable over Y_s, the composition $G_s(q_s) = U_s^{-1}(f_s(q_s))$ is differentiable over Q_s. We compute the derivative using the chain rule: $G'_s(q_s) = U_s^{-1'}(f_s(q_s))f'_s(q_s)$. The derivative of the inverse $U_s^{-1}(f_s(q_s))$ can be computed as

$$U_s^{-1'}(f_s(q_s)) = \frac{1}{U'_s(U_s^{-1}(f_s(q_s)))} > 0.$$

With the inequality $f'_s(\cdot) < 0$, we get $G'_s(q_s) < 0$, $q_s \in Q_s$. Hence, $G_s(q_s)$ is strictly monotone decreasing in Q_s, so its inverse $G_s^{-1}(x_s)$ exists on X_s. □

2.2 Equilibrium Structure and Second Order Utility Optimization

In this section we study the above model at equilibrium, i.e. we assume, that rates and prices are at fixed equilibrium values x^*, y^*, p^*, q^*. From the above model, we immediately have the relationships:

$$y^* = Rx^*, \quad q^* = Rp^*.$$

In equilibrium, the sending rates $x_s \in X_s$, $s \in S$ satisfy:

$$x_s^* = U_s^{-1}([f_s(q_s^*)]_{u_s^{min}}^{u_s^{max}}) = G_s(q_s^*). \tag{3}$$

Since q_s represents the congestion in the path $L(s)$, the sending rates will be decreasing at higher q_s, and increasing at lower q_s. Now we consider the inverse $G_s^{-1}(x_s)$ of the above function on the interval X_s, and construct the second order utility $F_s(x_s)$ as the integral of $G_s^{-1}(x_s)$. Hence, $F_s(\cdot)$ has the following form and property:

$$F_s(x_s) = \int G_s^{-1}(x_s)dx_s \quad \text{with} \quad F_s'(x_s) = G_s^{-1}(x_s). \tag{4}$$

Lemma 2. *The second order utility $F_s(\cdot)$ is a positive, continuous, strictly increasing, and strictly concave function of $x_s \in X_s$.*

Proof. This follows directly from Lemma 1 and the relation

$$F_s''(x_s) = G_s^{-1'}(x_s) = \frac{1}{G_s'(q_s)} < 0. \qquad \square$$

The construction of $F_s(\cdot)$ leads to the following property:

Lemma 3. *The equilibrium rate (3) is the unique solution of the optimization problem:*

$$\max_{x_s \in X_s} F_s(x_s) - q_s x_s. \tag{5}$$

Proof. The first order necessary optimality condition to problem (5) is:

$$F_s'(x_s) = q_s$$
$$\Leftrightarrow G_s^{-1}(x_s) = q_s$$
$$\Leftrightarrow \qquad x_s = U_s^{-1}([f_s(q_s)]_{u_s^{min}}^{u_s^{max}})$$

Due to the strict concavity of $F(\cdot)$ on X_s, the second order sufficient condition is also satisfied completing the proof. $\qquad \square$

The above optimization problem can be interpreted as follows. $F_s(x_s)$ is the second order utility a source receives, when sending at rate x_s, and $q_s x_s$ is the price per unit flow the network would charge. The solution to (5) is the maximization of individual utility profit at fixed cost q_s and exactly corresponds to the proposed source law (2). Now we turn to the overall system utility optimization problem. The aggregate prices q_s ensure that individual optimality does not collide with social optimality. An appropriate choice of prices $p_l, l \in L$ must guarantee that the solutions of (5) also solve the system utility optimization problem:

$$\max_{x \geq 0} \sum_{s \in S} F_s(x_s) \tag{6}$$

$$\text{s.t.} \quad Rx \leq c. \tag{7}$$

This problem is a convex program, similar to the convex programs in [1,5,7], for which a unique optimal rate vector exist. For solving this problem directly global knowledge about actions of all sources is required, since the rates are coupled through the shared links. This problem can be solved by considering its dual [7].

3   Dual Problem and Global Stability

In accordance with the approach in [1], we introduce the Lagrangian and consider prices $p_l, l \in L$ as Lagrange multipliers for (6),(7). Let

$$L(x,p) = \sum_{s \in S} F_s(x_s) - \sum_{l \in L} p_l(y_l - c_l) = \sum_{s \in S} F_s(x_s) - q_s x_s + \sum_{l \in L} p_l c_l$$

be the Lagrangian of (6) and (7). The dual problem can be formulated as:

$$\min_{p_l \geq 0} \sum_{s \in S} V_s(q_s) + \sum_{l \in L} p_l c_l, \tag{8}$$

where

$$V_s(x_s) = \max_{x_s \geq 0} F_s(x_s) - q_s x_s, \quad x_s \in X_s. \tag{9}$$

Due to the strict concavity of the objective and the linear constraints, at optimal prices p^*, the corresponding optimal x^* solving (9) is exactly the unique solution of the primal problem (6),(7). Note that (5) has the same structure as (9), so we only need to assure that the prices q_s given in (5) correspond to Lagrange multipliers q_s given in (9).

As shown in [7], a straightforward method to guarantee that equilibrium prices are Lagrange multipliers is the gradient projection method applied to the dual problem (8):

$$\frac{d}{dt} p_l(t) = \begin{cases} \gamma_l(p_l(t))(y_l(t) - c_l) & \text{if } p_l(t) > 0 \\ \gamma_l(p_l(t))[y_l(t) - c_l]^+ & \text{if } p_l(t) = 0, \end{cases} \tag{10}$$

where $[z] = max\{0, z\}$ and $\gamma_l(p_l) > 0$ is a nondecreasing continuous function. This algorithm can be implemented in a distributed environment. The information needed at the links is the link bandwidth c_l and the aggregate transmission rate $y_l(t)$, both of which are available. In equilibrium, the prices satisfy the complementary slackness condition, i.e. $p_l(t)$ are zero for non-saturated links and non-zero for bottleneck links. We conclude this section with a global convergence result of the dual algorithm (8) combined with the static source law (5) using Lyapunov techniques along the lines of [12]. It is only assumed that the routing matrix R is nonsingular. This guarantees that for any given $q_s \in S$ there exists a unique vector $(p_l, l \in L_s)$ such that $q_s = \sum_{l \in L_s} p_l$.

Theorem 1. *Assume the routing matrix R is nonsingular. Then the dual algorithm (10) starting from any initial state converges asymptotically to the unique solution of (6) and (7).*

The proof of this theorem can be found in [13]. For further analysis of the speed of convergence, we refer to [1].

4   Utility Proportional Fairness

Kelly et al. [2] introduced the concept of *proportional fairness*. They consider elastic flows with corresponding strictly concave logarithmic bandwidth utility functions. A proportional fair rate vector $(x_s, s \in S)$ is defined such that for any other feasible rate vector $(y_s, s \in S)$ the aggregate of proportional change is non-positive:

$$\sum_{s \in S} \frac{y_s - x_s}{x_s} \le 0.$$

This definition is motivated by the assumption that all users have the same logarithmic bandwidth utility function $U_s(x_s) = \log(x_s)$. By this assumption, a first order necessary and sufficient optimality condition for the system bandwidth optimization problem

$$\max_{x_s \ge 0} \sum_{x_s \ge 0} U_s(x_s) \quad \text{s.t.} \quad Rx \le 0$$

is

$$\sum_{s \in S} \frac{\partial U_s}{\partial x_s}(x_s)(y_s - x_s) = \sum_{s \in S} \frac{y_s - x_s}{x_s} \le 0.$$

This condition is known as the *variational inequality* and it corresponds to the definition of proportional fairness.

Before we come to our new fairness definition, we restate the concept of utility max-min fairness. It is simply the translation of the well known bandwidth max-min fairness applied to utility values.

Definition 2. *A set of rates $(x_s, s \in S)$ is said to be* utility max-min fair, *if it is feasible, and for any other feasible set of rates $(y_s, s \in S)$, the following condition hold: if $U_s(y_s) > U_s(x_s)$ for some $s \in S$, then there exists $k \in S$ such that $U_k(y_k) < U_k(x_k)$ and $U_k(x_k) \le U_s(x_s)$.*

Suppose we have a utility max-min fair rate allocation. Then, a user cannot increase its utility, without decreasing the utility of another user, which receives already a smaller utility. We further apply the above definition to a utility allocation of a single path.

Definition 3. *Consider a single path in the network denoted by a set of adjacent links $(l \in L_p)$. Assume a set of users $S_{L_p} \subset S$ share this path, i.e. $L(s) = L_p$ for $s \in S_{L_p}$. Then, the set of rates $x_s, s \in S$ is said to be* path utility max-min fair *if the rate allocation on such a path is utility max-min fair.*

Now we come to our proposed new fairness criterion, based on the second order utility optimization framework.

Definition 4. *Assume, all second order utilities $F_s(\cdot)$ are of the form (4). A rate vector $(x_s, s \in S)$ is called* utility proportional fair *if for any other feasible rate vector $(y_s, s \in S)$ the following optimality condition is satisfied:*

$$\sum_{s \in S} \frac{\partial F_s}{\partial x_s}(x_s)(y_s - x_s) = \sum_{s \in S} G_s^{-1}(x_s)(y_s - x_s)$$

$$= \sum_{s \in S} f_s^{-1}(U_s(x_s))(y_s - x_s) \leq 0 \tag{11}$$

The above definition ensures, that any proportional utility fair rate vector will solve the utility optimization problem (6), (7). If we further assume, all users have the same transformation function $f(\cdot) = f_s(\cdot), s \in S$, then we have the following properties of a utility proportional fair rate allocation, which are proven in Appendix A.

Theorem 2. *Suppose all users have a common transformation function $f(\cdot)$ and all second order utility functions are defined by (4). Let the rate vector $(x_s \in X_s, s \in S)$ be proportional utility fair, i.e. the unique solution of (6). Then the following properties hold:*

(i) *The rate vector $(x_s \in X_s, s \in S)$ is path utility max-min fair.*
(ii) *If $q_{s_1} \in Q_{s_1}, q_{s_2} \in Q_{s_2}$ and $q_{s_1} \leq q_{s_2}$ for sources s_1, s_2, then $U_{s_1}(x_{s_1}) \geq U_{s_2}(x_{s_2})$.*
(iii) *If source s_1 uses a subset of links that s_2 uses, i.e. $L(s_1) \subseteq L(s_2)$, and $U_{s_1}(x_{s_1}) < u_{s_1}^{max}$, then $U_{s_1}(x_{s_1}) \geq U_{s_2}(x_{s_2})$.*

It is a well-known property of the concept of proportional fairness that flows traversing several links on a route receive a lower share of available resources than flows traversing a part of this route provided all utilities are equal. The rationale behind this is that these flows use more resources, hence short connections should be favored to increase system utility. Transferring this idea to utility proportional fairness, we get a similar result. Flows traversing several links receive less utility compared to shorter flows, provided a common transformation function is used. If this feature is undesirable, since the path a flow takes is chosen by the routing protocol and beyond the reach of the single user, the second order utilities can be modified to compensate this effect. We show that an appropriate choice of the transformation functions $f_s(\cdot)$ will assure a utility max-min bandwidth allocation in equilibrium.

Theorem 3. *Suppose all users have the same parameter dependent transformation function $f_s(q_s, \kappa) = q_s^{-\frac{1}{\kappa}}, s \in S, \kappa > 0$. The second order utilities $F_s(x_s, \kappa)$, $s \in S$ are defined by (4). Let the sequence of rate vectors $x(\kappa) = (x_s(\kappa) \in X_s, s \in S)$ be utility proportional fair. Then $x(\kappa)$ approaches the utility max-min fair rate allocation as $\kappa \to \infty$.*

The proof of this theorem can be found in Appendix B.

5   Conclusion

We have obtained decentralized congestion control laws at links and sources, which are globally stable and provide a utility proportional fair resource allocation in equilibrium. This new fairness criterion ensures that bandwidth utility values of users (applications), rather than bit rates, are proportional fair in equilibrium. We further showed that a utility proportional fair resource allocation also ensures utility max-min fairness for all users sharing a single path in the network. As a special case of our model, we incorporate utility max-min fairness for all users sharing the network. To the best of our knowledge, this is the first paper dealing with resource allocation problems in the context of global optimization, that includes non-concave bandwidth utility functions.

We are currently working on ns-2 (Network Simulator) implementations of the described algorithms. First simulation results are promising. An open issue and challenge is to design the feedback control interval for real-time applications. There is clearly a tradeoff between the two conflicting goals: stability (delay) and minimal packet overhead (i.e. multicast) in the network. Nevertheless, we believe that this framework has a great potential in providing real-time services for a growing number of multimedia applications in future networks.

References

1. S. H. Low and D. E. Lapsley: Optimization Flow Control I. IEEE/ACM Trans. on Networking **7** (1999) 861–874
2. F. P. Kelly, A. K. Maulloo and D. K. H. Tan: Rate Control in Communication Networks: Shadow Prices, Proportional Fairness, and Stability. Journal of the Operational Research Society **49** (1998) 237–52
3. R. J. Gibbens and F. P. Kelly: Resource pricing and the evolution of congestion control. Automatica (1999) 1969–1985
4. S. H. Low: A duality model of TCP flow controls. In: Proceedings of ITC Specialist Seminar on IP Traffic Measurement, Modeling and Management. (2000)
5. S. H. Low, F. Paganini, J. Doyle: Internet Congestion Control. IEEE Control Systems Magazine **22** (2002)
6. S. Athuraliya, V. H. Li, S. H. Low and Q. Yin: REM: Active Queue Management. IEEE Network **15** (2001) 48–53
7. S. H. Low, F. Paganini, J. C. Doyle: Scalable laws for stable network congestion control. In: Proceedings of Conference of Decision and Control. (2001)
8. S. Shenker: Fundamental Design Issues for the Future Internet. IEEE JSAC **13** (1995) 1176–88
9. Z. Cao, E. W. Zegura: Utility max-min: An application-oriented bandwidth allocation scheme. In: Proceedings of IEEE INFOCOM'99. (1999) 793–801
10. J. Cho, S. Chong: Utility Max-Min Flow Control Using Slope-Restricted Utility Functions. Available at http://netsys.kaist.ac.kr/Publications (2004)
11. S. Floyd: TCP and Explicit Congestion Notification. ACM Comp. Commun. Review **24** (1994) 10–23
12. F. Paganini: A global stability result in network flow control. Systems and Control Letters **46** (2002) 165–172
13. T. Harks: Utility Proportional Fair Resource Allocation - An Optimization Oriented Approach. Technical Report ZR-04-32, Konrad-Zuse-Zentrum für Informationstechnik Berlin (ZIB) (2004)

Appendix A

Proof of Theorem 2:

To (i): if sources $s \in S_{L_p}$ share the same path, they receive the same aggregate congestion feedback in equilibrium $q_p = q_s$, $s \subset S_{L_p}$. Two cases are of interest.

(a) Suppose for all sources the following inequality holds: $f(q_p) < u_s^{max}$, $s \in S_{L_p}$. Hence, all sources adapt their sending rates according to the available utility $f(q_p) = U_s(x_s)$. This corresponds to the trivial case of path utility max-min fairness, since all sources receive equal utility.

(b) Suppose a set $s \in Q \subset S_{L_p}$ receives utility $U_s(x_s) = u_s^{max} < f(q_p)$. We prove the theorem by contradiction. Assume the utility proportional fair rate vector $(x_s, s \in S)$ is not path utility max-min fair with respect to the path L_p. By definition, there exists a feasible rate vector $y_s, s \in S$ with

$$U_j(y_j) > U_j(x_j) \text{ for } j \in S_{L_p} \setminus Q \tag{12}$$

such that for all $k \in S_{L_p} \setminus (Q \cup \{j\})$ with $U_k(x_k) \leq U_j(x_j)$ the inequality

$$U_k(y_k) \geq U_k(x_k) \tag{13}$$

holds. In other words, we can increase the utility of a single source rate $U_j(x_j)$ to $U_j(y_j)$ by increasing the rate x_j to y_j without decreasing utilities $U_k(y_k), k \in S_{L_p} \setminus (Q \cup \{j\})$ which are already smaller. We represent the rate increase of source j by $y_j = x_j + \xi_j$, where $\xi_j > 0$ will be chosen later on. Here again, we have to consider two cases:

(b1) Suppose, there exists a sufficiently small $\xi_j > 0$ that we do not have to decrease any source rate of the set $\{y_k, k \in S_{L_p} \setminus (Q \cup \{j\})\}$ to maintain feasibility. Hence, we can increase the system utility while maintaining feasibility. This clearly contradicts the proportional fairness property of x.

(b2) Suppose, we have to decrease a set of utilities $(U_k(y_k), k \in K)$, which are higher then $U_j(x_j)$, i.e. $U_k(y_k) < U_k(x_k)$ with $U_k(y_k) > U_j(x_j), k \in K \subset S_{L_p} \setminus (Q \cup \{j\})$. This correspond to decreasing the set of rates $y_k = x_k - \xi_k$, $k \in K$ with $\sum_{k \in K} \xi_k \leq \xi_j$. Due to the strict concavity of the objective functions of (6), we get the following inequalities:

$$F_j'(x_j) = f^{-1}(U_j(x_j)) > f^{-1}(U_k(y_k)) = F_k'(y_k), \ k \in K \subset S_{L_p} \setminus (Q \cup \{j\}).$$

Due to the continuity of $F_s'(\cdot), s \in S$, we can choose ξ_j with $y_j = x_j + \xi_j$ such that

$$F_j'(x_j + v_j \xi_j) > F_k'(y_k) \text{ for all } k \in K \subset S_{L_p} \setminus (Q \cup \{j\}) \text{ and } v_j \in (0,1).$$

Comparing the aggregate second order utilities of the rate vectors x and y using the mean value theorem, we get:

$$\sum_{s \in S} F_s(x_s) - \sum_{s \in S} F_s(y_s) = \sum_{k \in K} (F_k(x_k) - F_k(y_k)) + F_j(x_j) - F_j(y_j)$$

$$= \sum_{k \in K} (F_k(y_k + \xi_k) - F_k(y_k)) + F_j(x_j) - F_j(x_j + \xi_j)$$

$$= \sum_{k \in K} (F_k(y_k) + F_k'(y_k + v_k \xi_k)\xi_k - F_k(y_k))$$

$$+ F_j(x_j) - (F_j(x_j) + F_j'(x_j + v_j \xi_j)\xi_j$$

$$= \sum_{k \in K} F_k'(y_k + v_k \xi_k)\xi_k - F_j'(x_j + v_j \xi_j)\xi_j$$

$$\leq \sum_{k \in K} \xi_k \max_{k \in K}(F_k'(y_k + v_k \xi_k)) - F_j'(x_j + v_j \xi_j)\xi_j$$

$$\leq \xi_j (\max_{k \in K}(F_k'(y_k + v_k \xi_k)) - F_j'(x_j + v_j \xi_j))$$

$$< 0, \quad v_j \in (0,1), \ v_k \in (0,1), k \in K.$$

The last inequality shows that x is not the optimal solution to (6). Thus, x cannot be utility proportional fair. This contradicts the assumption and proves that x is path utility max-min fair.

To (ii): Assume $q_{s_1} \in Q_{s_1}$, $q_{s_2} \in Q_{s_2}$ and $q_{s_1} \leq q_{s_2}$ for sources s_1, s_2. Applying (1) to given q_{s_1}, q_{s_2}, we have $f(q_{s_1}) = U_{s_1}(x_{s_1}) \geq f(q_{s_2}) = U_{s_2}(x_{s_2})$ because of the monotonicity of $f(\cdot)$.

To (iii): From $L(s_1) \subseteq L(s_2)$ it follows, that $q_{s_1} \leq q_{s_2}$. Since the available utility $f(\cdot)$ is monotone decreasing in q_s and the bandwidth utility $U_{s_1}(x_{s_1}) < u_{s_1}^{max}$ of user s_1 is not bounded by its maximum value, it follows, that $f(q_{s_1}) = U_{s_1}(x_{s_1}) \geq [f(q_{s_2})]_{u_{s_2}^{min}}^{u_{s_2}^{max}} = U_{s_2}(x_{s_2})$. □

Appendix B

Proof of Theorem 3:

Since all elements of the sequence $x(\kappa)$ solve (6) subject to (7), the sequence is bounded. Hence, we find a subsequence $x(\kappa_p), p \in \mathbb{N}^+$, such that $\lim_{\kappa_p \to \infty} = x$. We show, that this limit point x is utility max-min fair. The uniqueness of the utility max-min fair rate vector x will ensure that every limit point of $x(\kappa)$ is equal x. This proves the convergence of $x(\kappa)$ to x.

Since all users $s \in S$ use the same transformation function $f_s(q_s) = q_s^{-\frac{1}{\kappa}}$, $s \in S$, the second order utility and its derivative applied to the rate vector $x_s(\kappa)$ have the following form:

$$F_s(x_s(\kappa)) = \int U_s(x_s(\kappa))^{-\kappa} dx_s(\kappa) \quad \text{with} \quad \frac{\partial F_s}{\partial x_s(\kappa)} = U_s(x_s(\kappa))^{-\kappa}, \ s \in S.$$

We assume that the limit point $x = (x_s \in X_s, \ s \in S)$ is not utility max-min fair. Then we can increase the bandwidth utility of a user j while decreasing the utilities of other users $k \in K \subset S \setminus \{j\}$ which are larger than $U_j(x_j)$. More formal, it exists a rate vector $y = (y_s \in X_s, \ s \in S)$ and an index $j \in S$ with $U_j(y_j) > U_j(x_j)$, $j \in S$ and $U_k(y_k) < U_k(x_k)$ with $U_k(y_k) > U_j(x_j)$ for a subset $k \in K \subset S \setminus \{j\}$. We choose κ_0 so large that for all elements of the subsequence $x(\kappa_p)$ with $\kappa_p > \kappa_0$ the inequalities $U_j(y_j) > U_j(x_j(\kappa_p))$, $j \in S$, and $U_k(y_k) < U_k(x_k(\kappa_p))$ with $U_k(y_k) > U_j(x_j(\kappa_p))$ for a subset $k \in K \subset S \setminus \{j\}$ hold. With the inequality $U_j(x_j(\kappa_p)) < U_k(x_k(\kappa_p))$, $k \in K$, we can choose $\kappa_1 > \kappa_0$ large enough such that

$$U_j(x_j(\kappa_p))^{-\kappa_p} > C \cdot U_k(x_k(\kappa_p))^{-\kappa_p}, \tag{14}$$

for all $k \in K$, $\kappa_p > \kappa_1$, and $C > 0$ an arbitrary constant. Hence, there exists a κ_1 large enough that the following inequality holds:

$$U_j(x_j(\kappa_p))^{-\kappa_p} > \sum_{k \in K} \underbrace{(x_k(\kappa_p) - y_k)}_{>0} \max_{k \in K} U_k(x_k(\kappa_p))^{-\kappa_p}, \ \kappa_p > \kappa_1. \tag{15}$$

We evaluate the variational inequality (11) given in the definition of utility proportion fairness for the candidate rate vector $(y_s \in X_s, \ s \in S)$ and $\kappa_p > \kappa_1$.

$$\sum_{s \in S} \frac{\partial F_s}{\partial x_s(\kappa_p)}(x_s(\kappa_p))(y_s - x_s(\kappa_p)) = \sum_{s \in S} U_s(x_s(\kappa_p))^{-\kappa_p}(y_s - x_s(\kappa_p))$$

$$= U_j(x_j(\kappa_p))^{-\kappa_p}(y_j - x_j(\kappa_p)) + \sum_{k \in K} U_k(x_k(\kappa_p))^{-\kappa_p}(y_k - x_k(\kappa_p))$$

$$> U_j(x_j(\kappa_p))^{-\kappa_p}(y_j - x_j(\kappa_p)) - \max_{k \in K} U_k(x_k(\kappa_p))^{-\kappa_p} \sum_{k \in K}(x_k(\kappa_p) - y_k)$$

$$> 0, \text{ using (15).}$$

Hence, the variational inequality is not valid contradicting the utility proportional fairness property of $x(\kappa_p)$. □

Packet Size Distribution: An Aside?

György Dán, Viktória Fodor, and Gunnar Karlsson

KTH, Royal Institute of Technology,
Department of Microelectronics and Information Technology
{gyuri,viktoria,gk}@imit.kth.se

Abstract. For multimedia traffic like VBR video, knowledge of the average loss probability is not sufficient to determine the impact of loss on the perceived visual quality and on the possible ways of improving it, for example by forward error correction (FEC) and error concealment. In this paper we investigate how the packet size distribution affects the packet loss process, the distribution of the number of packets lost in a block of packets and the related FEC performance. We present an exact mathematical model for the loss process of an $MMPP + M/E_r/1/K$ queue and compare the results of the model to simulations performed with various other packet size distributions (PSDs), among others, the measured PSD from an Internet backbone. We conclude that the packet size distribution affects the packet loss process and thus the efficiency of FEC. This conclusion is mainly valid in access networks where a single multimedia stream might affect the multiplexing behavior. The results show that analytical models of the PSD matching the first three moments (mean,variance and skewness) of the empirical PSD can be used to evaluate the performance of FEC in real networks. We also conclude that the exponential PSD, though it is not a worst case scenario, is a good approximation for the PSD of today's Internet to evaluate FEC performance.

1 Introduction

For flow-type multimedia communications, as opposed to elastic traffic, the average packet loss is not the only measure of interest. The burstiness of the loss process, the number of losses in a block of packets, has a great impact both on the user-perceived visual quality and on the possible ways of improving it, for example by error concealment and forward error correction.

Forward error correction (FEC) is an attractive means to decrease the loss probability experienced by delay sensitive traffic, such as real-time multimedia, when ARQ schemes can not be used to recover losses due to strict delay constraints. There are two main directions of FEC design to recover from packet losses. One solution, proposed by the IETF and implemented in Internet audio tools is to add a redundant copy of the original packet to one of the subsequent packets [1]. The other set of solutions, considered in this paper, use block coding schemes based on algebraic coding, e.g. Reed-Solomon coding [2]. The error correcting capability of RS codes with k data packets and c redundant packets is c if data is lost. Thus, the capability of FEC to recover from losses depends on the distribution of the number of packets lost in a block, e.g. the burstiness of the loss process.

M. Ajmone Marsan et al. (Eds.): QoS-IP 2005, LNCS 3375, pp. 75–87, 2005.

The burstiness of the loss process in the network can be influenced by three factors, the burstiness of the stream traversing the network, the burstiness of the background traffic and the packet size distribution. The effects of the burstiness of the stream traversing the network and the background traffic have been investigated before [2–5]. The effects of the packet size distribution are not clear however. It is well known that in an M/G/1 queue the average number of customers is direct proportional to the coefficient of variation (CoV) of the service time distribution, as given by the Pollaczek-Khinchine formula [6]. For the finite capacity M/G/1 queue there is no closed form formula to calculate the packet loss probability [7, 8], though we know from experience that a lower CoV of the service time distribution yields lower average loss probability. It is however unclear how the distribution of the service time affects the loss process in a finite queue and thus how the potential of using FEC changes. The packet size distribution (PSD) in the network can vary on the short term due to changes in the ongoing traffic and on the long term as new applications and protocols emerge. As individual applications can not control the PSD in the network, it is important to know how the PSD will affect their performance, e.g. how much gain can an application expect from FEC given a certain measured end-to-end average loss probability.

In this paper we present a model to analyze the packet loss process of a bursty source, for example VBR video, multiplexed with background traffic in a single multiplexer with a finite queue and Erlang-r distributed packet sizes. We model the bursty source by an L-state Markov-modulated Poisson process (MMPP) while the background traffic is governed by a Poisson process. We compare the results of the model to the results of a model with deterministic packet sizes and to various simulations performed with general PSDs, among them the measured PSD of an Internet backbone [9], and investigate the effects of the network PSD on the packet loss process and the efficiency of FEC.

It is well known that compressed multimedia, like VBR video, exhibits a self-similar nature [10]. Yoshihara et al. use the superposition of 2-state IPPs to model self-similar traffic in [11] and compare the loss probability of the resulting $MMPP/D/1/K$ queue with simulations. They found that the approximation works well under heavy load conditions and gives an upper bound on the packet loss probabilities. Ryu and Elwalid [12] showed that short term correlations have dominant influence on the network performance under realistic scenarios of buffer sizes for real-time traffic. Thus the MMPP may be a practical model to derive approximate results for the queuing behavior of long range dependent traffic such as real-time VBR video, especially in the case of small buffer sizes [13]. Recently Cao et al. [14] showed that the traffic generated by a large number of sources tends to Poisson as the load increases due to statistical multiplexing and hence justifying the Poisson model for the background traffic. Recent measurements indicate that Internet traffic can be approximated by a non-stationary Poisson process [15]. According to the results the change free intervals are well above 150 ms, the ITU's G.114 recommendation for end-to-end delay for real-time applications.

The paper is organized as follows. Section 2 gives an overview of the previous work on the modeling of the loss process of a single server queue. In Section 3 we describe our model used for calculating the loss probabilities in a block of packets. In Section 4 we evaluate the effects of the PSD on the packet loss process in various scenarios. We consider constant average load in Subsection 4.1, constant average loss probability in Subsection 4.2, and we isolate the effect of the PSD from other factors in Subsection 4.3. We conclude our work in Section 5.

2   Related Work

In [16], Cidon et al. presented an exact analysis of the packet loss process in an M/M/1/K queue, that is the probability of losing j packets in a block of n packets, and showed that the distribution of losses may be bursty compared to the assumption of independence. They also considered a discrete time system fed with a Bernoulli arrival process describing the behavior of an ATM multiplexer. Gurewitz et al. presented explicit expressions for the above quantities of interest for the M/M/1/K queue in [17]. In [18], Altman et al. obtained the multidimensional generating function of the probability of j losses in a block of n packets and gave an easy-to-calculate asymptotic result under the condition that $n \leq K + j + 1$.

Schulzrinne et al. [19] derived the conditional loss probability (CLP) for the $N *$ $IPP/D/1/K$ queue and showed that the CLP can be orders of magnitude higher than the loss probability. In [2] Kawahara et al. used an interrupted Bernoulli process to analyze the performance of FEC in a cell switched environment. The loss process of the $MMPP/D/1/K$ queue was analyzed in [20] and the results compared to a queue with exponential packet size distribution.

Models with general service time distribution have been proposed for calculating various measures of queuing performance [21, 22], but not to analyze the loss process. Though models with exponential and deterministic PSDs are available, a thorough analysis of the effects of the PSD on the packet loss process has not yet been done.

3   Model Description

Flows traversing large networks like the Internet cross several routers before reaching their destination. However, most of the losses in a flow occur in the router having the smallest available bandwidth along the transmission path, so that one may model the series of routers with a single router, the bottleneck [23, 24].

We model the network with a single queue with Erlang-r distributed packet sizes having average transmission time $1/\mu$. The Erlang-r distribution is the distribution of the sum of r independent identically distributed random variables each having an exponential distribution. By increasing r to infinity the variance of the Erlang-r distribution goes to zero, and thus the distribution becomes deterministic.

Packets arrive to the system from two sources, a Markov-modulated Poisson process (MMPP) and a Poisson process, representing the tagged source and the background traffic respectively. The packets are stored in a buffer that can host up to K packets, and are served according to a FIFO policy. Every n consecutive packets from the tagged source form a block, and we are interested in the probability distribution of the number of lost packets in a block in the steady state of the system. Throughout this section we use notations similar to those in [16].

We assume that the sources feeding the system are independent. The MMPP is described by the infinitesimal generator matrix Q with elements r_{lm} and the arrival rate matrix $\Lambda = diag\{\lambda_1,\ldots,\lambda_L\}$, where λ_l is the average arrival rate while the underlying Markov chain is in state l [25]. The Poisson process modeling the background traffic has average arrival rate λ. The superposition of the two sources can be described by a single MMPP with arrival rate matrix $\hat{\Lambda} = \Lambda \oplus \lambda = \Lambda + \lambda I = diag\{\hat{\lambda}_1,\ldots,\hat{\lambda}_L\}$, and infinitesimal generator $\hat{Q} = Q$, where $\oplus$ is the Kronecker sum. Packets arriving from both sources have the same size distribution. Each packet in the queue corresponds to r exponential stages, and the state space of the queue is $\{0,\ldots,rK\} \times \{1,\ldots,L\}$.

Our purpose is to calculate the probability of j losses in a block of n packets $P(j,n)$, $n \geq 1, 0 \leq j \leq n$. We define the probability $P_{i,l}^a(j,n), 0 \leq i \leq rK, l = 1\ldots L, n \geq 1, 0 \leq j \leq n$ as the probability of j losses in a block of n packets, given that the remaining number of exponential stages in the system is i just before the arrival of the first packet in the block and the first packet of the block is generated in state l of the MMPP. As the first packet in the block is arbitrary,

$$P(j,n) = \sum_{l=1}^{L} \sum_{i=0}^{rK} \Pi(i,l) P_{i,l}^a(j,n). \tag{1}$$

$\Pi(i,l)$, the steady state distribution of the exponential stages in the queue as seen by an arriving packet can be derived from the steady state distribution of the $MMPP/E_r/1/K$ queue as

$$\Pi(i,l) = \frac{\pi(i,l)\lambda_l}{\sum_{l=1}^{L} \lambda_l \sum_{i=0}^{rK} \pi(i,l)}, \tag{2}$$

where $\pi(i,l)$ is the steady state distribution of the $MMPP/E_r/1/K$ queue.

The probabilities $P_{i,l}^a(j,n)$ can be derived according to the following recursion. The recursion is initiated for n = 1 with the following relations

$$P_{i,l}^a(j,1) = \begin{cases} 1 & j=0 \\ 0 & j \geq 1 \end{cases} \qquad i \leq r(K-1),$$

$$P_{i,l}^a(j,1) = \begin{cases} 0 & j=0, j \geq 2 \\ 1 & j=1 \end{cases} \qquad r(K-1) < i. \tag{3}$$

Using the notation $p_m = \frac{\lambda_m}{\lambda_m + \lambda}$ and $\overline{p}_m = \frac{\lambda}{\lambda_m + \lambda}$, for $n \geq 2$ the following equations hold.

$$P_{i,l}^a(j,n) = \sum_{m=1}^{L} \sum_{k=0}^{i+r} Q_{i+r,lm}(k)\{p_m P_{i+r-k,m}^a(j,n-1) + \overline{p}_m P_{i+r-k,m}^s(j,n-1)\} \tag{4}$$

for $0 \leq i \leq r(K-1)$, and for $r(K-1) < i$

$$P_{i,l}^a(j,n) = \sum_{m=1}^{L} \sum_{k=0}^{i} Q_{i,lm}(k)\{p_m P_{i-k,m}^a(j-1,n-1) + \overline{p}_m P_{i-k,m}^s(j-1,n-1)\}. \tag{5}$$

$P_{i,l}^s(j,n)$ is given by

$$P_{i,l}^s(j,n) = \sum_{m=1}^{L} \sum_{k=0}^{i+r} Q_{i+r,lm}(k)\{p_m P_{i+r-k,m}^a(j,n) + \overline{p}_m P_{i+r-k,m}^s(j,n)\}, \tag{6}$$

for $0 \leq i \leq r(K-1)$, and for for $r(K-1) < i$

$$P_{i,l}^s(j,n) = \sum_{m=1}^{L} \sum_{k=0}^{i} Q_{i,lm}(k)\{p_m P_{i-k,m}^a(j,n) + \overline{p}_m P_{i-k,m}^s(j,n)\}. \tag{7}$$

The probability $P_{i,l}^s(j,n), 0 \le i \le rK, l = 1\ldots L, n \ge 1, 0 \le j \le n$ is the probability of j losses in a block of n packets, given that the remaining number of exponential stages in the system is i just before the arrival of a packet from the background traffic and the MMPP is in state l. $Q_{i,lm}(k)$ denotes the joint probability of that the next arrival will be in state m of the MMPP and that k exponential stages out of i will be completed before the next arrival from the joint arrival process given that the last arrival was in state l of the MMPP. A way to calculate $Q_{i,lm}(k)$ is shown in the Appendix.

The procedure of computing $P_{i,l}^a(j,n)$ is as follows. First we calculate $P_{i,l}^a(j,1), i = 0\ldots rK$ from the initial conditions (3). Then in iteration k we first calculate $P_{i,l}^s(j,k), k = 1\ldots n-1$ using equations (6) and (7) and the probabilities $P_{i,l}^a(j,k)$, which have been calculated during iteration $k-1$. Then we calculate $P_{i,l}^a(j,k+1)$ using equations (4) and (5).

4 Performance Analysis

In this section we show results obtained with the $MMPP + M/E_r/1/K$ model described in Section 3, the $MMPP + M/D/1/K$ model described in [20] and simulations. The average packet length of both the tagged and the background traffic is set to 454 bytes, which is the mean packet size measured on an Internet backbone [9]. Note that increasing the average packet length is equivalent to decreasing the link speed, and thus the particular fixed value of the average packet length does not limit the generality of the results presented here. The PDF, CoV (σ/m) and skewness ($\sum(X-m)^3/\sigma^3$) parameters of the twelve considered PSDs are shown in Table 1. The G1 distribution is the measured PSD on a 2.5 Gbps Internet backbone link as given by the Sprint IP Monitoring project [9]. The considered link speeds are 10 Mbps, 22.5 Mbps and 45 Mbps. The queuing delay is set to around 1.5 ms in all cases, resulting in queue lengths from 5 to 20 packets depending on the link speed. Both in the analytical models and in the simulations we consider a 3 state MMPP, with an average bitrate of 540 kbps, arrival intensities $\lambda_1 = 116/s, \lambda_2 = 274/s, \lambda_3 = 931/s$ and transition rates $r_{12} = 0.12594, r_{21} = 0.25, r_{23} = 1.97, r_{32} = 2$. These values were derived from an MPEG-4 encoded video trace by matching the average arrival intensities in the three states of the MMPP with the average frame size of the I,P and B frames. The simulations were performed in ns-2, the simulation time was between 40 thousand and 400 thousand seconds (5-50 million packets from the tagged source).

We use two measures to compare the packet loss process, the probability of loosing j packets in a block of n packets. The first one is a commonly used measure of closeness, the Kullback-Leibler distance [26] defined for two distributions as

$$d(p_1, p_2) = \sum_{j=0}^{n} P_1(j,n) log_2 \frac{P_1(j,n)}{P_2(j,n)}, \tag{8}$$

The Kullback-Leibler distance is the same as the relative entropy of p_1 with respect to p_2. It is not a true metric, as it is not symmetric and does not satisfy the triangle inequality, but it is always non-negative and equals zero only if $p_1 = p_2$.

Table 1. Considered packet size distributions: coefficient of variation, skewness, PDF and notation in the figures. $N(m, \sigma)$ denotes a normal distribution with mean m and variance σ^2. $E(r, 1/\mu)$ denotes an r-stage Erlang distribution with mean $1/\mu$.

Distribution	CoV	Skewness	PDF	Notation
General 1	1.2	1.07	$b(x)$ taken from [9], see Figure 1	G1
General 2	1.2	1.07	$b(x) = 0.74N(127, 20) + 0.26N(1366, 20)$	G2
Phase type	1.2	1.07	$b(x) = 0.54E(5, 26) + 0.46E(5, 956)$	G3
Exponential	1	2	$E(1, 454)$	M
General 4	1	$\sqrt{2}$	$b(x) = 0.79N(219, 1) + 0.21N(1331, 1)$	G4
General 5	$1/\sqrt{2}$	2	$b(x) = 0.85N(321, 1) + 0.15N(1229, 1)$	G5
Erlang-2	$1/\sqrt{2}$	$\sqrt{2}$	$E(2, 454)$	E2
General 6	$1/\sqrt{2}$	$\sqrt{0.4}$	$b(x) = 0.65N(219, 1) + 0.35N(892, 1)$	G6
General 7	$\sqrt{0.1}$	$\sqrt{2}$	$b(x) = 0.79N(379, 1) + 0.21N(731, 1)$	G7
Erlang-10	$\sqrt{0.1}$	$\sqrt{0.4}$	$E(10, 454)$	E10
General 8	$\sqrt{0.1}$	0	$b(x) = 0.5N(310, 1) + 0.5N(598, 1)$	G8
Deterministic	0	0	$b(x) = \delta_{454}(x)$	D

The second measure is based on the gain that can be achieved by using FEC. Given the probabilities $P(j, n)$ the uncorrected loss probability for an RS(k,c+k) scheme can be calculated as

$$P_{loss}^{k,c+k} = \frac{1}{c+k} \sum_{j=c+1}^{c+k} jP(j, c+k). \qquad (9)$$

Based on the uncorrected packet loss probability we define the FEC gain as the ratio of the average loss probability without the use of FEC and the uncorrected loss probability when using FEC: $f(k, c+k) = P_{loss}/P_{loss}^{k,c+k}$.

4.1 Constant Average Load Case

In this subsection we investigate the effects of the PSD on the packet loss process and the efficiency of FEC as a function of the average load in the network. Figure 2 shows the uncorrected packet loss probability for FEC(1,1), FEC(10,11) and FEC(20,22) on a 10 Mbps link for the G1, G2, G3, M and D distributions. Figures 3 and 4 show the same results on a 22.5 Mbps and a 45 Mbps link. The figures show that results obtained with the G1, G2 and G3 distributions are practically the same (the difference is less than 5%). This indicates that by matching the first three moments of a distribution one can derive accurate results in terms of average loss probability and FEC gain even for very low loss probabilities. In the following we will only use the G1 distribution out of these three distributions. Figures 5 and 6 show the Kullback-Leibler distance obtained with different PSDs on a 10 Mbps link for $P(j, 11)$ and $P(j, 22)$ respectively. Figures 7 and 8 show the FEC gain for the same scenarios. Comparing the figures we conclude that FEC(10,11) and FEC(20,22) are qualitatively similar, and thus in the following we will only show figures for FEC(20,22) for brevity. Comparing results obtained with PSDs having the same CoV but different skewness we can see that even though the skewness has an effect on the packet loss process (especially at low loss probabilities), the CoV

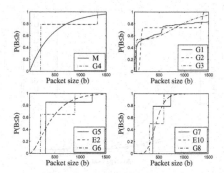

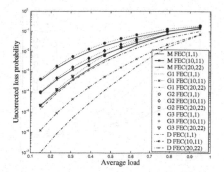

Fig. 1. Cumulative density functions of the considered packet size distributions.

Fig. 2. Average loss probability with and without FEC vs average load on a 10 Mbps link.

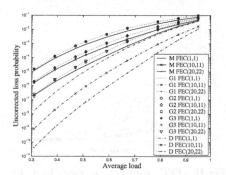

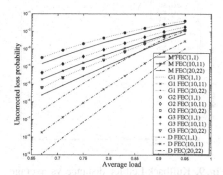

Fig. 3. Average loss probability with and without FEC vs average load on a 22.5 Mbps link.

Fig. 4. Average loss probability with and without FEC vs average load on a 45 Mbps link.

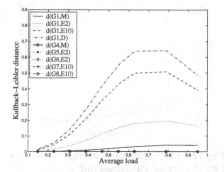

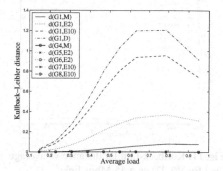

Fig. 5. Kullback-Leibler distance vs average load for P(j,11) on a 10 Mbps link.

Fig. 6. Kullback-Leibler distance vs average load for P(j,22) on a 10 Mbps link.

of the PSD has the biggest impact on the efficiency of FEC. We draw the same conclusion by examining Figures 9, 10, 11 and 12 which show the Kullback-Leibler distance and the FEC gain on a 22 Mbps and a 45 Mbps link as a function of the average load for $P(j, 22)$ and FEC(20,22) respectively. Thus, analytically tractable PSD models (for

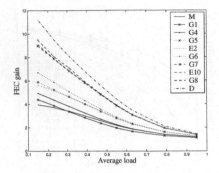

Fig. 7. FEC gain vs average load for FEC(10,11) on a 10 Mbps link.

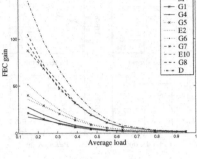

Fig. 8. FEC gain vs average load for FEC(20,22) on a 10 Mbps link.

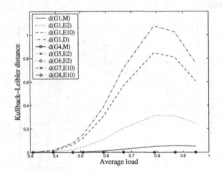

Fig. 9. Kullback-Leibler distance vs average load for P(j,22) on a 22.5 Mbps link.

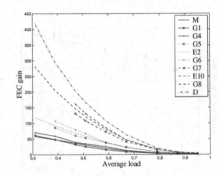

Fig. 10. FEC gain vs average load for FEC(20,22) on a 22.5 Mbps link.

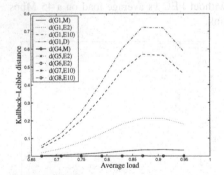

Fig. 11. Kullback-Leibler distance vs average load for P(j,22) on a 45 Mbps link.

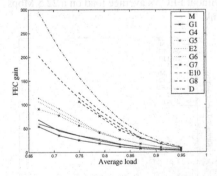

Fig. 12. FEC gain vs average load for FEC(20,22) on a 45 Mbps link.

example phase-type, which includes both the Erlang and the hyper-exponential distributions as special cases, and has an extensive literature [27–30]) can be used to derive approximate results for FEC performance by matching the first two, and accurate results by matching the first three moments of the empirical PSD. Furthermore as the CoV of the PSD in the network is bounded from above, applications can be given a lower bound on the achievable gain of using FEC independent of the packet size distribution in the network. Though for some networks the exponential PSD might fit, it is clear from the results that it does not represent a worst case scenario if the average packet size is not equal to the center of the domain of the PSD and thus the CoV of the PSD can exceed one. Nevertheless, the exponential PSD is a good approximation for the considered empirical PSD G1, and for other empirical PSDs to be found at [9]. This finding justifies the assumption of exponential service time distribution in earlier works on the efficiency of FEC [1, 16–18, 31].

The difference between the results obtained at a particular average load with distributions having different CoV values is significant, up to one order of magnitude in terms of FEC gain in the considered scenarios, a lower CoV value yielding a less bursty loss process. The difference however is partly due to the different average loss probabilities. We eliminate the effects of the average loss probability in the following subsection.

4.2 Constant Average Packet Loss Case

In this subsection we consider results with different PSDs as a function of the average loss probability. This enables us to investigate what an application (unaware of the network PSD) can expect from FEC given that it experiences a certain end-to-end average packet loss probability. In order to be able to compare the packet loss process at a certain average loss probability we take the results from simulations with the G1 PSD and increase the background traffic of the mathematical models to match the average packet loss probability given by the simulations.

Figure 13 shows the Kullback-Leibler distance between the results obtained with the different distributions as a function of the average loss probability on a 10 Mbps link for $P(j,22)$. The figure shows that the distance between the results obtained with different distributions decreased significantly (three orders of magnitude). Figure 14 shows the FEC gain for the same scenario. The effects of the PSD are significantly smaller compared to Figure 8.

Figures 15 and 16 show the FEC gain on a 22.5 Mbps and a 45 Mbps link respectively. Comparing the figures we can see that the difference between results with different PSDs in terms of FEC gain decreases as the link speed increases (from 10 Mbps to 45 Mbps). The reason for this is that the higher the link speed the less the background

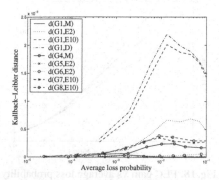

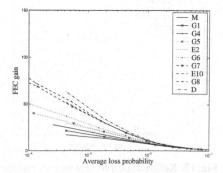

Fig. 13. Kullback-Leibler distance vs average loss probability for P(j,22) on a 10 Mbps link.

Fig. 14. FEC gain vs average loss probability for FEC(20,22) on a 10 Mbps link.

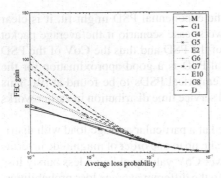

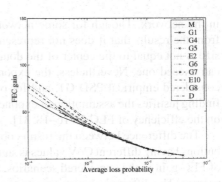

Fig. 15. FEC gain vs average loss probability for FEC(20,22) on a 22.5 Mbps link.

Fig. 16. FEC gain vs average loss probability for FEC(20,22) on a 45 Mbps link.

traffic has to be changed to keep the average loss probability constant, and thus the change in the level of statistical multiplexing decreases.

The observed difference in FEC gain can be due to the difference in the level of statistical multiplexing (the background traffic intensity was increased to maintain the average loss probability constant and as a result the packet loss process became more independent) and to the difference between the packet size distributions.

4.3 Isolating the Effects of the Packet Size Distribution

In this subsection we separate the effects of the level of statistical multiplexing and the PSD. We do it by changing the arrival intensity of both the background traffic and the tagged stream in the mathematical models in order to match the average loss probability given by the simulations with the G1 PSD, thus we keep the level of statistical multiplexing constant (doing so is equivalent to matching the average loss probability through decreasing the link speed). Figure 17 shows the Kullback-Leibler distance as a function of the average loss probability on a 10 Mbps link for $P(j, 22)$. Comparing this to Figure 13 we can see a further significant decrease in the distance of the distributions.

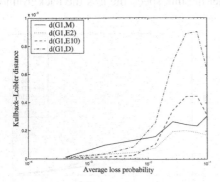

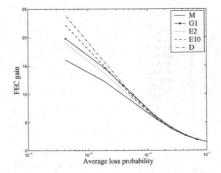

Fig. 17. Kullback-Leibler distance vs average loss probability for P(j,22) on a 10 Mbps link (same level of statistical multiplexing).

Fig. 18. FEC gain vs average loss probability for FEC(20,22) on a 10 Mbps link (same level of statistical multiplexing).

The same effect can be seen in Figure 18, which shows the FEC gain on a 10 Mbps link as a function of the average loss probability for FEC(20,22). Thus the difference in the FEC gain considering a fixed average loss (Section 4.2) is mainly due to the different levels of statistical multiplexing and in a lower extent to the different PSD. This is in accordance with the observation in Section 4.2 that the difference between the results with different PSDs decreases as the link speed increases.

5　Conclusion

In this paper we investigated the effects of the packet size distribution on the packet loss process and the related FEC performance in a single server queue with a finite buffer. We presented a mathematical model for the analysis of the packet loss process of the $MMPP + M/E_r/1/K$ queue and compared the results of simulations and mathematical models in different scenarios. Our results show that analytical models of the PSD matching the first three moments of the empirical PSD can be used to evaluate the performance of FEC in real networks, while the exponential PSD is a reasonable approximation for the PSD of today's Internet to evaluate FEC performance. Nevertheless the exponential PSD is not a worst case scenario, the PSD in today's networks has a higher CoV and thus shows slightly worse queueing performance. However, as the CoV of the packet size distribution in a real network is bounded from above, one can give a lower bound on the efficiency of FEC and thus predict its performance. The results show that the effects of the packet size distribution decrease as the link speed increases if one considers a particular average packet loss probability. Thus at a given average loss probability the actual network PSD does not influence the efficiency of FEC on a backbone link, however it has a big influence on it in access networks. At the same time applications can have a bigger influence on the packet size distribution in access networks and thus have an impact on the packet loss process of their traffic. The results presented here can serve as a basis for future research on the performance of end-to-end error control and facilitate the use of FEC in tomorrow's applications.

References

1. P. Dube and E. Altman, "Utility analysis of simple FEC schemes for VoIP," in *Proc. of Networking 2002*, pp. 226–239, May 2002.
2. K. Kawahara, K. Kumazoe, T. Takine, and Y. Oie, "Forward error correction in ATM networks: An analysis of cell loss distribution in a block," in *Proc. of IEEE INFOCOM*, pp. 1150–1159, June 1994.
3. E. Biersack, "Performance evaluation of forward error correction in ATM networks," in *Proc. of ACM SIGCOMM*, pp. 248–257, August 1992.
4. G. Dán and V. Fodor, "Quality differentiation with source shaping and forward error correction," in *Proc. of MIPS'03*, pp. 222–233, November 2003.
5. P. Frossard, "FEC performances in multimedia streaming," *IEEE Comm. Letters*, vol. 5, no. 3, pp. 122–124, 2001.
6. L. Kleinrock, *Queueing Systems*, vol. I. Wiley, New York, 1975.
7. E. Altman, C. Barakat, and V. M. Ramos, "On the utility of FEC mechanisms for audio applications," in *Proc. of Quality of Future Internet Services, LNCS 2156*, pp. 45–56, 2001.
8. J. W. Cohen, *The Single Server Queue*. North-Holland Publishing, Amsterdam, 1969.
9. Sprint IP Monitoring Project, "http://ipmon.sprint.com/."

10. J. Beran, R. Sherman, M. Taqqu, and W. Willinger, "Long-range dependence in variable-bit-rate video traffic," *IEEE Trans. on Communications*, vol. 43, no. 2/3/4, pp. 1566–1579, 1995.
11. T. Yoshihara, S. Kasahara, and Y. Takahashi, "Practical time-scale fitting of self-similar traffic with markov-modulated poisson process," *Telecommunication Systems*, vol. 17, no. 1-2, pp. 185–211, 2001.
12. B. Ryu and A. Elwalid, "The importance of long-range dependence of VBR video traffic in ATM traffic engineering: Myths and realities," in *Proc. of ACM SIGCOMM*, pp. 3–14, 1996.
13. P. Skelly, M. Schwartz, and S. Dixit, "A histogram-based model for video traffic behavior in an ATM multiplexer," *Trans. on Networking*, vol. 1, pp. 446–458, August 1993.
14. J. Cao, W. S. Cleveland, D. Lin, and D. X. Sun, "Internet traffic tends toward poisson and independent as the load increases," in *Nonlinear Estimation and Classification*, Springer, 2002.
15. T. Karagiannis, M. Molle, and M. Faloutsos, "A nonstationary poisson view of internet traffic," in *Proc. of IEEE INFOCOM*, pp. 1–1, March 2004.
16. I. Cidon, A. Khamisy, and M. Sidi, "Analysis of packet loss processes in high speed networks," *IEEE Trans. on Inform. Theory*, vol. IT-39, pp. 98–108, Jan. 1993.
17. O. Gurewitz, M. Sidi, and I. Cidon, "The ballot theorem strikes again: Packet loss process distribution," *IEEE Trans. on Inform. Theory*, vol. IT-46, pp. 2599–2595, November 2000.
18. E. Altman and A. Jean-Marie, "Loss probabilities for messages with redundant packets feeding a finite buffer," *IEEE Journal on Selected Areas in Comm.*, vol. 16, no. 5, pp. 779–787, 1998.
19. H. Schulzrinne, J. Kurose, and D. Towsley, "Loss correlation for queues with bursty input streams," in *Proc. of IEEE ICC*, pp. 219–224, 1992.
20. G. Dán, V. Fodor, and G. Karlsson, "Analysis of the packet loss process for multimedia traffic," in *Proc. of the 12th International Conference on Telecommunication Systems, Modeling and Analysis*, July 2004.
21. H. Heffes and D. M. Lucantoni, "A markov modulated characterization of packetized voice and data traffic and related statistical multiplexer performance," *IEEE Journal on Selected Areas in Comm.*, vol. 4, pp. 856–868, September 1986.
22. C. Blondia, "The N/G/1 finite capacity queue," *Commun. Statist. - Stochastic Models*, vol. 5, no. 2, pp. 273–294, 1989.
23. J. C. Bolot, "End-to-end delay and loss behavior in the Internet," September 1993.
24. O. J. Boxma, "Sojourn times in cyclic queues - the influence of the slowest server," *Computer Performance and Reliability*, pp. 13–24, 1988.
25. W. Fischer and K. Meier-Hellstern, "The markov-modulated poisson process MMPP cookbook," *Performance Evaluation*, vol. 18, no. 2, pp. 149–171, 1992.
26. S. Kullback, *Information Theory and Statistics*. Wiley, New York, 1959.
27. L. Le Ny and B. Sericola, "Transient analysis of the BMAP/PH/1 queue," *I.J. of Simulation*, vol. 3, no. 3-4, pp. 4–15, 2003.
28. W. Whitt, "Approximating a point process by a renewal process, i: Two basic methods," *Operations Research*, vol. 30, no. 1, pp. 125–147, 1982.
29. M. Neuts, *Matrix Geometric Solutions in Stochastic Models*. John Hopkins University Press, 1981.
30. E. P. C. Kao, "Using state reduction for computing steady state probabilities of queues of GI/PH/1 types," *ORSA J. Comput.*, vol. 3, no. 3, pp. 231–240, 1991.
31. O. Ait-Hellal, E. Altman, A. Jean-Marie, and I. A. Kurkova, "On loss probabilities in presence of redundant packets and several traffic sources," *Performance Evaluation*, vol. 36-37, pp. 485–518, 1999.
32. G. Dán, "Analysis of the loss process in an MMPP+M/M/1/K queue," TRITA-IMIT-LCN R 04:02, KTH/IMIT/LCN, January 2004.

Appendix

The probability $Q_{i,lm}(k)$ denotes the joint conditional probability that between two arrivals from the joint arrival stream there are k exponential stage completions out of i and the state of the MMPP at the moment of the arrival is m given that at the time of the last arrival the MMPP was in state l. $Q_{i,lm}(k)$ can be expressed as

$$
\begin{aligned}
Q_i^{l,m}(k) &= P^{l,m}(k) && if\, k < i \\
Q_i^{l,m}(k) &= \sum_{j=i}^{\infty} P^{l,m}(j) && if\, k = i,
\end{aligned}
\tag{10}
$$

where $P^{l,m}(k)$ denotes the joint probability of having k exponential stage completions between two arrivals and the next arrival coming in state m of the MMPP given that the last arrival came in state l.

The z-transform $P^{l,m}(z)$ of $P^{l,m}(k)$ is given by

$$
P^{l,m}(z) = \sum_{k=0}^{\infty} \left(\int_0^{\infty} \frac{(r\mu t)^k}{k!} e^{-r\mu t} f^{l,m}(t) dt \right) z^k = f^{l,m*}(r\mu - r\mu z),
\tag{11}
$$

where $f^{l,m}(t)$ is the joint distribution of the interarrival-time and the probability that the next arrival is in state m given that the last arrival was in state l of the MMPP. The Laplace transform of $f^{l,m}(t)$ is denoted with $f^{l,m*}(s)$ and is given by

$$
f^{l,m*}(s) = \mathscr{L}\left\{ e^{(\hat{Q}-\hat{\Lambda})x}\hat{\Lambda} \right\} = (sI - \hat{Q} + \hat{\Lambda})^{-1}\hat{\Lambda}.
\tag{12}
$$

The inverse Laplace-transform of (12) and thus the inverse z-transform of (11) can be expressed analytically by partial fraction decomposition as long as $L \le 4$, and has the form

$$
f^{l,m}(t) = \sum_{j=1}^{L} B_j^{l,m} e^{\beta_j t},
\tag{13}
$$

where β_j are the roots of $t(s) = det[sI - \hat{Q} + \hat{\Lambda}]$. Using the substitution $\alpha_j = 1 + \beta_j/\mu$ and $A_j^{l,m} = B_j^{l,m}/(\mu \alpha_j)$ one can calculate $P^{l,m}(k)$ based on (11)

$$
P^{l,m}(k) = \sum_{j=1}^{L} A_j^{l,m} \frac{1}{\alpha_j^k}.
\tag{14}
$$

Given the probability $P^{l,m}(k)$ one can express $Q_i(k)$ as

$$
Q_i(k) = \begin{cases} \sum_{j=1}^{L} A_j^{lm} \left(\frac{1}{\alpha_j} \right)^k & 0 \le k < i \\ \sum_{j=1}^{L} \frac{A_j^{lm}}{1-1/\alpha_j} \left(\frac{1}{\alpha_j} \right)^i & k = i. \end{cases}
\tag{15}
$$

A more detailed description of the calculation of $Q_{i,lm}(k)$ can be found in [32].

Synthesis and MAVAR Characterization
of Self-similar Traffic Traces from Chaotic Generators[*]

Stefano Bregni[1], Eugenio Costamagna[2], Walter Erangoli[1],
Lorenzo Favalli[2], Achille Pattavina[1], and Francesco Tarantola[2]

[1] Politecnico di Milano, Dept. of Electronics and Information
Piazza L. Da Vinci 32, 20133 Milano, Italy
Tel.: +39-02-2399.3503, Fax: +39-02-2399.3413
{bregni,erangoli,pattavina}@elet.polimi.it
[2] Università di Pavia, Dept. of Electronics, Via Ferrata 1, 27100 Pavia, Italy
Tel.: +39-0382-985200, Fax: +39-0382-422583
{Eugenio.Costamagna,Lorenzo.Favalli,Francesco.Tarantola}
@unipv.it

Abstract. Experimental measurements show that many relevant processes in telecommunication engineering exhibit self-similarity and long-range dependence (LRD) characteristics. Internet traffic is a significant example. A traditional parameter used to characterize self-similarity and LRD is the Hurst parameter H. Recently, the Modified Allan Variance (MAVAR) has been proposed to estimate the power-law spectrum and thus the Hurst parameter of LRD series. Chaotic generators have been introduced in the last years to mimic time series derived both from sampled Internet or packet video traffic. They are built by means of optimized weighted sums of geometric characteristic sampled from Lorenz strange attractors. The optimization of the structure is obtained observing both the variation coefficients (i.e., a traditional long term feature well matched to the analysis of the error gap processes, related to the Variance-Time plot and to the Hurst parameter too), and, more recently, the logscale diagram. Then, it was quite natural to explore the MAVAR characterization of time series derived from the above chaotic generators, and to compare it to that exhibited by the mimicked process, looking for a fruitful inclusion of MAVAR terms in the optimization cost function.

1 Introduction

In perspective, Internet is expected to become the main connecting infrastructure, enabling a whole variety of applications, and the notion of "quality of service" needs to be included in the Internet paradigm too. So, a large amount of efforts is currently being spent to determine suitable performance measurements, access schemes, and resource assignment procedures, to allow the sharing of this infrastructure among general users and privileged users, which need not to be trapped in congestions or suffer from sudden bottlenecks. The evaluation of such schemes requires the availability of accurate models of the traffic process.

[*] Work partially supported by the Italian Ministry of Education, University and Research (MIUR) under the FIRB project "TANGO" and the PRIN project "SHINES".

Traditionally, stochastic models (and specifically traffic models) are often implemented in the form of Markov processes, as they provide a flexible framework that can be customized to fit different system properties. They are also attractive since the statistical behavior of the system can be directly derived from the model. Their drawback is the definition of the state diagram and of its transition probabilities to properly describe the target system. This often leads to large and complex structures with several parameters to be tuned. Moreover, it has been shown that packet traffic exhibits long-range dependence (LRD) and selfsimilarity characteristics [1,2].

The discovery of these features forced traffic engineering to take them into account both in network design and in modeling theory and practice, to be able to generate models of traffic behavior accurate enough for the novel network paradigm. Several approaches have been adopted ranging from use of fractional brownian motion to heavy tailed and power-tailed distributions via superposition of on-off sources, or to the chaotic generators used in the following, as a tool to simulate the behavior of the error gap process in fading digital transmission channels. These generators are built by means of summing up geometric coordinates sampled from the trajectories of a suitable number of Lorenz attractors: both the weight of the sum and the velocities by which the trajectories are traversed are optimized, usually by means of a Nelder and Mead simplex procedure in the present case. The structure of the generators has been found effective to mimic time series derived from sampled Internet and packet video traffic too. The optimization of the structure is obtained observing both the variation coefficients (i.e., a traditional long term feature well matched to the analysis of the error gap processes, related to the Variance-Time plot and to the Hurst parameter too), and, more recently, the logscale diagram.

In this work we explore the Modified Allan Variance (MAVAR) characterization of time series derived from the above chaotic generators, and we compare it to that exhibited by the mimicked process, looking for a fruitful inclusion of MAVAR terms in the optimization cost function. Various sequences are generated after optimizing the generator parameters to mimic the variation coefficient behavior or the logscale diagram of some target time series, sampled from measured packet video or Internet traffic. Then, we use MAVAR to compare the statistical characteristics of the sequences, due to its superior sensitivity as spectral analysis tool.

1.1 The Modified Allan Variance

In time and frequency measurement theory, a well-known tool in the time domain for stability characterization of precision oscillators is the Modified Allan Variance (MAVAR) Mod $\sigma_y^2(\tau)$ [3-8]. This variance was proposed in 1981 by modifying the definition of the two-sample variance recommended by IEEE in 1971 for characterization of frequency stability [5]. Compared to the poor discrimination capability of the original Allan variance against white and flicker phase noise, the MAVAR discriminates effectively all power-law noise types recognized very commonly in frequency sources.

Given an infinite sequence $\{x_k\}$ of samples evenly spaced in time with sampling period τ_0, the MAVAR is defined as

$$\text{Mod } \sigma_y^2(n\tau_0) = \frac{1}{2n^2\tau_0^2}\left\langle\left[\frac{1}{n}\sum_{j=1}^{n}\left(x_{j+2n} - 2x_{j+n} + x_j\right)\right]^2\right\rangle, \tag{1}$$

where the observation interval is $\tau = n\tau_0$. In time and frequency stability characterization, the data sequence $\{x_k\}$ is made of samples of random time deviation $x(t)$ of the chronosignal under test. To summarize, modified Allan variance differs from basic Allan variance in the additional average over n adjacent measurements. For $n=1$ ($\tau = \tau_0$), the two variances coincide.

In practical measurements, given a finite set of N samples $\{x_k\}$, spaced by sampling period τ_0, an estimate of MAVAR can be computed using the standard estimator [7]

$$\text{Mod } \sigma_y^2(\tau) = \frac{1}{2n^4\tau_0^2(N-3n+1)}\sum_{j=1}^{N-3n+1}\left[\sum_{i=j}^{n+j-1}\left(x_{i+2n} - 2x_{i+n} + x_i\right)\right]^2, \tag{2}$$

with $n=1, 2,..., \lfloor N/3 \rfloor$. A recursive algorithm for fast computation of this estimator exists [3], which cuts down the number of operations to $O(N)$ instead of $O(N^2)$.

Following its definition, the MAVAR can be seen as the mean-square value of the signal output from a hypothetical filter, with proper impulse response shaped according to (1), receiving the data sequence $\{x_k\}$. Hence, the MAVAR can be also defined in the frequency domain [3][8], as the area under the spectral density of the signal output from such filter, i.e.,

$$\text{Mod } \sigma_y^2(\tau) = \int_0^\infty S_x(f)(2\pi f)^2 \frac{2\sin^6 \pi\tau f}{(n\pi f)^2 \sin^2 \pi\frac{\tau}{n}f}df, \tag{3}$$

where $S_x(f)$ is the power spectral density of input signal $x(t)$.

We are interested in power-law processes, whose spectral density behaves asymptotically for $f \to 0$ as (~ means "asymptotically of the order of")

$$S_x(f) \sim k_1 f^\alpha \tag{4}$$

for $-1 \leq \alpha \leq 0$ (LRD). Under this hypothesis (actually for the whole range $-4 \leq \alpha \leq 0$, $\alpha \in \Re$), the MAVAR obeys asymptotically to a power law of the observation time τ, as

$$\text{Mod } \sigma_y^2(\tau) \sim k_2\tau^\mu, \tag{5}$$

where $\mu = -3-\alpha$ [3][8].

1.2 Chaotic Models Derived from Lorenz Attractors

For the chaotic model, the approach is similar to that used in [10], and is derived almost straightforwardly from the experience gained in simulating error gap series in mobile radio channels and the attenuation process in satellite links. It stems from the observation that a chaotic attractor provides a description of a dynamical system joining a deterministic mechanism to an unpredictable behavior. This translates in a number of properties, the most important being the fact that small differences in sampling times along the trajectory result in completely different evolutions of sampled series, and the capability of exhibiting long term correlation properties after short term behaviors apparently uncorrelated with the past evolution. The modeling procedure is thoroughly described in [10] with reference to the error gap process in multipath transmission channels. It is based on a weighted sum of components derived from Lorenz strange attractors [11] as in the following formula:

$$F\left[f\Sigma_i(w_i\,u_i)\right] \tag{6}$$

where:

- $f(\bullet)$ is a polynomial or exponential function: we present here results obtained with exponentials;
- the u_i variables are selected coordinates or geometric distances sampled on the attractor trajectories (the x coordinate in the current case), and the w_i are suitable weights;
- explores the attractors: in these experiments the number of attractors is seven;
- weights and sampling distances are optimized to match the characteristics of the target sequence.

F($\bullet$) is a probability shaping function, used to match the distribution of the sample amplitudes derived from (6) to the target time series. This function is obtained, after some preliminary sample generation runs of the model, by means of a suitable inversion procedure for the cumulative probability function, providing good correspondence with the target probability distribution. This matching is only an ancillary duty: the whole modeling procedure is performed to mimic more interesting features of the target sequence, i.e., correlation and long term properties.

Note that we are not attempting to describe any *physical* characteristic of the underlying system, and that the generated time series are likely to represent realizations of the target stochastic process, but no local correspondence with any target sequence is obtainable. This is very different from the behavior of the Markovian models introduced for the sake of comparison.

To optimize the parameters in (6), various cost functions have been implemented, tailored to the actual problem. When mimicking error gap processes, the main features observed were the block error probabilities for various block lengths, windowed autocorrelation functions, and the variation coefficients already introduced in [12] and references quoted therein. Tabu search, metaheuristics or the Nelder and Mead simplex procedure [13] have been experimented, and the last is used here.

In the video traffic and packet data cases, emphasis was done to medium term moving autocorrelations, short term moving autocovariance functions and histograms, and to some mean differences, averaged over the sequence length, among samples taken at medium term distances, i.e., 20, 100 and 1000 samples. However, as in the error gap case, the main long term feature were the variation coefficients, derived for large intervals of the multisample order r: even if the original meaning of the multigaps, i.e., summed up lengths of consecutive gaps between errors is lost, the curves $K(r)$s computed following [12] keep their significance of comparison with a well defined reference process. Moreover, the well known Variance-Time plots are derived from very similar computations, the main difference with respect to the variation coefficient computations being to consider non overlapping intervals in the related aggregate processes instead of overlapping.

Good results have been obtained in several conditions, and some examples are described in [14-16], even if many problems concerning the optimization and the very structure of generators based on (6) are still open, deserving further work.

In [17], the comparison of the logscale diagrams [1] of the target and of the optimized time series was introduced in the cost function, both replacing the variation coefficients term or to do simultaneous optimization, and again the results were promising.

Notice that in [15] and [17], besides comparing the statistical features exhibited by the time series obtained from the optimized chaotic generators with the target ones, we try to compare the behaviors of the system at higher level, i.e., at the transmission network level: we observe the cumulative distributions of the interarrival delays for various link utilizations when the target and the modeled sequences are supplied to a congested node. In a way, the present work is providing further evaluation of the modeling results, based on statistical features not directly optimized.

Markov models have been applied since the beginning to describe the behavior of traffic events, and we use here hidden Markov models (HMM) [18,19] to provide comparison data, as done in [14-17]. In HMMs, there is no more correspondence between a state and a physical event and the properties are hidden in the model structure, as recalled in the name. From a mathematical point of view, an HMM can be described as a 5-tuple, $\lambda=(S,V,A,B,\Pi)$, where S is the set of the states of the system, with cardinality N; V is the set of observable values ("the alphabet"), with cardinality M; A is the matrix of state transition probabilities; B is the matrix of observable probabilities, and Π is the initial state vector. The parameters of the model are commonly provided using the method of moments, or gradient, or the Baum Welch (BW) procedure [14,15] used in this work. The BW algorithm is very robust in that it always converges, but there is no guarantee that it converges to a global maximum: consequently, the optimized parameters may be not the optimal ones in an absolute sense. Furthermore, there is no guidance in the best selection of the number of states, where the convergence speed is highly influenced by the size of the alphabet of symbols V.

Several parameter sets have been checked here for the HMM models: since previous results were not greatly improved for numbers of states larger than 10, and large numbers lead to very large computing times with long sequences as analyzed in this paper, 10 states have been used. The data sets need also to be discretized for the Baum-Welch algorithm to converge in reasonable time: the presented results are for a number of symbols in the alphabet equal to 50. With these numbers, a sequence of about 10^6 events is processed in a few hours on a 1.5 GHz PIV computer. Similar times are necessary to optimize the chaotic generators, depending on the difficulty of the task (and the skillfulness, or luckiness, of the user).

2 Models for a Packet Video Sequence

The time series we model is the Group of Pictures (GOP) sequence already considered in [14], sampling aggregate traffic derived from the "True Lies" movie, 13,533 GOP long. We provide a HMM sequence of the same length and two chaotic generator models, optimized in slightly different conditions for about 17,000 and 19,000 GOPs. We mimic windowed autocorrelation and short term autocovariance functions of the target, windowed histograms for both, but the variation coefficients $K(r)$s up to order $r=1,000$ for the first, and the logscale diagram of the target for the second. The variation coefficients curves and the logscale diagrams are reported for all sequences in figures 1 and 2. For the last, we follow the notation in [1]: abscissas are the octaves j and ordinates are the relative energies in logarithmic units.

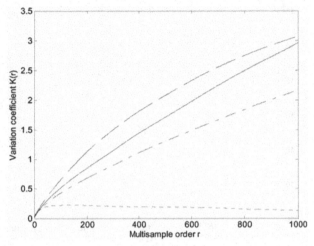

Fig. 1. The variation coefficient behaviors for the target "True Lies" GOP sequence and for some modeled time series. Solid line: the target; dashed line: the model obtained optimizing the variation coefficients; dashed-dotted line: the model obtained optimizing the logscale diagram; dots: the HMM sequence

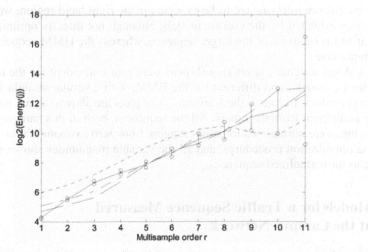

Fig. 2. The logscale diagram for the same target and modeled GOP time series as in Fig. 1, with the same line symbols. Confidence intervals are shown only for the target, and were similar for the three models

We see that the variation coefficients curves obtained mimicking the $K(r)$s them-selves or mimicking the logscale diagram are not very different. When optimized, the logscale diagram is similar to the target one. Otherwise the main differences are found in the right hand region, where the confidence intervals are very large. On the con-trary, both the variation curve and the logscale diagram computed for the HMM se-quence are very different from the target. The short term statistical features obtained for all models (not shown) were similar, and good in any case.

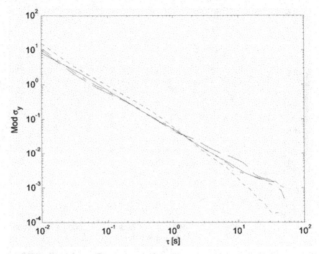

Fig. 3. The modified Allan variance characterization for the same target and modeled GOP time series in Fig. 1, with the same line symbols

The modified Allan variance behaviors are shown in Fig. 3, for $\tau_0 = 10^{-2}$ s: even if large confidence intervals are to be expected in the right hand region, we see that the behaviors exhibited by the chaotic models, although not directly optimized, are very similar to the behavior of the target sequence, whereas the HMM sequence leads to a different curve.

The Allan variance curves (not shown) were near coincident for the target and for the chaotic models, and different for the HMM, with a similar straight line behavior but larger values. In Fig. 4, the Variance-Time plots are shown: again, only the HMM curve is different from the others. All the sequences, both in this paragraph and in the following, were normalized to values ranging from zero to about 100 or 200, suitable for the optimization procedures, and all the variable magnitudes shown in the figures relate to the normalized sequences.

3   Models for a Traffic Sequence Measured at the Campus Network

A packet size sequence of aggregate Internet traffic was measured at the border gate-way of the Pavia University campus network, already used in [16], 500,000 packet long. We study here the modified Allan variances for this target sequence, two chaotic

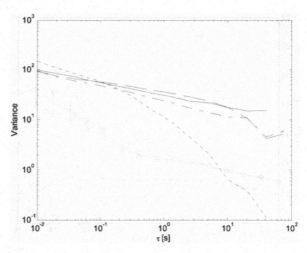

Fig. 4. The Variance-Time plots for the sequences in Fig. 1, with the same line symbols

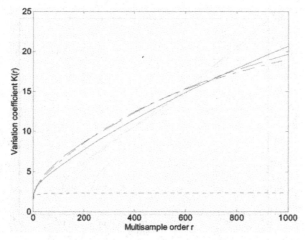

Fig. 5. The variation coefficients for a target measured aggregate Internet traffic sequence and for various models. Solid line: the target sequence; dashed line and dashed-dotted lines: a short and a long chaotic sequences, both optimized for the variation coefficients; dotted line: the HMM sequence

models of 500,000 and 53,000 packets, optimised in slightly different conditions for the variation coefficients and short term features, and a HMM model (500,000 packet long).

The variation coefficient behaviors are shown in Fig. 5: both the chaotic models are well mimicking the target curves, whereas the HMM is very different. The logscale diagrams are reported in Fig. 6, and, although not optimised, they look similar to the target for the chaotic generators, and again very different for the HMM. The same is true for the MAVAR diagrams and for the Variance-Time plots in figures 7 and 8, even if the curves for the short chaotic sequence cannot reach the right hand of the figure.

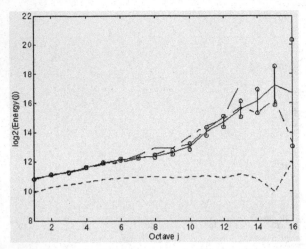

Fig. 6. The logscale diagrams for the time series in Fig. 5, with the same line symbols

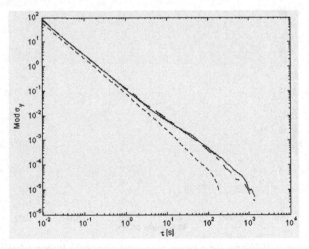

Fig. 7. The modified Allan variance characterization for the same target and modeled time series in Fig. 5, with the same line symbols. Notice that the curve for the short chaotic sequence ends at about $\tau = 10^3$ s; both the chaotic sequences exhibit curves very similar to the target, whereas the HMM curve is lower

Notice the very good match of the target and the chaotic MAVAR curves, maintained in the right hand region.

4 A Further Example: A MAWI Sequence

A further example is provided by a MAWI trans-Pacific T1 line traffic trace [20] and the models derived from it, i.e., a sequence optimized for the variation coefficients, a sequence optimized for the logscale diagram, a sequence optimized for both, and again a HMM, all 100,000 samples long.

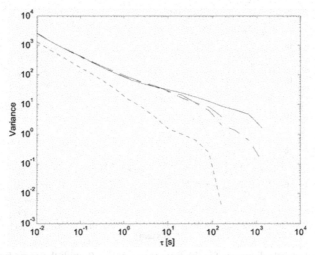

Fig. 8. The Variance-Time plots for the sequences in Fig. 5, with the same line symbols

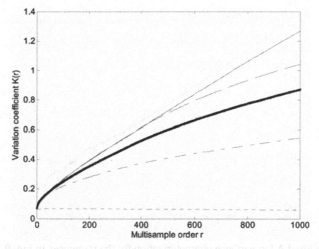

Fig. 9. The variation coefficients for the MAWI sequence and for various models. Solid line: the target sequence; dashed line, dashed-dotted line and heavy dots: three sequences optimized respectively for the variation coefficients, for the logscale diagram, and for both; dotted line: the HMM sequence

First of all, we observe that to optimize simultaneously the variation coefficient curve and the logscale diagram often seems to be an easy task, and the results in figures 9 and 10 are comforting.

Then, a conclusion similar to that derived for the previous examples holds, as all the chaotic models lead to MAVAR curves similar to the target, both for the MAVAR characterization and for the Variance-Time plot, where the HMM leads to different results, as seen in figures 11 and 12.

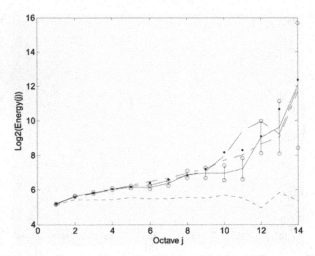

Fig. 10. The logscale diagrams for the time series in Fig. 9, with the same line symbols

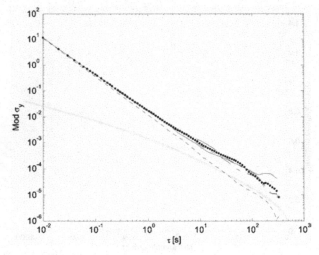

Fig. 11. The modified Allan variance characterization for the sequences in Fig. 9, with the same line symbols

5 Conclusions

Various sequences have been generated by means of chaotic generators based on Lorenz attractors, after optimizing the generator parameters to mimic the variation coefficient behavior or the logscale diagram of some target time series, sampled from measured packet video or Internet traffic. Then we analyzed the MAVAR characterizations of the sequences.

We observe that, although not directly used in optimizing coefficients, the MAVAR behavior exhibited by the chaotic sequences looks very similar to that of the target. The same holds for the Variance-Time plots. For the same targets, the behavior

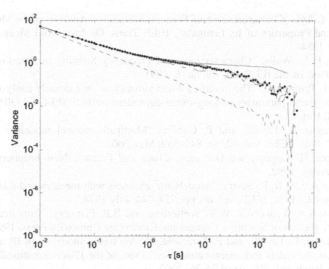

Fig. 12. The Variance-Time plots for the sequences in Fig. 9, with the same line symbols

obtained in sequences obtained by HMM looks quite different, emphasizing the significance of the comparison. For the sake of comparison, HMM sequences derived from a Baum-Welch procedure have been analyzed, and their MAVAR characteristics have been found very different from the target, as were the variation coefficient curves.

The next step in research will be direct optimization based on MAVAR characteristics, opening a new approach in chaotic modeling of traffic and more various time series. We expect to achieve an even closer approximation of the spectral characteristics of the target sequence, due to the superior sensitivity of MAVAR as spectral analysis tool.

References

1. K. Park and W. Willinger, Eds, Self Similar Network Traffic and Performance Evaluation. Wiley Interscience, 2000.
2. P. Abry, R. Baraniuk, P. Flandrin, R. Riedi, D. Veitch, "The Multiscale Nature of Network Traffic: Discovery Analysis and Modelling," IEEE Signal Processing Magazine, April 2002.
3. S. Bregni, Synchronization of Digital Telecommunication Networks. Chichester, UK: John Wiley and Sons, March 2002.
4. J. A. Barnes, A. R. Chi, L. S. Cutler, D. J. Healey, D. B. Leeson, T. E. McGunigal, J. A. Mullen Jr., W. L. Smith, R. L. Sydnor, R. F. C. Vessot and G. M. R. Winkler, "Characterization of Frequency Stability," IEEE Trans. on Instr. and Meas., vol. IM-20, no. 2, May 1971.
5. D. W. Allan, J. A. Barnes, "A Modified Allan Variance with Increased Oscillator Characterization Ability", Proc. of the 35th Annual Frequency Control Symposium, 1981.
6. L. G. Bernier, "Theoretical Analysis of the Modified Allan Variance," Proc. of the 41st Annual Frequency Control Symposium, 1987.

7. P. Lesage, T. Ayi, "Characterization of Frequency Stability: Analysis of the Modified Allan Variance and Properties of Its Estimate", IEEE Trans. On Instr. And Meas., vol. IM-33, no. 4, Dec. 1984.
8. J. Rutman, F. L. Walls, "Characterization of Frequency Stability in Precision Frequency Sources", Proc. of the IEEE, vol. 79, no. 6, 1991.
9. S. Bregni, L. Primcrano, "The modified Allan variance as time-domain analysis tool for estimating the Hurst parameter of long-range dependent traffic," IEEE GLOBECOM 2004, Dallas, TX, Usa.
10. E. Costamagna, L. Favalli, and P. Gamba, "Multipath channel modeling with chaotic attractors," Proc. IEEE, vol. 90, pp. 842-859, May 2002.
11. H.O. Peitgen, H. Jurgens, and D. Saupe, Chaos and Fractals, New Frontiers of Science, Springer-Verlag, 1992.
12. L. N. Kanal and A. R. K. Sastry, "Models for channels with memory and their application to error control," Proc. IEEE, vol. 66, pp. 724-744, July 1978.
13. W.H. Press, S.A. Teukolsky, W.T. Wetterling and B.P. Flannery, Numerical Recipes in FORTRAN, the Art of Scientific Computing, Cambridge University Press, 1992.
14. E. Costamagna, L. Favalli, and F. Tarantola, "Video traffic modeling in IP networks with hidden Markov models and chaotic attractors," Proc. of the 12th international Packetvideo Workshop, Pittsburgh, PA, April 24-26, 2002.
15. E. Costamagna, L. Favalli, and F. Tarantola, "Chaos equation and hidden Markov models for packet video traffic simulation," in E. Del Re, Ed., Proc. of EMPS 2002, Baveno/Stresa, Italy, September 25-26, 2002.
16. E. Costamagna, L. Favalli and F. Tarantola, Modeling and Analysis of Aggregate and Single Stream Internet Traffic," IEEE GLOBECOM 2003, San Francisco, CA, Dec. 1-5, 2003.
17. E. Costamagna, L. Favalli, P. Savazzi and F. Tarantola, "Chaos model for sample time series and wavelet analysis," WPMC 2004, Abano Terme, Italy, Sept. 12-15, 2004.
18. L. R. Rabiner, "A tutorial on hidden Markov models and selected applications in speech recognition," Proc. IEEE, vol. 77, 1989, pp. 257-286.
19. W. Turin, and M. M. Sondhi, "Modeling error sources in digital channels" IEEE J. Select. Areas Commun., vol. 11, 1993, pp. 340-347.
20. Traffic trace repository is accessible at http://tracer.csl.sony.co.jp (maintained by MAWI Working Group).

Coupled Kermack-McKendrick Models for Randomly Scanning and Bandwidth-Saturating Internet Worms*

George Kesidis[1], Ihab Hamadeh[2], and Soranun Jiwasurat[2]

[1] Department of Electrical Engineering, Department of Computer Science and Engineering,
Pennsylvania State University, University Park, PA 16802
kesidis@engr.psu.edu
[2] Department of Computer Science and Engineering,
Pennsylvania State University, University Park, PA 16802
{hamadeh,soranun}@cse.psu.edu

Abstract. We present a simple, deterministic mathematical model for the spread of randomly scanning and bandwidth-saturating Internet worms. Such worms include Slammer and Witty, both of which spread extremely rapidly. Our model, consisting of coupled Kermack-McKendrick equations, captures both the measured scanning activity of the worm and the network limitation of its spread, i.e., the effective scan-rate per worm/infective. We fit our model to available data for the Slammer worm and demonstrate its ability to accurately represent Slammer's total scan-rate to the core.

1 Introduction

The spread of worms is now a chronic problem affecting the performance of the entire Internet. Certain worms, such as Blaster, Slammer and Witty, had very costly effects on the Internetworking community. Their propagation activity congested network links thereby creating a temporary denial-of-access to the Internet for large population of end-hosts. At a minimum, this together with the required response to the worm (e.g., patching to inoculate or cure infected end-hosts against the worm), resulted in a significant direct expenditure and very significant aggregate loss of productivity.

In this paper, we focus on bandwidth-limited, random UDP-scanning worms like Slammer and Witty. These worms spread extremely rapidly in the wild: on Saturday, January 25, 2003, Slammer infected about 75 thousand SQL servers (nearly the entire population of susceptibles) in less than 10 minutes [7] and caused significant congestion in the stub-links connecting peripheral enterprise networks to the Internet core. In previous work [10], we conducted Slammer recreation experiments on an Internet-wide scale and demonstrated that the worm could be accurately "scaled-down" on the DETER cyber security testbed [4]. Our experiments were compared against Slammer data obtained from the University of Wisconsin's tarpit and reported in [7, 10]. This data is used in the following to validate our proposed model.

---

* This work is supported by both the NSF and DHS of the United States under NSF grant number 0335241.

M. Ajmone Marsan et al. (Eds.): QoS-IP 2005, LNCS 3375, pp. 101–109, 2005.
© Springer-Verlag Berlin Heidelberg 2005

In, e.g., [6, 11], a case was made for deployment of worm defenses (detection and response) in peripheral enterprise networks. Because of the need to realistically scale-down the simulation, we advocate an approach where a single enterprise network under test is simulated in much greater detail than the rest of the Internet to which it is connected. Therefore, a basic requirement for our testbed is the ability to realistically recreate a worm attack in this context; specifically, we need an accurate model of the worm probing (scanning) activity *from* the Internet *to* the enterprise network under test. Among our assumptions is that this quantity and the scans generated from the enterprise under test to the rest of the (much larger) Internet are negligibly dependent.

Such traffic generators can be formulated by using measured data from a particular worm (more precisely, extrapolations from measured data [7]) of total instantaneous scan-rate, $S(t)$, when this is available or by using a mathematical model whose parameters can be

- fit to the salient data of a given worm (again, if that data is available) or
- varied in an attempt to capture the behavior of actual worms for which measured Internet data is unavailable or set for hypothetical worms.

Under random scanning, the scan-rate from the Internet directed at the enterprise under simulation could be approximated as $(A/2^{32})S(t)$ where A is the size of the address space of enterprise network under simulation; alternatively, a similar but random thinning of $S(t)$ could be used. Individual scans would be directed to an end-system of the enterprise network that is chosen at random.

This paper is organized as follows. In Section 2, we briefly describe two salient characteristics of Slammer's spread that will subsequently be used to validate our model. In Section 3, we describe a homogeneous network model that accurately predicts the total scanning rate per worm (infective) but overestimates the total instantaneous scanning rate. A heterogeneous model is then presented in Section 4 that can be fit to both characteristics. We conclude with a summary and discussion of current and future work.

2   Slammer's Internet Spread and General Model Assumptions

The success of the simple Kermack-McKendrick model for certain Internet worms, e.g., Code Red, was demonstrated in [9, 12, 1, 13, 8]. Modeling Slammer and Witty is substantially more complex because network bandwidth limitations mitigated the spread of the worm, i.e., the worm's scanning activity saturated certain links. Beyond just spreading very quickly, Slammer was the first significant worm without a constant scanning rate.

Figure 1 [7] shows Slammer's total instantaneous scanning rate extrapolated from data measured at the University of Wisconsin's tarpit. Note that at around 175s (zero slope), there was a brief period in which tarpit data was unavailable. Dividing this data by an estimate of the number of infected end-systems (again, based on the tarpit data) gives the scan-rate per worm (Figure 2 [7]) which quickly declines from an initial peak of over 19,000 scans-per-second-per-worm to slightly over 800 scans-per-second-per-worm over the course of the infection. This behavior was not seen in previous worms. In [10], a similar signature was displayed for Witty and, therefore, it appears to be typical

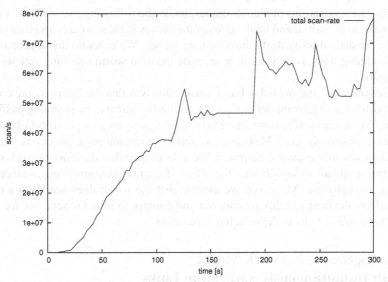

Fig. 1. Slammer's total scanning rate, as measured at the University of Wisconsin Tarpit Network.

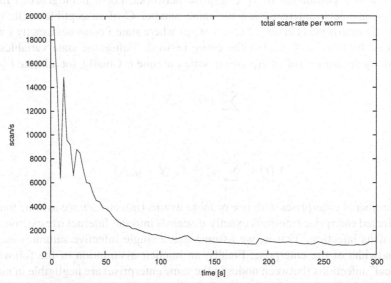

Fig. 2. Slammer's scan-rate per worm, as derived from the scans seen by the University of Wisconsin Tarpit Network.

of bandwidth-limited scanning worms. Because a rather complete data-set is available only Slammer, especially in the preliminary stages of the worm, we focus on Slammer herein.

It was argued in [7, 10] that the oscillations observed in these figures are due to measurement errors by the tarpit, errors that were amplified when the data was extrapolated to an estimate of the activity on the scale of the entire Internet. See [2] for a discussion of measurement variations of blackholes/tarpits. In the following, we will "fit" our proposed model to the initial and final values of the curves in these figures and then roughly check the model's fidelity to the intermediate values. We reiterate that a primary goal of our modeling work is to accurately recreate the total worm scanning activity to the core, Figure 1.

In the following and in [10], a basic assumption was that the Internet core connecting the peripheral enterprise networks only negligibly affects any scanning traffic they generate. Worm scan traffic is assumed to be limited only by a single stub-link connecting the enterprise to the core. Moreover, we ignore intra-enterprise infections (i.e., from an infective within the same enterprise). We also assume that the class of worms under consideration spread so rapidly that the effect of counter-measures (e.g., vulnerability patching) is negligible. Moreover, we assume that the worm does not harm a host so as to mitigate the host's ability to scan out and attempt to infect others. So, we do not model "removals" [3] from the infected population.

3   Homogeneous Network Model with Instantaneously Saturating Links

Consider now a population of N enterprise networks. For a homogeneous Internet model, assume each enterprise has the same number C of susceptible (SQL server) nodes. Each enterprise is in one of $C + 1$ states where state i connotes exactly i worms (infectives) for $0 \leq i \leq C$. For the entire network, define the state variables $y_i(t)$ representing the number of enterprises in state i at time t. Clearly, for all time $t \geq 0$

$$\sum_{i=0}^{C} y_i(t) = N. \tag{1}$$

Define

$$Y(t) \equiv \sum_{i=1}^{C} y_i(t) = N - y_0(t)$$

as the number of enterprises with one or more worms (infectives); we assume that each such infected enterprise transmits exactly σ scans/s into the Internet irrespective of the degree of its infection. That is, we assume that a single infective saturates the stub-link bandwidth of the enterprise. Finally, an implicit assumption of the following is that "local" infections (between nodes in the same enterprise) are negligible in number. Thus, the total rate of scanning (causing infection) into the Internet at time t is

$$S(t) = \sigma Y(t).$$

The likelihood that a particular susceptible is infected by a scan is $\eta = 2^{-32}$ (purely random scanning in the 32-bit IPv4 address space). The likelihood, therefore, that a scan causes an enterprise in state i at time t to transition to state $i + 1$ is $(C - i)\eta$ because there are $C - i$ susceptible but not infected nodes in the enterprise at time t. Thus, define the infection "rate" of an enterprise in state i by

$$\beta_i \equiv \sigma \eta (C - i).$$

The time-evolutions of the states y_i are governed by the following coupled Kermack-McKendrick equations: For times $t \geq 0$,

$$\dot{y}_C(t) = \beta_{C-1} y_{C-1}(t) Y(t), \tag{2}$$

$$\dot{y}_i(t) = (\beta_{i-1} y_{i-1}(t) - \beta_i y_i(t)) Y(t) \text{ for } 1 \leq i \leq C - 1 \tag{3}$$

$$\dot{y}_0(t) = -\beta_0 y_0(t) Y(t). \tag{4}$$

The total number of worms (infectives) at time t is clearly

$$\sum_{i=1}^{C} i y_i(t). \tag{5}$$

Thus, the scan-rate per worm (per infective) is

$$\frac{\sigma Y(t)}{\sum_{i=1}^{C} i y_i(t)} = \frac{\sigma \sum_{i=1}^{C} y_i(t)}{\sum_{i=1}^{C} i y_i(t)}. \tag{6}$$

Summing equations $i = 1$ to C yields the "standard" Kermack-McKendrick equation $dY/dt = \beta_0 y_0 Y = \beta_0(N - Y)Y$ whose solution is $Y(t) = NY(0)[Y(0) + (N - Y(0)) \exp(-\beta_0 N t)]^{-1}$.

For the homogeneous mathematical network model instantaneously (after just one infection) saturating links, we fitted just three parameters to measured data:

- The initial value of scan-rate per worm: $\sigma = 15000$
- The ratio of initial to final value of scan-rate per worm: $C = 18$
- The final value of total instantaneous scan-rate, $NC\sigma$ (or simply from the total number of initially susceptible (and ultimately infected) end-systems, $NC = 73,782$) giving $N = 4099$.

Numerically solving (2)-(4) with initial conditions $y_0(0) = N - 1$ and $y_1(0) = 1$ (i.e., one initially infected server) yielded the "homog-i"[1] curves given in Figure 3 and 4 over a 5 minute period. Note that Figure 4 is roughly similar to that which was extrapolated from measured data of Slammer's actual spread in the Internet, but the total instantaneous scan-rate "homog-i" curve of Figure 3 significantly overestimates that drawn from measured data. If we reduce the cluster size from 18 to, say, $C = 9$ (thereby doubling the number of enterprises to $N = 8198$), we would see that the total scan-rate curve will more accurately shift to the right, but the scan-rate per worm curve would also shift slightly right and its limiting value will increase (double in value if $C = 9$). This simple deterministic mathematical model of a homogeneous network with instantaneous link saturation yielded numerical results (reported in the next section) similar to those obtained by simulation of the "homogeneous clusters" model in [10].

---

[1] *homog*eneous enterprises with *i*nstantaneous (after just one infection) saturation of the stub-link.

4   Heterogeneous Network Model with Gradually Saturating Links

The previous model can be extended to allow for more gradual stub-link saturation as a function of the number of infectives and network heterogeneity can be introduced by creating different classes of enterprise networks where networks of class j:

 - have $C(j)$ susceptibles
 - have *maximum* scan-rate $\sigma_{j,C(j)}$

That is, the scan-rate to the Internet for class-j enterprise with $i \leq C(j)$ infectives is $\sigma_{j,i}$ where, under gradual link saturation, $\sigma_{j,i}$ is nondecreasing and subproportional to i, i.e., $\sigma_{j,i} \leq i\sigma_{j,1}$ for all $i \geq 1$ (but note that this model also accommodates scanning dynamics that do *not* saturate links, i.e., $\sigma_{j,i} = i\sigma_{j,1}$).

Let $y_{j,i}(t)$ be the number of class-j enterprises at time t with i infectives and let $N(j)$ be the total number of class-j enterprises. Thus,

$$\sum_j N(j) = N \text{ and } \sum_{i=0}^{C(j)} y_{j,i}(t) = N(j)$$

for all times t. Also, at time t, the total number of infected end-systems (worms, infectives) is

$$\sum_{j,i} iy_{j,i}(t),$$

the total instantaneous scan-rate is

$$S(t) \equiv \sum_{j,i} \sigma_{j,i}y_{j,i}(t), \tag{7}$$

and the scan-rate per worm is the ratio of these two quantities. Thus, a more general set of coupled Kermack-McKendrick equations modeling worm spread than those used for a homogeneous network is as follows: For times $t \geq 0$ and all classes j:

$$\dot{y}_{j,C(j)}(t) = \eta y_{j,C(j)-1}(t)S(t)$$
$$\dot{y}_{j,i}(t) = \eta[(C(j) - i + 1)y_{j,i-1}(t) - (C(j) - i)y_{j,i}(t)]S(t) \text{ for } 1 \leq i \leq C(j) - 1$$
$$\dot{y}_{j,0}(t) = -\eta(C(j) - 1)y_{j,0}(t)S(t)$$

where we recall $\eta = 2^{-32}$.

Fitting to data from the Slammer worm, we would require that the total number of susceptibles

$$\sum_j N(j)C(j) = 73,782. \tag{8}$$

Fitting to the final value of the scan-rate per infective curve,

$$\frac{\sum_j N(j)\sigma_{j,C(j)}}{\sum_j N(j)C(j)} \approx 15000/18. \tag{9}$$

Fitting to the initial value of scan-rate per infective ("in mean"):

$$\frac{\sum_j N(j)\sigma_{j,1}}{\sum_j N(j)} \approx 15000. \tag{10}$$

These three equations will determine three of the model parameters, leaving a number that can be used for potentially finer curve fitting to measured data, or simple model variations, as an experimenter would desire.

For example, we considered a model with two classes of enterprises, i.e., $j \in \{0, 1\}$. Under instantaneous saturation ($\sigma_{j,i}$ is a constant function of i), equations (8)-(10) would involve the following six unknowns: $N(0), N(1), C(0), C(1), \sigma_0, \sigma_1$. Under a gradual saturation, each "σ_j" would be replaced by up to $C(j)$ "$\sigma_{j,i}$" parameters. In our numerical example, we have simplified this by assuming that under gradual saturation, the quantities $\sigma_{j,i} - \sigma_{j,i-1} \equiv \delta_j$ are constant functions of i so that only two additional parameters are introduced: δ_0 and δ_1. In particular, for equation (10), $\sigma_{j,1} = \sigma_{j,C(j)} - (C(j) - 1)\delta_j$. So, the experimenter can stipulate 5 of 8 of these parameters and solve (8)-(10) for the remaining three parameters. For example, if we take $\sigma_{1,C(1)} = 25000, \sigma_{0,C(0)} = 6000, \delta_1 = 500, \delta_0 = 200$, and $C(0) = 5$; then the three unresolved parameters are computed $N(0) = 1008, N(1) = 2217, C(1) = 31$. The total instantaneous scan-rate and the scan-rate per infective (worm) for these model parameters were numerically computed and are depicted in Figure 3 and 4 ("hetero-g"[2] curves). Note that they agree closely with the curves extrapolated from the University of Wisconsin's tarpit data for Slammer.

Two other sets of parameters were numerically evaluated and the curves are depicted in these figures: "hetero-i" and "homog-g". For "hetero-i," the maximum scanning rates of the two enterprise classes were initialized to 8000 scans/s and 20000 scans/s and the number of enterprises in each class, $N(0)$ and $N(1)$, were computed to agree with (8)-(10) as above; also, the links were assumed to be instantaneously saturated. For "homog-g," there was a single enterprise class (class 0) with gradually saturating links; specifically, $C(0) = 18$ and $\sigma_{0,18} = 15000$ (as in Section 3), and $\delta_0 = 300$. In our preliminary numerical studies, we found that varying the parameters δ had the most significant effects on the resulting total instantaneous scan-rate curves.

5   Summary and Future Work

In summary, we proposed a system of coupled Kermack-McKendrick epidemic equations to model the spread of a bandwidth-limited, randomly scanning Internet worm. We applied this model to the available data for the Slammer worm. Specifically, we showed that the model can recreate Slammer's characteristic scan-rate-per-infective and its total scan-rate to the core. Our simulation code will be available at [5].

In [10], routeview data indicating the number of susceptible SQL servers with a given address prefix was reported (Figure 15). We are currently incorporating this data into our model and studying the results. When other relevant data becomes available (such as stub-link capacities *correlated* together with the number of susceptibles per

---

[2] *hetero*geneous enterprises with gradual saturation of the stub-link.

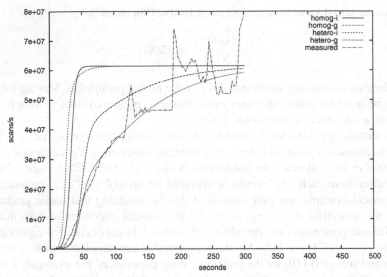

Fig. 3. Slammer total instantaneous scan-rate for homogeneous and heterogeneous network models.

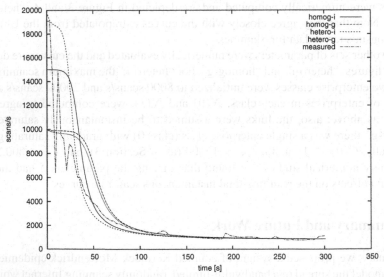

Fig. 4. Slammer scan-rate per infective (worm) for homogeneous and heterogeneous network models.

address prefix), we will select our model parameters accordingly. Such data is being pursued by participants of the DHS PREDICT project.

Finally, we are also exploring ways to incorporate into our model information about transmission delay variations of scanning packets through the core. Such latency issues are significantly more important for worms like Blaster and Sasser that propagated via TCP.

References

1. Z. Chen, L. Gao and K. Kwait. Modeling the spread of active worms", In *Proc. IEEE INFOCOM*, San Francisco, 2003.
2. E. Cooke, M. Bailey, Z.M. Mao, D. Watson, F. Jahanian, D. McPherson. Toward understanding distributed blackhole placement. In *Proc. ACM WORM*, Washington, DC, Oct. 29, 2004.
3. D.J. Daley and J. Gani. *Epidemic modeling, an introduction.* Cambridge University Press, 1999.
4. DETER project URL: http://www.isi.edu/deter
5. EMIST project URL: http://emist.ist.psu.edu
6. D. Moore, C. Shannon, Geoffrey M. Voelker, Stefan Savage Internet Quarantine: Requirements for Containing Self-Propagating Code. In *Proc. IEEE INFOCOM*, San Francisco, 2003.
7. D. Moore, V. Paxson, S. Savage, C. Shannon, S. Staniford and N. Weaver. Inside the Slammer worm. IEEE Security and Privacy, 2004, http://www.computer.org/security/v1n4/j4wea.htm
8. M. Liljenstam, D.M. Nicol, V.H. Berk and R.S. Gray, Simulating Realistic Network Worm Traffic for Worm Warning System Design and Testing. In *Proc. ACM WORM*, Washington, DC, Oct. 2003.
9. S. Staniford, V. Paxson, and N. Weaver. How to own the Internet in your spare time. In *Proc. USENIX Security Symposium*, pages 149–167, Aug. 2002.
10. N. Weaver, I. Hamadeh, G. Kesidis and V. Paxson, Preliminary results using scale-down using scale-down to explore worm dynamics. In *Proc. ACM WORM*, Washington, DC, Oct. 29, 2004.
11. N. Weaver, S. Staniford and V. Paxson, Very Fast Containment of Scanning Worms. In *Proc. 13th USENIX Security Symposium*, Aug. 2004.
12. C.C. Zou, W. Gong, and D. Towsley. Code red worm propagation modeling and analysis. In *Proc. 9th ACM Conference on Computer and Communication Security (CCS'02), Washington, DC*, Nov. 2002.
13. C.C. Zou, W. Gong, and D. Towsley. Worm propagation modeling and analysis under dynamic quarantine defense. In *Proc. ACM WORM*, Washington, DC, Oct. 2003.

On-Line Segmentation of Non-stationary Fractal Network Traffic with Wavelet Transforms and Log-Likelihood-Based Statistics*

David Rincón and Sebastià Sallent

Universitat Politècnica de Catalunya (UPC),
Av. Canal Olímpic s/n, Castelldefels, 08860 Barcelona, Spain
{drincon,sallent}@mat.upc.es

Abstract. Network traffic exhibits fractal characteristics, such as self-similarity and long-range dependence. Traffic fractality and its associated burstiness have important consequences for the performance of computer networks, such as higher queue delays and losses than predicted by classical models. There are several estimators of the fractal parameters, and those based on the discrete wavelet transform (DWT) are the best in terms of efficiency and accuracy. The DWT estimator does not consider the possibility of changes to the fractal parameters over time. We propose using the Schwarz information criterion (SIC) to detect changes in the variance structure of the wavelet decomposition and then segmenting the trace into pieces with homogeneous characteristics for the Hurst parameter. The procedure can be extended to the stationary wavelet transform (SWT), a non-orthogonal transform that provides higher accuracy in the estimation of the change points. The SIC analysis can be performed progressively. The DWT-SIC and SWT-SIC algorithms were tested against synthetic and well-known real traffic traces, with promising results.

1 Introduction

It is well known [1, 2] that network traffic exhibits some fractal properties (heavy tails, slow-decaying autocorrelation, and scaling, among others) that cannot be captured by traffic models based on Poissonian or Markovian stochastic processes. New fractal-aware algorithms, which exploit the properties of self-similarity, long-range dependence, and multifractality, have been developed for traffic analysis and generation. Few studies (to our knowledge, only [3], [4] and [5]) have explored the possibility of the non-stationarity of fractal parameters. Real traffic is expected to change its behavior as time goes on, and detecting the change points (i.e., the instants that bound the segments that show homogeneous fractal behavior) can be useful for some algorithms and network mechanisms that exploit the long-memory properties of network traffic. Some examples are the TCP congestion control described in [6], a predictive bandwidth control

---

* This work was supported in part by the Spanish Government under grant CICYT TIC2001-0956-C04-02, and the EuroNGI European Network of Excellence (IST-FP6).

M. Ajmone Marsan et al. (Eds.): QoS-IP 2005, LNCS 3375, pp. 110–123, 2005.

for MPEG sources [7], the effective bandwidth estimator described in [8] and the novel application of traffic fractality to the design of VLSI video decoders [9]. Apart from those applications, a better knowledge of traffic characteristics is needed when creating synthetic traces for simulation or testing purposes. Our long-term aim is to accurately describe the temporal and frequential evolution of fractal and multifractal traffic parameters, which, in turn, can provide us with a better understanding of network dynamics.

Among the fractal parameter estimation algorithms, those based on the use of the Discrete Wavelet Transform (DWT) exhibit a higher performance in terms of accuracy, computational efficiency, and adaptability. The Abry-Veitch estimator is widely accepted as the best and most efficient DWT-based LRD estimator, since it is capable of performing a joint estimation of α (the scaling parameter, related to the Hurst parameter) and c_f (related to the variance of the series under study), and computing the associated confidence intervals. The DWT decomposes the traffic trace in a multiresolution analysis (MRA) that turns the $1/f$ spectrum of fractal traces into a certain structure of the variance across scales, which can be easily detected. If the scaling parameter changes at any moment, so does the complete variance structure. Therefore, if a change point detection algorithm monitors the variance at every scale, the simultaneous detection of changes across scales will indicate a change in the scaling parameter. Changes must appear at several scales to be significant; a change in variance at a single scale only tells us of non-stationarity in the variance at that scale. A final phase of automatic selection and clustering of the candidate change points must be performed in order to infer the real changes of the fractal parameters.

We propose the use of the Schwarz Information Criterion (SIC) as the change point detection algorithm. This algorithm was developed from the Akaike Information Criterion (AIC) for model selection, widely used in statistical analysis. The SIC is based on the maximum likelihood function for the model, and can be easily applied to change point detection by the comparison of the likelihood of the null hypothesis (no changes in the variance series) against the opposite hypothesis (a change is present). The time-frequency characterization provided by the wavelet transforms and the SIC statistic is useful for off-line analysis, but can also be easily adapted to perform an on-line (in the sense of sequential, progressive) monitoring of the behavior of traffic. Both the DWT and SWT can be performed sequentially, and the SIC statistic can be updated appropriately.

This paper presents the SIC algorithm working with the DWT and its non-decimated version, the Stationary Wavelet Transform (SWT). The SWT provides higher temporal accuracy for the change point estimation, due to its inherent time redundancy. Both the DWT-SIC and SWT-SIC algorithms are described and applied to synthetic traces and real traffic. The rest of the paper is organized as follows: Section II gives a brief introduction to LRD parameters and their estimation; in Section III, the AIC and SIC algorithms are presented; Section IV explains how the DWT/SWT and the SIC statistic can work together, along with tests performed on synthetic traces; Section V shows the results of applying the algorithms to a real traffic sample; Section VI discusses the on-line performance of the analyzers; and Section VII concludes the paper.

2 Long-Range Dependence and Its Estimation

2.1 A Brief Review of Long-Range Dependence

A stationary stochastic process $x(t)$ is considered long-range dependent (LRD) if its autocovariance function decays at a rate slower than a negative exponential. Equivalently, LRD can also be defined in the frequency domain as a $1/f$-like spectrum around the origin: $S_x(f) \sim \frac{c_f}{|f|^\alpha}$ when $|f| \to 0$.

The LRD parameters are α and c_f. The scaling parameter α is related to the intensity of the LRD phenomenon (a qualitative measure) and is usually expressed as the Hurst parameter $H = (1 + \alpha)/2$, while c_f has dimensions of variance and can be interpreted as a quantitative measure of LRD. Though the Hurst parameter has received more attention in published literature, c_f is not negligible, since it appears in the expression of loss probability when LRD traffic is fed into a queue, and also in the variance of the sample mean of an LRD process (and thus determines the confidence intervals for its estimation) [10]. In the rest of the paper we will focus our attention on α and H, although our algorithms can be extended to an analysis of the evolution of c_f.

2.2 The Discrete Wavelet Transform

The Discrete Wavelet Transform is a powerful tool that allows a fast, efficient and precise estimation of LRD parameters. Given a signal $x(n)$, we denote by $d_x(j, k) = \langle x, \psi_{j,k} \rangle$ the coefficients of the DWT, where $\psi_{j,k}$ is a function with finite support, and j and k are respectively the scale and the location where the analysis is performed. The basis of the decomposition is actually a family generated from translated and dilated versions of the *mother wavelet* $\psi_{0,0}$: $\psi_{j,k}(n) = 2^{-j/2}\psi_{0,0}(2^{-t}n - k)$ for $j = 1 \ldots J, k \in \mathcal{Z}$.

The DWT can be understood as the result of a cascade of quadrature-mirror low-pass and high-pass filters ($h(n)$ and $g(n)$, respectively), in which the output of the high-pass filter (the details of the signal) are the coefficients of the DWT at each scale j, while the output of the low-pass filter (the approximation of the signal) is filtered again iteratively. The output of the filters is decimated in order to maintain orthogonality; therefore, the number of coefficients is halved at each iteration. In terms of the signal spectrum the output of the DWT is a decomposition in subbands that are halved at each step, giving rise to a multiresolution analysis in which the original signal is decomposed into a low-pass approximation at scale J, $a_x(J, k)$ and a set of high-pass details $d_x(j, k)$ for each scale $j = 1 \ldots J$. Figure 1 illustrates the filter bank interpretation.

2.3 DWT Analysis of LRD Traffic: The Abry-Veitch Estimator

Abry and Veitch developed the LogScale diagram [10], an unbiased and efficient estimator of LRD parameters. It estimates the power of the subbands of the wavelet decomposition as the sample variance of the coefficients at each branch of the filter, μ_j. Due to the power law,

$$\mu_j = E[d_x^2(j, k)] = 2^{j\alpha} c_f C(\alpha, \psi_0) \tag{1}$$

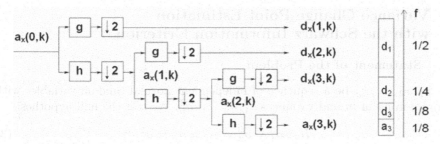

Fig. 1. On the left, the DWT as filter bank for a level 3 decomposition. On the right, the subbands of the normalized spectrum, with the approximation a_3 and details $d_{1...3}$.

and taking logarithms on both sides,

$$\log_2(\mu_j) = j\alpha + \log_2(c_f C) + g_j \tag{2}$$

where $g_j = \psi(n_j/2)/\ln 2 - log_2(n_j/2)$ are bias-correction terms that depend only on n_j (the number of wavelet coefficients at scale j), required because the expectation of the logarithm is not the logarithm of the expectation, and $\psi(x)$ is the Psi function. Parameters α and $c_f C$ can be estimated from expression (1) performing a weighted linear regression on μ_j, in which the weight is related to the number of coefficients available at the corresponding scale (decreasing by 2 for each increase in the scale). Assuming μ_j follows a Gaussian distribution, confidence intervals for the estimation can be derived.

2.4   Real-Time Estimation with the DWT

There exists an on-line version of the DWT estimator [4] that performs sub-band variance computation progressively, but it is accumulative, not dynamic: it returns updated estimations performed over all available samples from $t = 0$. The same authors studied the stationarity of the scaling exponent (α or H) and developed a statistical test that is capable of determining its constancy [5], but this study is limited to a dyadic temporal decomposition (i.e., the traffic traces are divided into non-overlapping segments whose lengths are powers of 2). We expect traffic to change at arbitrary points (*transition points*) where the trace changes its behavior (in terms of the distribution of its variance across scales).

The change point detection problem is a classical one in statistics, and several algorithms have been developed to address it. In a previous study [11] we used the iterated cumulated sum of squares (ICSS) algorithm, in conjunction with the stationary wavelet transform, to develop a method for the temporal segmentation of the trace into regions that show homogeneous LRD properties (constant Hurst parameter). The results of the ICSS technique are good, but it lacks flexibility in selecting the significance level (critical values are computed using Monte Carlo simulation), and it is also difficult to implement progressively. In this paper we use a more powerful approach, based on an information theory criterion, which can be computed on-line (in the sense of a progressive, sequential analysis). We follow the theory presented in [12] in the context of stock price analysis.

3   Variance Change Point Estimation with the Schwarz Information Criterion

3.1   Statement of the Problem

Let $x_1, x_2, \ldots, x_n$ be a sequence of independent normal random variables with a common mean m and variances $\sigma_1^2, \sigma_2^2, \ldots, \sigma_n^2$. We test the null hypothesis

$$H_0 : \sigma_1^2 = \sigma_2^2 = \cdots = \sigma_n^2 = \sigma^2 \tag{3}$$

versus the alternative

$$H_1 : \sigma_1^2 = \cdots = \sigma_{k_1}^2 \neq \sigma_{k_1+1}^2 = \cdots = \sigma_{k_q}^2 \neq \sigma_{k_q+1}^2 = \cdots = \sigma_n^2 \tag{4}$$

where q is the unknown number of change points and $1 \leq k_1 < k_2 < \cdots < k_q < n$ are the unknown positions of the change points. Following the binary segmentation procedure suggested in [12] we reduce the problem to the detection of a single change point, and then iterate the process in the two subsequences that surround the change point. If a change is detected in the subsequence, split again and iterate until no more changes are found in any of the subsequences. Our problem is then reduced to testing the null hypothesis against

$$H_1 : \sigma_1^2 = \cdots = \sigma_k^2 \neq \sigma_{k+1}^2 = \cdots = \sigma_n^2 \tag{5}$$

3.2   Information-Based Criteria: AIC and SIC

Now we turn to the discussion about the statistic that will allow us to decide whether a change point exists. One of the most used statistics for change point detection is the Akaike information criterion (AIC) for model selection [13]. Following Akaike's work, other authors have applied information theory criteria in other fields. Schwarz [14] defined the SIC statistic as:

$$SIC(k) = -2 \log L(\hat{\theta}) + p \log k \tag{6}$$

where $L(\hat{\theta})$ is the maximum likelihood function for the model, p is the number of free parameters, and k is the sample size. In our context there are two different models, which correspond to the null and alternative hypotheses, respectively. Our decision will be taken following the principle of minimum information; that is, do not reject H_0 if $SIC(n) \leq \min_k SIC(k)$, reject H_0 if $SIC(n) > SIC(k)$ for some k, and estimate the position of the change point by $\hat{k}$ such that $SIC(\hat{k}) = \min_{1 \leq k < n} SIC(k)$ where $SIC(n)$ is the SIC statistic under H_0 and $SIC(k)$ is the SIC under H_1 for $k = 1 \ldots n - 1$. Then, (6) becomes:

$$SIC(n) = n \log 2\pi + n \log \hat{\sigma}^2 + n + \log n \tag{7}$$

$$SIC(k) = n \log 2\pi + k \log \hat{\sigma_1}^2 + (n - k) \log \hat{\sigma_2}^2 + n + 2 \log n \tag{8}$$

where $\hat{\sigma}^2$, $\hat{\sigma_1}^2$ and $\hat{\sigma_2}^2$ are the biased estimators of the variances of the sequence and both subsequences $1 \ldots k$ and $k + 1 \ldots n$, respectively. These expressions limit our change detection range to $2 \leq \hat{k} \leq n - 1$. Reference [12] gives a proof that $\hat{k}$ is a consistent estimator of the true change point, and it also gives the expression for the computation of the significance level. The authors define the critical level c_α, which modifies the criterion: accept H_0 if $SIC(n) < \min SIC(k)$ for some k, and estimate the position of the change point by $\hat{k}$ such that

$$SIC(\hat{k}) = \min_{2 \leq k \leq n-2} SIC(k) + c_\alpha \qquad (9)$$

The expression of the critical level c_α is derived from the asymptotic null distribution of the statistic [12]. The same paper discusses other versions of the SIC criterion, such as an unbiased estimator and the unknown mean cases. The unbiased estimator yields a slight increase in accuracy at the cost of a high increase in computation. In our study we used only definition (9), which provides acceptable results at a lower computational cost.

4 DWT-SIC and SWT-SIC Algorithms for Detecting Changes in the Scaling Parameter α

4.1 Connecting the DWT and SIC

The main contribution of our work is to apply the SIC change point estimator to the output of each of the branches of the wavelet filter bank. If we find the same variance change position across all or a significant number of scales, it will signal a change point in the scaling (or Hurst) parameter. The number of samples available at each scale is not the same, since each branch of the wavelet bank suffers a different number of decimations. For example, if the DWT filter bank shown in Fig. 1 is fed with a traffic trace of 1024 samples, the first branch will output 512 $d_x(1, k)$ samples, the second branch will output 256 $d_x(2, k)$ samples, and the third branch will output only 128 $d_x(3, k)$ samples. There will still be a temporal relationship between them: the second sample of the lowest frequency subband ($d_x(3, 2)$), the third and fourth samples of the middle subband ($d_x(2, 3)$ and $d_x(2, 3)$), and the $5^{th}, 6^{th}, 7^{th}$ and 8th samples of the highest subband ($d_x(1, 5) \ldots d_x(1, 8)$) are all related to a certain time segment, which corresponds to samples 9 to 16 of the input trace. Figure 2 illustrates the relationship between samples across scales. In order to provide a good estimation of the change point at all available scales, a *phase correction* is applied to higher scales, with a scale-dependent delay as illustrated in the right side of Fig. 2. This correction aligns the position of the wavelet coefficients with their *zone of influence*. After the change point detection a clustering-and-decision step is needed, in order to decide if a change is present in enough scales. This step has not yet been fine-tuned, but we expect to provide a good solution with pattern-matching algorithms; for the moment, we rely on a visual heuristic technique, and therefore our results for the change points are approximate.

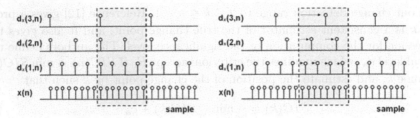

Fig. 2. On the left, the temporal relationship between the coefficients of the DWT. On the right, the *phase-corrected* coefficients.

4.2   The Stationary Wavelet Transform and SIC

Another approach to dealing with the phase-correction problem is the use of a non-orthogonal wavelet transform, known as stationary or maximum-overlap wavelet transform (SWT, MODWT). This transform is essentially identical to the DWT except in the decimation step, which is not performed in the SWT. Therefore, the variance change point can be accurately located at each scale. The expression for the estimation of the Hurst parameter is similar to (1), although the constant term is not the same, due to time redundancy.

$$\mu_j = E[d_x^2(j,k)] = 2^{j\alpha} c_f' C'(\alpha, \psi_0) \tag{10}$$

The bias-correcting terms are essentially the same as those developed for the DWT, except that their value is constant across scales, due to their exclusive dependence only on n_j (number of coefficients available at each scale. The Gaussian confidence intervals are also constant, since the expectation and variance of the logscale values depend only on n_j. Thus, the SWT produces a constant-weight estimation along the logscale diagram, which simplifies and speeds up regression computation. The main advantage, however, is increased accuracy at higher scales, where long-range dependence is detected (the DWT generates few coefficients at higher scales).

The SIC statistic is used with the SWT in the same way as with the DWT. The only difference is that with the SWT all the branches of the filter bank produce the same number of samples.

4.3   Testing the Algorithms with Synthetic Traces

Both algorithms (the DWT-SIC and SWT-SIC) were validated using synthetic traces before they were tried on real traffic traces. The tests were performed using fractional Gaussian noise (FGN) traces with different values for the Hurst parameter, which ranged from $H = 0.5$ (no LRD) to $H = 0.9$ (high LRD). Several tests were performed, although we only present one of them. The trace had 131072 samples, the first half of which were FGN with $H = 0.8$, followed by two segments of 32768 samples with $H = 0.9$ and $H = 0.7$. Therefore, the change points were located at $n = 65536$ and $n = 98304$. All the segments had mean 1024 and variance 1 (the distribution of the variance across scales depends on the scaling parameter). All the tests were performed using the Haar wavelet.

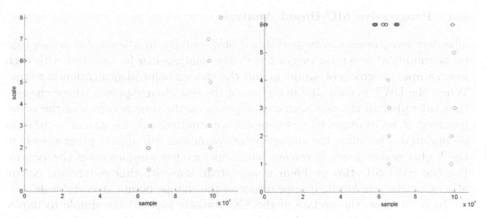

Fig. 3. Change point candidates at each scale. On the left, DWT-SIC at 8 levels and significance=0.001. On the right, SWT-SIC at 8 levels and significance=10^{-6}.

Figure 3 shows, on the left, the result of the phase-corrected 8-level DWT-SIC analysis of the test trace. At the lower scales both change points are clearly identified. At higher scales the DWT lacks of temporal accuracy, and the change point located at 65536 disappears. If the decomposition is performed with a lower significance level, the number of fake change point candidates increases, adding some "noise" to the diagram. The reason for this phenomenon is the slight difference between the values of H for the segments separated by the change point. At those scales, a higher temporal resolution is needed in order to distinguish the two variance structures of the segments with $H = 0.8$ and $H = 0.9$. A fake candidate appears immediately at the beginning of the sequence, but can be easily avoided since it is detected in just one subband. The DWT's lack of accuracy is also responsible for the deviation of the second change point (at position 98304) at lower frequencies (higher subbands), such as at the eighth level. Finally, although the second change point is detected in almost all the scales, subband 4 is missing. This is because scale 4 is the level at which the variance structures of both segments cross, i.e., the values of the variance of both segments at level 4 are the same (or very similar). It is the same phenomenon that causes the disappearance of the candidates for the first change point at higher levels, but in the second change point the variances are so separated at higher levels that even the DWT is capable of distinguishing them.

In order to compare the results with the non-decimated wavelet transform, the same SIC analysis is performed with the SWT. The results are plotted in the right side of Fig. 3. Both change points are clearly detected, although scales 4 and 5 suffer from the *coincidence of variances*. Generally speaking, the SWT-SIC returns better results, but it requires more memory and more computations (because of the increased number of SIC operations) than the DWT-SIC alternative.

4.4   Progressive SIC-Based Analysis

The decision process can be performed progressively, in a *sequential* sense; that is, beginning at a certain origin $t = 0$, the analysis can be updated with each new sample (or group of samples) and the change point identification is re-run. When the DWT is used, the detection of the new change points (those changes that take place in the new segment) depends on the time accuracy of the scales involved. If, for example, 64 new samples are acquired, only 5 scales ($2^6 = 64$) will be updated. Therefore, the change point candidates will appear progressively in the higher scales (lower frequency subbands) as new samples enter the system. For the SWT-SIC this problem is inexistent, since its time-redundant nature allows for the *synchronized* appearance of the change points at every scale.

In both cases, the update of the SIC statistic is relatively simple to implement, since it only requires the variance of the new samples at each scale to be added and is therefore scalable. The recalculation of the decision for the change points can be more difficult, since the number of computations is intrinsically variable due to its iterative nature: if more changes are found, more calculations are needed, and vice versa. In any case, it is reasonable to opt for updating the change point estimation at a lower rate than the update of the variances. Of course, the actual rate that is needed must be selected by the application: a routing decision process is slow and runs at the scale of minutes, while an access control mechanism must work faster and run at the scale of seconds or faster.

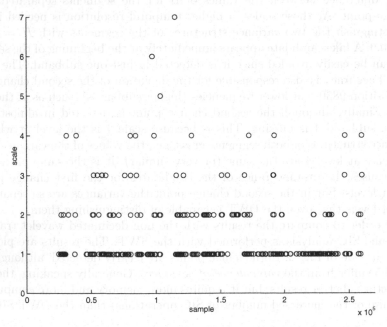

Fig. 4. Bellcore trace, DWT-SIC at 7 levels, significance=10^{-5}.

5   SIC-Based Analysis of Real Traffic Traces

We applied our algorithms to a well-known traffic trace: the pAug89 Bellcore trace, studied extensively in [1]. This series has usually been studied as a whole, without any investigation of the possible changes in the long-range dependence parameters. The Bellcore trace is composed of 314283 samples, which correspond to the amount of bytes transported by an Ethernet network during each 10ms segment. The trace is known to have a Hurst parameter of around 0.8. The Abry-Veitch estimator returns a value of $H = 0.803$, with 95 percent confidence intervals $[0.792, 0.813]$. Figure 4 and Table 1 show the results of the DWT-SIC algorithm applied to Bellcore trace. Only the first 262144 samples were used, due to current implementation limitations. It is quite clear in Fig. 4 that an important change point is present around $n = 98000$. The Abry-Veitch code gives an estimation of $H = 0.824$ for the left segment ($n = 1$ to 98000) and $H = 0.803$ for the rest of the trace. Several other change points can be identified at (approximately) $n = 40000, 80000, 125000, 160000, 190000$ and 245000. The results, shown in Table I, are similar to those found by the Iterative Cumulative Sum of Squares (ICSS) statistic described in [11]. The dynamic range of the results is important: H ranges from 0.720 to 0.873 (± 10 percent of the estimated value of 0.803). The trace seems to oscillate above 0.8 for the first 245000 samples, and then abruptly changes to a much lower value of 0.720. This is coherent with the "average" value of 0.803 returned by the estimator for the whole trace.

Table 1. DWT-SIC results for pAug89 trace.

Segment (initial and final sample)	Hurst parameter
1-40000	0.814
40000-80000	0.805
80000-98000	0.857
98000-125000	0.873
125000-160000	0.815
160000-190000	0.804
190000-245000	0.829
245000-262144	0.720

6   On-Line SIC-Based Analysis of Real Traffic

One of the main reasons behind developing the segmentation algorithm is the future possibility it provides for performing an on-line analysis of traffic. Both wavelet transforms, the DWT and SWT, can be performed progressively; that is, as new samples are fed into the filter bank, the transforms give the new samples corresponding to the convolution of the new data and the *memory* of the filter. The SIC statistic can also be computed sequentially, since we only have to update the partial sums of squares of the wavelet coefficients with the new outputs from the transforms. However, there is a step that cannot be predicted: the binary segmentation procedure. The computation of this step depends completely on the existence of a variance change point in the studied data. There is no way to predict the number of times that the binary segmentation (and re-run of the SIC statistic on the subsequences) must be performed, since we do not actually know the number and location of the change points. That is why we cannot ensure that the complete process can be performed in real time, although the algorithms allow some truncations, such as setting the minimum segment size, in which we consider the Hurst parameter to be locally stationary. For example, all the SIC tests in this paper were performed with a minimum segment of 128 samples. The actual value must be chosen taking into account the physical meaning, in time, of the minimum segment size, and its impact on the application. For example, the use of 128 samples with a trace like pAug89, in which each sample represents 10 ms of data, means that we are considering that fractal parameters do not change faster than once every 1.28 seconds. This can be a good choice for a routing algorithm, but not for a queue scheduling algorithm.

Several tests were performed to evaluate the behavior of the SIC statistic when a new change point is included in the updated sample to be analyzed. Progressive versions of the DWT-SIC and SWT-SIC were tested with the same traces used in Sections IV and V. A significant drawback to the DWT-based algorithm is that its output rate diminishes as the scale increases, due to the decimation step (as explained in Section IV). Since the SIC statistic needs some extra samples after a change point before detecting the change, the detection at higher scales of the DWT suffers from delays. There is also the possibility that two close change points are perfectly resolved at lower scales, but are merged into just one change point at higher scales. The SWT-SIC is better suited, providing good resolution at every scale and faster detection than the DWT-SIC.

Figure 5 shows a three-dimensional diagram in which the x-axis represents the analyzed time (as new samples are added), the y-axis is the location of the change points, and the z-axis is the scale at which the change is detected. These diagrams are just the generalization of the time-scale (x- and y-axis, respectively) plots shown in Section IV. In Fig. 5 we can see that the changes at positions $n = 65536$ and $n = 98304$ are clearly detected at more than one scale, and what is more important, that once the statistic has detected it for the first time, the detection is maintained as the time progresses. In this sense, the algorithm is *coherent*. This is an important feature, because random false change candidates appear (usually over the $y = x$ line at the first or second scales, as we will mention later), but these are easily detected because they disappear as the statistic progresses with new data. Therefore, the SIC statistic in some way corrects its previous incorrect estimation as it advances in the analysis. We can also see a delay in the detection of change points at higher scales. Take, for example, scales 5 and 6 for the $n = 98304$ change point. Since the DWT outputs 16 (24) or 32 (25) times less samples at those scales that at scale 1, due to the decimation step, this has an impact on the detection capacity of the SIC statistic, which works *slower* (in the sense that many more samples have to be processed before the change points are detected). Nevertheless, even at those high scales, the estimation is still coherent. We can also see a change candidate at $n = 38$ that gives a very coherent appearance, but which is circumscribed at the sixth scale. This signals an isolated variance change point, which is probably caused by an imperfection in the synthetic trace generator.

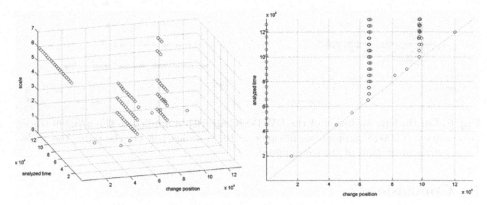

Fig. 5. On the left, the evolution of the DWT-SIC algorithm when applied to the synthetic trace. Granularity = 5000 samples, significance = 0.001, scales 1-6. On the right, the *time-versus-time* projection of the same figure.

Finally, there is a curious phenomenon that has been mentioned before: the spurious false change points on the $x = y$ line. The right side of Fig. 5, which is merely the projection of the 3D diagram onto the x-y plane, clearly shows that these points are located precisely at the end of each of the intervals that are being studied. Our interpretation is that the SIC statistic can be somehow fooled by some kind of boundary effect (this could also be the reason for the change candidate at $n = 38$, which is located very close to the beginning of the trace, but its coherence within all the estimations makes us think it is a real variance change rather than a boundary effect). In any case, these fake boundary candidates are easily detected and rejected, since they disappear rapidly.

The *time-versus-time* projection in Fig. 5 also allows us to evaluate the detection delay, which is defined not in terms of actual computation time, but rather on the quantity of samples located after the change point that have to enter the system, before the change point is detected. The ideal instantaneous detection would put the points over the $x = y$ bisectrix. Our tests show that the typical detection delay is under 100 samples, and in most cases 50 samples is enough, although a more detailed study is needed.

The on-line DWT-SIC algorithm has also been applied to the Bellcore trace, which yielded the results shown in Fig. 6. We can again see the coherence of the change candidates, which are the same as were found in Section V. The projection on the y-z plane (right side of Fig. 6) is practically identical to Fig. 4. The x-y fake change points are difficult to distinguish, due to the high density of candidates at the lower scales. We conclude that the on-line version of our estimation algorithm is as reliable as the static version.

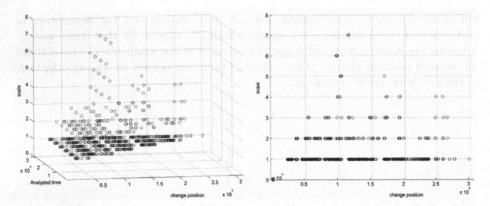

Fig. 6. On the left, evolution of the DWT-SIC algorithm applied to the pAug89 trace. Granularity = 50000 samples, significance = 10^{-6}, scales 1-8. On the right, the *scale-versus-time* projection of the same figure.

7   Conclusions

The paper describes the Schwarz Information Criterion and its application to the study of the variations of the scaling parameter of long-range dependence network traffic, using the Abry-Veitch wavelet-based estimator. Two different algorithms were presented, the DWT-SIC and the SWT-SIC, which exhibit different features: SWT-SIC is more accurate and can provide fast updates of the change points, while DWT-SIC is faster and less computationally intensive. Some results from the study of synthetic traces are given, along with preliminary results from an on-line version of the estimators. The algorithms were applied to real traffic traces from the Bellcore dataset. We found an important result: the scaling (Hurst) parameter of real traces is highly variable over time. Until today, the LRD analysis of real traffic traces assumed, at least implicitly, a constant value of the scaling parameter. We have shown that this is not true. Since LRD characteristics can by and large affect network dynamics, it is very important to use a method that is capable of characterizing the evolution of LRD parameters over time, as the SIC-based algorithms can.

There are several issues that require further investigation, among which the most important are the following: computing the confidence intervals of the estimations; evaluating the computational cost of the real-time implementation of the algorithms; studying how departures from the ideal situation (non-Gaussianity, non-stationarity of the mean) influence on the change point detection; determining the influence of the wavelet family used in the analysis and the significance level of the estimation, and its role in a network environment. Our efforts are focused on providing a complete, progressive, time-frequency segmentation and characterization of network traffic. Some of the algorithms presented here are being implemented in a traffic analyzer/generator described in [15].

References

1. Leland, W.e.a.: On the self-similar nature of ethernet traffic. IEEE/ACM Transactions on Networking **2** (1994) 1–15
2. Park, K., Willinger, W., eds.: Self-similar traffic and network performance. John Wiley & Sons (2000)
3. Duffield, N., Lewis, J., O'Connell, N., Russell, R., Toomey, F.: Statistical issues raised by the Bellcore data. In: Proceedings. of the 11th IEE UK Teletraffic Symposium (1994)
4. Roughan, M., Veitch, D., Abry, P.: On-line estimation of the parameters of long-range dependence. In: Proceedings of Globecom 98. Volume 6. (1998) 3716–3721
5. Veitch, D., Abry, P.: A statistical test for the time constancy of scaling exponents. IEEE Transactions on Signal Processing **49** (2001) 2325–2334
6. He, G., Gao, Y., Hou, J., Park, K.: A case for exploiting self-similarity of network traffic in TCP congestion control. In: Proceedings of the 10th IEEE International Conference on Network Protocols. (2002) 34–43
7. Ouyang, Y.C., Yeh, L.B.: Predictive bandwidth control for MPEG video: a wavelet approach for self-similar parameters estimation. In: Proceedings of ICC 2001. Volume 5. (2001) 1551–1555
8. Yu, Xiang Yu; Li-Jin Thng, I., Jiang, Y.: Measurement-based effective bandwidth estimation for long range dependent traffic. In: Proc. of IEEE Region 10 Intl. Conf. on Electrical and Electronic Technology (TENCON). Volume 1. (2001) 359–365
9. Varatkar, G., Marculescu, R.: On-chip traffic modeling and synthesis for MPEG-2 video applications. IEEE Transactions on VLSI Systems **12** (2004) 108–118
10. Veitch, D., Abry, P.: A wavelet-based joint estimator of the parameters of long-range dependence. IEEE Transactions on Information Theory **45** (1999) 878–897
11. Rincón, D., Sallent, S.: Characterizing fractal traffic with redundant wavelet-based transforms. Preprint (2004)
12. Chen, J., Gupta, A.: Testing and locating variance change points with application to stock prices. Journal of the American Statistical Association **92** (1997) 739–747
13. Akaike, H.: Information theory and an extension of the maximum likelihood principle. In: Proc. of the 2nd Intl. Symposium on Information Theory. (1973) 267–281
14. Schwarz, G.: Estimating the dimension of a model. The Annals of Statistics **6** (1978) 461–464
15. Rincón, D., Martinez, S., Cano, C., Sallent, S.: Synthesis and analysis of fractal LAN traffic at high speeds. In: Procs. of IEEE LANMAN 2004. (2004) 259–264

Implementation of Virtual Path Hopping (VPH) as a Solution for Control Plane Failures in Connection Oriented Networks and an Analysis of Traffic Distribution of VPH

Manodha Gamage, Mitsuo Hayasaka, and Tetsuya Miki

The Department of Electronics, Faculty of Engineering,
The National University of Electro-Communications,
1-5-1, Chofugaoka, Chofu-Shi, Tokyo, 182-8585, Japan
manodha@ice.uec.ac.jp

Abstract. A connection oriented next generation Internet could guarantee the QoS preferred by the emerging real time applications. These connection oriented networks are inherently more prone to network failures: link/path failures and degraded failures. The control and the data plane of connection oriented networks such as MPLS are logically separated. Therefore a failure in the control plane should not always immediately affect the communications in the data plane. Control plane failures are usually detected using *Timers*. If any of the *Timers* expire, the control plane session is terminated resulting failures in the corresponding data plane. This paper discusses in detail the implementation of VPH concept that intent to eliminate the degraded type failures in the data plane and its advantages. Also this evaluates a more efficient, non-periodic VPH concept. The results of the computer simulations show VPH is a proactive technique to eliminate degraded type failures and improve the availability of the networks.

1 Introduction

Initially the Internet was introduced with a limited set of services such as email, file transfer, remote terminal emulation etc. In the recent years the growth of the Internet users has been enormous and these users have come to expect highly available telecommunications services supporting high quality voice, video and accurate data. IP technology offers multiple grades of voice and video quality at lower cost. The focus of this paper is to address the reliability, availability and QoS guarantees of high-end real-time multimedia services. In most of these future applications such as telemedicine, the consequences of losing the QoS guarantees can be very severe. The durations of the sessions of most applications such as remote lecturing, multiparty video conferencing etc. will be very long. It is our belief that the connection oriented networks such as Multi Protocol Label Switching (MPLS) [1] could fulfill the requirements of future applications of the internet, much better than the conventional connectionless, best effort networks. These connection oriented networks are inherently more prone to network failures and therefore the need for an efficient scheme to overcome network failures in these networks is high. Furthermore these schemes

M. Ajmone Marsan et al. (Eds.): QoS-IP 2005, LNCS 3375, pp. 124–135, 2005.

should be able to provide 100% restorability even for dual failure situations, as the assumption of single failure at any given point of time may not be valid for rapidly expanding future Internet.

MPLS has its control plane logically separated from the data plane. Therefore the failures in the control plane do not immediately disconnect the data plane communications, but they will create temporary interruptions in the data plane due to loss of maintenance functions of the control plane and if the control plane failure is not recovered in a specified time, the data communication session is disconnected. These types of control plane failures are defined as degraded type failures according to the RFC 3469 [2] and they account for about 50% of total network failures [5]. To the best of our knowledge, fast re-routing techniques [9-12, 15] so far proposed, do not distinguish the degraded type failures from the normal link/path failures and they always perform a re-routing when the communication session in the data plane is disconnected due to degraded type failures.

This paper discusses the implementation of our proposed, VPH concept [3] in detail and analyses the traffic distributions aspects of it. The main objective of implementing VPH is to eliminate the degraded type failures due to control plane failures and minimize the number of re-routings that causes service outages. The VPH has the ability to dynamically distribute the traffic over the network. It is also capable of handling dual failure situations because it forms a VP-pool, where only one VP is active at any given time and others can be used as backup paths. The rest of this paper is organized as follows. In the next section basic knowledge of MPLS to understand the implementation of VPH concept is discussed briefly. The application of the VPH concept is not limited to MPLS and it is discussed here as a well known architecture eases the understanding of this concept. In the section 3, the problem analysis and related work are discussed. The implementation of VPH concept and its advantages are discussed in section 4. A numerical analysis is done in the section 5 and the simulated results are analyzed in Section 6. Finally the paper is concluded in Section 7.

2 Briefly on MPLS

MPLS [1] provides connection-oriented services to the inherently connectionless IP networks. A router supporting MPLS is a Label Switched Router, or LSR. An ingress is a node by which a packet enters the MPLS network, an egress is a node by which a packet leaves the MPLS network. The main functions of ingress are calculating the path through the MPLS network, initiating label switched path, classifying inbound traffic into Forward Equivalence Classes (FEC) etc. An FEC represents the binding of a group of packets or flows that require the same handling. The requests with the same destination egress and same QoS requirements are mapped to the same FEC by default, if no local policy is stated otherwise.

Each LSR maintains the "Next Hop Label Forwarding Entry" (NHLFE) in its routing tables, to forward labeled packets. Mapping packets to an FEC is done only once when they enter the MPLS network at the ingress. Therefore every ingress node maintains a FEC to NHLFE (FTN) to map each FEC to one or a set of NHLFEs, when unlabeled packets arrive at them. The ingress will always label them before they are dispatched. All other LSRs will maintain an "Incoming Label Map" (ILM) instead a FTN to map incoming labeled packets to NHLFEs.

MPLS supports Constraint-based Routing (CR). LDP and RSVP-TE are two main control signaling protocols used in MPLS [4, 7]. RSVP-TE uses UDP in its transport layer where as most of the LDP uses TCP. Since RFC 3468 [21] encourages RSVP-TE over CR-LDP, RSVP-TE is used in this discussion even though the simulations were performed for both CR-LDP and RSVP-TE. The VPH concept can be equally applied for both these control plane protocols.

3   Problem Analysis and Related Work

According to the RFC 3469 [2], the failures in connection oriented networks such as MPLS are classified into two types, namely link/path failures and degraded failures. As per the definitions of [2], a link/path failure means a situation where the actual connectivity of the path between the ingress and egress is lost whereas a degraded failure is a situation where the links at lower layers are not in suitable quality for data transmission and the applications feel temporary interruptions regularly. Studies done on actual ISP networks have shown almost 50% of total network failures are of degraded type [5]. The main reason for degraded type failures is the control plane failures.

The control plane failures do not immediately disconnect the communication sessions in the data plane as the two planes are logically separated, but they cause temporary interruptions for the communication sessions. This is because the corresponding LSP in data plane loses its maintenance functions, when the control plane session is failed. These control plane failures are detected using *timers (T)* of control plane peers. *RSVP Hello State Timer* in RSVP-TE [7] and the *Keep Alive Timer/cleanup timer* in LDP [4] of the control plane of MPLS are two such examples. The threshold values of these *timers* are usually in the range of 30-40s depending on the equipment used and the requirements of the network. Each LSP will have a corresponding control plane session. Each LSP will have a corresponding control plane session and the control plane peers of active sessions exchange Protocol Data Units (PDU) continuously. Whenever a PDU is received by a peer it resets its *timer*. If this *timer* expires without receiving any PDU, then it is assumed that the neighboring peer of the control plane session is failed and the corresponding LSP is disconnected as it cannot be maintained without the underlying control plane session. These degraded failures can be avoided and data plane communications can be continued without failure, if the control plane can start a new session instead of the failed session before these *timers* expire. This is the goal of implementing the VPH concept.

Among many related work in finding suitable restoration strategies for high-end real-time communication services, it is worth while to mention the recent work of [9-12] and [15]. References [9] and [11] describe fast restoration strategies to minimize the spare capacity that are efficient and effective compared to the conventional 1+1 and 1:1 protection. These back up bandwidth (BBW) sharing strategies are not very effective unless back up paths are dynamically revised as in [12] due to the variations of the network conditions. Also the restoration times of all these strategies are comparatively high. On the other hand [10] and [15] explain the strategies to have back up paths (BP) with guaranteed BW. They make sure the BP always has enough BW in the event of a failure. They give faster restoration times but less efficient with respect to the resource utilization. Furthermore almost all these studies consider only single fault situations with the assumption that any failure could be repaired before the next failure occurs. Very little studies done on the dual failure scenarios have revealed that the currently popular Backup Band-Width (BBW) sharing schemes with 100% restoration for single faults could in average recover about 60-70% failures in dual faults situations but it can be as low as 20% [20]. The expansion of networks and increased durations of applications require future networks to have 100% restorability even for dual fault situations. The other major draw back of these existing strategies is, they treat both degraded and link/path failures in the same manner. In other words they do not treat degraded failures before they become link/path failures even though there is a possibility to avoid them. Since 50% of the failures are of this nature, a solution similar to VPH that recovers degraded failures before they disconnects data plane communications could improve the QoS and availability drastically.

4   Implementation of Virtual Path Hopping and Its Advantages

The main objective of the proposed VPH is to eliminate degraded failures due to control plane failures of connection oriented networks. The term *re-routing* in this article refers to 'the change of data transmission from Active Path (AP) to a Backup Path (BP), *after* a failure in the data plane'. On the other hand VPH refers to changing the active VP in the data plane to a different VP in the VP-pool, *before* it fails. Any restoration by re-routing after a failure will have an outage of service and therefore reducing number of re-routings will improve QoS of the network. In VPH, when a request for a communication session is arrived at the ingress, a link disjoint *VP-pool*, which contains the candidate VPs that participate in VPH is decided for the ingress and egress pair, as shown in Fig 1. This VP-pool should contain at least three link-disjoint VPs for the effectiveness of this concept and with the increment of number of VPs in the VP-pool, the effectiveness will increase. The number of VPs in the VP-pool depends on several factors such as the congestion of the network, Service Level Agreements (SLA) of clients, failure analysis and failure distributions of the network, availability of resources etc. There are many algorithms proposed in the literature [16-19] to find link disjoint paths between a pair of ingress and egress. It is beyond the scope of this paper to discuss them in great detail, but if it is not possible to find completely link disjoint VPs for the VP-pool, it is good enough to find VPs with

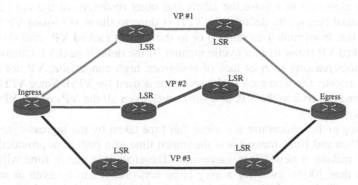

Fig. 1. The concept of Virtual Path Hopping with a VP-pool of 3 VPs

minimum overlaps, in other words best suitable VPs for the intended traffic type. The previous studies have shown that there is a high probability of finding such VPs [17]. Also instead of using Shortest Path First (SPF) only, to determine the VP-pool, it is necessary to use QoS routing along with SPF as proposed in [16] and [18].

In this discussion, VP-pool of 3 VPs is considered for simplicity. All VPs in the VP-pool should be ranked according to their suitability to the traffic type/class they serve. In this concept, always only one VP in the VP-pool is activated at any given time and therefore resources are allocated and consumed only by one VP at a particular instant of time. Another VP in the VP-pool can be always used as a BP to overcome link/path failures. At the beginning of the session ingress will always start the communication via the rank-1 VP. Then it will hop periodically to rank-2, rank-3, back to rank-1 and so on. The VPH period is a vital parameter and to avoid all the control plane failures this period should be less than the threshold of T. If the proposed VPH is done with a period less than the threshold value of T of control plane peers, it is possible to continue the communication session with a *smooth transition of LSP*. In other words in VPH the active VP in the data plane is changed periodically by changing the corresponding session in the control plane.

The ingress nodes of MPLS or any other connection oriented network will have to play a major role in the implementation of VPH concept. Once the rank-1 VP/LSP is decided, it will originate a Path message (of RSVP-TE) as usual and forward it to the egress using explicit routing. Then the egress will send a Resv message (of RSVP-TE) in the return path. When the FEC and the labels to be used in this rank-1 VP are decided based on the destination IP address, the traffic class, QoS requirements etc. and the resources are reserved, the ingress will update the FTN with the FEC and outgoing label information. Then the data transmission is started. The VPH is to be implemented using the 'make before break' policy [14] and it will make sure that there is no packet loss when VPH is done. Therefore when the VP/LSP is to be hopped, first of all the ingress will send a Path message explicitly to the already decided nodes of the next ranked VP/LSP of the VP-pool. Then the ingress will receive a Resv message with a new label for the same FEC through the return path. Just at this moment there will be two label entries for the same FEC in the FTN at the ingress. The LSP can be now changed without loss of any traffic, by activating the FEC: new-label entry of FTN and deactivating the FEC: old-label entry. Then the RSVP Teardown messages (PathTear and ResvTear) are exchanged between ingress and egress explicitly to release the labels and other resources of the old LSP. As it was mentioned before, by default, the VPH is done to the next ranked VP from any active VP (i.e. from rank-1 to rank-2) or to the highest ranked VP (rank-1) from the lowest ranked VP (rank-n) in a cyclic manner. If the default next VP cannot be used due to various reasons such as lack of resources, high congestion, VP not available due to failures etc. the next rank in the sequence is used for VPH. Since VPH uses the existing standards to hop the VP, this works well even all the VPs of the VP-pool are not link disjoint.

According to this procedure it is clear that time taken by the ingress-egress pair to exchange Path and Resv messages is the transit time of a path hop, provided the first attempt of making a new VP is successful. Therefore this transit time will be very small, less than 200ms even for a very large network. It can be even as smaller as 20ms for very high speed all-optical network in the future. It is possible to set priority as *Key* to all the Path and Resv messages involved in VPH to reduce the transition time further.

Since the ingress is mainly controlling VPH a simple buffering at the ingress could implement the VPH smoothly. Also since different paths have different transmission delays, caching at the egress will overcome any delay jitter due to VPH. 'Actively reserved bandwidth architecture' [8] can be used for fast allocation of resources and minimize the rejections due to lack of resources. It is necessary to maintain a *VP/LSP-pool Table* at the ingress and such a table essentially should include information such as LSP/VP ID, rank of LSP and status of LSP (active or inactive) apart from the information of a conventional routing table of MPLS. Whenever a VPH is performed according to the above algorithm, there will be some overhead traffic added to the network. If the number of hops in VPs can be minimized without affecting the performances of VPH, it is possible to minimize this overhead.

Therefore it is suggested to implement a not periodic VPH instead of a periodic VPH. It is called non-periodic VPH here onwards. In non-periodic VPH, path hopping is triggered by a 2nd timer, called the VPH-Timer at each control plane peer. Here a VPH is done only if the VPH-Timer expires and not periodically. This VPH-Timer is reset to zero whenever a VPH is done or a PDU is received. In a similar way to the RSVP Hello State Timer is decided in RSVP [23], the VPH-Timer values also can be decided by each control plane peer that participates in VPH and they should be less than T. Whenever the VPH-Timer expires, a VPH is performed before T expires and T is reset to zero. This will eliminate unnecessary VP hops. Hence a vast reduction of additional overhead due to VPH can be achieved. VPH-Timer value should be decided so that there is enough time to inform the ingress about the expiration of the VPH-Timer and to carry out the VP hop before data communication is affected. It is possible to use the *Lost* state of *Hello* messages to inform the expiration of VPH-Timer to the ingress.

Another very important phenomenon in the implementation of VPH is its ability to dynamically distribute traffic throughout the network. This will reduce the stress on links and nodes and improves the efficiency of the network resources. Further more in this VPH concept, a traffic engineered LSP from the VP-pool is used as a back up LSP. Therefore any re-routing will not damage the TE model.

5 Numerical Analysis

In a network, the links differ widely in their failure characteristics and a link failure model should account for it [5]. In a general model considered here, it is assumed, there are n different link disjoint VPs in each VP-pool and each VP has $H_1, H_2 \ldots \ldots H_n$ number of links respectively. The probability of failure is considered as $p_{ij} = p_{ij}^f + p_{ij}^d$; where i ($1 \leq i \leq n$), j ($1 \leq j \leq H_i$), p_{ij}^f and p_{ij}^d are path number, link number, probability of failure due to link/node failures and probability of failure due to degraded failures respectively. Since about 50% of the total failures are of degraded type, p_{ij}^f and p_{ij}^d will be approximately equal to each other. The probability of having a failure in n^{th} VP is given by;

$$1 - \prod_{j=1}^{H_n} (1 - (p_{nj}^f + p_{nj}^d)) \tag{1}$$

If VPH is not implemented and no failure occurs, any communication session will use the same VP throughout the whole session. The probability of path failure in such a session, P_{No_VPH} is given by (1). On the other hand if periodic VPH is implemented, where there are K VP hops during a session, the time average of probability of path failure P_{VPH} over a the period of communication session duration is given by:

$$\frac{1}{K} \sum_{i=1}^{K} \left[1 - \prod_{j=1}^{H_i} \left(1 - (p_{ij}^f + p_{ij}^d)\right) \right] \text{ for } K<n \text{ and}$$

$$\frac{1}{n} \sum_{i=1}^{n} \left[1 - \prod_{j=1}^{H_i} \left(1 - (p_{ij}^f + p_{ij}^d)\right) \right] \text{ for } K \geq n$$

If $K > n$ the VPH is done in cyclic manner and therefore each path is used more than once as explained in the previous section. The VPH concept will almost eliminate degraded type failures, making p_{ij}^d very small (almost zero).

Therefore $P_{VPH} < P_{No\_VPH}$. This shows an improved reliability and availability of the network due to VPH concept.

6   Performance Evaluation

The performance of VPH concept and its contributions to the traffic engineering were evaluated by way of computer simulations. Different network topologies with different number of nodes and links as shown in Table 1 were simulated. These network topologies were derived using random graphs. Table 2 indicates the different failure combination scenarios of networks. Each topology in Table 1 was simulated for all failure combinations in Table 2. These simulations were performed for periodic VPH, non-periodic VPH and without VPH scenarios. All simulations of Topology A were carried out for duration of 1 year and that of topologies B and C were simulated for a period of 6 months. In these simulations all the link/path failures in the data plane were recovered by a re-routing to a BP and always BP was selected from the VP-pool. Therefore number of re-routings performed was used as a performance measure that indicates the number of failures in the data plane.

6.1   Simulation Model

Three links disjoint VPs were formed for all the VP-pools between the 10 different ingress-egress pairs in all simulations, using the Dijkstra's [13] shortest path algorithm and QoS routing. Randomly allocated link costs in terms of bandwidth were used to find the shortest path because the link characteristics of a WAN vary from link to link. The links that are not suitable in terms of QoS were eliminated before finding the shortest path. Once the best VP is decided those links were eliminated when finding the 2nd best VP and so on. If it is impossible to find link disjoint VPs for the VP-pool, the least overlapped best VPs were decided in a similar way to the algorithm in [16]. Results shown here are for communication sessions arrive as per to a Poisson distribution with an average of 60s. The session durations were randomly decided based on an exponential distribution with an average of 3600s. Three major types of failures namely, hardware and software failures of equipment, failures due to burst losses and congestion, and control plane failures were considered. Their arrivals were assumed to be distributed exponentially and the averages were decided for each simulation according to the values of Tables 1 and 2. The bandwidth of the sessions was decided randomly with an average of 10Mbps. According to many simulations performed the parameters of the distribution of repair times of failed links were found to have no effects to the performance of VPH. Therefore the repair time after a link failure was assumed to be a constant. The main reason for this is the ability of VPH to provide 100% restorability for dual failures due the formation of VP-pool with many VPs. The threshold of T in periodic VPH was set to 40s. Since T was set to 40s, it is found, as expected, the periodicity of the periodical VPH should be less than 40s and therefore it is set to 30s [3]. VPH-Timer values were randomly decided for each peer to be a multiple of 10s in the range of 30s-80s.

6.2 Simulated Results and Traffic Analysis

Almost all simulation scenarios could obtain an average reduction of 48% in re-routings, as expected and that accounts for almost 50% of control plane failures in the network that could be avoided by VPH. Fig. 2 shows the variations of the number of re-routings with respect to time, without the VPH, with the VPH (periodic), and with the VPH (non-periodic) are applied. Fig 2a and 2b show graphs for two network topologies with 20 and 50 nodes respectively. A re-routing is necessary when ever there is a failure in the network as the traffic on the lost LSP should be recovered by a back up LSP. Therefore the number of re-routings was counted as a measure of link/path failures. In the event of a re-routing there will be a data loss and by reducing the number of failures it is possible to reduce this data loss due to re-routings. This could minimize the VPHs, and thereby minimize the overheads added to the network due to the VPH concept. The Fig. 2 shows that the non-periodic VPH delivers almost the same performance compared to periodic VPH and this simulation showed the non-periodic VPH could reduce the number of VP hops by 75%.

Table 1. Simulated network topologies

Network	No. of Nodes	No. of Links (Bidirectional)
A	20	50
B	40	115
C	50	135

Table 2. Simulated failure combinations

Combination	Failures/node/month	Failures/link/month
I	0.08	0.1
II	0.08	0.2
III	0.1	0.1
IV	0.1	0.2
V	0.1	0.3

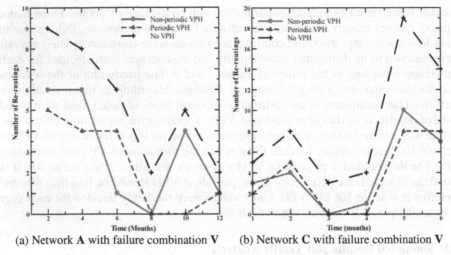

(a) Network **A** with failure combination **V**   (b) Network **C** with failure combination **V**

Fig. 2. Variation of Number of Re-routings with respect to Time

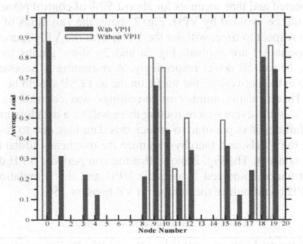

Fig. 3. Variation of Average Load with respect to Node Number

Traffic distribution analysis was done by monitoring the traffic related to a particular ingress-egress pair over a period of 24 days. Simulations were performed to compare the traffic distribution due to non-periodic VPH concept with No-VPH condition. According to the bar chart in Fig. 3, the traffic of the selected ingress-egress pair traveled through 8 nodes without VPH. When VPH is implemented it was distributed over 14 nodes reducing the load of each node and utilizing the network resources efficiently. Similarly Fig. 4 shows the traffic distribution that can be achieved over many links by implementing the VPH concept. In other words, instead of over using few links and not using many other links at all, VPH distributes the traffic among many links reducing the load of each link. Figure 5 shows the variation of the average traffic load per link with respect to time over a period of 24 hours. The vast reduction of average traffic load per link, with the VPH concept implemented is due to the dynamic traffic distribution over many links in the network that can be achieved with VPH. The traffic distribution will reduce the stress of the network and will improve the robustness to failures of the network.

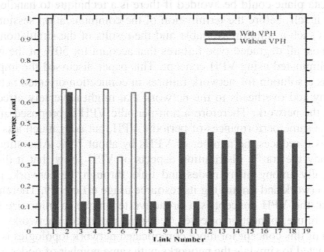

Fig. 4. Variation of Average Load with respect to Link Number

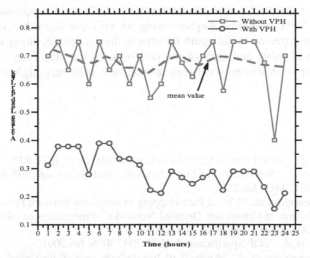

Fig. 5. Variation of Average Load per Link with respect to Time

7 Conclusions

The QoS, reliability and availability preferred by future Internet applications could easy to be satisfied by the connection oriented networks such as MPLS. These networks are inherently more vulnerable for network failures. According to the RFC

3469, network failures in connection oriented networks such as MPLS are of two types: link/path failures and degraded failures. Degraded failures are mainly due to control plane failures and the control plane of MPLS is logically separated from the data plane of MPLS. Therefore these control plane failures affecting the communications in the data plane could be avoided if there is a technique to handle them in the control plane it self, before the termination of the communication session in the data plane. VPH is such a solution available and the results of the simulations performed show that almost all degraded type failures that account for 50% of the network failures can be eliminated using VPH concept. This paper discussed the implementation of the VPH as a solution for network failures in connection oriented networks. Periodic VPH may add overheads to the network that might adversely affect the traffic conditions of the network. Therefore a non-periodic VPH is proposed and evaluated here. It has the same performances of periodic VPH, but adds much less overhead to the network as it reduces the number of VPHs by about 75%. Also simulations performed to study the traffic distribution aspects of VPH show that it distributes the load dynamically among many nodes and links through the network, reducing the stress of the network and improving its resource usage efficiency. Therefore it can be concluded that the VPH concept is a proactive technique to minimize the network failures in the future connection oriented, real-time multimedia networks.

Evaluation of this concept for more complicated network topologies is in progress. Also it is expected to simulate the networks with same number of nodes and links, but with different connectivity to make more general and statistical conclusions on the performances of VPH. Also it is worth while to investigate further any degradation that could occur due to the implementation of the VPH and the optimum buffer size to implement VPH smoothly. Implementing an efficient algorithm to recover the communications affected by link/path failures in the data plane, along with VPH will improve the availability and QoS of the networks. In the near future we should be able to come up with this new network architecture with very high availability and QoS.

References

1. E. Rosen et al., "Multi Protocol Label Switching Architecture", IETF RFC 3031, Jan 2001.
2. V. Sharma et al., "Framework for Multi-Protocol Label Switching (MPLS)-based Recovery", IETF RFC 3469, Jan 2003.
3. Manodha Gamage et al., "Virtual Path Hopping to overcome Network Failures due to Control Plane Failures in Connection Oriented Networks", Proceedings of joint conference of the 10th APCC and the 5th MDMC, Aug. 2004.
4. L. Anderson et al., "LDP Specification", IETF RFC 3036, Jan 2001.
5. Gianluca Iannaccone et al., "Analysis of link failures in a IP backbone" Proceedings of Internet Measurement Workshop 2002, http://www.icir.org/vern/imw-2002/imw2002-papers/202.pdf
6. Jing Wu, Delfin Y. Montuno, Hussein T. Mouftah, Guoqiang Wang and Abel C. Dasylva, "Improving the Reliability of the Label Distribution Protocol" Proceedings of the 26th Annual IEEE conference on Local Computer Networks, 2001.
7. D. Awduche et al., "RSVP-TE: Extensions to RSVP for LSP Tunnels", IETF RFC 3209, Dec. 2001

8. Geng-Sheng Kuo and C. T. Lai, "A new architecture for transmission of MPEG-4 video on MPLS networks", ICC 2001 - IEEE International Conference on Communications, no. 1, June 2001 pp. 1556-1560.
9. Yijun Xiong and Lorne G. Mason, "Restoration strategies and spare capacity requirements in self-healing ATM networks", IEEE/ACM Transactions on Networking, no. 1, Feb 1999 pp. 98-110.
10. Murali Kodialam and T. V. Lakshman, "Dynamic routing of restorable bandwidth-guaranteed tunnels using aggregated network resource usage information", IEEE/ACM Transactions on Networking, no. 3, Jun 2003 pp. 399-410
11. Dahai Xu, Chunming Qiao and Yizhi Xiong, "An Ultra-fast Shared Path Protection Scheme-Distributed Partial Information, Part II", Proceedings of the 10th ICNP '02- IEEE International Conference on Network Protocols, June 2002.
12. Yu Liu, David Tipper and Peerapon Siripongwutikorn, "Approximating Optimal Spare Capacity Allocation by Successive Survivable Routing", Proceedings of IEEE INFOCOM 2001.
13. Dijkstra's shortest path algorithm, http://en.wikipedia.org/wiki/Dijkstra%27s_algorithm
14. Ping Pan, George Swallow and Alia Atlas, "Fast Reroute Extensions to RSVP-TE for LSP Tunnels" http://www.ietf.org/internet-drafts/draft-ietf-mpls-rsvp-lsp-fastreroute-07.txt
15. Li Li, Milind M. Buddhikot, Chandra Chekuri and Katherine Guo, "Routing Bandwidth Guaranteed Paths with Local Restoration in Label Switched Networks", Proceedings of the 10th IEEE International Conference on Network Protocols 2002.
16. S. D. Nikolopoulos, A. Pitsillides and D. Tipper, "Addressing Network Survivability Issues by Finding the K-best Paths through a Trellis Graph", Proceediings of IEEE INFOCOM 1997.
17. B. Szviatovszki, A. Szentesi and A.Juttner, "On the Effectiveness of Restoration Path Computation Methods", http://www.cs.elte.hu/~alpar/publications/proc/RestoPath.pdf
18. Yuchun Guo, Fernando Kuipers and Piet Van Mieghem, "Link-Disjoint Paths for Reliable QoS Routing", http://www.nas.its.tudelft.nl/people/Piet/papers/dimcra.pdf
19. Y. Bejerano et al., "Algorithms for Computing QoS paths with Restoration", http://citeseer.ist.psu.edu/cache/papers/
20. cs/29147/http:zSzzSztiger.technion.ac.ilzSz~spalexzSzpubzSzTM-restoration.pdf/bejerano02algorithms.pdf
21. M. Clouqueur and W. D. Grover, " Availability Analysis of Span-Restorable Mesh Networks", IEEE Jurnal on Selected Areas in Communications, vol. 20, no. 4, May 2002.
22. L. Andersson and G. Swallow "RSVP The Multiprotocol Label Switching (MPLS) Working Group decision on MPLS signaling protocols", IETF RFC 3468, Feb. 2003.
23. R. Braden et al., "Resource ReServation Protocol (RSVP)", IETF RFC 2205, Sept. 1997

Experimental Comparison of Fault Notification and LSP Recovery Mechanisms in MPLS Operational Testbeds

Roberto Albanese[1], Daniele Ali[1], Stefano Giordano[2], Ugo Monaco[1],
Fabio Mustacchio[2], and Gregorio Procissi[2]

[1] INFO-COM dept. of University "La Sapienza" of Rome, Italy
{albanese,ali,monaco}@infocom.uniroma1.it
[2] Dept. of Information Engineering University of Pisa, Italy
{s.giordano,fabio.mustacchio,g.procissi}@iet.unipi.it

Abstract. This paper reports on the comparison of recovery strategies in MPLS-TE experimental testbeds. We focus on alternative notification mechanisms and compare different end-to-end recovery techniques. A measurements campaign has been performed in two different trials based on commercial routers and PC/Linux boxes. In the former an inbuilt signaling-based mechanism provides remote fault notification to the edge routers while in the latter a prototype of IGP flooding-based mechanism has been implemented. We investigate the impact of the alternative fault notification schemes and present a performance evaluation distinguishing different components of the overall recovery time. We believe that the ideas and experimental insight contained in this work will be helpful to other people involved in standardization and implementation of MPLS-related recovery techniques.

1 Introduction

The definition and design of an advanced dynamic control plane for future MPLS network has attracted a large interest in the nowadays networking arena. This challenge is addressed by extending and integrating existing signaling and routing protocols into the so called MPLS-TE protocol suite [1, 2]. We are currently involved in the design, implementation and testing of experimental/commercial testbeds – on PC/Linux platform and commercial routers – with a dynamic control plane based on MPLS-TE. This activity is part of a broader research project called TANGO [3]. In this paper we present an experimental comparison and report on our experience raised during a measurements campaign on two different MPLS trials. In particular, we focus on LSP recovery mechanisms and provide experimental results exploiting alternative fault notification solutions based on RSVP-TE signaling and OSPF-TE flooding.

The rest of the paper is organized as follows. In Section 2 we briefly overview MPLS-based recovery techniques. In Section 3 we focus on end-to-end recovery strategies and discuss about alternative fault notification mechanisms. In Section

M. Ajmone Marsan et al. (Eds.): QoS-IP 2005, LNCS 3375, pp. 136–149, 2005.

4 we present detailed information about the two testbeds and define the measurements scenario. In Section 5 we report the experimental results and provide a performance comparison of the proposed fault recovery mechanisms. Finally, in Section 6 we conclude.

2   MPLS-Based Recovery

MPLS-based protection of traffic (called MPLS-based recovery) may be motivated by the notion that there are limitations to improving the recovery times of current routing algorithms. In fact, although IP routing algorithms are robust and survivable, the amount of time they take to recover from a fault can be significant, in the order of 10's of seconds (for interior gateway protocols (IGPs)) or minutes for exterior gateway protocols, such as the Border Gateway Protocol (BGP)), causing disruption of service for some applications in the interim. This is unacceptable in situations where the aim is to provide a highly reliable service, with recovery times that are in the order of seconds down to 10's of milliseconds.

IP routing may also not be able to provide bandwidth recovery, where the objective is to provide not only an alternative path, but also bandwidth equivalent to that available on the original path. MPLS, on the other hand, by integrating forwarding based on label-swapping of a link local label with network layer routing allows flexibility in the delivery of new routing services. MPLS allows for using such media-specific forwarding mechanisms as label swapping. This enables some sophisticated features such as quality-of-service (QoS) and traffic engineering (TE) to be implemented more effectively.

The most important ability of MPLS-based recovery is to increase network reliability by enabling a faster response to faults than it is possible with traditional Layer 3 (or IP layer) approaches alone while still providing the visibility of the network afforded by Layer 3. We can observe that to reduce restoration times is only one of the goals of MPLS-based recovery. Other conflicting objectives such as optimal use of resources, applicability of traffic protection at various granularities and different scopes, minimization of the degradation of the unprotected traffic have to be reached and a trade off may exist between them. This leads to the definition of several recovery mechanisms and the choice between them will often involve engineering compromises based on a variety of factors such as cost, end-user application requirements, network efficiency, complexity involved and revenue considerations.

In [4], [5], [6] and [7] a taxonomy of different recovery mechanisms is proposed. This classification is also useful to describe the recovery mechanisms we are going to employ in our experiments. A first distinction among different recovery mechanisms can be made observing that recovery can be applied at various levels throughout the network (recovery scope):

- *Local (Span-Level) Recovery* refers to the recovery of an LSP over a link between two nodes. In this case a failure is notified and solved at intermediate nodes, next to the failed resource.

– *End-to-End (Path-Level) recovery* refers to the recovery of an entire LSP from its source (ingress node end-point) to its destination (egress node end-point). These mechanisms require that a failure notification is propagated till the end nodes of the LSP and there solved.

Because of the sub-optimality of the resulting backup paths, span level strategies are prone to waste resources in the network, whereas end-to-end recovery strategies are more efficient, because they provide the computation of the best end-to-end backup paths. Despite of the recovery scope, two recovery mechanisms categories can be identified based on the resource allocation done during the recovery LSP/span establishment:

– *Protection Strategies:* in this paradigm one or more dedicated protection LSP(s)/span(s) is/are fully established in advance to protect one or more LSP(s)/span(s). This implies a pre-computation and pre-allocation of backup resources and no signalling takes place to establish the protection LSP/span when a failure occurs;
– *Restoration Strategies:* in this paradigm the complete establishment of the restoration LSP/span occurs only after a failure of the working LSP/span, and requires some additional signalling. The distinction between different types of restoration is made based on the level of route computation, signalling and resource allocation done during the restoration LSP/span establishment. For example, backup resources could be allocated on-demand only at time of failure (On-the-fly) or could be pre-computed and only booked for a future restoration (Fast Restoration). We can observe that restoration fits better the dynamical assignment/release of the network resources with respect to protection; but, in case of a fault, a higher blocking probability for the restoring traffic might be experimented, due to the failure handling by control plane mechanisms instead of hardware ones (e.g. detection, notification and mitigation).

Different recovery types can be further identified depending on the number of recovery LSPs/spans that are protecting a given number of working LSPs/spans. Five recovery types could be distinguished as described in [4]. The choice of the recovery mechanisms to be implemented in the experimental testbeds has been strictly influenced by the network topology. Some constraints, due to the utilization of a metropolitan testbed equipped with commercial routers, have led to the comparison of the recovery mechanisms in a ring topology. In such a scenario path-level techniques are preferable in order to limit the waste of resources in the network. Both end-to-end protection and restoration dedicated recovery with extra traffic mechanisms (see [4]) have been implemented to evaluate the impact of the fault notification mechanism and the traffic recovery operation process.

3   Fault Notification Mechanisms

The end-to-end protection approach requires some mechanism to convey the fault notification from the internal node(s) detecting the fault to the edge nodes which are in charge of switching the traffic onto the backup LSPs. In the MPLS-TE standards no solution has been given to address this point. Instead, this topic is currently being discussed in the IETF working groups in the context of GMPLS [8]. Basically two approaches are being compared: signaling-based and flooding-based. For the Linux-based testbed we have preferred the latter, and we have implemented a prototype of such functionality in the OSPF-TE daemon [9]. In sections 4 and 5 a comparison with a commercial router inbuilt notification mechanism – based on RSVP-TE-signaling – is provided and experimental measurements on the performances of different recovery techniques are presented.

3.1 The Fault Notification Dilemma: Signaling-Based or Flooding-Based?

The signaling-based approach foresees the fault notification message be carried by the RSVP-TE protocol, from the detecting node upstream along the LSP to the ingress edge node. This approach has several drawbacks. For instance the number of messages generated upon a failure equals the number of impacted LSPs, say N. Thus, the detecting node will generate N different messages, while each of its reachable neighbours (say k) will have to process N/k different messages on average. In order to diminish the processing overhead associated with this approach, one could think to the aggregation of such messages, similarly to the proposal made in [10]. On the other hand, this would require additional capabilities at the RSVP-TE daemon, namely message aggregation and branching, which are not currently foreseen in the protocol. Also, this solution would lead the signaling-based dissemination process to closely reassemble a sort of "partial flooding" along the network subgraph constituted by the upstream LSP tree, an approach that does not participate of the additional advantages of the complete flooding scheme discussed below. In any case, the signaling-based approach would require the detecting node to parse the entire set of supported LSPs so as to identify those impacted by the fault (as made explicit in [11]), and this would add processing burden to the internal router.

With the flooding-based approach, the fault notification message is flooded throughout the network. An attractive possibility would be to reuse the existing IGP (OSPF-TE or ISIS-TE) flooding process of Opaque LSAs, a capability that is intrinsically present in the MPLS-TE model. Recall that the Opaque LSAs are flooded transparently through the network, without triggering the recomputation of the routing/forwarding tables, and are used to disseminate link load information in MPLS-TE. With this approach, there is no need for the detecting node to parse the entire set of LSPs. Also, the per-node processing load after a failure is limited to one single message.

Two additional advantages of the flooding-based approach are the *minimum notification delay* and *complete dissemination*. In fact, by assuming an equal processing time of the Opaque LSA and RSVP-TE messages, and that each node floods/forwards the received message immediately, it is easy to recognize that *IGP flooding always achieves the minimum possible notification delay towards any edge node*, while in the signaling-based approach it depends on the actual length of the impacted LSP paths. This point was already suggested in [12]: "[...] notification message exchanges through a GMPLS control plane may not follow the same path as the LSP/spans for which these messages carry the status. In turn, this ensures a fast, reliable [...] and efficient [...] failure notification mechanism".

As an additional advantage, because of the immediate dissemination of failure notification, that is failure notification reaches all the edge nodes and not only those responsible for the impacted LSPs, flooding-based approach is helpful in preventing from erroneous route selections in the aftermath of the failure. In fact, with the signaling-based approach, those edge nodes not having any LSP crossing the failed link l will be not notified the failure and erroneous route selections could occur for new LSPs resulting in fastidious signaling overhead and re-computation procedure; this propability is even higher if we consider multiple contemporary failures. In fact, assume two generic links along the path of a generic LSP, say l_1 and l_2, the former being upstream with respect to the latter. If both links fail simultaneously, the propagation of l_2 failure will not reach the edge node, due to upstream interruption of the path and there is a potential for this edge node to include the failed link l_2 in the new selected routes. With flooding-based approach such cases are eliminated as all edge nodes are immediately notified any failure. Finally the usage of link state IGP for fault notification is compliant with RFC 3272 [13]: "network state information may be distributed by link state advertisements also under exceptional conditions".

As an alternative to IGP flooding, in the context of GMPLS for Optical Transport Network it has been proposed to add flooding capabilities to the Link Management Protocol (LMP) [14]. Despite this approach has the undoubtedly advantage to avoid further modifications to the IGP platform, on the other hand it would fatally lead to the duplication of many flooding-related mechanisms between IGP and LMP, and at the same time breaks the local nature of the LMP protocol. Our preference for IGP flooding was dictated by the fact that LMP does not belong to the MPLS-TE suite for packet network. A comparative view on the experience from practical implementations of both approaches can be found in [15].

4   Network Scenario

The comparison of different recovery techniques has been carried out on two operational MPLS networks that consist of four nodes in ring topology configuration:

1. Juniper Gigabit network with a signaling-based fault notification mechanism;
2. PC/Linux routers FastEthernet network equipped with a prototypical flooding-based fault notification scheme.

End-to-end (Path Level) recovery techniques have been considered, in particular we focus on protection and restoration (with pre-computed backup path) dedicated recovery with extra traffic strategies (see [4]). In order to evaluate the impact of the number of LSPs established on the overall recovery time, measurements have been carried out in two different network conditions: (i) 1 protected LSP established with data traffic load and (ii) 10 protected LSPs established and 1 LSP with data traffic load. In the latter condition the loaded LSP on witch measurements have been performed is the last to be recovered. The key difference beetween the considered scenarios are the fault notification mechanism and the actions made by the ingress Label Edge Router (LER) to recover the protected traffic. In the following we define *notification time* the time between the fault occurrence and the receiving of the fault notification message (RSVP-TE PATH_ERR message or OSPF-TE Opaque_LSA) from downstream router on the ingress LER and *recovery operation time* the time needed to perform recovery operation and traffic recovery.

4.1 Commercial Testbed

The first testbed on which the experimental activities have been carried out is depicted in Fig 1(a) and is represented by a Gigabit Ethernet network fiber optical ring that permanently interconnects three remote sites all across the town of Pisa: the Department of Information Engineering of the University of Pisa, the CNIT National Laboratory of Photonic Networks and the Institute for Informatics and Telematics (IIT) of the Italian National Research Council. The experimental testbed is equipped with four M10 Juniper routers, running JunOS 6.0 (see [16] and [17] for details), with two Gigabit Ethernet interfaces at 1 Gbit/s and a variable number of FastEthernet interfaces connecting the router to the internal LANs. A set of recovery mechanisms are available on M10 routers including both path and span level techniques. Considering the first ones, that are the focus of this paper, two different mechanisms can be distinguished:

1. Secondary Path Standby (Path Level Protection). Backup LSP resources are pre-established and pre-reserved at primary LSP setup time;
2. Secondary Path No-Standby (Path Level Restoration). Backup LSP resources are pre-calculated but not signalled (Backup LSP RSVP-TE signalling takes place only after a fault is detected).

When the fault occurs the router that experiences the failure, on both mechanisms, notifies the network impairment to the ingress LER by means of the following actions:

– A Flexible PIC Concentrator (FPC) informs the Packet Forwarding Engine (PFE) that a fault is detected;
– This notification is delivered to the RSVP State Machine that, in turn, sends a PATH_ERR message to the PFE and changes state into down;
– PFE forwards a PATH_ERR message toward the ingress LER.

The notification is completed when the PATH_ERR message is received by the RSVP State Machine on the Ingress LER. Subsequent actions, defined as recovery operations, depend on the mechanism configured. In the Standby scheme backup resources are pre-reserved (Backup LSP is signalled and resources are allocated in advance) and only to switch traffic from Primary LSP to Backup one is needed. In the No-Standby scheme, at first, RSVP signalling along with resource reservation is needed and the switch operation takes place only when the RESV message is received on the Ingress LER and the Backup LSP is established. Finally, traffic switch-over on the Backup LSP is performed.

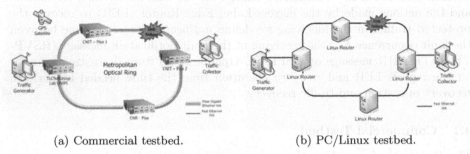

(a) Commercial testbed.   (b) PC/Linux testbed.

Fig. 1. Operational MPLS networks.

4.2 Linux Testbed

The second testbed on which the experimental activities have been carried out is depicted in Fig 1(b). It is made-up by four PC 2.0 Ghz CPU, 512 MB RAM, Linux 2.4.20 kernel interconnected on a 100 Mbit Ethernet ring. All Linux boxes are DiffServ aware MPLS-TE routers with on-demand end-to-end service capabilities. Fig.2 shows the logical organization of the software modules at the generic router. The Node Manager (NM) is in charge of receiving connection requests from the external and coordinate the information flow between the modules. Therefore, for each incoming connection request, the NM will first compute the end-to-end routes of both the working and backup LSPs (arrow a). The route computation is local to the edge node (source routing) and run by a separate module called Route Selection Engine (RSE). The route selection is based (arrow b) on the information about network topology and residual link bandwidth that is disseminated by OSPF-TE through the flooding of Opaque LSAs (arrow c), and maintained at the local Network State Database.

The route selection algorithm running at the RSE is the one described in [18] and [19]. Basically it jointly selects the routes of the working and backup LSP for the new connection taking into account i) the disjointedness constraint, ii) the available bandwidth constraint and iii) a bandwidth minimization / balancing objective (achieved by simply minimizing a link-cost metric that is inversely proportional to the residual reservable bandwidth for the specific Diffserv class). Additionally, the RSE algorithm has been also extended to route demands with time-varying bandwidth profile (see [20]) and supports Shared-Risk Link Groups (SRLG) and SRLG-disjointedness for protected demands (see [19] for details).

After selected the routes, the NM triggers the RSVP-TE daemon (arrow d) to start the setup signaling procedures for the working and backup LSPs. During

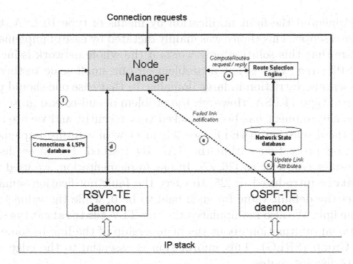

Fig. 2. Structure of logical modules at the generic Linux router.

the signaling phase, each node along the path enforces Admission Control in order to check current bandwidth availability. This step is necessary because the link load information available at the edge node may be not synchronized with the current network state. This can be due to the intrinsic delays in the flooding process and/or to the adoption of some conservative link-state update policy aimed at controlling the flooding overhead. Examples of flooding reduction mechanisms can be found in [21–23]; we adopted the algorithm proposed in [23] (see also [18]), applied in a multi-class bandwidth allocation environment.

Upon occurrence of link failure, the edge node must be notified the event and promptly switch the incoming packets from the working to the backup LSP. Therefore, a mechanism is needed to convey the failure notification from the internal node to the edge nodes. According to what stated in Section 3 we preferred the flooding-based approach and we implemented in OSPF-TE flooding of fault notification messages via Opaque LSA; in the following some details on the prototypical implementation are reported.

Details of the Notification Mechanism Implementation. In this subsection we report on some details of our prototypical implementation of flooding-based fault notification through OSPF-TE Opaque LSAs.

The Opaque LSA option (O-LSA for short) has been defined in [24], where three types of O-LSAs are defined with different scope: type 9 (link-local, flooded only within the subnetwork), type 10 (area-local, flooded within the associated area) and type 11 (flooded throughout the entire AS). The usage of O-LSAs for traffic engineering purposes has been described in [25], that specifically prescribes to use type 10 LSAs. These are considered for the dissemination of link attributes, included currently unreserved bandwidth, and introduces a number of nine message elements associated to the link. These are included in the O-LSA, and are called the sub-TLVs of the "Link TLV". All such features were already supported by the OSPF implementation [9] that was used in the testbed.

We implemented the fault notification by means of type 10 LSA, therefore with area-level scope. This choice was mainly dictated by ease of implementation. We are aware that this solution only works if the whole network is included in a single OSPF area. This is not a problem on our small scale testbed, but it might be a serious restriction in large domains. In that case one should probably consider use of type 11 LSA. However, the problem of end-to-end protection in a multi-region environment has been attacked very recently, and we are currently supporting the discussion in IETF (see [8]) in view of a future implementation of multi-region protection within the TANGO testbed. For more details the interested reader can refer to [26, 27]. In our implementation, we used the sub-TLV `TE-metric` introduced in [25] to carry the fault notification semantic: we arbitrary set the default value for such field to be 0, while the value 1 indicate failure of the link. We also manipulated the sub-TLV `Administrative-Group` to carry additional information about the membership of the link to some Shared-Risk Link Group (SRLG). This information is essential to the edge nodes to select SRLG-disjoint routes.

A key point we had to cope with is the presence of hold-down timers in OSPF. In OSPF [28] two timers are present to enforce a minimum spacing between consecutive LSAs referred to the same link. The first one, `MinLSInterval`, inhibits the generation of new LSAs for 5 seconds after transmitting a LSA. The second one, `MinLSArrival`, imposes the discarding of any new LSA received within 1 second since the last received LSA. With such timers, the flooding of an O-LSA advertising the failure of link j could be delayed in the case that the previous Opaque LSA for the same link was generated within the latter 5 seconds – for example to advertise a change in the reserved bandwidth. In order to eliminate this possibility, we introduced a mechanism called "timer forcing". That means the generation/reception of a new O-LSAs carrying a fault notification semantic (i.e., `TE-metric` set to 1) can force the hold-down timer to expire immediately and be reset. The timer is associated to a flag variable, which is set to 1 when the timer is forced, and returns back to its default value 0 when the timer expires normally. The arriving O-LSA can force the timer only if $i)$ it carries fault notification semantic and $ii)$ if the flag variable is set to 0. This again enforces a minimum spacing between O-LSA with a failure notification content. With this mechanism we eliminated the interspacing between a generic O-LSA and the first fault notification O-LSA, but apart this "exception" the timer behavior remains compliant with the standard. We remark that this solution does not introduce any additional processing burden, since in any case the content of O-LSAs has to be processed before forwarding. This mechanism can be considered as a simple example of content-based routing message processing [29].

5   Experimental Results

In order to evaluate the recovery time that can be obtained by means of different recovery mechanisms in the two considered scenario a measurement methodology has been defined. On PC/Linux routers a network monitor software measure with

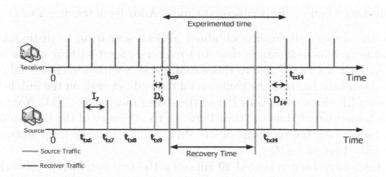

Fig. 3. Estimation of the overall recovery time.

ms precision recovery events occurrence. Unfortunately Juniper M10 routers timestamp the events with a resolution of 1 s, this means that the router's log file can't be used for recovery time evaluation. So, to introduce in the testbeds two more network elements (Linux PCs) acting respectively as traffic generator (TG) and traffic collector (TC) is needed.

In both the scenarios the TG is directly connected with the ingress LER and a Constant Bit Rate (CBR) traffic is generated by means of BRUTE (a software tool developed by UniPi [30]) and mapped onto the primary LSP. Data flow rate has been chosen to load the networks about 0,3% of the total links' capacity (such a choice is also motivated by synthetic traffic generation constraints):

- 3 Mbit/s UDP traffic flow with packet size 64 byte and 163 μs inter-departure time has been mapped into the primary LSP in the Juniper testbed;
- 300 kbit/s UDP traffic flow with packet size 64 byte and 1,63 ms inter-departure time has been loaded in the PC/Linux network.

On the other side traffic is terminated and collected on the TC, directly connected with the Egress LER. Traffic traces, with accurate packet timestamping, are collected on both PCs keep synchronized by means of GPS receivers. The experiment begins when a failure is forced to occur on the primary LSP and recovery mechanism starts. The overall recovery time is estimated by the analysis of collected traffic traces. Since is not possible to timestamp network impairment and switch-over operation completion in the commercial testbed, they are approximated with the time at which the last packet before the failure (t_{rx9} in Fig. 3) and the first packet after the recovery (t_{rx14}) are received on the TC. The difference between t_{rx9} and t_{rx14} is defined as Experimented Time (ET). We can observe that this time is influenced by the difference between the One-Way Delay (D) experimented after the recovery and the one before the failure, referred as Additive Latency (AL). This value can be computed from the synchronized collected traces and has to be subtracted to ET. So, the measurement error, induced in this scheme by the discrete packet sending, depends only on the packet inter-departure time (ID) and is up to $2 * ID$. This time has been reduced (163 μs) in order to not influence the recovery time estimation (10's of ms).

$$\text{Recovery\_time} = \text{Experimented\_time} - \text{Additive\_latency} \pm 2 * ID$$

In order to compare the results obtained in both scenarios, to distinguish the components such as *notification time* and *recovery operation time* a network analyzer (AdTech AX4000), able to timestamp packets with a microsecond resolution, has been inserted into the commercial testbed network on the link between the ingress LER (located at UniPI) and the router located at CNIT. *Notification time* has been estimated as the time between the delivery of the last packet before the failure and the receiving of the PATH_ERR message from downstream router on the Ingress LER.

The tests have been repeated 20 times on the two considered networks, for each mechanism in both network conditions, in order to compute a *Mean Recovery Time*. The results are collected in Tab.1. With only 1 protected LSP established we can observe that in the commercial testbed the most relevant contribution on the overall recovery the time is represented by *notification time* while in PC/Linux network the fault notification process takes much less (the mean value is ten times lower). Such unexpected delays that potentially could be caused by hold off timers, should be investigated more in depth (even if the manufacturer doesn't provide any information about). On the other hand a longer *recovery operation time* is needed by the ingress LER, after the reception of the flooded notification message, to identify LSPs to protect and start recovery operations and traffic switch-over – a list of protected LSPs with corresponding routing paths should be maintained at each ingress LER. Unexpectedly the mean *notification time* measured for Restoration strategy is longer with re-

Table 1. Measurements results: mean values (standard deviation) of the notification time and recovery operation time for different recovery techniques.

		End-to-end Recovery Mechanism	Notification Time (ms)	Recovery Operation Time (ms)	Overall Recovery Time (ms)
Commercial Testbed	1 LSP	*Protection*	400,753 (231,428)	27,463 (2,884)	428,216
		Restoration	462,770 (276,290)	31,320 (5,920)	494,090
	10 LSP	*Protection*	388,610 (240,310)	86,000 (9,930)	474,610
		Restoration	456,730 (246,400)	118,600 (22,270)	575,330
PC/Linux Testbed	1 LSP	*Protection*	39,600 (20,298)	68,700 (15,649)	108,300
		Restoration	40,500 (21,280)	101,300 (28,900)	141,800
	10 LSP	*Protection*	41,800 (17,570)	113,500 (23,140)	155,300
		Restoration	45,400 (18,330)	138,200 (24,930)	183,600

spect to the time obtained with Protection recovery; infact the intermediate router that performs fault notification is totally blind of the adopted recovery mechanism. Because of the high standard deviation values of measured times an higher number of data should be collected to better address this issue. From the comparison beetween Protection and Restoration mechanisms also emerges the faster capability to setup LSP by commercial routers (about 3 ms) with respect to PC/Linux boxes (about 30 ms).

When 10 protected LSPs are established after the fault occurs, 10 RSVP-TE PATH_ERR messages are sent to the ingress LER in the commercial testbed while handling a larger protected LSPs database is required by the ingress router to perform recovery operation in the PC/Linux network. In this case advisable differences could be noted for the *recovery operation time* in both the testbeds, indeed a longer time is needed to find the LSPs to recover by the PC/Linux router but also considerable time is spent by commercial router to handle all notification messages.

6 Conclusions

In this work we proposed a comparison of recovery strategies in MPLS-TE experimental testbeds. We focus on alternative notification mechanisms and compare different end-to-end recovery techniques. A measurements campaign has been performed in two different trials based on commercial routers and PC/Linux boxes. In the former an inbuilt signaling-based mechanism provides remote fault notification to the edge routers while in the latter an original prototype of IGP flooding-based mechanism has been realized. From the comparison of different recovery schemes appears that the proposed flooding-based approach is efficient and fast, moreover we showed how it can be implemented within the existing protocols for MPLS traffic engineering, particularly OSPF-TE. Besides, the results presented here suggest that also for the commercial routers, where ad-hoc hardware/software design improve processes operation times and separate control and forwarding plane, network condition – in particular the number of established LSPs – considerably affects the recovery operation time.

We investigated the impact of the alternative fault notification schemes and ran tests in order to assess a performance metric, namely the *achievable overall recovery time* in terms of *notification* and *recovery operation* components. We stress that such results were obtained in ideal conditions: a small unloaded network, without packets nor LSPs other than those under test. Collectively they confirm the expectation that the reference of 50 ms of MPLS Fast Rerouting can not be accomplish by end-to-end mechanisms. On the other hand, they seem to be encouraging towards the possibility of achieving end-to-end MPLS recovery in the order of few hundreds of milliseconds in operational conditions.

Our current efforts are directed to repeat the measurements in heavy-load network conditions, in order to assess the robustness of the considered fault recovery schemes. Admittedly, our results are limited to quasi-ideal conditions, particularly about the choice of the traffic load and the simple topology. We recognize that a conclusive quantitative assessment of our analysis necessitates a traffic/topology scenario closer to reality. We believe that the ideas and experimental insight contained in this work will be helpful to other people involved in standardization and implementation of MPLS-related recovery techniques.

Acknowledgments

This work was funded by the TANGO project of the FIRB programme of the Italian Ministry for Education, University and Research.

References

1. D. Awduche et al.: Requirements for Traffic Engineering Over MPLS. RFC 2702 (1999)
2. X. Xiao et al.: Traffic Engineering with MPLS in the Internet. IEEE Network (2000)
3. Project, T.: (Homepage: http://tango.isti.cnr.it)
4. D. Papadimitriou, E. Mannie eds.: Recovery (Protection and Restoration) Terminolgy for Generalized Multi-protocol Label Switching (GMPLS). draft-ietf-ccamp-gmpls-recovery-terminology-05.txt. Work in progress (2004)
5. P. Lang, B. Ragjagopalan: Generalized Multi-protocol Label Switching (GMPLS) Recovery functional Specification. draft-ietf-ccamp-gmpls-recovery-functional-02.txt. Work in progress (2004)
6. G. Mohan, C.S.R.M.: Lightpath restoration in WDM optical network. IEEE Network **14** (2000) 24–32
7. V. Sharma, F. Hellstrand: Framework for Multi-Protocol Label Switching (MPLS)-based Recovery. RFC 3469 (2003)
8. Mail archive of the CCAMP working group: (http://ops.ietf.org/lists/ccamp)
9. Zebra Home Page: (http://www.zebra.org/)
10. C. Huang, V. Sharma, K. Owens, S. Makam: Building Reliable MPLS Networks Using a Path Protection Mechanism). IEEE Communications Magazine (2002)
11. J. P. Lang, B. Rajagopalan eds.: Generalized MPLS Recovery Functional Specification. draft-ietf-ccamp-gmpls-recovery-functional-02.txt. Work in progress (2004)
12. D. Papadimitriou, E. Mannie eds.: Analysis of Generalized MPLS-based Recovery Mechanisms (including Protection and Restoration). draft-ietf-ccamp-gmpls-recovery-analysis-03.txt. Work in progress (2004)
13. D. Awduche et al.: Overview and Principles of Internet Traffic Engineering. RFC 3272 (2002)
14. T. Soumiya, R. Rabbat eds.: Extensions to LMP for Flooding-based Fault Notification. draft-soumiya-lmp-fault-notification-ext-02. Work in progress (2003)
15. R. Rabbat, T. Soumiya, S. Kanoh, V. Sharma, F. Ricciato, R. Albanese: Implementation and Performance of Flooding-based Fault Notification. draft-rabbat-ccamp-perf-flooding-notification-exp-00.txt. Work in progress (2004)
16. A. Garrett, G.D.C., Networks, J.: Field Guide and Reference. Addison-Wesley (2002)
17. J. Doyle, M.K.: Juniper Network Routers: The Complete Reference. (McGraw-Hill/Osborne)

18. F. Ricciato, M. Listanti, A. Belmonte, D. Perla: Performance Evaluation of a Distributed Scheme for Protection against Single and Double Faults for MPLS. 2nd Int'l Workshop on Quality of Service in Multiservice IP Networks (QoS-IP 2003), Milano (2003) Published in Lecture Notes in Computer Science, vol. 2601, Springer, pp. 218-232.
19. F. Ricciato, S. Salsano, M. Listanti: An Architecture for Differentiated Protection agains Single and Double Faults in GMPLS. (to appear in Photonic Networks Magazine, June/July 2004)
20. F. Ricciato, U. Monaco: On-Line Routing of MPLS Tunnels with Time-Varying Bandwidth Profiles. High Performance Switching and Routing HPSR04 , Phoenix (2004)
21. A. Shainkh, J. Rexford, K. G. Shin: Evaluating the Overheads of Source-Directed Quality-of-Service Routing. Int'l Conference on Network Protocols (ICNP) (1998)
22. G. Apostolopoulos, R. Guerin, S. Kamat, S.K.Tripathi: Quality of Service Based Routing: A Performance Perspective. SIGCOMM (1999)
23. A. Botta, P. Iovanna, M. Intermite, S. Salsano: Traffic Engineering with OSPF-TE and RSVP-TE: Flooding Reduction Techniques and Evaluation of Processing Cost. CoRiTeL Report. Submitted. (2003)
24. R. Coltun: The OSPF Opaque LSA Option. RFC 2370 (1998)
25. D. Katz, K. Kompella, D. Yeung: Traffic Engineering (TE) Extensions to OSPF Version 2). RFC3630 (2003)
26. F. Ricciato, U. Monaco, A. D'Achille: A novel scheme for end-to-end protection in a multi-area network. (Proc. of 2nd International Workshop on Inter-domain Performance and Simulation, IPS04, Budapest, Hungary, 22-23 March 2004.)
27. A. D'Achille, M. Listanti, U. Monaco, F. Ricciato, V. Sharma: Diverse Inter-Region Path Setup/Establishment . draft-dachille-diverse-inter-region-path-setup-01.txt (2004)
28. J. Moy: OSPF Version 2. RFC 2328 (1998)
29. G. Iannaccone, J. Chen-Nee Chuch, S. Bhattachoryya, C. Diot: Feasibility of IP Restoration in a Tier-1 Backbone. IEEE Networks (2004)
30. BRUTE Home Page: (http://netgroup-serv.iet.unipi.it/brute)

NPP: A Facility Based Computation Framework for Restoration Routing Using Aggregate Link Usage Information[*]

Faisal Aslam, Saqib Raza, Fahad Rafique Dogar,
Irfan Uddin Ahmad, and Zartash Afzal Uzmi

Lahore University of Management Sciences, Pakistan
{faisal,saqibr,fahad,irfank,zartash}@lums.edu.pk

Abstract. We present NPP – a new framework for online routing of bandwidth guaranteed paths with local restoration. NPP relies on the propagation of only aggregate link usage information [2,9] through routing protocols. The key advantage of NPP is that it delivers the bandwidth sharing performance achieved by propagating complete per path link usage information [9], while incurring significantly reduced routing protocol overhead. We specify precise implementation models for the restoration routing frameworks presented in [1] and [2] and compare their traffic placement characteristics with those of NPP. Simulation results show that NPP performs significantly better in terms of number of LSPs accepted and total bandwidth placed on the network. For 1000 randomly selected LSP requests on a 20-node homogenous ISP network [8], NPP accepts 775 requests on average compared to 573 requests accepted by the framework of [2] and 693 requests accepted by the framework of [1]. Experiments with different sets of LSP requests and on other networks indicate that NPP results in similar performance gains.

1 Introduction

The destination based forwarding paradigm employed in plain IP routing does not support routing along explicit routes determined through constraint based routing [12]. The emergence of Multi-Protocol Label Switching (MPLS) has overcome this limitation of traditional shortest path routing, by presenting the ability to establish a virtual connection between two points on an IP network, maintaining the flexibility and simplicity of an IP network while exploiting the ATM-like advantage of a connection-oriented network [13]. Ingress routers of an MPLS network classify packets into forwarding equivalence classes and encapsulate them with labels before subsequently forwarding them along pre-computed paths [15]. The path a packet takes as a result of a series of label switch operations in an MPLS network is called a label switched path (LSP). LSPs may be routed through constraint based routing that adapts to current network state information (e.g., link utilization) and selects explicit routes that satisfy a set of

---

[*] This work was supported by a research grant from Cisco Systems, San Jose, CA.

M. Ajmone Marsan et al. (Eds.): QoS-IP 2005, LNCS 3375, pp. 150–163, 2005.
© Springer-Verlag Berlin Heidelberg 2005

constraints. The ability to explicitly route network traffic using constraint based routing enables service providers to provision quality of service and also leads to efficient network utilization [14].

An important application of constraint based routing is provisioning of bandwidth guaranteed LSPs [6–8]. Furthermore, many real-time applications require that the guaranteed bandwidth remains available when network facilities[1] fail. When recovery mechanisms are employed at the IP layer, restoration may take several seconds which is unacceptable for real-time applications [11]. In contrast, MPLS *local restoration* meets the requirements of real-time applications with recovery times comparable to those of SONET rings [6,10]. In local restoration, each LSP passing through a facility is protected by a backup path which originates at the node immediately upstream to the facility. This node, which redirects the traffic onto the preset backup path in case of failure, is called the Point of Local Repair (PLR). Since this decision to redirect traffic is strictly local, faster recovery is possible. In this paper, we consider routing bandwidth guaranteed paths with local restoration.

There are two distinct approaches to local restoration: In *one-to-one* backup approach [6–8], the PLRs maintain separate backup paths for each LSP passing through a facility. The backup path terminates by merging back with the primary path at a node called the Merge Point (MP). In one-to-one backup approach, the MP can be any node downstream the protected facility. Maintaining state information for backup paths protecting individual LSPs, as in the one-to-one approach, is a significant resource burden for the PLR. Moreover, periodic refresh messages[2] sent by the PLR, in order to maintain each backup path, may become a network bottleneck [4]. On the other hand, in *many-to-one* approach, a PLR maintains a single backup path to protect a set of primary LSPs traversing the triplet (PLR, facility, MP)[3]. Thus, fewer states need to be maintained and refreshed which results in a scalable solution. Many-to-one backup approach, also called *facility backup*, is illustrated in Fig. 1. Note that in this approach, the MP should be the node immediately downstream to the facility.

Backup provisioning requires bandwidth reservation along the backup paths, thereby reducing the total number of LSPs that can otherwise be placed on the network. This reduction is significant if resources along the backup paths are not shared. Since it is reasonable to assume that different network facilities will not fail simultaneously [1, 6–8], backup paths protecting different facilities should share bandwidth.

In this paper, we present NPP – a new framework for facility backup, which relies on the propagation of aggregate link usage information through routing protocols [2, 17]. The key advantage of NPP is that it delivers the bandwidth sharing performance achieved by propagating complete per path link usage in-

---

[1] The term facility refers to either a node or a bidirectional link.

[2] Local Protection primarily uses RSVP-TE extensions, which is a soft-state protocol and requires periodic refresh messages to maintain its states.

[3] Throughout this paper, an LSP traversing the triplet (PLR, Facility, MP) refers to a primary LSP that passes through the PLR, the facility, and the MP in that order.

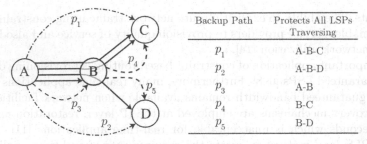

Backup Path	Protects All LSPs Traversing
p_1	A-B-C
p_2	A-B-D
p_3	A-B
p_4	B-C
p_5	B-D

Fig. 1. Many-to-one or Facility Backup.

formation [9], while incurring significantly reduced routing protocol overhead. We compare the traffic placement characteristics of NPP with two existing facility backup frameworks, described in [1] and [2], and present results based on the total number of LSP demands accepted and the bandwidth placed on the network.

2   Background

2.1   Routing Paradigm

The optimality of primary and backup paths computed to serve an LSP request is a function of network state information available during path computation. Two path computation scenarios are possible in this context: Centralized and Distributed. In centralized path computation, one central entity makes all the routing decisions. This central entity can achieve optimal backup bandwidth sharing as it maintains complete network information. However, a centralized path computation server incurs additional costs in terms of high processing power and high bandwidth control channels [5]. Distributed control entails autonomous path computation by distributed nodes, based on the node's view of the network state. The state maintained by a node, in the distributed path computation scenario, is a function of its local state and the network state information periodically distributed by other routers via link state routing protocols. We consider restoration routing frameworks that use distributed path computation, such that the primary path is computed by the LSP ingress node and backup paths are computed locally. The primary and backup paths are signaled using protocols like RSVP-TE [3] and CR-LDP [18]. In case of primary paths, the ingress node initiates the signaling, while each backup path is signaled by its PLR. Each node along the signaled route reserves the required bandwidth, before forwarding the signaling request to the next node along the path.

2.2   Fault Model

Failure of a network facility along an LSP results in the need to divert the LSP traffic onto a preset backup path. Failure of network nodes and links is a low probability event, and network measurements reveal that chances of multiple failures in the network are even lower. Furthermore, upon failure along the primary path new reoptimized primary and backup paths may be provisioned, with local restoration serving only as a temporary measure [1]. The probability of additional failures during the setup of reoptimized paths is negligible. Therefore, a more realistic restoration objective is to provide protection against the failure of a single facility. We consider the fault model wherein backup paths provision restoration in the event of a single link or single node failure. We refer to this fault model as the *single element protection* fault model.

In order to elucidate local recovery for the single element protection fault model we distinguish between two types of backup paths: *next-hop* paths and *next-next-hop* paths.

Definition 1. *A next-hop path that spans a link* $(i,j)^4$ *is a backup path which:*

a) *originates at node* i,

b) *merges with the primary LSP(s) at node* j, *and*

c) *provides restoration for one or more primary LSPs that traverse* (i,j), *if* $\{i,j\}$ *fails.*

Definition 2. *A next-next-hop path that spans a link* (i,j) *and a link* (j,k) *is a backup path which:*

a) *originates at node* i,

b) *merges with the primary LSP(s) at node* k, *and*

c) *provides restoration for one or more primary LSPs that traverse* (i,j), *if* $\{i,j\}$ *or node* j *fails.*

Fig. 2 depicts local restoration with respect to a single primary path according to the single element protection fault model. The figure shows that setting up next-next-hop paths along the primary path provides restoration in event of single node failure. Note that such a configuration also protects against the failure of all except the last link. In order to provision single element protection, an additional next-hop backup path spanning the last link is setup.

3   Problem Definition

We consider a network with n nodes and m bidirectional links. LSP requests arrive one by one at the ingress node, and the routing algorithm has no a priori knowledge of future requests. An LSP request is characterized by the LSP ingress node, the LSP egress node, and an associated bandwidth demand b.

In order to serve an LSP request, a bandwidth guaranteed primary path must be setup along with locally restorable backup paths that provide protection against the failure of facilities along the primary path. If the routing algorithm

---

[4] A bidirectional link between two nodes constitutes a single facility. However, traffic traverses a link in a specific direction. We, therefore, use $\{i,j\}$ to represent the bidirectional link between node i and node j, and use the ordered pair (i,j) when direction is significant. Thus, (i,j) refers to the directed stem of $\{i,j\}$ from node i to node j. Note that failure of the facility $\{i,j\}$ implies failure of (i,j) and (j,i).

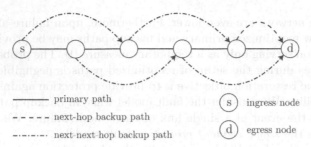

Fig. 2. Local Restoration for Single Element Failure.

is able to find sufficient bandwidth in the network for the requisite primary and backup paths, the paths are setup, and the LSP request is accepted; otherwise, the LSP request is rejected. The next LSP request arrives only after the current LSP request has either been accepted or rejected. Our goal is to optimize network utilization. To this end, we wish to minimize the bandwidth reserved for the primary and backup paths for each LSP request.

Suppose we are serving a new LSP request with bandwidth demand b. Further suppose that the computed primary LSP for this request traverses (i,j). A backup path, that provisions restoration for this primary LSP along (i,j), is a next-hop path if j is the egress node, and is a next-next-hop path otherwise. Each backup path protects one or more facilities and has a merge point. The PLR for both types of backup paths is node i. In sum, we are trying to protect a primary LSP that traverses the triplet (PLR, facility, MP). It is possible that a backup path $\wp_{old}$ already exists, that protects previously routed primary LSPs traversing the same triplet. If such a backup path exists, restoration for the new request may be provisioned by reserving additional bandwidth along that path. However, some links along $\wp_{old}$ may not have the capacity for the requisite additional reservation. Therefore, if the primary LSP belonging to the current LSP request traverses the triplet (PLR, facility, MP), the PLR re-computes a new backup path that provisions restoration for all primary LSPs traversing that triplet, including the new primary LSP. We refer to this new backup path as $\wp$. If the cumulative bandwidth reserved for the previous LSPs traversing the triplet (PLR, facility, MP) was b_{old}, the bandwidth required for $\wp$ is given by $b_{new} = b + b_{old}$. The details for path computation, bandwidth sharing, and subsequent setup of $\wp$ depend upon the restoration routing framework and are explained in the following sections.

4   Backup Bandwidth Sharing

Failure of a protected facility (link or node) results in the activation of a set of backup paths. We refer to the set of backup paths that are simultaneously activated, if a facility fails, as the *activation set* for that facility. PLRs detect[5]

---

[5] Mechanisms exist that allow the PLR to distinguish between link and node failure; see [16] for details.

link or node failure and subsequently activate all backup paths that protect the failed facility. The following enumerates the backup paths included in the activation sets for the facilities $\{i,j\}$ and node j:

Activation Set for $\{i,j\}$: Recall from section 2 that a next-hop path protects against failure of a link, and a next-next-hop path protects against failure of both a link and a node. Incase $\{i,j\}$ fails all next-hop and next-next-hop paths protecting $\{i,j\}$ are activated. This activation set for $\{i,j\}$ comprises the following backup paths (see Fig. 3):

- next-hop path that spans (i,j)
- next-hop path that spans (j,i)
- next-next-hop paths that span (i,j) and $(j,x), \forall x \neq i$
- next-next-hop paths that span (j,i) and $(i,x), \forall x \neq j$

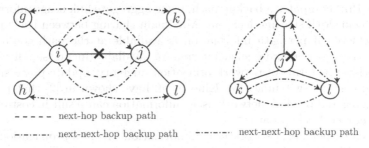

```
- - - - -  next-hop backup path
-··-··-··-  next-next-hop backup path      -··-··-·  next-next-hop backup path
```

Fig. 3. Paths in the Activation Set for link $\{i,j\}$ (left) and for node j (right).

Activation Set for Node j: Recall from section 2, only next-next-hop paths protect against failure of a node. Thus, the activation set for node j comprises the next-next-hop paths that span (x,j) and $(j,y), \forall x \neq y$. Fig. 3 shows the set of backup paths that are activated incase node j fails.

Since only a single link or a single node may fail at a time, two backup paths will not be simultaneously active unless they are in the same activation set. It follows that such backup paths can share bandwidth with each other. In contrast, backup paths that are simultaneously active must make bandwidth reservations that are exclusive of each other. Consequently, a next-hop path that spans (i,j) is activated if $\{i,j\}$ fails, and therefore, cannot share bandwidth with other backup paths belonging to the activation set of $\{i,j\}$. Similarly, a next-next-hop path that spans (i,j) and (j,k) is activated if either $\{i,j\}$ or node j fails, and hence cannot share bandwidth with backup paths belonging to the activation sets of either $\{i,j\}$ or node j.

5 Restoration Routing Frameworks

The extent of bandwidth sharing is governed by a number of parameters that include the distribution of path computation, the amount of network state information propagated through routing protocols, and the signaling mechanisms used for path setup. In this section, we present two existing restoration routing frameworks. We then introduce NPP, our own restoration routing framework, and describe how it draws upon the advantages of the other two frameworks to achieve more efficient bandwidth sharing.

5.1   Kini's Framework

In this framework, each backup path is computed by its PLR. The PLR relies on network state information propagated through routing protocols to decide how much bandwidth a backup path can share on any given link. Kini's framework involves propagation of aggregated per-link network usage information as proposed in [2]. Kini's framework is characterized by two distinct stages: a suboptimal path computation stage, and a corrective signaling stage.

Suboptimal Path Computation Stage: In the suboptimal path computation stage, the PLR computes a backup path, using aggregate link usage information, to make bandwidth sharing decisions. Maximum sharing between backup paths can be achieved if per-path information is available at the path computation server [9]. However, making such information available incurs significant protocol overhead in terms of network utilization, memory, update-processing and associated context switching [14]. Kini, et al. have shown in [2] that propagation of aggregated per-link network usage information can result in cost-effective sharing between backup paths.

Aggregated per-link network usage information involves propagation of the following values for each link (i, j) in the network:

F_{ij} : Bandwidth reserved on (i, j) for primary LSPs
G_{ij} : Bandwidth reserved on (i, j) for backup LSPs
R_{ij} : Residual bandwidth on (i, j)

The above network state information (i.e., F_{ij}, G_{ij}, and R_{ij}) can be propagated using traffic engineering extensions to existing link state routing protocols [17]. The following describes how Kini's framework makes use of this aggregate link usage information to make bandwidth sharing decisions.

In order to route $\wp$, we seek to find the amount of bandwidth reserved on a link (u, v) that can be shared by $\wp$, for every (u, v) in the network. To this end, we consider the following two cases:

Case 1: $\wp$ Is a Next-Hop Path That Spans a Link (i, j). Recall from section 2 that a next-hop path that spans (i, j) is activated when $\{i, j\}$ fails. As described in section 4, the activation set of $\{i, j\}$ consists of next-hop and next-next-hop backup paths that provision restoration for primary LSPs that traverse $\{i, j\}$. The total amount of bandwidth for these LSPs is $F_{ij} + F_{ji}$. Thus, a link (u, v) can have a maximum of $F_{ij} + F_{ji}$ units of backup bandwidth reserved for backup paths that are activated when $\{i, j\}$ fails. However, b_{old} units of bandwidth, out of $F_{ij} + F_{ji}$, are reserved for $\wp_{old}$. Therefore, the amount of bandwidth that is simultaneously active with $\wp$ is given by $F_{ij} + F_{ji} - b_{old}$. It follows that S_{uv}, the bandwidth available for sharing by $\wp$ on (u, v), is given by:

$$S_{uv} = \max(0, G_{uv} - F_{ij} - F_{ji} + b_{old}) \tag{1}$$

Case 2: $\wp$ Is a Next-Next-Hop Path That Spans a Link (i, j) and a Link (j, k).
Recall from section 4 that a next-next-hop path that spans (i, j) and (j, k) is
activated if either $\{i, j\}$ or node j fails. As explained in Case 1, $F_{ij} + F_{ji} - b_{\text{old}}$ is
the worst case bandwidth reserved for backup paths that will be simultaneously
active with $\wp$ when $\{i, j\}$ fails. Similarly, the activation set of node j comprises
next-next-hop paths that protect primary LSP traffic traversing all links (x, j),
such that node x is adjacent to node j. The maximum amount of such traffic
is given by $\sum_x F_{xj}$. Thus, a link (u, v) can have a maximum of $\sum_x F_{xj}$ units
of backup bandwidth reserved for backup paths that are activated when node j
fails. However, b_{old} units of bandwidth, out of $\sum_x F_{xj}$, are reserved for $\wp_{\text{old}}$.
Therefore, the maximum amount of bandwidth that is simultaneously active with
$\wp$, when node j fails, is given by $\sum_x F_{xj} - b_{\text{old}}$. Since $\wp$ is activated if either node j
or $\{i, j\}$ fails, a link (u, v) can have up to $\max(F_{ij} + F_{ji} - b_{\text{old}}, \sum_x F_{xj} - b_{\text{old}})$
units of bandwidth reserved for backup paths that are simultaneously active
with $\wp$. It follows that S_{uv}, the bandwidth available for sharing by $\wp$ on (u, v),
is given by:

$$S_{uv} = \max(0, G_{uv} - \max(F_{ij} + F_{ji} - b_{\text{old}}, \sum_x F_{xj} - b_{\text{old}})) \qquad (2)$$

For both the above cases, the additional bandwidth reservation required on
(u, v), if $\wp$ traverses (u, v), is given by $\max(0, b_{\text{new}} - S_{uv})$. The PLR computes
a route for $\wp$ such that the least amount of additional backup bandwidth is reserved. This culminates the suboptimal path computation phase of Kini's framework.

Corrective Signaling Stage: Once the PLR has computed a route for $\wp$,
it signals for the path to be setup. Note that in the previous stage the route
computed for $\wp$ is suboptimal. Since the PLR makes routing decisions on the
basis of aggregate link usage information, for every link (u, v), it assumes that
all backup paths in the activation set of a facility protected by $\wp$ traverse (u, v).
It is possible that only a subset of these paths traverse (u, v), and therefore, the
full extent of backup bandwidth sharing possible on (u, v) is obscured.

The corrective signaling stage can partially compensate for the suboptimal
routing of the first stage, during signaling of the path. Although the route computed for $\wp$ remains suboptimal, maximum bandwidth sharing along links in the
route computed for $\wp$ is ensured. This is accomplished as follows:

The head-end of each link (u, v) maintains a form of state which we refer to
as Link to Facility Incidence Map (LTFIM). The LTFIM for (u, v) contains a list
of all facilities that are protected by backup paths that traverse (u, v). For each
such facility, the LTFIM for (u, v) maintains b_{facility}, which is the total bandwidth
required on (u, v) by backup paths protecting that facility. Once a reservation
request arrives at the head end of a link (u, v) along the route computed for
$\wp$, the entries corresponding to the facilities protected by $\wp$ are located in the
LTFIM for (u, v).[6] Recall that the total bandwidth reserved for backup paths on

---

[6] In Kini's framework, installing $\wp$ and tearing off $\wp_{\text{old}}$ require LTFIM updates. In
practice, explicit tearing off is avoided because RSVP uses soft states.

(u, v) is given by G_{uv}. Since the bandwidth required by $\wp$ is b_{new}, G_{uv} must be at least equal to $b_{\text{new}} + b_{\text{facility}}$ for each facility protected by $\wp$. In case it is not so, we increase G_{uv} so that G_{uv} is equal to $b_{\text{new}} + b_{\text{facility}}$ for that facility. Note that for any link (u, v), additional bandwidth is only reserved when necessary, and therefore, $\wp$ benefits from maximum possible bandwidth sharing on (u, v).

5.2   Facility-Based-Computation (FBC) Framework

The key idea behind the FBC framework is that backup path computation is performed by a node that can make optimal bandwidth sharing decisions for that path. This is accomplished by maintaining a form of local state called Facility to Link Incidence Map (FTLIM). Every node j maintains FTLIM for each link adjacent to it and for itself. Each entry in an FTLIM for a facility corresponds to a link (u, v) and contains b_{uv}, which is the amount of bandwidth reserved on (u, v) by backup paths that belong to the activation set for that facility.

Furthermore, in FBC, each link of the topology maintains, logically disjoint, backup and primary pools [1]. A predetermined[7] percentage of each link is reserved for use by the primary paths and the remaining bandwidth is available for use by the backup paths. We refer to the amount of bandwidth constituting the backup pool on a link (u, v) as B_{uv}. We further define two indicator variables: I_{uv}^{new} which equals 1 if $\wp$ traverses (u, v), and is zero otherwise; and I_{uv}^{old} which equals 1 if $\wp_{\text{old}}$ traverses (u, v), and is zero otherwise. The following provides details of computing a route for $\wp$:

Case 1: $\wp$ Is a Next-Hop Path That Spans a Link (i, j). Recall from section 4 that a next-hop path that spans (i, j) is activated if $\{i, j\}$ fails. The FTLIM corresponding to link $\{i, j\}$ is maintained at both node i and node j. In the FBC framework, node j computes the route for $\wp$. It checks the FTLIM for $\{i, j\}$ at node j, and finds out b_{uv}, which is the bandwidth reserved on (u, v) by backup paths that belong to the activation set of $\{i, j\}$. Observe that it is feasible for $\wp$ to traverse (u, v) if and only if B_{uv} is greater than or equal to $b_{\text{new}} + b_{uv} - I_{uv}^{\text{old}} b_{\text{old}}$. Node j computes a route for $\wp$ using feasible links.

Upon routing $\wp$, some FTLIM entries also need to be updated. When $\wp$ is a next-hop path that spans (i, j), node j locates the FTLIM for the facility $\{i, j\}$, and increments[8] the entry corresponding to (u, v) by $I_{uv}^{\text{new}} b_{\text{new}} - I_{uv}^{\text{old}} b_{\text{old}}$. This update must be made to the FTLIM for $\{i, j\}$ maintained at node i. This is accomplished during signaling of $\wp$. Since node i is the PLR for $\wp$, node j must communicate the route computed for $\wp$ to node i, so that node i can signal $\wp$. At the same time, node i can update its FTLIM for $\{i, j\}$.

Case 2: $\wp$ Is a Next-Next-Hop path That Spans a Link (i, j) and a Link (j, k). Recall from section 4 that a next-next-hop path that spans (i, j) and (j, k) is activated if either $\{i, j\}$ or node j fails. The FTLIM for node j is maintained at

---

[7] The primary and backup pools are statically assigned by the network administrator and do not change afterwards.

[8] Since I_{uv}^{new} and I_{uv}^{old} can be 0 or 1 independently, FTLIM update may result in increase or decrease in the value of the entry.

node j and the FTLIM for $\{i, j\}$ is maintained at both node i and node j. In the FBC framework, node j computes the route for $\wp$. The FTLIMs for node j and $\{i, j\}$ contain values of b_{uv} for every link (u, v). Let $b_{uv}^\star$ be the maximum of all such values. Therefore, $\wp$ can traverse (u, v) if and only if the backup pool on (u, v) is greater than or equal to $b_{new} + b_{uv}^\star - I_{uv}^{old} b_{old}$. Node j uses this information to compute a route for $\wp$ that comprises only feasible links.

As in Case 1, node j increments the entries corresponding to (u, v) by an amount $I_{uv}^{new} b_{new} - I_{uv}^{old} b_{old}$, in the FTLIMs for the facilities $\{i, j\}$ and node j. Once again, such updates are also made to the FTLIM for $\{i, j\}$ maintained at node i during signaling of $\wp$.

An important feature of FBC is that no explicit bandwidth reservations are made on the links included in the route computed for a backup path. Since the node computing $\wp$ is aware of the bandwidth reservations of backup paths that may be simultaneously active with $\wp$, it ensures that there is no overbooking of bandwidth. Note that this implicitly results in optimal sharing of bandwidth between backup paths that belong to different activation sets.

5.3   NPP: Facility Based Computation Using Aggregate Information

We now present NPP, a new restoration routing framework that draws upon the advantages of Kini's framework and the FBC framework to achieve more efficient bandwidth sharing.

Recall that in Kini's framework, we use aggregate link usage information to compute a route for $\wp$. That is, for this computation, we assume that all backup paths in the same activation set as $\wp$ will traverse a given link (u, v). In actuality, however, only a subset of such paths will be traversing (u, v). Thus, we make a conservative estimate of the bandwidth that can be shared by $\wp$ on (u, v). Therefore, the route computed for $\wp$ is suboptimal. Note that the corrective signaling stage can partially compensate for the suboptimal routing of the initial stage. This is accomplished by reserving the precise amount of bandwidth required by $\wp$ during its signaling. This route is still suboptimal: since the route selection was based on aggregate information, another route that would have minimized the additional bandwidth reservation may have been rejected.

In contrast, the FBC framework computes optimal backup paths using local state. Moreover, perfect bandwidth sharing along the optimally computed paths is also ensured. However, a major disadvantage of the FBC framework is the static allocation of active and backup pools. Even carefully selected pools can remain under-utilized. This under-utilization represents a significant resource wastage and results in a greater number of LSP requests being rejected.

NPP does not suffer from the disadvantages associated with Kini's framework and the FBC framework. Similar to the FBC framework, it shifts the computation of a backup path from the PLR to the node that can make optimal bandwidth sharing decisions for that path. The NPP framework is similar to Kini's framework in that it makes use of aggregate link usage information. The residual capacity of every link in the network represents a single pool of bandwidth in this framework, as opposed to being divided into logically disjoint primary and backup pools. Furthermore, in the NPP framework nodes are required to maintain the LTFIM and FTLIM as they did in section 5.1 and section 5.2, respectively.

To explain routing a backup path $\wp$, as in the following, we use the same definitions for b_{uv}, I_{uv}^{old}, and I_{uv}^{old} as in the FBC framework:

Case 1: $\wp$ Is a Next-Hop Path That Spans a Link (i, j). Recall from section 4 that a next-hop path that spans (i, j) is activated if $\{i, j\}$ fails. The FTLIM corresponding to link $\{i, j\}$ is maintained at both node i and node j. In NPP, node j computes the route for $\wp$. It checks the FTLIM for $\{i, j\}$ at node j, and finds out b_{uv}, which is the bandwidth reserved on (u, v) by backup paths that belong to the activation set of $\{i, j\}$. Observe that it is feasible for $\wp$ to traverse (u, v) if and only if G_{uv} is greater than or equal to $b_{new} + b_{uv} - I_{uv}^{old} b_{old}$. Node j computes a route for $\wp$ using feasible links.

As in the FBC framework, FTLIM entries need to be updated. Node j increments the entries corresponding to (u, v) by an amount $I_{uv}^{new} b_{new} - I_{uv}^{old} b_{old}$, in the FTLIM for $\{i, j\}$. As in the FBC case, such updates are also made to the FTLIM for $\{i, j\}$ maintained at node i during signaling of $\wp$.

Case 2: $\wp$ Is a Next-Next-Hop Path That Spans a Link (i, j) and a Link (j, k). Recall from section 4 that a next-next-hop path that spans (i, j) and (j, k) is activated if either $\{i, j\}$ or node j fails. The FTLIM for node j is maintained at node j and the FTLIM for $\{i, j\}$ is maintained at both node i and node j. In the FBC framework, node j computes the route for $\wp$. The FTLIMs for node j and $\{i, j\}$ contain values of b_{uv} for every link (u, v). Let $b_{uv}^{\star}$ be the maximum of all such values. Therefore, $\wp$ can traverse (u, v) if and only if G_{uv} is greater than or equal to $b_{new} + b_{uv}^{\star} - I_{uv}^{old} b_{old}$. Node j uses this information to compute a route for $\wp$ that comprises only feasible links.

Once again, node j increments the entries corresponding to (u, v) by an amount $I_{uv}^{new} b_{new} - I_{uv}^{old} b_{old}$, in the FTLIMs for the facilities $\{i, j\}$ and node j. Such updates are also made to the FTLIM for $\{i, j\}$ maintained at node i during signaling of $\wp$. Thus, the mechanism to update FTLIMs in NPP framework is quite similar to the mechanism used in FBC framework.

In NPP, the bandwidth reservations on links, along the route computed for $\wp$, are made exactly like those in Kini's framework. That is, for each link (u, v), the head-end node maintains and updates the LTFIM. Once a reservation request arrives at the head end of a link (u, v), along the route computed for $\wp$, the entries corresponding to the facilities protected by $\wp$ are located in the LTFIM for (u, v). The total bandwidth reserved for backup paths on (u, v) is given by G_{uv}. Since the bandwidth required by $\wp$ is b_{new}, G_{uv} must be at least equal to $b_{new} + b_{facility}$ for each facility protected by $\wp$. In case it is not so, we increase G_{uv} so that G_{uv} is equal to $b_{new} + b_{facility}$ for that facility. Note that for any link (u, v), additional bandwidth is only reserved when necessary, and therefore, $\wp$ benefits from maximum possible bandwidth sharing on (u, v).

6 Simulation Experiments

In this section, we describe the simulation experiments that depict the benefits of NPP over existing frameworks. We conduct a set of experiments and compare the total number of accepted LSP requests and the total bandwidth placed in NPP, FBC and Kini's frameworks. Results for these statistics are presented for element protection. For simulations, a homogeneous network topology is adapted from the network used in [8]. It represents the Delaunay triangulation for the twenty largest metros in continental United States [8]. All the links in the network are symmetric with each link having a capacity of 120 units. Each node in the network may be an LSP ingress or egress node. Therefore, there are 380 possible ingress-egress pairs in the network. LSP requests arrive one by one, where the LSP ingress and egress nodes are chosen randomly from amongst all ingress-egress pairs. The bandwidth demand for an LSP request is uniformly distributed between 1 and 6 units, and the call holding time for each LSP request is infinite. For each LSP request, if it is possible to route the requisite primary and locally restorable facility backup paths, the LSP request is accepted and the associated bandwidth reservations are made on the network; otherwise, the LSP request is rejected. We calculate the number of LSP requests that are successfully placed and the total bandwidth demand associated with these LSPs. We conducted 100 experiments with randomly selected ingress-egress pairs. In each experiment, the primary paths are computed using link costs given by the inverse of residual bandwidth available for primary paths on respective links.

We computed the average of the total number of LSPs and the total bandwidth associated with placed LSPs in these hundered experiments. Fig. 4 shows the number of LSPs placed in NPP in comparison with FBC and Kini's frameworks. It is expected that efficient bandwidth sharing results in better network utilization and hence a greater number of accepted LSP requests. Therefore, NPP – that performs better bandwidth sharing – accepts 775 requests on average compared to 693 requests accepted by FBC framework and 573 requests accepted by Kini's framework. Moreover, the sum of bandwidths for these accepted LSP requests is higher in NPP, as shown in Fig. 4. Experiments on other networks with different sets of LSP requests also indicate similar performance gains when NPP is used.

In our simulations for FBC, a parameter of particular interest is the proportion of bandwidth allocated for primary and backup pools. The value of this parameter is fixed throughout and affects the LSP placement in the network. We used a value for this parameter calculated in the following manner: twenty evenly distributed backup pool percentages between 5 and 100 are used and the one that gives maximum placement of LSPs requests is selected for all the experiments.

7 Conclusions

We investigated the problem of online routing of bandwidth guaranteed LSPs with local restoration. We presented NPP, a new framework for restoration

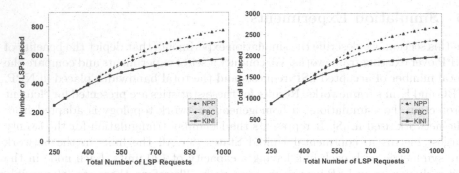

Fig. 4. Total LSPs and total Bandwidth Placed on the Network.

routing using facility backup, and described it in the backdrop of two existing frameworks for restoration routing: Kini and FBC. Kini's framework computes suboptimal routes, since it relies on aggregate rather than complete link usage information to make bandwidth sharing decisions. However, once a path has been computed Kini's framework allowed precise bandwidth reservations to be made during the signaling phase. On the other hand, the FBC framework makes possible optimal path computation by shifting path computation to a node that keeps track of the bandwidth sharing characteristic for that backup path. However, the FBC framework had the disadvantage of statically allocating primary and backup pools, resulting in unutilized bandwidth on certain links. We have shown how NPP draws upon the advantages of both the Kini and FBC frameworks, while avoiding their disadvantages. The key advantage of NPP is that it delivers the bandwidth sharing performance achieved by propagating complete per path link usage information [9], while incurring the significantly reduced routing protocol overhead. Simulation results show that NPP performs significantly better in terms of number of LSPs accepted and total bandwidth placed on the network. For 1000 randomly selected LSP requests on a 20-node homogenous ISP network [8], NPP accepts 775 requests on average compared to 573 requests accepted by Kini's framework of [2] and 693 requests accepted by FBC framework of [1]. Experiments with different sets of LSP requests and on other networks indicate that NPP results in similar performance gains.

References

1. Vasseur, J.-P., Charny, A., Le Faucheur, F., Achirica, J., Leroux, J.-L.: MPLS Traffic Engineering Fast Reroute: Bypass Tunnel Path Computation for Bandwidth Protection. IETF draft. February 2003.
2. Kini, S., Kodialam, M., Sengupta, S., Villamizar, C.: Shared Backup Label Switched Path Restoration. IETF draft. May 2001.
3. Pan, P., Swallow, G., Atlas, A. (Editors): Fast Reroute Extensions to RSVP-TE for LSP Tunnels. IETF draft. August 2004.
4. Vasseur, J.-P., Pickavet, M., Demeester, P.: Network Recovery: Restoration and Protection of Optical, SONET-SDH, IP and MPLS. Morgan Kaufmann. Elsevier. 2004.

5. Awduche, D., Chiu, A., Elwalid, A., Widjaja, I., Xiao, X.: Overview and Principles of Internet Traffic Engineering. RFC 3272. May 2002.
6. Kodialam, M., Lakshman, T. V.: Dynamic Routing of Locally Restorable Bandwidth Guaranteed Tunnels using Aggregated Link Usage Information. Proceedings of IEEE Infocom. pp. 376–385. 2001.
7. Li, L., Buddhikot, M. M., Chekuri, C., Guo, K.: Routing Bandwidth Guaranteed Paths with Local Restoration in Label Switched Networks. Proceedings of IEEE ICNP. pp 110–120. 2002.
8. Norden, S., Buddhikot, M. M., Waldvogel, M. , Suri, S.: Routing Bandwidth Guaranteed Paths with Restoration in Label Switched Networks. Proceedings of IEEE ICNP. pp 71–79. 2001.
9. Kodialam, M., Lakshman, T. V.: Dynamic Routing of Bandwidth Guaranteed Tunnels with Restoration. Proceedings of IEEE Infocom. pp 902–911. 2000.
10. Swallow, G.: MPLS Advantages for Traffic Engineering. IEEE Communications Magazine. pp 54–57. vol. 37. no. 12. December 1999.
11. Iannaccone, G., Chuah, C. N., Bhattacharrya, S., Diot, C.: Feasibility of IP Restoration in a Tier-1 Backbone. IEEE Network. pp. 13–19. vol. 18. no. 2. March 2004.
12. Davie, B., Rekhter, Y.: MPLS Technology and Applications. Morgan Kaufmann. San Francisco, CA. 2000.
13. Alcatel: Traffic Engineering Solutions for Core Networks. White Paper. July 2001.
14. Apostolopoulos, G., Guerin, R., Kamat, S., Tripathi, S. K.: Quality of Service Routing: A Performance Perspective. The Proceedings of ACM SIGCOMM. pp 17–28. 1998.
15. Rosen, E., Viswanathan, A., Callon, R.: Multi-Protocol Label Switching (MPLS) Architecture. RFC 3031. January 2001.
16. Charny, A., Vasseur, J.-P.: Distinguish a link from a node failure using RSVP Hellos extensions. IETF draft. October 2002.
17. Katz, D., Kompella, K., Yeung, D.: Traffic Engineering (TE) Extensions to OSPF Version 2. RFC 3630. September 2003.
18. Jamoussi, B. (Editor): Constraint-Based LSP Setup using LDP. RFC 3212. January 2002.

An Efficient Backup Path Selection Algorithm in MPLS Networks

Wook Jeong[1], Geunhyung Kim[1], and Cheeha Kim[2]

[1] Technology Network Laboratory, Korea Telecom (KT),
463-1 Jeonmin-dong, Yusung-gu, Daejeon, 305-811, Korea
{wjeong,geunkim}@kt.co.kr
[2] Department of Computer Science and Engineering,
Pohang University of Science and Technology(POSTECH),
San 31 HyoJa-Dong, Nam-Gu, Pohang, 790-784, Korea
chkim@postech.ac.kr

Abstract. The rapid growth of real-time and multimedia traffic over IP networks makes not only QoS guarantees but also network survivability more critical. This paper proposes an efficient algorithm which supports end-to-end path-based connection restoration in MPLS networks. We review previous related work. This includes SPR (Shortest Path Restoration), PIR (Partial Information Restoration) and CIR (Complete Information Restoration). The objective of backup path selection algorithms is to minimize the total network bandwidth consumed due to backup paths. Backup path bandwidth usage can be reduced by sharing backup paths among disjoint service paths. In CIR, since a path selection algorithm uses per-LSP information, backup path sharing can be optimized. However, the large amount of information, which each node advertises and maintains, makes it impractical. In the case of PIR, some sharing of backup paths is possible while using the aggregated service bandwidth and backup bandwidth used on each link. We think it is reasonable to increase backup path sharing using aggregated information as with PIR. In this paper we propose an efficient backup path selection algorithm to outperform PIR while using aggregated information. Simulation results show that our algorithm uses less total backup bandwidth compared to PIR.

1 Introduction

The rapid growth of real-time and multimedia traffic over IP networks makes not only QoS guarantees but also network survivability more critical. The current Internet has a degree of survivability because dynamic routing protocols will react to faults detected from routing information updates and compute alternate routes. Currently, Multi-protocol label switching (MPLS) is rapidly becoming a key technology for use in core networks. MPLS networks are more vulnerable to failures because of their connection-oriented nature. MPLS uses a technique known as label switching to forward data through the network. A small, fixed-format label is inserted in front of each data packet on entry into the MPLS network. At each hop across the network, the packet is routed based on the

M. Ajmone Marsan et al. (Eds.): QoS-IP 2005, LNCS 3375, pp. 164–175, 2005.

value of the incoming interface and label, and dispatched to an outwards interface with a new label value. The path that the data follow through a network is defined by the transition in label values, as the label is swapped at each LSR (Label Switch Router). Such a path is called a label switched path (LSP). LSPs are set up using signaling protocols, such as a resource reservation protocol with traffic engineering (RSVP-TE) or constraint routed label distribution protocol (CR-LDP) [5]. Several MPLS recovery mechanisms have been proposed to ensure continuity of service after network impairments. MPLS recovery methods can be classified into two recovery models: restoration and protection switching [12]. In the case of restoration backup LSPs are established on-demand after the detection of a failure. Since the calculation of new routes and the signaling and resource reservation of a new LSP are time-consuming, restoration is considerably slower than protection mechanisms. In the case of protection switching, the backup LSP is pre-established and pre-reserved realizing the shortest disruption of traffic in the case of a failure. Both 1+1 and 1:1 protections are possible. With 1+1 protection, packets are forwarded simultaneously on a service and backup path. In the case of a failure on the service path, the downstream side simply selects packets from the backup path. In the case of 1:1 protection the packets are forwarded on a predefined alternative path only in the case of a network failure [5]. If a 1:1 resource allocation is used, the recovery LSP may additionally carry low-priority, preemptable traffic when no failure is present in the network. This preemptable traffic must be dropped if the LSP is needed for the recovery of a failed LSP. It is also possible to share the backup path. In this paper we propose an efficient backup path selection algorithm. The paper is organized as follows. In section 2, we review previous related work. In section 3, we present our path selection algorithms for restoration. Next, in section 4, we provide simulation results and evaluate the performance of the proposed algorithm for restoration.

2   Previous Works

The objective of the backup path selection algorithm is to minimize the total amount of bandwidth consumed by the backup paths. Backup path sharing must be exploited to minimize the total amount of bandwidth consumed by the backup paths. The amount of backup path sharing depends on the information available to the routing algorithm. There are three cases with the information model. They are NI (No Information), PI (Partial Information), and CI (Complete Information) [3]. In the case of NI, required information is only the residual bandwidth in each link. In the case of PI, the required information is the residual bandwidth, the aggregated service bandwidth, and the backup bandwidth in each link. In the case of CI, per-LSP information is required. In this section, we explain the concept of the backup path sharing and introduce three previous path-based backup path selection algorithms which are SPR (Shortest Path Restoration) [1, 6], PIR (Partial Information Restoration) [1, 3, 4], and CIR (Complete Information Restoration) [3, 4]. Each backup path selection algorithms exploits NI, PI, and CI, respectively.

2.1 Notations

In this paper the following notations will be used.

- $G(E, V)$: network G with link set E and node set V
- $(s{\rightarrow}d, b)$: connection request asking for b units of bandwidth from s to d
- S_p : service path
- B_p : backup path
- $S[i]$: the total amount of bandwidth used by service paths on each link i
- $B[i]$: the total amount of bandwidth used by backup paths on each link i
- $R[i]$: residual bandwidth on each link i
- $D[u]$: $max\{\delta_i^u, i{\in}S_p, u{\in}E\}$
- δ_i^u : the amount of bandwidth required on link u to restore all service paths if link i fails ($i{\in}E$), i.e., $\delta_i^u = \sum_{k{\in}\phi_i^u} D[k]$
- A_i : the set of service paths that use link i
- B_i : the set of backup paths that use link i
- b_k : the amount of bandwidth required by request k
- ϕ_i^u : $A_i{\cap}B_u$
- θ_i^u : additionally required bandwidth to restore link i on the link $u(u{\in}E)$ if link i is used in S_p
- M : maximum service bandwidth over all links along the selected service path, i.e., $M = max\, S[i], i{\in}S_p$
- l_M : the link whose service bandwidth is M in S_p
- $S_{INTER}[i]$: the total amount of bandwidth used by inter-service path on link i ($i{\in}S_p$)
- $B_{INTRA}[u]$: the total amount of backup bandwidth used by intra-backup paths on link u ($u{\in}E$)
- $E[u]$: $S_{INTER}[i] + B_{INTRA}[u]$
- $w[u]$: the weight to link u ($u{\in}E$) to compute the backup path

2.2 Assumption

We are only interested in online routing that routes LSP requests that arrive one by one. We assume that the service path is selected before computing the backup path. The service path selection procedure is composed of two steps. In the first step, remove the links with insufficient residual BW for request bandwidth b. In the second step, a service path is selected by Dijkstra's algorithm. We assume there are single link failures in a network. A single link failure means that no other failure occurs in the network until the current failure is repaired. We also assume that the connection request is rejected if the service path or backup path cannot be selected due to insufficient residual bandwidth on the network.

2.3 Backup Path Selection Algorithms

The amount of backup path sharing depends on the information available to the routing algorithm. In this subsection we introduce three previous backup path selection algorithms – SPR [1, 6], PIR [1, 3, 4] and CIR [3, 4] – which use NI, PI, and CI, respectively.

Shortest Path Restoration (SPR). The SPR algorithm assumes that only NI information is available. Since no information is known other than R[i], the path selection algorithm does not know the backup bandwidth currently used on the different links. Hence, it cannot consider backup path sharing. Thus, request bandwidth b units have to be reserved on each link in the service path as well as the backup path. After the service path is selected, the ingress node removes the links of the service path, as well as any link with insufficient available bandwidth, from the network topology. The ingress node then uses Dijkstra's algorithm to select the backup path.

Complete Information Restoration (CIR). In this algorithm, it is assumed the sets A_i and B_i are known for all links i. Kodialam presented a CIR solution [3, 4]. Kodilalam defined the quantity θ_i^u is the cost of using link u on the backup path if link i is used in the service path. Kodialam also defined $\phi_i^u = A_i \cap B_u$. This is the set of demands that use link i on the service path and link u on the backup path. The sum of all the demand values in the set ϕ_i^u is represented by $\delta_i^u = \sum_{k \in \phi_i^u} b_k$

$$\theta_i^u = \begin{cases} 0 & \text{if } b \leq B[u] - \delta_i^u \text{ and } i \neq u \\ \delta_i^u + b - B[u] & \text{if } b > B[u] - \delta_i^u \text{ and } i \neq u \\ \infty & \text{otherwise} \end{cases}$$

where b is the size of the bandwidth request.

The quantity δ_i^u is the amount of bandwidth needed on link u to backup the service paths currently traversing link i. Therefore, also taking the current request into account, a total of $\delta_i^u + b$ units of backup bandwidth are needed on link u if the current request were to traverse link i and use link u for backup. Then the current request can be backed up on link u without reserving any additional bandwidth if $b \leq B[u] - \delta_i^u$. If $b > B[u] - \delta_i^u$, an additional reservation of $\delta_i^u + b - B[u]$ units is necessary. After selecting the service path S_p, compute the $D[u]$, where $D[u] = \max\{\delta_i^u, i \in S_p, u \in E\}$. The ingress node then assigns a weight to each link in the network:

$$w[u] = \begin{cases} \varepsilon & \text{if } b \leq B[u] - D[u] \\ D[u] - b - B[u] & \text{if } b > B[u] - D[u] \text{ and } i \neq u \\ \infty & \text{otherwise} \end{cases}$$

where ε is a small number.

The ingress node removes any link with insufficient available bandwidth, from the network topology. In this step, if $R[u] < b$, $B[u] - D[u] > 0$ and $B[u] - D[u] + R[u] > b$ link u is not removed. This means sharable bandwidth is considered when the any link is removed from the network topology, then uses Dijkstra's algorithm to select the backup path B_p, that minimizes $\sum w[u]$ for u in B_p.

In CIR, since a path selection algorithm uses per-LSP information, backup path sharing can be optimized. However, the very large amount of information that must be advertised from each node and maintained at each node makes it impractical.

Partial Information Restoration (PIR). The PIR algorithm assumes that PI information – aggregated service bandwidth and backup bandwidth on each link – is available. The basic idea of PIR is to weight each link using an estimate of the additional bandwidth that needs to be reserved if a particular backup path is selected. After the service path S_p is selected, the ingress node computes the maximum service bandwidth M over all links along the service path, i.e., $M = \max\{S[i], i \in S_p\}$. Then the ingress node assigns a weight to each link in the network:

$$w[u] = \begin{cases} \min\{b, M + b - B[u]\} & \text{if } b > B[u] - M \text{ and } i \notin S_p \\ \varepsilon & \text{if } b \leq B[u] - M \text{ and } i \notin S_p \\ \infty & \text{if } i \in S_p \end{cases}$$

The ingress node removes any link with insufficient available bandwidth, from the network topology and then uses Dijkstra's algorithm to select the backup path Bp, that minimizes $\sum w[u]$ for u in B_p. For a potential link u on the backup path, if $B[u] > M$ and $u \notin S_p$, it is estimated that link u has sharable bandwidth. This is because PIR only uses aggregated bandwidth and service bandwidth as routing information. Using only this aggregated information, the ingress node cannot know the distribution of the backup paths for service paths composing M. M assumes that when a failure occurs along S_p, all the connections that routed over the failed link would indeed reroute onto link u. If $b \leq B[u] - M$, no additional bandwidth needs to be reserved on the backup path because any link failing on the service path generates a bandwidth need of at most $M + b$ on the links of the backup path. If $b > B[u] - M$, only an additional reservation of $\min\{b, M + b - B[u]\}$ units is necessary.

3   Proposed Algorithm

In this section we present our research motivation and bring up the problems in previous backup path selection algorithms and propose an efficient backup path selection algorithm in order to alleviate these problems. We also introduce a previous scheme for exact reservation for the backup path. We apply this exact reservation scheme to our scheme to further reduce the amount of total backup bandwidth.

3.1   Motivation

In section 2, we studied three previous backup path selection algorithms for restoration which are SPR, PIR, and CIR respectively. In the case of SPR, since it does not know the backup bandwidth currently used on each link in a network, backup path sharing is impossible. In the case of CIR, since it uses per-LSP information, the backup path sharing can be optimized. However, the very large amount of information that needs to be advertised from each node and maintained at each node makes it impractical. In the case of PIR, some sharing of backup paths is possible while using the aggregated service bandwidth and backup bandwidth used on each link. PIR cannot optimize backup path sharing, but it is easy to maintain and use routing information in a distributed fashion. So we think it is reasonable to increase backup path sharing using PI. In this section, we propose an efficient backup path selection algorithm to outperform PIR while using PI.

3.2   The Proposed Backup Path Selection Algorithm

In this subsection we propose an efficient backup path selection algorithm. In PIR, some sharing of backup paths is possible using only the aggregated service bandwidth and backup bandwidth on each link. For sharing of backup path the value of M was defined. It is estimated that link u which satisfies the condition $B[u] - M > 0$ and $u \notin S_p$ has sharable bandwidth. However, the M assumes that when a failure occurs along S_p, all the connections that routed over the failed link would indeed be rerouted onto link u and is therefore a crude estimate of the restoration bandwidth needed due to link failures along S_p. This approach may overestimate the bandwidth that needs to be reserved on some links.

If we can determine partial backup paths distribution of link i whose service bandwidth is M, we can increase the amount of sharing bandwidth in comparison with PIR. In fact, each ingress node knows some distribution information about service paths and backup paths. In other words, the ingress node knows the information of the service paths and backup paths, which are started from itself. We divide the service path set which passes the l_M. One is the intra-service path set; the other is inter-service path set. The intra-service path set is composed of service paths, which are started from the ingress node, and the inter-service path set is composed of service paths, which are initiated from other nodes. The ingress node also knows the distribution of backup paths for intra-service paths. The intra-backup path set is composed of these backup paths which are started from the ingress node, while the inter-backup path set is composed of backup paths which are started from other nodes.

We define estimator $E[u]$ in order to use path information of the ingress node for backup path selection: $E[u] = S_{INTER}[i] + B_{INTRA}[u]$ where link i is l_M, $S_{INTER}[i]$ is the total bandwidth of inter-service paths on link i, and $B_{INTRA}[u]$ is the total bandwidth of intra-backup paths on link u. The ingress node assigns a weight to each link in the network for computing the backup path as follows:

$$w[u] = \begin{cases} \min\{b, E[u] + b - B[u]\} & \text{if } b > B[u] - E[u] \text{ and } u \notin S_p \\ \varepsilon & \text{if } b \leq B[u] - E[u] \text{ and } u \notin S_p \\ \infty & \text{if } u \in P_s \end{cases}$$

The ingress node removes any link with insufficient available bandwidth from the network topology. In this step, if $R[u] < b$, $B[u] - E[u] > 0$ and $B[u] - E[u] + R[u] > b$, link u is not removed. This means that sharable bandwidth is considered when the any link is removed from the network topology. Next, the ingress node uses Dijkstra's algorithm to select the backup path B_p, that minimizes $\sum w[u]$ for u in B_p.

The ingress node does not know the distribution of the inter-service and inter-backup paths but knows the distribution of intra-service and intra-backup paths.

Therefore, we cannot assign an exact weight to each link from the viewpoint of inter-service and inter-backup paths but can assign an exact weight to each link from the viewpoint of intra-service and intra-backup paths. So the value of estimator E[u] is the summation of $S_{INTER}[i]$ and $B_{INTRA}[u]$, where i is the M link. From this, we can determine $E[u] \leq M$.

The term $\min\{b, E[u]+b-B[u]\}$ means the amount of additional bandwidths needed if, upon failure of the service path S_p, the connection is routed over a backup path that contains link u. As we assign an exact weight to each link from the viewpoint of intra-service and intra-backup paths we can further increase the amount of shared bandwidth over PIR. For a potential link u on the backup path, if $B[u] > E[u]$ and $u \notin S_p$, it is estimated that the link u has sharable bandwidth. If $b \leq B[u] - E[u]$, no additional bandwidth needs to be reserved on the backup path because any link failing on the service path generates a bandwidth need of at most $E[u] + b$ on the links of the backup path. If $b > B[u] - E[u]$, only an additional reservation of $\min\{b, E[u] + b - B[u]\}$ units is necessary.

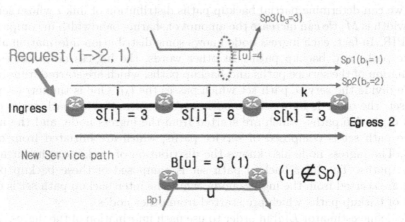

Fig. 1. Sharable bandwidth in PIR and the proposed algorithm.

Fig. 1 shows an example illustrating our proposal. Suppose a connection request asking for one unit of bandwidth from Ingress 1 to Egress 2 arrives at Ingress 1. We suppose the Ingress 1 selects $i - j - k$ for the service path. In this example M link is the j and the value of M is 6. Ingress 1 knows intra-service path S_p1 ($b_1=1$) and S_p2 ($b_2=2$) is passing the link j and only B_p1 for S_p1 among these intra-service paths is passing the link u. We suppose a inter-service path, $S_p3(b_3=3)$, is passing the link j and $B[u] = 5$. In the case of PIR, u is not a sharable link because $M(= 6) > B[u](= 5)$. However, in the case of proposal, u is a sharable link because $E[u] = S_{INTER}[j] + B_{INTRA}[u] = 3+1 = 4 < B[u] = 5$. And $E[u] + b = 5 \leq B[u]$. Therefore, the weight of u is ε for backup path selection.

In the case of PIR, link u will not be selected as a potential link in the backup path, but in the case of the proposal, link u will be selected as a potential link in the backup path. From this example we can determine that our proposal can further increase the amount of shared bandwidth over PIR.

3.3 Exact Reservation

In PIR scheme, the amount of bandwidth reservation for backup path is decided by the value of $M + b - B[u]$ on each link. In PIR scheme, the required amount of additional bandwidth on each link along the backup path is as follows:

$$\begin{cases} \min\{b, M + b - B[u]\} & \text{if } M + b - B[u] > 0 \text{ and } u \in B_p \\ 0 & \text{if } M + b - B[u] \leq 0 \text{ and } u \in B_p \end{cases}$$

In the same way, our scheme requires the amount of additional bandwidth on each link along the backup path as follows:

$$\begin{cases} \min\{b, E[u] + b - B[u]\} & \text{if } E[u] + b - B[u] > 0 \text{ and } u \in B_p \\ 0 & \text{if } E[u] + b - B[u] \leq 0 \text{ and } u \in B_p \end{cases}$$

This is because our scheme uses aggregated information and path information of the ingress node for backup path selection and for computation of the amount of additional bandwidth on each link along the backup path.

Kodialam proposed an exact reservation scheme [3]. The basic idea of this scheme is that the set of links in the service path is conveyed to every link in the backup path during the signaling phase and each node maintains this information. In other words, each node maintains δ_i^u ($i \in E$, u is a link of each node). Then, the total backup bandwidth of link u equals to $B[u] = \max \delta_i^u, i \in E$ ensures that there is sufficient bandwidth reserved to protect against any single link failure. The node of link u along the backup path updates the δ_i^u as follows: $\delta_i^u \leftarrow \delta_i^u + b$, if link i is on the service path and b is the requested bandwidth of the connection. The node of link u along the backup path updates the δ_i^u as follows: $\delta_i^u \leftarrow \delta_i^u - b$, if link i is on the deletion required and b is the requested bandwidth of the connection.

We can apply this scheme to our scheme for exact bandwidth reservation along the backup path. If we apply this exact reservation scheme to our scheme we can further reduce the amount of total backup bandwidth.

4 Simulation Results

This section compares the performance of SPR, PIR and proposal. We performed simulation on the network which is the same as that used in [3, 4] and has 15 nodes and 56 links. Core nodes are 2, 5, 6, 9, and 10. The network is shown in Fig. 2. Each undirected link in Fig. 2 represents two directed links. In this simulation two performance metrics are used. One is network load and the other is the drop ratio.

4.1 Network Load

Simulation A (Random Request Generation). For this simulation the link capacities are set to infinity. Connection requests arrive one at a time at the network. The source and destination nodes for the requests are chosen at random, except for the core nodes. The bandwidth request is uniformly distributed

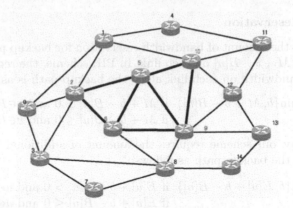

Fig. 2. Network topology (random request generation).

between 100 and 200 units. Since the capacity is infinite, there is no request rejection. The objective, therefore, is to compare how much total bandwidth the proposed algorithm uses relative to the PIR and SPR. Service path selection uses Dijkstra's algorithm. There are five algorithms run on the same data for backup path selection. They are SPR, PIR, proposal, PIR with exact reservation and proposal with exact reservation. Each algorithm is compared with the same number of total connection requests from 200 to 2000 in increments of 200.

Fig. 3 shows the simulation results of the total number of connections vs. total bandwidth units required for backup paths depending on the five different algorithms. The total used backup bandwidth after 2000 requests for SPR is 1,046,727 units, for PIR is 794,459 units and for proposal 708,034 units. And for PIR with exact reservation is 611,618 units and for proposal with exact reservation is 576,639 units. We can say the proposal provides about 11% improvement in comparison with PIR and proposal with exact reservation provides about 6%

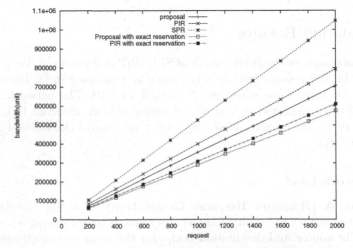

Fig. 3. Total used bandwidth (random request generation).

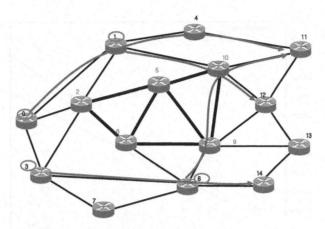

Fig. 4. Network topology (limited request generation).

improvement in comparison with PIR with exact reservation. We can conclude that performance improvement is caused by path information of the ingress node used.

Simulation B (Limited Request Generation). In this simulation, the source and destination node for the Requests are chosen among the special set. The selected special set is $(0 \rightarrow 11)$, $(3 \rightarrow 14)$, $(8 \rightarrow 11)$, $(1 \rightarrow 12)$. Every service path for the elements in the set will be link disjointedly selected from each other by Dijkstra's algorithm (Fig. 4).

Fig. 5 shows the simulation results of the total number of connections vs. total bandwidth units required for backup paths depending on the four different algorithms. These algorithms are SPR, PIR, CIR and proposal. The total used backup bandwidth after the 2000 request for SPR is 1,193,570 units; for PIR is 1,033,464 units; and for the proposal and CIR is 811,104 units. No difference in the backup bandwidth usage between the proposal and CIR is caused by the absence of inter service pathes in this scenario. It is evident that without inter service paths the ingress node in the proposal clearly has enough information to compute the path to maximize backup path sharing as much as in CIR. The proposed algorithm and CIR provide about 21% improvement in comparison with PIR. In fact, this scenario is extremely profitable for our proposal. This result is meaningful, however. If service paths are selected as disjointedly from each other as possible, the proposed algorithm can obtain greater backup path sharing in comparison with PIR and approximates to the case of CIR as the link-disjointedness among service paths increases. This is feasible when only some paths require setup of a backup path, according to the provider policy.

4.2 Simulation with Dropped Requests

This simulation studies the behavior of the algorithms with respect to the number of dropped requests when the network is overloaded. In this simulation, we

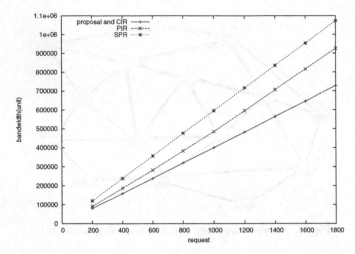

Fig. 5. Total used bandwidth (limited request generation).

set each link capacity to 20000 units of bandwidth. Except for this point other simulation scenarios are the same as Simulation A. Table 1 shows the numbers of requests rejected after 2000 demands are loaded on the network using different backup path selection algorithms. If there is a lack of available capacity to choose either the service path or the backup path, then the request is rejected. Table 1 shows that our proposal performs better than SPR and PIR, and the proposal with exact reservation performs better than PIR with exact reservation.

Table 1. Rejected requests.

Methods	number of rejected requests
SPR	863/2000
PIR	607/2000
Proposal	539/2000
PIR with exact reservation	527/2000
Proposal with exact reservation	503/2000

5   Conclusion

The objective of the backup path selection algorithms is to minimize the total network bandwidth consumed due to backup paths. Backup path bandwidth usage can be reduced by sharing backup paths among disjoint service paths. We regarded it as reasonable to increase backup path sharing using aggregated information similar to PIR. Therefore, we proposed an efficient backup path selection algorithm to outperform the PIR while using aggregated information.

PIR does not use the path information of the ingress node for selection of the backup path. We defined estimator $E[u]$ in order to use path information of the ingress node for backup path selection. In simulation A, our proposal provided about 11% improvement in comparison with PIR and the proposal with exact reservation provided about 6% improvement in comparison with PIR with exact reservation. In simulation B, the proposal provided about 21% improvement in comparison with PIR. This is the same result of CIR. We can conclude that performance improvement was caused by the ingress path information used and emphasize that the proposal does not require any extra routing information in comparison with PIR.

Scenarios of Simulation B were extremely profitable for our proposal since inter service paths did not exist in this scenario. However, this result is meaningful. It is evident that without inter service path the ingress node in the proposal clearly has enough information to compute the path to maximize backup path sharing as much as in CIR. If service paths are selected as disjointedly from each other as possible, the proposed algorithm can obtain more backup path sharing in comparison with PIR and approximates to the case of CIR as the link-disjointedness among service paths increases. This is feasible when only some paths require setup of a backup path, according to the provider policy.

Future work will develop an algorithm which can minimize interference among service paths while maintaining an acceptable path distance. If this algorithm is applied to the proposed algorithm, a greater performance improvement will be provided.

References

1. G. Li and D. Wang, Efficient distributed path selection for shared restoration connections, IEEE INFOCOM 2002.
2. S. Norden and Milind M. Buddhikot, Routing bandwidth guaranteed paths with restoration label switched networks, ICNP 2001.
3. M. Kodialam and T.V. Lakshman, Restorable dynamic quality of service routing, IEEE Communication Magazine, June 2002.
4. M. Kodialam and T.V. Lakshman, Dynamic routing of bandwidth guaranteed tunnels with restoration, IEEE INFOCOM 2000.
5. V. Sharma and F. Hellstrand, Framework for MPLS-based recovery", Internet Draft, July 2002.
6. R. Bhandari, Survivable networks: algorithms for diverse routing, Kluwer Academic Publisher, 1999
7. C. Qiao and D. Xu, Distributed partial information management (DPIM) schemes for survivable networks, IEEE INFOCOM 2002.
8. M. Kodialam and T.V. Lakshman, Dynamic routing of locally restorable bandwidth guaranteed tunnels using aggregated link usage information, IEEE INFOCOM 2001.
9. R. Doverspike and J. Yates, Challenges for MPLS in optical network restoration, IEEE Communication Magazine, February 2001.
10. C. Huang and V. Sharma, Building reliable MPLS networks using a path protection mechanism, IEEE Communication Magazine, March 2002.
11. G. Iannaccone and C.-N. Chuah, Analysis of link failures in an IP backbone, Internet Measurement Workshop 2002.
12. A. Autenrieth and A. Kirstadter, RD-QoS : The integrated provisioning of resilience and QoS in MPLS-based networks, ICC 2002.

Planning Multiservice VPN Networks:
An Analytical/Simulative Mechanism
to Dimension the Bandwidth Assignments

Raffaele Bolla, Roberto Bruschi, and Franco Davoli

DIST – Department of Communications, Computer and Systems Science
University of Genoa, Via Opera Pia 13, 16145 Genova, Italy
{raffaele.bolla,roberto.bruschi,franco.davoli}@unige.it

Abstract. The planning of a Virtual Private Network, carrying both Quality of Service (QoS) and Best-Effort (BE) traffic is considered in the paper, and a hybrid analytical/simulation tool for link bandwidth allocation is proposed and tested. The tool works in three phases, related to assessing the capacities needed to satisfy requirements for QoS and BE traffic individually, and to take into account the effects of multiplexing gains arising from traffic mixing, respectively. The analytical part is based on a simplified version of call-level (circuit-switched) traffic models; the simulation model is a fast one, based on a fluid representation of TCP traffic aggregates. The tool is tested both on simple networks and on a more complex real national backbone.

1 Introduction

Starting from the WWW explosion, data services have become a more and more strategic element in the economic/industrial world. In the past years, this kind of services grew exponentially and represents now a large part of the overall traffic carried by telecommunication networks. As a consequence, Public Data Networks, i.e., mainly the Internet, have so enlarged to become a suitable platform for both new data services and Quality of Service (QoS) applications (voice, video, ...). One of the most interesting "new" services which involves potentially all traffic types and has a lot of customer-attractive aspects is that of Virtual Private Networks (VPNs), which in fact has attracted the attention of both the standard bodies and the scientific community (see, among others, [1-4], and [5]).

In the last decade a wide range of network architectures and technologies has been proposed (and used) to support this kind of service. Today, most VPNs are realized by using layer 2 (Frame Relay, ATM, MPLS) and layer 3 technologies (IP over IP, GRE, IPSEC) [6]. But it should be noted that the same type of service may be build also by using layer 1 technologies (as SDH or WDM) [7]. In particular, we refer to the new layer 1 protocols able to guarantee bandwidth allocations on the network links with a minimum acceptable level of granularity and efficiency, e.g., the SDH Generic Framing Procedure (GFP) [8], which provides leased links with a bandwidth allocation up to 10 Gbps and a minimum granularity of about 1 Mbps. Note that all these technological solutions (layer 1, 2 and 3) can be used to realize VPNs with guaranteed bandwidth allocations and support for QoS algorithms (see. e.g., [9] and [10]).

M. Ajmone Marsan et al. (Eds.): QoS-IP 2005, LNCS 3375, pp. 176–190, 2005.

The problem approached in this paper is the bandwidth dimensioning of a fully integrated data and multimedia VPN with heterogeneous traffic patterns and requirements. The planning/design phase is always a very complex and difficult task in (data or integrated) packet networks ([11], [12]), but it assumes a specific role in this context, where it has to be solved a number of times, at least at each new VPN activation upon customer's request. For what concerns the need of precise sizing of a VPN, it can be noted that, nowadays, backbone and local networks are often characterized by capacities so high to be heavily under-utilized: as reported by the traffic measurements of many Internet Service Providers (ISPs) the bandwidth occupation peaks are about 10% of the overall available capacity. But, in spite of all that, bandwidth on-demand is still a quite expensive resource and then the studies about the optimal dimensioning for VPN planning are a strongly topical issue.

In this kind of context our effort is to propose a set of analytical and simulation-based mechanisms and tools to optimize the link bandwidth allocations for fully integrated data and multi-service IP VPNs. In the scientific literature, some different approaches can be found to solve this kind of problems. One of the most interesting is that based on the "hose" model, which has been proposed and recently developed [12], [13]. This technique does not need a complete traffic matrix among the connected customer nodes. As underlined in [14], the hose model is characterized by some key advantages with respect to the others (as the ease of specification, the flexibility or the multiplexing gain), but, especially in a static capacity provisioning environment, this kind of model might over-dimension the network resources. Other approaches, based on different types of technologies, can be found in [15], [16] and [17], among others.

In our approach we define a dimensioning mechanism for static VPN bandwidth allocation. This mechanism uses a partially analytic and partially simulative model. The simulation part concerns mainly Best Effort (TCP) services and is based on the fluidic TCP traffic model that is described in [18]. This type of approach allows us to impose very specific and accurate constraints on the traffic performance, but it requires the knowledge of the traffic matrix, in terms of mean generation rates per traffic classes. The proposed mechanism can be quite easily extended to provide a complete VPN planning (including the topology design).

The rest of the paper is organized as follows. The main hypotheses and the basic characteristics of the approached problem and scenario are summarized in Section II. In Section III the proposed dimensioning mechanism is described. Section IV shows some simulation results obtained by the tool's application, while Section V reports the conclusions.

2 Network, Traffic and Problem Description

As previously outlined, in this work we consider the sizing of VPNs for IP services with QoS support. In a multi-service environment, it is reasonable to suppose that the network can carry at least two main sets of traffic: a first set composed by all the multimedia streams generated by applications that require QoS constraints (i.e., VoIP, videoconference, or similar) indicated as QoS traffic in the following, and another set that groups together all the standard best effort data traffic (WWW, e-mail, FTP, Remote login, etc.). We assume that the bandwidth is allocated to a VPN statically and that it is not shared with other VPNs. The bandwidth allocated to the VPN is shared by both QoS and Best Effort traffic. Moreover, to achieve the desired end-to-end service level for all the traffic classes, we suppose that the ISP provides the needed set of infrastructure for QoS support on its backbone and, specifically, that this is achieved within a Differentiated Services approach. In fact, as is well described in [9], the DiffServ environment suits VPN applications very well for many and different reasons; first of all, DiffServ handles traffic aggregates and it is able to differentiate the service levels per VPN and/or per traffic types. The DiffServ network architecture aims at satisfying common-class QoS requirements, by controlling network resources at the nodes (Per Hop Behavior, PHB). For example, this is done by setting suitable queue service disciplines at the Core Routers (CR), by limiting the amount of traffic with QoS requests in the network, by assigning different priorities, and by performing Call Admission Control (CAC) at the Edge Routers (ER) of the provider's network. Thus, to achieve the desired service level, the provider's Core Routers need to apply a suitable active queue management mechanism that forwards the QoS traffic packets with a certain priority, or, better, a different PHB, with respect to the BE ones. This differentiation should be done on two levels: among VPNs, and inside each VPN among different traffic types. In this sense, we can suppose the VPN bandwidth to be allocated by using global VPN MPLS pipes; so, our proposed tool tries to find the optimal size of these pipes. Inside the VPN we suppose each QoS flow, i.e., QoS traffic streams between the same pair of ISP border routers, to be activated dynamically and subjected to Call Admission Control (CAC). More specifically, for what concerns the QoS traffic, we suppose that each stream has to ask for bandwidth during the setup phase and, if it is accepted (by CAC), its bandwidth is in principle completely dedicated to it. To realize this type of allocation, the QoS streams are treated as Connection Oriented (CO) flows and routed by using a CO static routing. This means that every time there is a new QoS stream activation request, a static path is computed and along this path the availability of the requested bandwidth is verified. If the bandwidth is available, it is reserved along the whole path (e.g., by a second level MPLS path setup) and the traffic is sent along that route. The BE traffic uses all the VPN bandwidth not utilized by the QoS streams and it is routed by using datagram adaptive routing, where the link weights are computed as the inverse of the bandwidth available on the link for the BE. In our mechanism we provide a bandwidth threshold for the QoS traffic on each link, which avoids that the QoS streams occupy all the available resources of the VPN. In summary, our planning problem is the following. We suppose to have the traffic matrix of QoS traffic, in terms of average stream activation request rate and duration. We suppose to have one traffic matrix for the BE traffic, with the average data burst arrival rate and length. Moreover, we define QoS constraints at call level, in terms of the maximum blocking probability for the QoS traffic stream requests. We fix a constraint with respect to the BE traffic in terms of the maximum number of "congestion periods" (to be precisely defined in the next Section) with a certain duration acceptable during a day. The latter is defined as a stochastic constraint (i.e., the maximum number with a certain probability). With this initial data and with the topology of the VPN we find the minimum value of the VPN link capacities and CO thresholds that can allow the satisfaction of the constraints. Note that the CO and BE traffic flows are modeled at the edge nodes of the Provider's network as aggregates: this means that the same node can generate more CO connections or BE flows. In particular, we think of an aggregate as composed by all the connections that traverse the network between the same pair of nodes. The CO flows are generated at each source with exponentially distributed and independent inter-arrival and service times; the BE ones follow exponentially distributed and independent inter-arrival times, whereas the dimension of the data blocks is modeled as a Pareto process.

3 The Dimensioning Mechanism

The proposed dimensioning mechanism is composed by three different parts: the first two (QoS module and BE module) are essentially used to identify a set (one for each link) of small capacity ranges, inside which the optimal solution should be contained; the last one finds the final solution by searching the optimal assignments inside these reduced ranges. More in particular, the QoS module dimensions the network by considering only the QoS traffic, while the other (BE module) does the same action by considering only the BE traffic. The QoS module finds the link capacities needed to maintain the blocking probability of QoS stream requests under a certain threshold. The other one, by using a simulative flow model, finds the link capacities needed to maintain the maximum (over the links) average utilization around 80%; this value has been shown as a threshold above which the TCP traffic starts to suffer bad performance. From these two modules, which can operate in parallel, we obtain two capacity values for each link; their sum represents an upper bound on the link bandwidth dimensioning, when the two traffic types are carried by the same network. In this case, we neglect the resource gains generated by the partial superposition of the two traffic types.

Consider now the VPN dimensioned with the above described upper bound (for each link, the sum of the two capacities found by the two modules). First of all, note that the capacity values found for the QoS traffic are also the QoS thresholds, i.e., the maximum capacity value usable by QoS traffic on a link. This means that the considered network correctly respects the performance constraints on the QoS traffic, i.e., the network is correctly dimensioned with respect to it. However, one main problem remains: we have done an over-provisioning for BE traffic. There are two reasons why this happens. First, because this kind of traffic can utilize both the capacity computed by the BE modules and the amount of bandwidth left unused by the QoS traffic. Second, the constraint imposed to the BE traffic by the BE module is quite rough and quite conservative. To overcome this drawback, the final module first computes the occurrence distribution histogram of the capacity used by QoS traffic per link; this is done by simulation and for each link of the network dimensioned by the QoS modules. By using this result, we can obtain the value of each link capacity occupied by QoS traffic for more than δ% of the time (e.g., 90%). For each link, we add this value to the one computed by the BE module to identify a bandwidth lower bound. At this point, we have the capacity ranges for the final solution. Thus, we define more precise performance constraint for BE traffic: we want to obtain a minimum capacity dimensioning that assures a maximum of χ BE congestion periods per day with probability β, where a congestion period is considered as every time interval larger than τ seconds, in which at least one network link exhibits 100% utilization value for the BE traffic. Finally, we operate by considering the two traffic types active and by using the two bounds described above and the QoS threshold (computed by the QoS module) to find the minimum capacities needed to assure the last BE constraint. The reduced range is required because of the relatively high complexity of this last computation. Let us now describe in detail the operation of the three modules that compose the proposed dimensioning mechanism.

3.1 QoS Traffic Dimensioning

As previously described, in summary, we consider the QoS traffic as composed by streams, which are activated dynamically at the edge nodes of the VPN and which are subject to CAC at those nodes. The hypotheses are that each stream is routed as a virtual circuit (the path is decided at the setup) and during the set-up phase it asks for an amount of bandwidth. If this bandwidth is available along the source-destination path identified by the routing, the stream is accepted in the VPN, otherwise it is refused. Each stream in progress occupies all its assigned capacity. With these assumptions, we implicitly suppose the QoS problem at the packet level to be already solved by the node schedulers; then, we model only the dynamics of stream activations and completions, and we take into account quality constraints at the stream level only. With this approach, for what concerns the QoS traffic, we can use a product-form loss network model identical to that described in [19]. We consider a VPN with J links that supports K stream classes. Every class-k stream has $\lambda_{QoS}^{(k)}$ [stream/s] arrival rate, $\mu_{QoS}^{(k)}$ [s^{-1}] departure rate, a bandwidth requirement $b^{(k)}$ and a route $R^{(k)}$, and the k-class offered load is

$$\rho^{(k)} = \lambda_{QoS}^{(k)} / \mu_{QoS}^{(k)} \tag{1}$$

Note that, owing to the use of a static shortest path routing, we have only one path for each source-destination pair. Moreover, we suppose the amounts of required bandwidth to belong to a finite size set $B=\{b^{(r)}, r=1,...,R\}$ of admissible sizes. The stream arrival and departure instants are supposed to be generated by a Poisson process (both arrival and duration times follow an exponential distribution). An arrival class-k request is admitted in the network if and only if there are at least $b^{(k)}$ free capacity units along each link of $R^{(k)}$. Our objective in this context is to find the minimum capacity assignment $\tilde{C}_j^{QoS}$, $\forall j=1,...,J$, needed to respect the following constraint on the blocking probability

$$Pb^{(k)} \le \overline{Pb}^{(k)}, k=1,...,K \tag{2}$$

where $Pb^{(k)}$ is the blocking probability of the streams of class k and $\overline{Pb}^{(k)}$ is its upper-bound constraint. The equilibrium distribution of this loss network and the corresponding blocking probabilities can be computed by using a stochastic knapsack problem solution. But this product form computation of $Pb^{(k)}$, $k=1,...,K$, requires the solution of a NP-complete problem (see [19]), which can be computationally heavy in this context (we have to use it inside the minimization procedure). To find $\tilde{C}_j^{QoS}$, $\forall$ $j=1,...,J$, we use an approach that follows the same philosophy as the whole mechanism: at first we apply an approximate fast analytical method to find a starting tempo-rary solution not far from the optimal one and finally we improve it by using a simu-lative method. The first step is to define a simple method to find $Pb^{(k)}$, $k=1,...,K$, given the C_j^{QoS}, $j=1,...,J$. The idea is to use an approximate computation link by link by supposing, in principle, the setup requests and CAC functions to be realized for each link along the stream path. For the sake of simplicity, let us suppose to have only one class rate, i.e., $b^{(k)}=1$ bandwidth unit, $\forall k=1,...,K$. Indicating by $\bar{\rho}_j$ the total offered traffic to link j, the blocking probability per link can be computed as

$$Pb_j = ER[\bar{\rho}_j, C_j^{QoS}] = \frac{(\bar{\rho}_j)^{C_j^{QoS}}/C_j^{QoS}!}{\sum_{s=0}^{C_j^{QoS}}(\bar{\rho}_j)^s/s!} \quad (3)$$

Equation (3) may be used to determine the $\widetilde{C}_j^{QoS}$, $\forall j=1,...,J$, if we are able to trans-form the end to end probability constraints (2) into the following link constraints

$$Pb_j^{(k)} \leq \overline{Pb}_j^{(k)}, k=1,...,K \quad (4)$$

So, now the problem is to find $\overline{Pb}_j^{(k)}$, $k=1,...,K$, and we solve it by fixing

$$\overline{Pb}^{(k)} = \sum_{j \in R^{(k)}} \overline{Pb}_j \quad (5)$$

This choice represents an upper-bound in the single rate case, and in general it results in a capacity assignment mostly rather close to the optimum one. Then, we use the following algorithm, by which we "distribute" the $\overline{Pb}^{(k)}$ probabilities along the links.

Let us introduce some notations:
- $\overline{Pb}_j^{max}$ is the maximum blocking probability constraint associated to link j;
- $R_{res}^{(k)}$ is the set of links in the residual path for class k;
- $Pb_{res}^{(k)}$ is the residual blocking probability for class k;
- $Pb_{eq}^{(k)}$ is the value of the blocking probability of class k if $\overline{Pb}^{(k)}$ would be equally spread on the path links;
- $\widetilde{K}$ is the set of the classes k with $R_{res}^{(k)} = \varnothing$.

The proposed algorithm works as follow:

1) Let $Pb_{res}^{(k)} = \overline{Pb}^{(k)}$, $k=1,...,K$, $\widetilde{K} = \varnothing$ and $R_{res}^{(k)} = R^{(k)}$, $k=1,...,K$

2) compute $Pb_{eq}^{(k)}$ as $Pb_{eq}^{(k)} = Pb_{res}^{(k)}/\left|R_{res}^{(k)}\right|, \forall k \notin \widetilde{K}$ ($|X|$ denotes the cardinality of the set X)

3) choose the class $\widetilde{k}$ (the class with the most restrictive constraint):
   $$\widetilde{k} = \underset{k}{\text{argmin}}\left\{Pb_{eq}^{(k)}\right\}, \forall k \notin \widetilde{K} \text{ and } \widetilde{K} \leftarrow \widetilde{K} \cup \{\widetilde{k}\}$$

4) for each link j that belongs to $R_{res}^{(\widetilde{k})}$:
   a) $\overline{Pb}_j = Pb_{eq}^{(\widetilde{k})}$

b) for each traffic class $k \notin \widetilde{K}$, if the link j belongs to $R_{res}^{(k)}$ then do:

$$R_{res}^{(k)} \leftarrow R_{res}^{(k)} - \{j\} \quad \text{and} \quad Pb_{res}^{(k)} \leftarrow Pb_{res}^{(k)} - \overline{Pb}_j$$

5) if $|\widetilde{K}| = K$ then stop, else return to 2.

By using the $\overline{Pb}_j$ $j=1,...,J$, computed by the above algorithm, to find a suboptimal value of the link capacities $\hat{C}_j^{QoS}$ $j=1,...,J$, we can adopt (3) in the case of a single class rate, and the per class blocking probabilities resulting from a stochastic knapsack in the multirate case. These values result in upper-bounds in the single rate case (this can be demonstrated by using the "Product Bound" theorem [18]) and in general they are rather close to the optimum quantities if the constraints are very restrictive (≤ 0.01) and the path's maximum length is not too large. To obtain the final result, we start with the $\hat{C}_j^{QoS}$, $j=1,...,J$, network configuration and apply an optimization procedure by using a simulative approach. The stream births and deaths are generated to find the exact (within a given confidence interval) end-to-end blocking probabilities, and if they are more restrictive than the constraints the capacities on the links are reduced (or increased if they do not respect the constraints) and the simulation is restarted in an interactive procedure. This kind of simulation is quite fast and the number of iterations needed are very few, so a near to optimal solution can be found in a short time.

3.2 BE Traffic Dimensioning

This phase of the mechanism is substantially based on the simulative model presented in [18]. We suppose the BE traffic is composed only by TCP packets (UDP should be used for the QoS traffic), and we model the aggregate TCP streams as flows. The model adopted is able to determine the average throughput for each link, where the time average can be obtained over any time period; namely, we can compute both the throughput value second by second or its global average value over a very long period. Different simulation results and measures have shown that for a link loaded with TCP traffic composed by many TCP connections, when the global throughput increases above 80%, the performance of each single connection's end-to-end throughput decreases very quickly. So, a first BE traffic dimensioning is done by trying to find the minimum capacities $\widetilde{C}_j^{BE}$, $\forall j=1,...,J$, for which the maximum average throughput results to be around 80%.

The model is based on the concept of aggregate flow of TCP connections. A flow models all the TCP data exchanges (TCP connections) between the same source and destination pair as a "fluid". This flow representation does not take into account the packet dynamics. The incoming traffic is represented as blocks of data with random size, which become available (for transmission) at a source node at a random instant. Each source is modeled as a tank, and the network links represent the pipes that drain the tank, while the data blocks fill it. Where more than one "fluid data" source (either coming directly from a tank or from a previous link) has to share a single pipe, this sharing is regulated by a specific law. This law, roughly speaking, divides the capacity of the link in proportion to a sort of "pressure" of the different flows. The source-destination path of each flow is decided by the routing algorithm. The rate (the number of bits per second) associated to each flow is fixed by the link along the path, which assigns the smallest available bandwidth to that specific flow (the bottleneck link). These rates are always dynamic ones; they can change when a tank in the network becomes void, or when it is void and receives a block of data, or when the routing changes one or more paths. For the notational aspects, we indicate with N the set of the nodes in the network and with F the set of flows f between each node pair (i,j): $i \neq j$, $i,j \in N$. Indicating by t_k, $k = 0,1,2,...$, the significant event occurrence instants (i.e., the instants when there is a new block arrival, or a buffer empties or the routing changes a routing table), $r_{t_k}^{(f)}$ is the rate associated to the flow $f \in F$ during the period $[t_k, t_{k+1})$.

To implement the data sources we have to generate two quantities: the arrival instants of the data blocks and their sizes. We use a Poisson process to generate the arrival instants, i.e., we use an inter-arrival exponential distribution with arrival rates $\lambda_{BE}^{(f)}$, $f \in F$. Concerning the size of the data blocks, many studies and measures reported in the literature suggest the use of heavy-tailed distributions, which originate a self-similar behavior, when traffic coming from many generators is mixed in the network links. We have chosen a Pareto-Levy distribution, i.e., a probability density function:

$$f_X^{(f)}(x) = \alpha^{(f)} \left(\Delta_{INF}^{(f)} \right)^{\alpha^{(f)}} x^{-\left(1 + \alpha^{(f)}\right)}, \; x \geq \Delta_{INF}^{(f)} \quad f \in F \tag{6}$$

where x is the block size in bits, $\alpha^{(f)}$ is the shape parameter and $\Delta_{INF}^{(f)}$ the location parameter, i.e., the minimum size of a block. Actually, in the model we utilize a truncated version of equation (6), both to obtain stable average performance indexes, and to represent short connections (mice), which do not exit the slow-start phase of TCP control flow, as well as very long data transfers (elephants).

The model acts at each event instant t_k, $k = 1, 2, 3,...$ by re-computing the value of the rate $r_{t_k}^{(f)}$ for every aggregate flow $f \in F$. Thus, the core of the procedure is represented by the computation algorithm for these rates, which is substantially based on the *min-max* rule described in [20]. It is worth noting that, though the min-max sharing is known not to hold when considering the behavior of individual TCP connections, it can represent fairly well the way the bandwidth is subdivided among aggregates of TCP connections between the same pairs. The critical and original part of this procedure is related to the law by means of which the different flows compete for the bandwidth on the bottleneck links. As stated and shown in [18], we have observed that the average bandwidth sharing has a stationary point proportional to the offered load (if we suppose that the instantaneous throughput of each connection be the same, on average). So, we have defined a sharing parameter $\rho^{(a,f)}$, which represents the maximum bandwidth portion that an aggregate flow f on a link j can use, proportional to the offered load of the different active flows on the link, i.e.:

$$\rho_{BE}^{(j,f)} = \frac{\lambda_{BE}^{(f)} \overline{x}_{BE}^{(f)}}{\sum_{l \in F^{(j)}} \lambda_{BE}^{(l)} \overline{x}_{BE}^{(l)}} \tag{7}$$

where $F^{(j)}$ represents the set of aggregate flows traversing the link j and $\overline{x}_{BE}^{(f)}$ is the average data block size for flow f. It is worth noting that this simulative mechanism has shown to be really very fast and remarkably precise in the determination of the throughput per link [18]. The flow simulator is used inside a descent procedure, which stops when the bandwidth on each network link is such that the average throughput satisfies the imposed constraint. For the notational aspects we define:

- $C_j^{BE(i)}$ the bandwidth assignment to link j at step i;
- L^i the set of links traversed by the BE traffic at step i;
- g_{BE} the minimum bandwidth granularity;
- ρ_j^i the throughput on link j at step i, measured by the simulative process;
- γ the required link bandwidth utilization (we always use 0.8 in this paper);
- Δ_j^i the maximum bandwidth tolerance range for link j at step i.

The bandwidths of all the network links are initialized with a value over the sum of all the BE traffic offered loads.

The algorithm at the generic step i works as follows:

1) $\forall j \in L^i$ do $\Delta_j^i = g_{BE} / C_j^{BE(i)}$
2) if $\rho_j^i < \gamma - \Delta_j^i$ then $C_j^{BE(i+1)} = C_j^{BE(i)} - g_{BE}$
3) if $\rho_j^i > \gamma + \Delta_j^i$ then $C_j^{BE(i+1)} = C_j^{BE(i)} + g_{BE}$
4) if $\gamma - \Delta_j^i < \rho_j^i < \gamma + \Delta_j^i$ $\forall j \in L^i$ then stop
5) run another simulation with the new bandwidth allocation and return to step 1.

3.3 The Final Dimensioning

As already stated in Section II, the output of the previous two steps gives us:

- an upper-bound capacity assignment represented by the values

$$C_j^{max} = \widetilde{C}_j^{QoS} + \widetilde{C}_j^{BE}, \quad \forall j=1,\dots,J \tag{8}$$

- the thresholds for the QoS traffic CAC (i.e., $\widetilde{C}_j^{QoS}$, $\forall j=1,\dots,J$).

We recall that the capacity assignment (8) represents an upper bound, because it corresponds to a very conservative configuration with respect to BE traffic. Thus, the mechanism runs a simple network simulation only with the CO flows' birth and death events and, during this simulation, it builds an occurrence distribution histogram for the bandwidth occupancy on each link, i.e., it computes the percentage of time $\Delta_j(C)$ during which link j has C bandwidth units occupied by QoS traffic. This information gives us the possibility to identify the capacity $\widetilde{C}_{j,\xi}^{QoS,min}$, below which the QoS utilization of link j remains for ξ % of the time, i.e.,

$$\widetilde{C}_{j,\xi}^{QoS,min} = min\left\{ C : \sum_{i=0}^{C} \Delta_j(i) > \xi \right\} \tag{9}$$

Thus we consider

$$C_{j,\xi}^{min} = \widetilde{C}_{j,\xi}^{QoS,min} + \widetilde{C}_j^{BE} \tag{10}$$

as the lower bound assignment, and

$$G_{j,\xi}^{max} = \tilde{C}_j^{QoS} - \tilde{C}_{j,\xi}^{QoS,min} \tag{11}$$

as the maximum gains that we can obtain when the two traffic types are carried on the same network. In fact, it is reasonable to think that in a network dimensioned with the $C_{j,\xi}^{min}$ capacities, when the CO bandwidth occupancy on one or more network links begins to utilize more than $\tilde{C}_{j,\xi}^{QoS,min}$ bandwidth units, the BE connections that cross those links might not receive the minimum capacities needed for an efficient average operation. Moreover, as the CO dynamics are generally very slow with respect to the BE ones, these periods of resource lack for the BE might result excessively long, and they might generate sensible deterioration of the user-perceived performance: the largest part of the TCP connections during this kind of periods suffers packet drops or expired timeouts; so, they apply the congestion avoidance mechanism, lowering their throughputs and increasing their transfer times. An example of this condition is shown in Fig. 1, which reports some measures obtained by ns2 (network simulator 2, [21]) simulations. The large difference between BE and QoS traffic models' dynamics is the main reason why the global simulation can be computationally very heavy. So, in this last step we have realized a sort of "smart" simulation tool, which operates in the following way. The algorithm is iterative and it executes one simulation at each step. During the general iteration step i, the main procedure simulates the activations and terminations of the QoS traffic streams. Only when at least on a link j the bandwidth occupied by the QoS traffic overcomes a threshold $T_j^{(i)}$, the aggregate flow BE simulator is activated and the BE performance is monitored and measured. Obviously, the BE simulator is stopped when all the QoS traffic bandwidth occupancies decrease below $T_j^{(i)}$. At step $i{=}0$, we initialize $C_j^{(0)} = \tilde{C}_{j,\xi}^{min}$ and $T_j^{(0)} = \tilde{C}_{j,\xi}^{QoS,min}$.

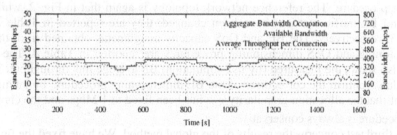

Fig. 1. Bandwidth occupation obtained by means of a ns2 simulation. The figure shows the decay in BE traffic performance (average throughput per TCP connection [Kbps]), when the bandwidth left free by the QoS traffic goes below a certain threshold

While the BE traffic is injected into the network, we collect the BE average bandwidth utilization on all the links every τ seconds. Every time we find a link's utilization equal to 100% on a generic link j, we count a congestion event. If we collect more than χ congestion events on the link j every Γ hours (e.g., 24 hours) with a probability greater than β and if $C_j^{(i)} < C_j^{max}$, then we increase the bandwidth assignment on such link of a bandwidth unit and we update the parameters as follows:

$$C_j^{(i+1)} = C_j^{(i)} + 1, \quad T_j^{(i+1)} = T_j^{(i)} + 1 \quad \text{and} \quad i{=}i{+}1$$

The dimensioning mechanism stops when the simulation yields stable values from the last link assignment change.

4   Numerical Results

In this Section the results obtained in the validation test sets are reported. All tests, except the last one, have been realized by using the simple network in Fig. 2. In the last test we have applied the global method on a complex network, which closely matches a national backbone. In the first test, we have used the traffic matrix reported in Table 1. To easily verify the correctness of the results, we have utilized the same end-to-end blocking probability constraint for all of the four QoS traffic classes. Note that, for this reason and since the A-B, B-E, C-D and D-E links (with reference to Fig. 2) are traversed by traffic flows with identical offered loads, they should result in the same capacity allocation, and this is what happens at the end of the procedures. Our first aim is to verify the QoS dimensioning procedure. Fig. 3 shows the blocking probabilities measured on the network, dimensioned both by the analytical procedure only and by the whole QoS phase (i.e., including the simulation-based refinement described at the end of sub-section III.A). It can be seen that the difference between the two dimensioning results grows with the constraint value, and it is also quite evident that the approximate analytical solution always remains a lower bound. On the other hand, by observing Figs. 4 and 5, where the corresponding computed capacities versus different values of the desired blocking probability are reported, it can be noted that even a significant difference in blocking probabilities determines a relatively small difference in capacity allocation. This means that in very complex networks, the analytical procedure might be considered precise enough to determine the upper bounds and the QoS thresholds. The aim of the second test is checking the BE dimensioning procedure. The reference network topology is again that in Fig. 2, while the QoS and BE traffic matrices have been changed; they are reported in Table 2 and Table 3, respectively. The BE and QoS capacity assignments obtained from the BE dimensioning and QoS dimensioning procedures are reported in Table 4. The same table also shows the measured (by ns2 simulation) BE average link throughput. These last values confirm the correctness of the BE dimensioning procedure, by taking into account that the allocation granularity g_{BE} has been fixed to 1 Mbps and the choice of the procedure is always conservative.

The third test concerns the results of the global method. We have fixed the final BE constraint by imposing Γ =24 hours and β =95%. (i.e., the constraint is imposed on the maximum number of BE congested periods per day and with probability 0.95). Fig. 6 reports the obtained final global capacity reductions (bandwidth gains) with respect to the initial upper-bound computation. These results are shown for an increasing number of allowed congestions per day (χ =10, 15, 20, 25, 30 and 35) and also for different durations of what is considered to be a congestion period (τ = 30, 60, 90 and 120 seconds). It can be seen that the advantage obtained by using the final procedure with respect to the first computed upper-bound assignment increases by progressively relaxing the constraints. In the extreme cases the gain reaches the maximum value forecast by the procedure (i.e., 25 Mbps); this means that the first BE constraint (80% utilization) is here more stringent than the other one.

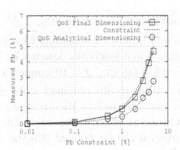

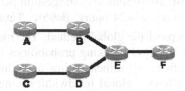

Fig. 2. The network topology used in the first validation test sessions

Fig. 3. Measured blocking probabilities versus imposed constraints with both analytical and complete QoS procedure capacity assignments

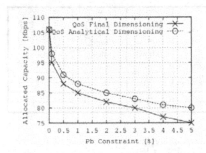

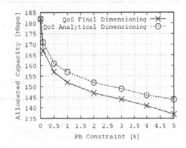

Fig. 4. QoS capacity assignment for links A-B, B-E, C-D and D-E, computed by only the analytical part and by the whole procedure versus the blocking probability constraint

Fig. 5. QoS capacity assignment for link E-F, computed by only the analytical part and by the whole procedure versus blocking probability constraints

Table 1. QoS traffic matrix for the first test

Class ID k	Source Node	Dest. Node	$\lambda_{QoS}^{(k)}$	$\mu_{QoS}^{(k)}$	$b^{(k)}$
			[streams/s]	[streams/s]	[Mbps]
1	A	F	0.1	0.01	1
2	A	F	0.1	0.01	2
3	C	F	0.2	0.002	1
4	C	F	0.2	0.002	2

Table 2. QoS traffic matrix for the second test

Class ID k	Source Node	Dest. Node	$\lambda_{QoS}^{(k)}$	$\mu_{QoS}^{(k)}$	$b^{(k)}$	$\overline{Pb}^{(k)}$
			[streams/s]	[streams/s]	[Mbps]	[%]
1	A	F	0.01	0.01	1	0.75
2	A	F	0.002	0.002	2	0.75
3	C	F	0.01	0.01	1	1
4	C	F	0.002	0.002	2	0.75

The last test has been realized to dimension a large VPN on the national (Italian) backbone, shown in Fig. 7, which represents substantially the network of the Italian academic and research network GARR [22]. The traffic matrix used represents a fully meshed traffic exchange among all the nodes. More in detail, each node has the following three traffic streams/flows sources for each possible destination: **QoS sources A:** Class rate 1 Mbps, $\lambda_{QoS}^{(k)} = 0.01$ streams/s, $\mu_{QoS}^{(k)} = 0.01$ streams/s, blocking probability constraint = 1% for $\forall k \in A$; **QoS sources B:** Class rate 2 Mbps, $\lambda_{QoS}^{(k)} = 0.02$ streams/s, $\mu_{QoS}^{(k)} = 0.02$ streams/s, blocking probability constraint = 1% for $\forall k \in B$;

BE sources: $\lambda_{BE}^{(f)} = 1$ flows/s, $\overline{x}_{BE}^{(f)} = 0.24$ MB, $\alpha^{(f)}=2$ for $\forall f \in BE$. Note that, in this case, (3) must be substituted by the expressions of the blocking probabilities corresponding to a stochastic knapsack. The final BE constraints are: congestion period duration $\square = 60$ sec, $\chi = 10$ congestion events every $\Gamma = 24$ hours, $\beta=95\%$. Table 5 reports all the results obtained at the different steps of the global method. The measures realized by ns2 of BE average throughput and QoS blocking probabilities (not reported here) confirm the correctness of the first two dimensioning procedures. Moreover, it can be noticed that the final step allows a global bandwidth saving of 299 Mbps.

Table 3. BE traffic matrix for the second test

Flow ID f	Source Node	Dest. Node	$\lambda_{BE}^{(f)}$ [conn/s]	$\overline{x}_{BE}^{(f)}$ [MB]	$\alpha^{(f)}$
1	A	F	1	0.24	2.5
2	B	F	1	0.24	2.5
3	C	F	1	0.24	2.5
4	D	F	1	0.24	2.5

Table 4. Capacity assignments for BE and QoS dimensioning and measured BE average throughput

Link	QoS Assigned Capacity [Mbps]	BE Assigned Capacity [Mbps]	Measured Average Through-put [Mbps]
A-B	14	3	2.0167
C-D	14	3	1.9983
B-E	14	5	3.940
D-E	14	5	3.975
E-F	20	10	7.915

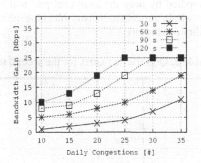

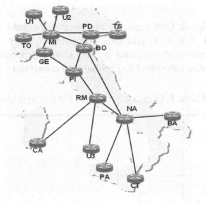

Fig. 6. Bandwidth gains (capacity reduction with respect to the upper bound) versus allowable number of daily BE congestion periods, with respect to different values of the congestion period duration

Fig. 7. Network topology used in the last test

5 Conclusions

A complete set of methodologies for bandwidth dimensioning in a VPN has been proposed and analyzed in the paper. The aim of the whole structure is to obtain a reasonable balance between precision and computational complexity. In particular, the presence of both QoS and BE traffic has been explicitly taken into account, and

Table 5. Last tests dimensioning results

Link	QoS Bandwidth [Mbps]	BE Bandwidth [Mbps]	Gain Range [Mbps]	Allocated Bandwidth [Mbps]	Link	QoS Bandwidth [Mbps]	BE Bandwidth [Mbps]	Gain Range [Mbps]	Allocated Bandwidth [Mbps]
MI-RM	186	199	25	365	RM-NA	144	143	25	269
MI-PD	51	36	16	74	NA-RM	78	0	0	78
MI-PI	51	36	14	77	NA-CT	81	36	22	107
MI-BO	51	0	0	51	CT-NA	81	36	18	109
PD-BO	63	0	0	63	NA-BA	81	36	21	108
PI-BO	42	0	0	42	BA-NA	81	36	14	108
BO-NA	96	0	0	96	NA-PA	81	36	18	107
RM-MI	123	59	23	187	PA-NA	81	36	19	106
PD-MI	51	36	14	77	RM-CA	78	36	15	104
PI-MI	51	36	13	78	CA-RM	78	36	18	104
BO-MI	135	143	23	261	MI-GE	75	36	17	106
BO-PD	42	0	0	42	GE-MI	75	36	17	106
BO-PI	42	0	0	42	PI-GE	15	0	0	15
NA-BO	156	143	19	285	GE-PI	15	0	0	15
MI-U1	81	36	23	101	MI-TO	81	36	18	107
U1-MI	81	36	20	104	TO-MI	81	36	20	107
MI-U2	81	36	18	104	PD-TS	15	0	0	15
U2-MI	81	36	17	107	TS-PD	42	0	0	42
RM-U3	78	36	15	104	MI-TS	75	36	16	106
U3-RM	78	36	17	101	TS-MI	54	36	10	84

two dimensioning procedures, tailored on the specific traffic models, have been devised. The final dimensioning considers the effect of the multiplexing gain, arising when the different traffic types share the assigned link capacities. This methodology has been tested on specific networks, and the effects of the approximations made have been examined, by comparison with ns2 detailed simulations. Globally, even though further testing and refinements are being planned, the proposed tool seems to be capable of responding to the needs of a network operator with reasonable accuracy and the possibility of realizing significant savings in bandwidth activation.

References

1. Y. Maeda, "Standards for Virtual Private Networks (Guest Editorial)", IEEE Commun. Mag., June 2004, pp. 114-115.
2. M. Carugi, J. De Clercq, "Virtual Private Network Services: Scenario, Requirements and Architectural Construct from a Standardization Perspective", IEEE Commun. Mag., June 2004, pp. 116-122.
3. C. Metz, "The latest in Virtual Private Networks: Part I", IEEE Internet Computing, Jan-Feb 2003, pp. 87-91.
4. C. Metz, "The latest in Virtual Private Networks: Part II", IEEE Internet Computing, May-June 2003, pp. 60-65.
5. J. De Clercq, O. Paridaens "Scalability Implications of Virtual Private Networks", IEEE Commun. Mag., May 2002, pp. 151-157.
6. P. Knight, C. Lewis, "Layer 2 and 3 Virtual Private Networks: Taxonomy, Technology, and Standardization Efforts", IEEE Commun. Mag., June 2004, pp. 124-131.
7. T. Takeda, I. Inoue, R. Aubin, M. Carugi, "Layer 1 Virtual Private Networks: Service Concepts, Architecture Requirements, and Related Advances in Standardization", IEEE Commun. Mag., June 2004, pp. 132-138.

8. S. S. Gorshe. T. Wilson, "Transparent Generic Framing Procedure (GFP): A Protocol for Efficient Transport of Block-Coded Data through SONET/SDH Networks", IEEE Commun. Mag., May 2002, pp. 88-95.
9. J. Zeng, N. Ansari, "Toward IP Virtual Private Network Quality of Service: A Service Provider Perspective", IEEE Commun. Mag., April 2003, pp. 113-119.
10. I. Khalil, T. Braun, "Edge Provisioning and Fairness in VPN-DiffServ Networks", J. Network and Syst. Manag., vol. 10, no. 1, March 2002, pp. 11-37.
11. A. Olsson, "Understanding Telecommunications", Chapter A.10, 1997-2002, Ericsson, STF, Studentlitteratur.
12. A. Kumar, R. Rastogi, A. Silberschatz, B. Yener , "Algorithms for Provisioning Virtual Private Networks in the Hose Model", IEEE/ACM Trans. Networking, vol. 10, no. 4, Aug. 2002, pp. 565-578.
13. N N. Duffield, P. Goyal, A. Greenberg, P. Mishra, K. Ramakrishnan, J. van der Merwe, "Resource Management with Hoses: Point-to-Cloud Services for Virtual Private Networks", IEEE/ACM Trans. Networking, vol. 10, no. 10, Oct. 2002, pp. 679-691.
14. A. Jàüttner, I. Szabó, Á. Szentesi "On Bandwidth Efficiency of the Hose Model in Private Networks", Proc. IEEE Infocom 2003, San Francisco, CA, April 2003.
15. W. Yu, J. Wang, "Scalable Network Resource Management for Large Scale Virtual Private Networks", Simulation Modelling Practice and Theory 12, Elsevier, June 2004, pp. 263–285.
16. R. Isaacs, I. Leslie, "Support for Resource-Assured and Dynamic Virtual Private Networks", IEEE J. Select. Areas Commun., vol. 19, no. 3, March 2001, pp. 460-472.
17. R. Cohen, G. Kaempfer, "On the Cost of Virtual Private Networks", IEEE/ACM Trans. Networking, vol. 8, no. 6, Dec. 2000, pp. 775-784.
18. R. Bolla, R, Bruschi, M. Repetto, "A Fluid Simulator for Aggregate TCP Connections", Proc. 2004 Internat. Symp. on Performance Evaluation of Computer and Telecommunication Syst. (SPECTS'2004), San Diego, CA, July 2004, pp. 5-10.
19. K. W. Ross, Multiservice Loss Models for Broadband Telecommunication Networks, Springer, Berlin, 1995.
20. D. Bertsekas, R. Gallager, Data Networks, 2nd Ed., Prentice-Hall, 1992.
21. The Network Simulator – Ns2. Documentation and source code from the home page: http://www.isi.edu/nsnam/ns/.
22. GARR backbone, network topology and statistics homepage: http://www.garr.it/

Topological Design of Survivable IP Networks Using Metaheuristic Approaches*

Emilio C.G. Wille[1], Marco Mellia[2], Emilio Leonardi[2], and Marco Ajmone Marsan[2]

[1] Dipartimento di Elettronica, Politecnico di Torino, Italy
[2] Departamento de Eletrônica, Centro Federal de Educação Tecnológica do Paraná, Brazil

Abstract. The topological design of distributed packet switched networks consists of finding a topology that minimizes the communication costs by taking into account a certain number of constraints such as the end-to-end quality of service (e2e QoS) and the reliability. Our approach is based on the exploration of the solution space using metaheuristic algorithms (GA and TS), where candidate solutions are evaluated by solving CFA problems. We propose a new CFA algorithm, called GWFD, that is capable of assign flow and capacities under e2e QoS constraints. Our proposed approach maps the end-user performance constrains into transport-layer performance constraints first, and then into network-layer performance constraints. A realistic representation of traffic patterns at the network layer is considered as well to design the IP network. Examples of application of the proposed design methodology show the effectiveness of our approach.

1 Introduction

Today, with the enormous success of the Internet, packet networks have reached their maturity, and all enterprises have become dependent upon networks or networked computations applications. In this context the loss of network services is a serious outage, often resulting in unacceptable delays, loss of revenue, or temporary disruption. In an usual network-design process, were networks are designed to optimize a performance measure (e.g., delay, throughput) without considering possible network failures, the network performance can degrade drastically at an element failure. A network-design process based only on reliability considerations (e.g., connectivity, successful communication probability) does not necessarily avoid network performance degradation, despite the connections guarantees during the failure of one or several stations. To avoid loss of network services, communications networks should be designed so that they remain operational and maintain as high performance level as feasible, even in presence of network component failures.

The packet network design methodology that we propose in this paper is quite different from the many proposals that appeared in the literature. Those focus almost invariably on the trade-off between total cost and average performance (expressed in terms of average network-wide packet delay, average packet loss probability, average

---

* This work was supported by the Italian Ministry for Education, University and Research under project TANGO. The first author was supported by a CAPES Foundation scholarship from the Ministry of Education of Brazil.

M. Ajmone Marsan et al. (Eds.): QoS-IP 2005, LNCS 3375, pp. 191–206, 2005.
© Springer-Verlag Berlin Heidelberg 2005

link utilization, network reliability, etc.). This may lead to situations where the average performance is good, but, while some traffic relations obtain very good quality of service (QoS), some others suffer unacceptable performance levels. On the contrary, our packet network design methodology is based on user-layer QoS parameters, and explicitly accounts for each source/destination QoS constraint.

From the end user's point of view, QoS is driven by e2e performance parameters, such as data throughput, web page latency, transaction reliability, etc. Matching the user-layer QoS requirements to the network-layer performance parameters is not a straightforward task. For example, the QoS perceived by end users in their access to Internet services is mainly driven by TCP, whose congestion control algorithms dictate the latency of information transfer. Indeed, it is well known that TCP accounts for a great amount of the total traffic volume in the Internet [1, 2].

According to the Internet protocol architecture, at least two QoS mapping procedures should be considered; the first one translates the application-layer QoS constraints into transport-layer QoS constraints, and the second translates transport-layer QoS constraints – such as *file transfer latency* (L_t), or *file transfer throughput* (T_h) – into network-layer QoS constraints, such as *Round Trip Time* (RTT) and *packet loss probability* (P_{loss}). In [3, 4], the authors describe *QoS translators* that are able to do this mapping process. While the first procedure is an ad-hoc one, the second is based on the numerical inversion of analytic TCP models presented in the literature.

The representation of traffic patterns inside the Internet is a particularly delicate issue, since it is well known that IP packets do not arrive at router buffers following a Poisson process [5], but a higher degree of correlation exists, which can be partly due to the TCP control mechanisms. This means that the usual approach of modeling packet networks as networks of M/M/1 queues [6–8] is not appropriate. In this paper we adopt a refined IP traffic modeling technique, already presented in [9], that provides an accurate description of the traffic dynamics in multi-bottleneck IP networks loaded with TCP mice and elephants. The resulting analytical model is capable of producing accurate performance estimates for general topology IP networks loaded by realistic TCP traffic patterns, while still being analytically tractable.

In addition, given that the network reliability depends on its components, it is necessary to consider the failure of one or several components in the design of such networks. In general, two disjoint physical paths p_1 and p_2 are associated with each existing path. In normal conditions, only path p_1 is used to information exchange; however, enough capacity is reserved on p_2 to be able to restore communication when one of the physical links (or nodes) belonging to p_1 fails. Protection and/or restoration schemes normally guarantee the resilience to the failure of a single physical link (or node); thus, the capacity to be reserved in the network must be sufficient to successfully restore all the paths disrupted by a link (or node) failure. In this paper, the concept of 2-connectivity is used as a reliability constraint.

The challenge in the area of network design is to determine the less expensive solution to interconnect nodes, and assign flow and capacities, while satisfying the reliability and end-to-end (e2e) QoS constraints. This problem is called the TCFA (Topological, Capacity and Flow Assignment) problem. This is a combinatorial optimization problem classified as NP-complete. Polynomial algorithms which can find the optimal solution for this problem are not known. Therefore, heuristic algorithms are applied, searching for solutions. In this paper we suggest two metaheuristic approaches: the *Genetic Algorithm* (GA) [10], and the *Tabu Search* (TS) algorithm [11] to address the topological design problem, where traffic is mostly due to TCP connections. GAs are heuristic search procedures which applies natural genetic ideas such as natural selection, mutations and survival of the fittest. The TS algorithm is an evolution of the classical Steepest Descent method, however thanks to an interior mechanism that provides to accept also worse solutions than the best solution found so far, it is less subject to local optima entrapments.

The evaluation of each candidate inside each metaheuristic algorithm is done considering the cost of the network obtained from the solution of its related Capacity and Flow Assignment (CFA) problem. In this paper, we present a nonlinear mixed-integer programming formulation for the CFA problem and propose an efficient heuristic procedure called Greedy Weight Flow Deviation (GWFD) to obtain CFA solutions.

When explicitly considering TCP traffic it is also necessary to tackle the Buffer Assignment (BA) problem, for which we propose an algorithm that computes solutions to the droptail case BA problem.

The rest of the paper is organized as follows. Section 2 outlines the problem and the solution approach adopted in this paper, In particular, the CFA and BA formulations are presented and the GWFD method is described. Section 3 describes fundamentals of GAs, their operating techniques, and synthesizes the adaptation and implementation details of the GAs applied to the topological design problem; it also presents the TS algorithm. Section 4 presents and analyzes computational results. Finally, Section 5 summarizes the main results obtained in this research.

2   Problem Statement

The topological design of packet networks can be formulated as a TCFA problem as follows: given the geographical location of the network nodes on the territory, the traffic matrix, the capacity costs; minimize the total link cost, by choosing the network topology and selecting link flows and capacities, subject to QoS and reliability constraints. As reliability constraint we consider that all traffic must be exchanged even if a single node fails (2-connectivity). There is a trade off between reliability and network cost; we note that more links between nodes imply more routes between each node pair, and consequently the network is more reliable; on the other hand, the network is more expensive. Finally, the QoS constrains correspond to maintain the e2e packet delay for each network source/destination pair below a maximum tolerable value.

Our solution approach is based on the exploration of the solution space (i.e., 2-connected topologies) using the metaheuristic algorithms (GA and TS). As the goal is to synthesize a network that remains connected despite one node failure, for each topology evaluation, actually, we construct n different topologies, that are obtained from the topology under evaluation by the failure of a node each time, and then for each topology we solve the CFA problem. Link capacities are set to the maximum capacity value find so far considering the set of topologies. Using these capacity values the objective function (network cost) is evaluated.

2.1 Traffic and Queueing Models

The representation of traffic patterns inside the Internet is a particularly delicate issue, since it is well known that IP packets do not arrive at router buffers following a Poisson process, see [5], but a higher degree of correlation exists, which can be partly due to the TCP control mechanisms. This means that the usual approach of modeling packet networks as networks of M/M/1 queues as discussed in [6–8, 12, 13] is not appropriate. In this paper we adopt a refined IP traffic modeling technique, already presented in [9], that provides an accurate description of the traffic dynamics in multi-bottleneck IP networks loaded with TCP mice and elephants. The resulting analytical model is capable of producing accurate performance estimates for general topology IP networks loaded by realistic TCP traffic patterns, while still being analytically tractable.

We choose to model the increased traffic burstiness induced by TCP by means of batch arrivals, hence using $M_{[X]}/M/1/B$ queues. A wide range of investigations performed in [9] certify the accurate network layer performance estimates by considering $M_{[X]}/M/1/B$ models. The batch size varies between 1 and W with distribution $[X]$, where W is the maximum TCP window size expressed in segments. The distribution $[X]$ is obtained considering the number of segments that TCP sources send in one RTT [9]. Additionally, our choice of using batch arrivals following a Poisson process has the advantage of combining the nice characteristics of Poisson processes (analytical tractability in the first place) with the possibility of capturing the burstiness of the TCP traffic [9].

Considering the $M_{[X]}/M/1/\infty$ queue, the average packet delay (queueing and transmission time) is given by:

$$E[T] = \frac{K}{\lambda} \frac{\rho}{1-\rho} = \frac{K}{\mu} \frac{1}{C-f} \tag{1}$$

with K given by:

$$K = \frac{m'_{[X]} + m''_{[X]}}{2m'_{[X]}} \tag{2}$$

where f is the average data flow on the link (bps), C is the link capacity (bps), the utilization factor is given by $\rho = f/C$, the packet length is assumed to be exponentially distributed with mean $1/\mu$ (bits/packet), $\lambda = \mu f$ is the arrival rate (packets/s), $m'_{[X]}$ and $m''_{[X]}$ are the first and second moments of the batch size distribution $[X]$.

2.2 Network Model

In the mathematical model, the IP network infrastructure is represented by a graph $G = (V, E)$ in which V is a set of nodes (with cardinality n) and E is a set of edges (with cardinality m). A node represents a network router and an edge represents a physical link connecting one router to another. The output interfaces of each router is modeled by a queue with finite buffer.

Each network link is characterized by a set of attributes which principally are the flow, the capacity and the buffer size. For a given link (i, j), the flow f_{ij} is defined as the effective quantity of information transported by this link, while its capacity C_{ij} is a measure of the maximal quantity of information that it can transmit. Flow and capacity are both expressed in bits per second (bps). Each buffer can accommodate a maximum of B_{ij} packets, and d_{ij} is the link physical length.

The average traffic requirements between nodes are represented by a traffic matrix $\hat{\Gamma} = \{\hat{\gamma}_{sd}\}$, where the traffic $\hat{\gamma}_{sd}$ between a node pair (s, d) represents the average number of bps sent from source s to destination d. We consider as traffic offered to the network $\gamma_{sd} = \hat{\gamma}_{sd}/(1 - P_{loss})$, thus accounting for the retransmissions due to the losses that flows experience along their path to the destination. The flow of each link that composes the topological configuration depends on the traffic matrix. We consider that for each source/destination pair (s, d), the traffic is transmitted over exactly one directed path in the network (a minimum-cost routing algorithm is used). The routing and the traffic uniquely determine the vector $\bar{f} = (f_1, f_2, ..., f_m)$ where $\bar{f}$ is a multicommodity flow for the traffic matrix; it must obey the law of flow conservation. A multicommodity flow results from the sum of single commodity flows f_{sd} with source node s and destination node d.

2.3   The Capacity and Flow Assignment Problem

Different formulations of the CFA problem result by selecting i) the cost functions, ii) the routing model, and iii) the capacity constraints; different methodologies must be applied to solve them. In this paper we focus on Virtual Private Networks (VPNs), in which common assumptions are i) linear costs, ii) non-bifurcated routing, and iii) continuous capacities.

The CFA problem requires the joint optimization of routing and link capacities. Our goal is to determine a route for the traffic that flows on each source/destination pair and the link capacities in order to minimize the network cost subject to the maximum allowable e2e packet delay. Let δ_{ij}^{sd} be a decision variable which is one if link (i, j) is in path (s, d) and zero otherwise. Thus the CFA problem is formulated as the following optimization problem:

$$Z_{CFA} = min \sum_{i,j} g(d_{ij}, C_{ij}) \tag{3}$$

subject to:

$$\sum_j \delta_{ij}^{sd} - \sum_j \delta_{ji}^{sd} = \begin{cases} 1 & \text{if } s = i \\ -1 & \text{if } t = i \\ 0 & \text{otherwise} \end{cases} \quad \forall (i, s, d) \tag{4}$$

$$K_1 \sum_{i,j} \frac{\delta_{ij}^{sd}}{C_{ij} - f_{ij}} \leq RTT_{sd} - K_2 \sum_{i,j} \delta_{ij}^{sd} d_{ij} \quad \forall (s, d) \tag{5}$$

$$f_{ij} = \sum_{s,d} \delta_{ij}^{sd} \gamma_{sd} \quad \forall (i, j) \tag{6}$$

$$C_{ij} \geq f_{ij} \geq 0 \quad \forall (i, j) \tag{7}$$

$$\delta_{ij}^{sd} \in \{0, 1\} \quad \forall (i, j, s, d) \tag{8}$$

The objective function (3) represents total link cost, which is a linear function of both the capacity C_{ij} and the physical length d_{ij} of link (i, j), i.e., $g(d_{ij}, C_{ij}) = d_{ij}C_{ij}$.

Constraint set (4) enforce flow conservation, defining a route for the traffic from a source s to a destination d. Non-bifurcated routing model is used where the traffic will follow the same path from source to destination. Equation (5) is the e2e packet delay constraint for each source/destination pair. It says that the total amount of delay experienced by all the flows routed on a path should not exceed the maximum RTT[1] minus the propagation delay of the route. Equation (6) defines the average data flow on the link. Constraints (7) and (8) are non-negativity and integrity constraints. Finally, $K_1 = K/\mu$ and K_2 is a constant to convert distance in time.

We notice that this problem is a nonlinear non convex mixed-integer programming problem. Other than the nonlinear constraint (5), it is basically a multicommodity flow problem. Multicommodity flow belongs to the class of NP-hard problems for which no known polynomial time algorithms exist. In addition, thanks to its non convex property there are in general a large number of local minima solutions. Therefore, in this paper we only discuss CFA sub-optimal solutions.

2.4   The Flow Deviation Method

Before presenting a new, fast CFA algorithm, we briefly review the classical Flow Deviation (FD) method [14], which allows to determine the routing of all the flows entering the network at different source/destination pairs; thus it solves the Flow Assignment (FA) problem.

It is well known that optimal routing directs traffic exclusively along paths which are shortest with respect to some *link weights* that depend on the flows carried by the links. This suggests that suboptimal routing can be improved by shifting the flow, for each source/destination pair, from a path to another with lower link weights values. This is the key idea in the FD method. The method starts from a given feasible path flow vector $\bar{f}$ and, using a set of link weights, finds a new path for each source/destination pair. The weight L_{ij} of the link (i, j) is obtained as the partial derivative of the average total network delay T (based on $M/M/1$ formulations) with respect to the flow rate f_{ij} which is traversing the link (i, j) evaluated at the current flow assignment f_{ij}. With this metric, the weight of a link represents the change in the objective function value if an incremental amount of the traffic is routed on that link. Then, the new flow assignment is determined by using the shortest path algorithm with the set of link weights L. The flow assignment is incrementally changed along the descent direction of the objective function. By incrementally changing the previous flow assignment into the new one, the optimal flow assignment is determined.

The FD method can be used, in a properly modified form, to solve the CFA problem; however it leads to local optima [14].

2.5   The Greedy Weight Flow Deviation Method

In this section we present a greedy heuristic to obtain solutions to the CFA problem presented in subsection 2.3. The key idea is to use a modified flow deviation method to find the flow distributions that minimize the network cost. The natural approach would

---

[1] RTT values are obtained by using the QoS translators given in [3, 4].

be to obtain the optimal values of the capacities C_{ij} as a function of the link flows f_{ij}, to substitute them in the cost function and finally (using partial derivatives) to obtain the link weights L_{ij}. Unfortunately, no closed form expression for the optimal capacities can be derived from our CFA formulation. However, we use a modified FD method in the following way.

First, it is straightforward to show that the link weights in the Kleinrock's method are given by:

$$L_{ij} = \frac{d_{ij}C_{ij}}{f_{ij}} \qquad (9)$$

Second, in order to enforce e2e QoS delay performance constraints, the link capacities C_{ij} must be obtained using an e2e QoS CA problem solver (in this paper we use the approximate CA solution presented in [3]). As our new method relies on the greedy nature of the CA solver algorithm to direct computations toward a local optima, we called it the Greedy Weight Flow Deviation (GWFD) method.

As noted before, the CFA problem has several local optima. A way to obtain a more accurate estimate of the global optima is restart the procedure using random initial flows. However, we obtained very good results setting as initial trail $L_{ij} = d_{ij}$.

2.6   The Buffer Assignment Problem

As final step in our methodology, we need to dimension buffer sizes, i.e., to solve the following problem:

$$Z_{BA} = \min \sum_{i,j} B_{ij} \qquad (10)$$

Subject to:

$$\sum_{ij} \delta_{ij}^{sd} p(B_{ij}, C_{ij}, f_{ij}, [X]) \leq P_{loss}, \quad \forall \, (s,d) \qquad (11)$$

$$B_{ij} \geq 0, \quad \forall \, (i,j) \qquad (12)$$

The objective function (10) represents the total buffer cost, which is the sum of the buffer values. Equation (11) is the loss probability constraint for each source/destination node pair. Where $p(B_{ij}, C_{ij}, f_{ij}, [X])$ is the average loss probability for $M_{[X]}/M/1/B$ queue, which is evaluated by solving its Continuous Time Markov Chain (CTMC). Constraints (12) are non-negativity constraints.

In the previous formulation we have considered the following upper bound on the value of loss probability for path (s,d) (constraint (11)).

$$\hat{P}_{loss} = 1 - \prod_{i,j} \left(1 - \delta_{ij}^{sd} p(B_{ij}, C_{ij}, f_{ij}, [X])\right)$$
$$\leq \sum_{ij} \delta_{ij}^{sd} p(B_{ij}, C_{ij}, f_{ij}, [X]) \qquad (13)$$

Consequently, the solution of the BA problem is a conservative solution.

The proof that the BA problem is a convex optimization problem is not a straight-forward task. The difficulty derives from the need of showing that $p(B, C, f, [X])$ is convex. Since, to the best of our knowledge, no closed form expression for the station-ary distribution is known, no closed form expression for $p(B, C, f, [X])$ can be derived. However, we conjecture that the BA problem is a convex optimization problem by con-sidering that: (i) for an $M/M/1/B$ queue, $p(B, C, f)$ is a convex function; and (ii) approximating $p(B, C, f, [X]) = \sum_{i=B}^{\infty} \pi_i$, where π_i is the stationary distribution of an $M_{[X]}/M/1/\infty$ queue, the loss probability is a convex function of B.

We can thus classify the BA problem as a multi-variable constrained convex mini-mization problem; therefore, the global minimum can be found using convex program-ming techniques. We solve the minimization problem applying first a constraints reduc-tion procedure which reduces the set of constraints by eliminating redundancies. Then, the solution of the BA problem is obtained via the *logarithm barrier method* [15]

2.7 Numerical Examples: 40-Node Networks

To evaluate the performance of the new GWFD method, we run a large number of nu-merical experiments and computer simulations. We present results obtained considering several fixed topologies (40–node, 160–link each), which have been generated using the BRITE topology generator [16] with the router level option. Our goal is to obtain rout-ing and link capacities. For each topology, we solved both the CFA and BA problems using the approaches described in previous sections.

Link propagation delays are uniformly distributed between 0.5 and 1.5ms, i.e., link lengths vary between 100 and 300 Km. Random traffic matrices were generated by picking the traffic intensity of each source/destination pair from a uniform distribution. The average source/destination traffic requirement was set to $\gamma = 5$ Mbps. The file size follows a distribution, which is derived from one-week long measurements [2] (we recall that K and $p(B, C, f, [X])$ are related to the file size distribution).

For all source/destination pairs, the target QoS constraints are: i) latency $L_t \leq 0.2s$ for TCP flows shorter than 20 segments, ii) throughput $T_h \geq 512$ kbps for TCP flows longer than 20 segments, iii) $P_{loss} = 0.001$. Using the transport-layer QoS translator (presented in [3, 4]), we obtain the equivalent network-layer performance constraint $RTT \leq 0.032s$ for all source/destinations node pairs.

In Fig. 1, we compare network costs, considering 10 random topologies, obtained with four different techniques: i) Lagrangian relaxation (LB); ii) primal heuristic with logarithmic barrier CA solution (PH); iii) logarithmic barrier CA solution with mini-mum-hop routing (MinHop + CA); and iv) Greedy Weight Flow Deviation method (GWFD). A lower bound to the network cost is obtained by using the Lagrangian relax-ation; using the second technique, feasible solutions are obtained from a sub-gradient optimization technique; the third technique just ignores the routing optimization when solving the CA problem (see [3, 4]).

We can observe that the GWFD solutions, for all considered topologies, always fall rather close to the lower bound (LB). The gap between GWFD and LB is about 13%. In addition, the GWFD algorithm is faster than the sub-gradient approach presented in [3, 4] (only 5 seconds of CPU time are needed to solve an instance with 40 nodes), while it obtains very similar results.

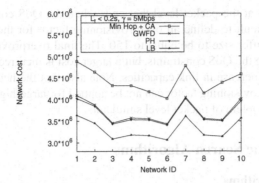

Fig. 1. Network Cost for 40–node Network Random Topologies.

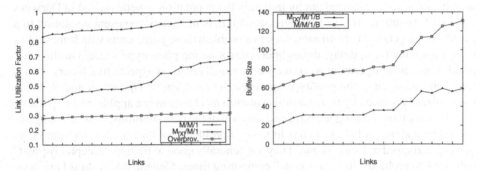

Fig. 2. Link Utilization Factors and Buffer Sizes for the 10–node network.

Avoiding to optimize the flow assignment subproblem results in more expensive solutions, as shown by the "Min Hop" routing associated with an optimized CA problem. This underlines the need to solve the CFA problem rather then a simpler CA problem.

To complete the evaluation of our methodology, we compare the link utilization factors and buffer sizes obtained when considering the classical $M/M/1$ queueing model [14] instead of the $M_{[X]}/M/1$ model. Fig. 2 shows the link utilizations (top plot) and buffer sizes (bottom plot) obtained with our method and with the classical model. It can be immediately noticed that considering the burstiness of IP traffic radically changes the network design. Indeed, the link utilizations obtained with our methodology are much lower than those produced by the classical approach, and buffers are much longer.

It is important to observe that the QoS perceived by end users in a network dimensioned using the classical approach cannot be tested by using simulations, because the loss probability experienced by TCP flows is so high that retransmissions cause the offered load to become larger than 1 for some links. This means that the network designed with the classical approach is not capable of supporting the offered load and therefore cannot satisfy the QoS constraints.

In addition in Fig. 2, we also compare our results to those of an overprovisioned network, in which the capacities obtained by using the traditional $M/M/1$ model are multiplied a posteriori by the minimum factor which allows the QoS constraints to be met. The overprovisioning factor was estimated by a trial and error procedure based on path simulations at the packet level that ends when the QoS constraints are satisfied. Since it is difficult to define an overprovisioning factor for the BA problem, we fixed a priori the buffer size to be equal to 150. The final overprovisioned network is capable of satisfying the QoS constraints, but a larger cost is incurred, which is directly proportional to the increase in link capacities. Note also that the heuristic used to find the minimum overprovisioning factor can not be applied for large/high-speed networks, due to scalability problem of packet level simulators.

3   Metaheuristic Search Algorithms

3.1   Genetic Algorithms

Genetic Algorithms (GAs) are stochastic optimization heuristics in which explorations in solution space are carried out by imitating the population genetics stated in Darwin's theory of evolution. To use a genetic algorithm, is necessary represent a solution to a problem as a genome (or *chromosome*). In a problem those parameters which are subject to optimization (e.g., delay, throughput) constitute the phenotype space. On the other hand, the genetic operators work on abstract mathematical aspects like binary strings (e.g., chromosomes), the genotype space. *Selection*, *Genetic Operation* and *Replacement*, directly derived by from natural evolution mechanisms are applied to a population of solutions, thus favoring the birth and survival of the best solutions. GAs are not guaranteed to find the global optimal solution to a problem, but they are less susceptible to getting entrapped at local optima. They are generally good at finding "acceptably good" solutions to problems in "reasonable" computing times. Genetic Algorithms have been successfully applied to many NP-hard combinatorial optimization problems, in several application fields such as business, engineering, and science. But, during the last decade, the application of GAs to the telecommunication problem domain has started to receive significant attention.

The population comprises a group of N_I individuals (chromosomes) from which candidates can be selected for the solution of a problem. Initially, a population is generated randomly. The *fitness* values of the all chromosomes are evaluated by calculating the objective function in a decoded form (phenotype). A particular group of chromosomes (parents) is selected from the population to generate the offspring by the defined genetic operations (*mutation* and *crossover*). The fitness of the offsprings is evaluated in a similar way to their parents. The chromosome in the current population are then replaced by their offspring, based on a certain replacement strategy. Such a GA cycle is repeated until a desired termination criterion is reached (for example, a predefined number of generations N_G is produced). If all goes well throughout this process of simulated evolution, the best chromosome in the final population can became a highly evolved solution to the problem.

In the following paragraphs, we describe various techniques that are employed in the GA process for encoding, fitness evaluation, parent selection, genetic operation, and replacement.

Encoding Scheme: Its should be noted that each chromosome represents a trial solution to the problem setting. To enhance the performance of the algorithm, a chromosome representation that stores problem-specific information is desired. The chromosome is usually expressed in a string of variables, each element of which is called a gene. The variable can be represented by binary, real number, or other forms and its range is usually defined by the problem.

Manipulation of topologies with the genetic algorithm requires that they are represented in some suitable format. Although more compact alternatives are possible, the characteristic matrix of a graph is quite an adequate representation. A graph is represented by an $n \times n$ binary matrix, where n is the number of nodes of the graph. A "1" in row i and column j of the matrix stands for an arc from node i to node j and a "0" represents that node i and node j are not connected.

Fitness Evaluation: The objective function is a main source to providing the mechanism for evaluating the status of each chromosome. It takes a chromosome (or phenotype) as input and produces a number or list of numbers (objective value) as a measure to the chromosome's performance. The evaluation of the objective function is usually the most demanding part of a GA program because it has to be done for every individual in every generation. In this paper, a fitness function, which is an estimation of the goodness of the solution for the topological design problem, is inversely proportional to the objective function value (cost). The lower the cost the better the solution is.

Parent Selection: Parent selection emulates the survival-of-the-fittest mechanism in nature. The choice of parents to produce offspring is somewhat more challenging than it might appear. Simply choosing the very best individual and breeding from that will not generally work successfully, because that best individual may have a fitness which is at a local, rather than a global optimal. Instead, the GA ensures that there is some diversity among the population by breeding from a selection of the fitter individuals, rather than just from the fittest.

There are many ways to achieve effective selection, including *proportionate schemes*, *ranking*, and *tournament* [10]. In this paper tournament selection is used. In this approach pairs of individuals are picked at random and the one with the higher fitness (the "winner of the tournament") is used as one parent. The tournament selection is then repeated on a second pair of individuals to find the other parent from which to breed.

Genetic Operations: Crossover is a recombination operator that combines subparts of two parent chromosomes to produce offspring that contain some parts of both parents' genetic material. The probability, p_c, that these chromosomes are recombined (mated) is a user-controlled option and is usually set to a high value (e.g., 0.95). If they are not allowed to mate, the parents are placed into the next generation unchanged. A number of variations on crossover operations are proposed, including *single-point*, *multipoint*, and *uniform* crossover [10]. In this paper single-point crossover is used. In this case, two parents are selected based on the above mentioned selection scheme. A crossover point is randomly selected and the portions of the two chromosomes beyond this point are exchanged to form the offspring.

Mutation is used to change, with probability p_m, the value of a gene (a bit) in order to avoid the convergence of the solutions to "bad" local optima. As a population evolves, there is a tendency for genes to become predominant until they have spread to all members. Without mutation, these genes will then be fixed for ever, since crossover alone cannot introduce new gene values. If the fixed value of the gene is not the value required at the global optimal, the GA will fail to optimize properly. Mutation is therefore important to "loosen up" genes which would otherwise become fixed. In our experiments good results were obtained using a mutation operator that simply changes one bit, picket at random, for each produced offspring.

Repair Function: Simple GAs assume that the crossover and mutation operators produce a high proportion of feasible solutions. However, in many problems, simply concatenating two substrings of feasible solutions, or modifying single genes do not produce feasible solutions. In such cases, there are two alternatives. If the operators produce sufficient number of feasible solutions, it is possible to let the GA destroy the unfeasible ones by assigning them low fitness values. Otherwise, it becomes necessary to modify the simple operators so that only feasible individuals result from their application. In our crossover operation we employ a simple repair function which aims to produce 2-connected (feasible) topologies. If the repair operation is unsuccessful the parents are considered as crossover outputs.

Replacement Strategies: After generating the subpopulation (offsprings), two representative strategies can be proposed for old generation replacement: *Generational-Replacement* and *Steady-State reproduction* [10]. We use Generational-Replacement. In this strategy each population size N_I generates an equal number of new chromosomes to replace the entire population. It may make the best member of the population fail to reproduce offspring in the next generation. So the method is usually combined with an *elitist strategy* where once the sons' population has been generated, it is merged with the parents' population according to the following rule : only the best individuals present in both sons' population and parents' population enter the new population. The elitist strategy may increase the speed of domination of a population by a super chromosome, but on balance it appears to improve the performance.

Population Size: Although biological evolution typically takes place with millions of individuals in the population, GAs work surprisingly well with quite small populations. Nevertheless, if the population is too small, there is an increased risk of convergence to a local optimal; they cannot maintain the variety that drives a genetic algorithm's progress. As population sizes increase, the GA discovers better solutions more slowly; it becomes more difficult for the GA to propagate good combinations of genes through the population and join them together.

A simple estimate of appropriate population size for topological optimization is given by $N_I \geq \frac{\log(1-p^{\frac{1}{q}})}{\log(1-\frac{q}{n(n-1)})}$ where N_I is the population size, n is the number of nodes, q is the number of links, and p the is probability that the optimal links occurs at least once in the population.

The choice of an appropriate value for q is based on the behavior of the topological optimization process, which can change the network number of links in order to reduce the network cost. We, therefore, set q as the minimum number of links that potentially can maintain the network 2–connectivity.

3.2   Tabu Search Algorithm

The second heuristic we propose relies on the application of the Tabu Search (TS) methodology. The TS algorithm can be seen as an evolution of the classical local optimal solution search algorithm called Steepest Descent. However, thanks to the interior mechanism that provides to accept also worse solutions than the best solution found so far, it is not subject to local minima entrapments. TS is based on a partial exploration of the space of admissible solutions, finalized to the discovery of a good solution. The exploration starts from an initial solution that is generally obtained with a greedy algorithm and when a stop criterion is satisfied (e.g., a number maximum of iterations N_T), the algorithm returns the best visited solution. During the search in the space of the admissible solutions, it is possible the utilization of some exploration criterion called *intensification* and *diversification* criterion. The first one it used when a very careful exploration of a small part of the space solution is required, while the second one is used to jump into another place of the space solution to better cover that.

For each admissible solution, a class of neighbor solutions is defined. A neighbor solution, is defined as a solution that can be obtained from the current solution by applying an appropriate transformation of that called *move*. The set of all the admissible moves uniquely defines the *neighborhood* of each solution. The dimension of the neighborhood is N_N. In this paper a simple move is considered, it either removes an existing link or adds a new link between network nodes.

At each iteration, all solutions in the neighborhood of the current one are evaluated, and the best is selected as the new current solution. A special rule, the *tabu list*, is introduced in order to prevent the algorithm to deterministically cycle among already visited solutions. Tabu list tracks the last accepted moves so that a stored move in it cannot be used to generate a new move for a duration of a certain number of iterations. Therefore it may happen that TS continuing the search will select an inferior solution because better solutions are tabu. The choice of the tabu list size is important: a small size could cause the cyclic visitation of the same solutions while a big one could block for many iterations the optimization process, avoiding a good visit of the space solution.

3.3   Time Complexity

We now discuss the complexity of the heuristics that were described before. The GA time complexity, considering a simple GA, is $O(N_G.[N_I.[\psi_{ps} + \psi_{go} + \psi_{of}] + \psi_{rp}])$. Where ψ_{ps}, ψ_{go}, ψ_{of} are time complexities related to the following GA tasks: parent selection, genetic operators, and individual (objective function) evaluation, respectively. These tasks are done for each of the N_I individuals. Afterwords the replacement, with complexity ψ_{rp}, is required. Finally, all these operations is made for each of the N_G generations.

The complexity of the tabu search is evaluated as follows. At each iteration the neighborhood generation and evaluation of N_N solutions are necessary; this requires $O(\psi_{ng}+N_N.\psi_{of})$ operations, since the generation of the neighborhood has complexity ψ_{ng}, and the evaluation of each solution has complexity ψ_{of}. If the number of iterations is N_T, the resulting complexity is $O(N_T.[\psi_{ng} + N_N.\psi_{of}])$.

4   Selected Results

In this section we present some selected numerical results considering network designs obtained with the metaheuristic approaches. We consider the same mixed traffic scenario where the file size follows the distribution shown in [2].

The first set of experiments is aimed at investigating the performance of GA and TS algorithms. As a first example, we applied the proposed methodology to set up a 10–node VPN network over a given 25–node, 140–link physical topology. The target QoS constraints for all traffic pairs are: i) file latency $L_t \leq$ 1s for TCP flows shorter than 20 segments, ii) throughput $T_h \geq$ 512 Kbps for TCP flows longer than 20 segments. Selecting $P_{loss} = 0.01$, we obtain a network-level design constraint equal to $RTT \leq$ 0.15s for all source-destination pairs. Each traffic relation offers an average aggregate traffic equal to $\gamma = 2$ Mbps. Link lengths vary between 25 Km and 760 Km.

In left plot of Fig. 3 we show the network cost versus computational time, in seconds, considering both the GA and TS algorithms (for different values of population and tabu list sizes). In order to improve further the performance of the TS algorithm we used a diversification criterion which corresponds to restart the procedure, with a new random initial solution. As we can see, in this case, the final solutions differ by at most 2%; but the GA can reach better solutions quicker than the TS. At each iteration the TS evaluates $N_N = 140$ solutions (recall that a move corresponds to add or remove a link) and the GA evaluates "only" $N_I = 50$ (or 100) solutions. Obviously, the complexity of each evaluation task (that is considered dominant regarding the other algorithm tasks) is the same for both GA and TS. Thus, in the studied cases, the GA will be faster than TS. As expected, for $N_I = 50$ individuals the GA converges faster than runs with $N_I = 100$ individuals, being the solution (after 30 minutes) almost the same.

In the second example, we set up a 15–node VPN network over a given 35–node, 140–link physical topology. The target QoS constraints for all traffic pairs are the same for the first case; and the average aggregate traffic is equal to $\gamma = 3$ Mbps. Link lengths vary between 15 Km and 300 Km (average = 225 Km). In right plot of Fig. 3 we report the network cost versus computational time. We notice that, after a period of 1.38 hours, the solution values differ by at most 2.5%. The best solution is given by the GA with $N_I = 160$ individuals (in contrast, the same solution value was reached by the TS, with $TL = 20$ moves, after a period of 12 hours).

The GA with a small population quickly stagnates, and its solution value is "relatively poor". Using a population size near the value suggested by the estimate ($N_I = 160$) the GA make slower progress but consistently discover better solutions. If the population size increases further, the GA becomes more slowly, without however, being able to improve the solutions (not show in the figure).

Coming back to the 10–node VPN network case, we also compare the network cost to those of a dimensioning where the 2–connectivity constraint is relaxed, i.e., in this case, the failure of a node can interrupt the information exchange for some origin/destination pairs of the network. Fig. 4 presents the network costs, from the GA solutions, for different values of the average traffic. The plots show that 2–connectivity networks have an extra cost of about 75%.

In addition, by considering physical topologies with different number of links, we can analyze the behavior of network costs and computational times. In this case we

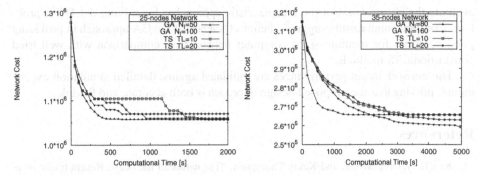

Fig. 3. Network Cost versus GA and TS Computational Time: 25–node network (left plot) and 35–node network (right plot).

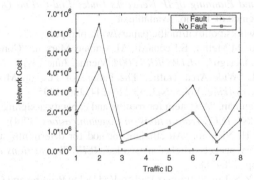

Fig. 4. Network Costs for Different Traffics Scenarios (10–node VPN network).

considered initially the case of 150 links, and further increased the number of links randomly (until an all-connected case). Initial solutions to the GA are obtained by considering 100 random topologies for each case (number of links).

5   Conclusion

In this paper, we have considered the QoS and reliability design of packet networks, and in particular the joint Topology, Capacity and Flow Assignment problem where the link placements, routing assignments and capacities are considered to be decisions variables. By explicitly considering TCP traffic, we also need to consider the impact of finite buffers, therefore facing the Buffer Assignment problem.

Our solution approach is based on the exploration of the solution space using metaheuristic algorithms (GA and TS), where candidate solutions are evaluated by solving CFA problems. In order to obtain a practical, useful TCFA solver, it is fundamental a fast CFA problem solver. For this scope, we have proposed a new CFA algorithm, called GWFD, that is capable of assign flow and capacities under e2e QoS constraints.

Examples of application of the proposed design methodology to different networking configurations have been discussed; also experiments were performed in order to evaluate the effectiveness of the metaheuristic approaches for solving the TCFA problem. Computational results suggest a better efficiency of the GA approach in providing good solutions for medium-sized computer networks, in comparison with well tried conventional TS methods.

The network target performances are validated against detailed simulation experiments, proving that the proposed design approach is both accurate and flexible.

References

1. K. Claffy, Greg Miller, and Kevin Thompson, "The nature of the beast: Recent traffic measurements from an Internet backbone". In *Proceedings of INET '98*, July 1998.
2. M. Mellia, A. Carpani, R. Lo Cigno, "Measuring IP and TCP behavior on Edge Nodes", *IEEE Globecom 2002*, Taipei, TW, Nov. 2002.
3. E. Wille, *Design and Planning of IP Networks Under End-to-End QoS Constraints*, PhD dissertation, Politecnico di Torino, Available at
   http://www.tlc-networks.polito.it/mellia/papers/wille_phd.pdf
4. E.Wille, M.Garetto, M.Mellia, E.Leonardi, M.Ajmone Marsan, "Considering End-to-End QoS in IP Network Design", *NETWORKS 2004*, Vienna, June 13-16
5. V.Paxson, S.Floyd, "Wide-Area Traffic: The Failure of Poisson Modeling," *IEEE/ACM Transactions on Networking*, Vol.3, N.3, pp.226–244, Jun. 1995.
6. B. Gavish and I. Neuman. "A system for routing and capacity assignment in computer communication networks". *IEEE Transactions on Communications*, 37(4):360–366, April 1989.
7. K. Kamimura and H. Nishino. "An efficient method for determining economical configurations of elementary packet-switched networks". *IEEE Transactions on Communications*, 39(2), Feb. 1991, pp. 278–288.
8. K. T. Cheng and F. Y. S. Lin, "Minimax End-to-End Delay Routing and Capacity Assignment for Virtual Circuit Networks", *Proc. IEEE Globecom*, pp. 2134-2138, 1995.
9. M.Garetto, D.Towsley, "Modeling, Simulation and Measurements of Queuing Delay under Long-tail Internet Traffic", *ACM SIGMETRICS 2003*, San Diego, CA, June 2003.
10. D. E. Goldberg, *Genetic Algorithm in Search, Optimization, and Machine Learning*, Addison Wesley Publishing Company, 1989.
11. Fred Glover and Manuel Laguna, *Tabu Search*, Kluwer Academic Publishers, 1997.
12. B. Gavish. "Topological design of computer communication networks – the overall design problem". *European Journal of Operational Research*, 58, (1992) 149–172.
13. T. T. Mai Hoang and W. Zorn, "Genetic Algorithms for Capacity Planning of IP-Based Networks", *Proceedings of the 2001 Congress on Evolutionary Computation CEC2001*, 1309–1315, 2001.
14. Gerla, M., L. Kleinrock, "On the Topological Design of Distributed Computer Networks", *IEEE Transactions on Communications*, Vol.25, pp.48-60, Jan. 1977.
15. Wright, M.; "Interior methods for constrained optimization", *Acta Numerica*, Vol.1, pp. 341–407, 1992.
16. A.Medina, A.Lakhina, I.Matta, J.Byers, "BRITE: Boston university representative internet topology generator", Boston University, http://cswww.bu.edu/brite, April 2001.

Replicated Server Placement
with QoS Constraints

Georgios Rodolakis[1], Stavroula Siachalou[2], and Leonidas Georgiadis[2]

[1] Hipercon Project, INRIA Rocquencourt, 78153 Le Chesnay Cedex, France
Georges.Rodolakis@inria.fr
[2] Aristotle Univ. of Thessaloniki, Faculty of Engineering,
School of Electrical and Computer Engineering, Telecommunications Dept.
Thessaloniki, 54124, Greece
{ssiachal,leonid}@auth.gr

Abstract. The problem of placing replicated servers with QoS constraints is considered. Each server site may consist of multiple server types with varying capacities and each site can be placed in any location among those belonging to a given set. Each client can de served by more than one locations as long as the request round-trip delay satisfies predetermined upper bounds. Our main focus is to minimize the cost of using the servers and utilizing the link bandwidth, while serving requests according to their delay constraint. This is an NP-hard problem. A pseudopolynomial and a polynomial algorithm that provide guaranteed approximation factors with respect to the optimal for the problem at hand are presented.

1 Introduction

Communications networks have become a widely accepted medium for distributing data and all kinds of services. In networks such as the Internet, the world's largest computer network, users and providers demand response times that satisfy QoS requirement and optimization of the system performance. This is achieved by providing access response times that satisfy QoS requirements and by minizing the operating cost of the network, that is the cost of utilizing the link bandwidth and placing the servers. Improvement of system service performance can by achieved by placing replicated servers at appropriately selected locations.

The problem of maximizing the performance and minimizing the cost of a computing system has been addressed in the past in several papers. Krishnan et al [1] developed polynomial optimal solutions to place a given number of servers in a tree network to minimize the average retrieval cost of all clients. Li et al [2] investigated the placement of a limited number of Web proxies in a tree so that the overall latency for accessing the Web server is minimized. In [3] two objectives were studied: minimization of the overall access cost by all clients to access the Web site and minimization of the longest delay for any client to access the Web site. The problem was reduced to the placement of proxies in a set of trees whose root nodes are the server replicas. Qiu et al [4] also assumed a

M. Ajmone Marsan et al. (Eds.): QoS-IP 2005, LNCS 3375, pp. 207–220, 2005.

restricted number of replicas and no restriction of the number of requests served by each replica. A client could be served by a single replica and the cost for placing a replica was also ignored. The objective was to minimize the total cost for all clients to access the server replicas, while the cost of a request was defined as the delay, hop count or the economic cost of the path between two nodes. They compared several heuristic solutions and found that a greedy algorithm had the best performance. Chen et al [5] tackled the replica placement problem from an other angle: minimizing the number of replicas when meeting clients' latency constraints and servers' capacity constraints by using a dissemination tree. In [6] the authors considered the problem of placing a set of mirrors only at certain locations such that the maximum distance from a client to its closest mirror (server replica), based on round trip time, is minimized. They assumed no cost for placing a mirror and showed that placing mirrors beyond a certain number offered little performance gain. Sayal et al presented some selection algorithms to access replicated Web servers in [7]. The algorithms found the closest replicated server for a client based on different metrics such as hop counts, round trip time and the HTTP request latency. In [8] the objective was to minimize the amount of resources, storage and update, required to achieve a certain level of service. They assumed that all servers in the network are organized into a tree structure rooted at the origin server.

In this paper we approach the problem of replicated server placement with QoS constraints from a system administrator's perspective. Thus we are interested in serving the users' requests so that their delay requirements are satisfied, while at the same time minimizing the total cost of placing the servers and using the link bandwidth. We consider that each server site may be comprised of a number of different server types with varying processing capacities. We also assume that a server site may by placed on any of a given set of locations and each client can be served by more than one location. Our objective is to select optimally the locations where the servers must be placed, the types that comprise each server, the proportion of traffic that must be routed by each client to each of the servers and the routes that will be followed by the requests issued by a client to a server. The problem is NP-hard and therefore an optimal solution is not likely to be found. We present a pseudopolynomial approximation algorithm and a polynomial time algorithm that provide guaranteed approximation factors with respect to the optimal for the problem at hand.

Due to space limitation proofs are omitted. For a complete version of the paper the interested reader is referred to the site http://genesis.ee.auth.gr/georgiadis/english/public/QoS_IP2005.pdf.

2   Problem Formulation

Let $G(V, E)$ represent a network with node set V, and link set E. Let also H be a subset of V. We are interested in placing servers at some of the nodes in H, that will serve requests originated by any of the nodes in V. We assume that the servers contain the same information and hence any node may obtain the requested information by accessing any of the servers.

With link (i, j) there is an associated delay d_{ij}. Requests should be obtained in a timely fashion, and hence there is a bound D on the time interval between the issuing of the request and the reception of the reply. We refer to this bound as the "round-trip delay bound". Note that the processing time of a request at the server can be incorporated in this model by replacing D with $D + d_p$, where d_p is an upper bound on the request processing time at the server.

The load in requests per unit of time originated by node $i \in V$ is g_i. To transfer an amount of x requests per second, it is required to reserve bandwidth αx on the links traversed by the requests. The transfer of server replies corresponding to the x requests back to the requesting node, requires the reservation of βx units of bandwidth on the links traversed by the replies.

The cost of transferring 1 unit of bandwidth on link (i, j) is e_{ij}. Hence the cost of transferring x requests per second on link (i, j) is $\alpha e_{ij} x$ while the cost of transferring the replies to these requests is $\beta e_{ij} x$. A node i can split its load g_i to a number of servers and routes as long as the delay bound D between the issuing of the request and the reception of the reply is satisfied. At each node $j \in H$ there is a set S_j of server types that can be selected. Server type s, $1 \leq s \leq K_j$, ($K_j = |S_j|$) costs f_j^s units and can process up to U_j^s requests per second.

Our objective is to determine,

1. the locations (subset of the nodes in H) where the servers will be placed,
2. the amount of traffic (in requests per unit of time) that will be routed by each node to each of the selected locations,
3. the routes through which the node traffic will be routed to each location,
4. the number and type of servers that should be opened at each location,

so that,

1. the round-trip delay bound for each request is satisfied,
2. the total cost of using the servers and utilizing the bandwidth is minimized.

Notice that in the current setup we do not consider link capacities. In effect we assume that the network links have enough bandwidth to carry the requested load by the network nodes. This is a reasonable assumption in an environment where the server requests are a small portion of the total amount of information carried by the network. The general problem where link capacities are also included, is a subject of further research.

2.1 Optimization Problem Formulation

A feasible solution to the problem consists of the following:

- A set of locations $F \subseteq H$ where the servers will be placed.
- A subset of server types $G_j \subseteq S_j$ that should be opened at location $j \in F$.
- The number n_j^s of server types $s \in G_j$ to be opened at location $j \in F$.
- A set of round-trip routes R_{ij} between node $i \in V$ and facility $j \in F$. A round-trip route, denoted $r_{ij} = (p_{r_{ij}}, q_{r_{ij}})$, consists of two simple paths, $p_{r_{ij}}$ and $q_{r_{ij}}$, used for transferring requests and replies respectively.
- The amount of requests per unit of time, $x_{r_{ij}}$, accommodated on route r_{ij}.

The constraints of the problem are the following:

- The request load of each node must be satisfied: $\sum_{j \in F} \sum_{r \in R_{ij}} x_r = g_i$, $i \in V$.
- The round-trip delay of each round-trip route should be at most D. That is, $\sum_{l \in p} d_l + \sum_{l \in q} d_l \leq D$, for $r = (p, q) \in R$, where R is the set of all round-trip routes, and the summation is over all links of the corresponding paths.
- The total server capacity at server location $j \in H$ should be at least as large as the request rate arriving at location j. That is,

$$\sum_{i \in V} \sum_{r \in R_{ij}} \sum_{s \in G_j} n_j^s U_j^s, \ j \in H. \tag{1}$$

The objective cost function to be minimized is

$$\sum_{j \in F} \sum_{s \in G_j} n_j^s f_j^s + \sum_{l \in E} e_l (\alpha \sum_{\substack{r=(p,q) \in R \\ l \in p}} x_r + \beta \sum_{\substack{r=(p,q) \in R \\ l \in q}} x_r). \tag{2}$$

The first term in (2) corresponds to the cost of opening the servers, while the second term corresponds to the cost of reserving bandwidth (for transmitting requests and replies) on the network links in order to satisfy the node requests.

For the rest of the paper we assume that the node loads g_i, $i \in V$ are nonnegative integers and that splitting of these loads to a number of server locations may occur in integer units. In practice this is not a major restriction, since usually the load is measured in multiples of a basic unit.

2.2 Problem Decomposition

In this section we decompose the problem defined in Sect. 2.1 into three independent subproblems. As will be seen all three problems are NP-hard.

For a round-trip route $r = (p, q)$, define the cost $C_r = \alpha \sum_{l \in p} e_l + \beta \sum_{l \in q} e_l$. Let r_{ij}^* be a minimum-cost round-trip route between node i and server location j, satisfying the round-trip delay D. It can be easily seen that it suffices to restrict attention to solutions that assign all the load from node i to server location j on r_{ij}^*. Hence, setting $c_{ij} = C_{r_{ij}^*}$, the second term in (2) becomes $\sum_{i \in V} \sum_{j \in F} c_{ij} x_{ij}$.

Consider now the first term in (2). Let $f_j(y)$ be the minimum cost server type assignment at location j, under the assumption that the request load at that location is y. By definition, the feasible solution that assigns this minimum cost server assignment at location j for request load $y_j = \sum_{i \in V} \sum_{r \in R_{ij}} x_r$, is at least as good as a solution inducing the same load at location j. Hence, we may replace this term with $\sum_{j \in F} f_j(y_j)$.

For our purposes, it is important to observe that the function $f_j(y)$ defined in the previous paragraph is subadditive, i.e., it satisfies the inequality

$$f_j(y_1) + f_j(y_2) \geq f_j(y_1 + y_2) \text{ for all } y_1 \geq 0, \ y_2 \geq 0. \tag{3}$$

To see this, note that if $S(y_1)$, $S(y_2)$ are the sets of servers achieving the optimal costs $f_j(y_1)$, $f_j(y_2)$ respectively, then the set of servers $S(y_1) \cup S(y_2)$ provides a feasible solution for request load $y_1 + y_2$, with cost $f_j(y_1) + f_j(y_2)$. Since $f_j(y_1 + y_2)$ is by definition the minimum cost server assignment with request load $y_1 + y_2$, (3) follows.

From the above it follows that we need to solve the following problems.

Problem 1. Given a graph, find a round-trip route with minimum cost, satisfying the round-trip delay bound for any node $i \in V$ and server location $j \in H$. This determines c_{ij}, $i \in V$, $j \in H$.

Problem 2. Given a set of server types S_j and a required load y at node $j \in H$, find the optimal selection of server types and the number of servers of each type so that the load is accommodated. That is, determine $n_j^s(y)$ so that $\sum_{s \in G_j} n_j^s(y) U_j^s \geq y$ and $f_j(y) = \sum_{s \in G_j} n_j^s(y) f_j^s$ is minimal.

Problem 3. Given non-decreasing subadditive functions $f_j(y)$, costs c_{ij}, integer node loads $g(i) \geq 0$, $i \in V$, solve

$$\min \sum_{j \in H} f_j(y) + \sum_{i \in V} \sum_{j \in H} c_{ij} x_{ij}$$
$$\text{subject to: } \sum_{j \in H} x_{ij} = g(i), \ i \in V, \ \sum_{i \in V} x_{ij} = y, \ j \in H, \ x_{ij} \geq 0.$$

The decision problems associated to Problems 1 and 2 are NP-hard. Indeed, when $\beta = 0$, the associated decision problem to Problem 1 is reduced to the Shortest Weight-Constrained Path problem which is known to be NP-hard [9]. Also, the associated decision problem to Problem 2 amounts to the Unbounded Knapsack Problem which is NP-hard [10]. However for both problems pseudopolynomial algorithms exist (see Sect. 3) and, as will be discussed in Sect. 4, fully polynomial time approximation algorithms can be developed. Regarding Problem 3, there is an extensive work in the literature under various assumptions on the function $f_j(y)$ and on the costs c_{ij} (see [11, 12]). Most of the work is concentrated on the case of "metric" costs, i.e., it is assumed that costs satisfy the inequality $c_{ij} + c_{jk} \geq c_{ik}$. However, this inequality does not hold in our case.

In the next section, by combining algorithms for the three problems discussed above, we provide a pseudopolynomial time approximation algorithm for the problem addressed in this paper. The algorithm for Problem 3 is based on the algorithm proposed in [12] and uses the fact that $f_j(y)$ is subadditive step function. In Sect. 4 we modify the algorithm in order to obtain a polynomial time algorithm with approximation factor close to the best possible (unless NP$\subseteq$DTIME($n^{O(\log \log n)}$)). The performance of the algorithm in simulated networks is studied in Sect. 5.

3   Pseudopolynomial Algorithm

In this section we discuss pseudopolynomial algorithms for Problems 1, 2 and 3, which combined give a pseudopolynomial algorithm for the problem at hand.

3.1   Pseudopolynomial Algorithm for Problem 1

Let $F_{ij}(d)$ be the minimum cost path from node i to j with delay at most d, and $B_{ij}(d)$ the minimum cost path from node j to i with delay at most d. Then it can be easily seen that, $c_{ij} = \min_{0 \leq d \leq D} \{F_{ij}(d) + B_{ij}(D - d)\}$. Hence c_{ij} can be determined provided $F_{ij}(d)$ and $B_{ij}(d)$ are known. There are fully polynomial time algorithms for computing these quantities, however in this section we will concentrate on efficient pseudopolynomial algorithms that work well in practice [13, 14]. We provide the discussion for $F_{ij}(d)$, since the same holds for $B_{ij}(d)$. The algorithms in [13, 14] are based on the fact that $F_{ij}(d)$ is a right continuous non-increasing step function with a finite number of jumps. Hence, in order to compute $F_{ij}(d)$ one needs only to compute its jumps, which in several practical networks are not many. Another useful feature of these algorithms is that in one run they compute $F_{ij}(d)$ from a given node $j \in H$ to all other nodes in V. The jumps of $F_{ij}(d)$ and $B_{ij}(d)$ can be computed using the algorithm in [13] in $O\left(|V| D\left(|V| \log|V| + |E| \log|V|\right)\right)$ time. Thus a pseudopolynomial algorithm for Problem 1 can be developed with the same worst case running time.

3.2   Pseudopolynomial Algorithm for Problem 2

We restate Problem 2 in its generic form, to simplify notation.

Problem 2. Given a set of server types S, server capacities U^s, server costs $f^s > 0$ and a required load y, find the optimal selection of server types G and the number of servers of each type so that the load is satisfied. That is, determine $n^s(y)$ so that $\sum_{s \in G} n^s(y) U^s \geq y$ and $f(y) = \sum_{s \in G} n^s(y) f^s$ is minimal.

Problem 2 is similar to the Unbounded Knapsack Problem (UKP) [10]. The difference is that in UKP the inequality constraint is reversed and maximization of the cost $\sum_{s \in G_j} n^s(y) f^s$ is sought. A pseudopolynomial algorithm for Problem 2 can be developed in a manner analogous to the one used for UKP. Specifically, number the servers from 1 to $|S|$ and define $A(f, i)$ to be the largest load achievable by using some among the first i servers so that their total cost is exactly f. The entries of the table $A(f, i)$ can be computed in order of increasing i and f using the dynamic programming equation

$$A(f, i + 1) = \min\{A(f, i), U^{i+1} + A(f - f^{i+1}, i + 1)\}, \tag{4}$$

with $A(f, 0) = 0$ for all f, $A(f, i) = -\infty$ if $f < 0$, and $A(0, i) = 0$ for all $0 \leq i \leq K$. The optimal server selection cost is then determined as $f(y) = \min\{f \mid A(f, K) \geq y\}$. By keeping appropriate structures one can also determine the server types and the number of servers of each type in the optimal solution.

Function $f(y)$ is a right continuous non-decreasing step function. Based on (4) and using an approach similar to [15], an efficient pseudopolynomial algorithm can be developed for finding the jump points of $f(y)$. Again, in one run, all jump points of $f(y)$ up to an upper bound can be determined. The running time of this approach is bounded by $O\left(|S| y\right)$, where $|S|$ is the number of server types.

3.3   Pseudopolynomial Algorithm for Problem 3

In [12] a polynomial time algorithm is provided for Problem 3 in the case of concave facility cost functions. It is assumed that the cost $f_j(y)$ of placing servers at node $j \in H$ to accommodate load y can be computed at unit cost and that all nodes have unit loads. It is shown that the proposed algorithm achieves an approximation factor of $\ln|V|$ compared to the optimal. In our case we have arbitrary integer node loads g_i and the functions $f_j(y)$ are subadditive and can be computed exactly only in pseudopolynomial time. As observed in [11] the assumption of unit loads can be removed by considering a modified network where node i is replaced with g_i nodes each having the same links (and link costs). However, now the algorithm becomes pseudopolynomial (even assuming unit costs for computing $f_j(y)$) since the number of nodes in the modified network can be as large as $|V g_{max}|$, where $g_{max} = \max_{i \in V}\{g_i\}$.

The approximability proof for general costs c_{ij} in [12] carries over without modification if $f_j(y)$ are subadditive rather than concave functions. Hence the approximability factor in our case becomes $\ln|V g_{max}|$. Moreover, the worst case running time of the algorithm in [12] is among the best of the proposed algorithms. Hence, we will use this algorithm as the basis for our development. We present it below (Algorithm 1) adapted to our situation. For the moment we assume that $f_j(y)$ can be computed exactly at unit cost.

Algorithm 1 Generic algorithm for solving Problem 3.

Input: Graph G, the array c with the costs of the routes and the Knapsack list
Output: Locations and types of servers, load assigned from each node to the locations.

1. For $j \in H$ set $load_j = 0$;
2. For $i \in V$ set $\psi(i, j) = 0$;
3. While there is an unassigned node do
4.   For $j \in H$ do
5. $t(j) = \min_k \dfrac{f_j(load_j + k) - f_j(load_j) + \sum_{s=1}^{n_j(k)} r(i_s)c_{i_s j} + l_j(k)c_{i_{n_j(k)+1}j}}{k}$;
6. $k(j) = \arg\min_k \dfrac{f_j(load_j + k) - f_j(load_j) + \sum_{s=1}^{n_j(k)} r(i_s)c_{i_s j} + l_j(k)c_{i_{n_j(k)+1}j}}{k}$;
7.   Let $j^* = \arg\min_{j \in H}\{t(j)\}$; Set $load_{j^*} \leftarrow load_{j^*} + k(j^*)$;
8.   For $1 \leq s \leq n_{j^*}(k^*)$ do
9. $\psi(i_s, j^*) \leftarrow \psi(i_s, j^*) + r(i_s)$; $r(i_s) = 0$;
10. $\psi(i_{n_{j^*}(k^*)+1}, j^*) = l_{j^*}(k^*)$; $r(i_{n_{j^*}(k^*)+1}) \leftarrow r(i+1) - l_{j^*}(k(j^*))$;

The algorithm performs a number of iterations. At each iteration a node j^* in H is selected and the load of some of the nodes in V is assigned to j^*. Let matrix $\psi(i, j)$ represent the total load from node i assigned to server location j at the beginning of an iteration (i.e., the beginning of the while loop at Step 3). Hence the load of node i remaining to be assigned is $r(i) = g(i) - \sum_{j \in H} \psi(i, j)$.

A node such that $r(i) > 0$ is called unassigned. For server location $j \in H$ consider the unassigned nodes arranged in non-decreasing order of their costs $c_{i_s j}$, i.e., $c_{i_1 j} \leq c_{i_2 j} \leq \ldots \leq c_{i_m j}$. Let $R_j(n) = \sum_{s=1}^{n} r(i_s)$, $1 \leq n \leq m$, and $n_j(k) = \max\{n : R_j(n) \leq k\}$. Define also $l_j(k) = k - R_j(n_j(k))$.

The variable $load_j$ holds the total load assigned to node $j \in H$ at the beginning of an iteration. In step 5, the most economical (cost per unit of assigned load) load assignment for each of the server locations is computed. In steps 7, the server location with the minimum economical assignment is selected and the associated load is placed on this location. In steps 8 to 10 the remaining loads of the nodes that will place their load on the selected location are updated.

The average running time of this algorithm can be improved by taking advantage of the fact that $f_j(y)$ is a step function. Specifically, in order to compute the minimum in step 5, one needs to do the computation only for values of k such that $load_j + k$ is a jump point of $f_j(y)$, or $k = R_j(n)$ for some n. The running time of Algorithm 1 is $O(|V|^3 g_{max}^2)$.

Letting $|S|$ be the maximum number of server types in any of the server locations, taking into account that the maximum load on any facility is $|V| g_{max}$ and that we may need to compute at most $|V|$ functions $f_j(y)$, we conclude that the worst case computation time of the complete algorithm is

$$O(|V| D (|V| \log |V| + |E| \log |V|) + |S| |V|^2 g_{max} + |V|^3 g_{max}^2).$$

The algorithm presented above works well in practice as will be seen in Sect. 5. However, it is pseudopolynomial and it is theoretically important to know whether there exists a polynomial time algorithm that can provide a guaranteed approximation factor with respect to the optimal. In the next section we will show that this can be done based on the algorithm presented above.

4   Polynomial Algorithm

In this section, by generalizing the approach in [12] we provide a polynomial time approximation algorithm for arbitrary integer node loads and non-decreasing subadditive functions that are not necessarily computable exactly in polynomial time. Note that a concave function is also subadditive and hence our results carry over to concave functions. However, as will be seen, for concave functions the approximation constants can be made smaller.

The approach we follow is to provide polynomial time approximation algorithms for each of Problems 1, 2 and 3. By combining these algorithms, we get a polynomial time algorithm for the problem at hand with guaranteed performance factor compared to the optimal.

In the previous section the costs c_{ij} and the functions $f_j(y)$ were computed exactly using pseudopolynomial algorithms for Problems 1 and 2 respectively. The use of polynomial time approximation algorithms for these problems provides only approximate values for c_{ij} and $f_j(y)$. That is, we can only ensure that for any $\varepsilon > 0$, we provide in polynomial time values $\overline{c}_{ij}$ and $\overline{f}_j(y)$ (for fixed y) such that $c_{ij} \leq \overline{c}_{ij} \leq (1+\varepsilon)c_{ij}$ and $f_j(y) \leq \overline{f}_j(y) \leq (1+\varepsilon)f_j(y), y \geq 0$. Replacing c_{ij} and $f_j(y)$ with $\overline{c}_{ij}$ and $\overline{f}_j(y)$ in Problem 3 and providing an α-approximate solution for the resulting instance, provides also an $(1 + \varepsilon)\alpha$- approximate solution for the original problem. This important observation was used in [11] and we present it here in the next lemma.

Lemma 1. *Consider the problems*

$$\min_{x \in A} g(\mathbf{x}), A \subseteq R^n, \tag{5}$$

$$\min_{x \in A} \overline{g}(\mathbf{x}), A \subseteq R^n, \tag{6}$$

where $g(x) \geq 0$. If for $x \in A$, $g(x) \leq \overline{g}(x) \leq \beta g(x)$, then an $\alpha-$approximate solution for problem (6), $\alpha \geq 1$, is a $\beta\alpha$-approximate solution for (5).

Using Lemma 1 we can proceed as follows.

- Compute in polynomial time approximate values $\overline{c}_{ij}$,
- Compute in polynomial time approximate values $\overline{f}(y)$ (for a given $y \geq 0$),
- Provide an approximation algorithm for Problem 3, based on Algorithm 1, using the approximate values $\overline{c}_{ij}$, $\overline{f}(y)$.

Difficulties arise in the approach outlined above for the following reasons. First, to compute the minimum in Step 5 of Algorithm 1, $\overline{f}(y)$ must be computed for all values of y in the worst case, and the number of these computations is bounded by $\sum_{i \in V} g_i$, i.e., it is not polynomial in the input size, even if $\overline{f}(y)$ is computable at unit cost. Second, the load assigned to a node in H at each iteration of the while loop at step 9 can be 1 in the worst case and hence the number of iterations of the while loop may be again $\sum_{i \in V} g_i$ in the worst case. Third, while $f(y)$ is subadditive, it cannot be guaranteed that $\overline{f}(y)$ is subadditive as well, and hence the approximation factors cannot be guaranteed a priori. Hence, the straightforward application of Algorithm 1 will result in pseudopolynomial running time and will not provide guaranteed performance bounds. However, we can modify the approach so that the resulting algorithm runs in polynomial time at the cost of a small increase in the approximation factor.

4.1 Polynomial Algorithms for Problem 1 and 2

A fully polynomial time approximation algorithm for the problem of finding the minimum constrained path from a source to a given destination was developed in [16]. The algorithm can be adapted for the development of a fully polynomial time approximation algorithm for Problem 1. The resulting algorithm runs in $t = O(|E||V|(\log \log |V| + 1/\varepsilon))$ for each route.

A fully polynomial time approximation algorithm for Problem 2 can be developed by paralleling the approach for the UKP [10, Section 8.5]. The resulting algorithm has a worst case running time of $T = O(\frac{1}{\varepsilon^2}|S|\log|S|)$, where $|S|$ is the number of server types.

4.2 Polynomial Algorithm for Problem 3

We now address the main problem of Sect. 4, i.e., the development of a polynomial algorithm for Problem 3, using the approximate costs $\overline{c}_{ij}$ and $\overline{f}_j(y)$. As explained above, we intend to use Algorithm 2 as the basis for the development. Assume for the moment that c_{ij} and $f_j(y)$ are computable exactly. As was mentioned above the fact that the node loads are general nonnegative integers in our case, renders the algorithm pseudopolynomial even under this assumption. However, if the functions $f_j(y)$ are concave, then the algorithm becomes polynomial. This is due to the fact that for concave functions Algorithm 1 can assign all the load of each node to a single server. This is shown in the next lemma. Recall that a function $f(y)$ defined for integer y is called concave if for all y in its domain of definition it holds, $f(y+1) - f(y) \leq f(y) - f(y-1)$.

Lemma 2. *If the functions $f_i(y)$ are concave and Algorithm 1 is applied, then the load of each node in V can be assigned to a single server.*

Lemma 2 implies that when c_{ij} and $f_j(y)$ are computable in polynomial time and $f_j(y)$ are concave, the algorithm runs in polynomial time, and the approximation factor is $\log(|V| g_{max})$, where $g_{max} = \max_{i \in V} \{g_i\}$.

We now return to the problem at hand. In our case, $f_j(y)$ is subadditive rather than concave and is not computable exactly in polynomial time. Hence the results above cannot be applied directly to obtain a polynomial time algorithm. The approach we follow is to construct in polynomial time a concave function $\widetilde{f}_j(y)$, such that for any y in its domain of definition $\widetilde{f}_j(y)$ is computed in polynomial time and, $f_j(y) \leq \widetilde{f}_j(y) \leq \alpha f_j(y)$. Then, by applying Lemmas 1 and 2 we get a polynomial time algorithm. To proceed we need some definitions.

Consider a nonnegative function $\phi : \{0, 1, \ldots, W\} \rightarrow \mathbb{Q}^+$ ($\mathbb{Q}^+$ is the set of nonnegative rationals) and let A be the convex hull of the set of points $S = \{(y, \phi(y)), y = 0, 1, \ldots, W\} \cup \{(0, 0), (W, 0)\}$. Recall that the convex hull of a set of points S is the smallest convex set that includes these points. In two dimensions it is a convex polygon. The vertices of the polygon correspond to a subset of S, of the form $S' = \{(y_k, \phi(y_k)), k = 1, .., K\} \cup \{(0, 0), (W, 0)\}$ where $y_k \in \{0, 1, \ldots, W\}$, $y_1 = 0$, $y_K = W$, and $y_k < y_{k+1}$ for all k, $1 \leq k \leq K - 1$.

Consider the piecewise linear function $\phi_2(y)$ with break points the set S', i.e., for $y_k \leq y < y_{k+1}$, $\phi_2(y)$ is defined as,

$$\phi_2(y) = \phi(y_k) + \frac{\phi(y_{k+1}) - \phi(y_k)}{y_{k+1} - y_k}(y - y_k), \tag{7}$$

The function $\phi_2(y)$ is concave. We call $\phi_2(y)$, the "upper hull" of $\phi(y)$. If $\phi(y)$ is non-decreasing, the same holds for $\phi_2(y)$. By construction it holds for all $y \in \{0, 1, \ldots, W\}$, $\phi(y) \leq \phi_2(y)$.

If the function $\phi(y)$ is subadditive and non-decreasing, then it also holds that its upper hull is smaller than twice the function value.

Lemma 3. *If a function $\phi : \{0, 1, \ldots, W\} \rightarrow \mathbb{Q}$ is subadditive and non-decreasing, then it holds for its upper hull $\phi_2(y)$, $\phi_2(y) \leq 2\phi(y)$.*

Consider now the subadditive function $f(y)$ of interest in our case (we drop the index j for simplicity). As a consequence of the approximate solution to Problem 2, for a given $\varepsilon > 0$ and a given $y \in \{0, 1, \ldots, W\}$, $W = |V| g_{max}$, we can construct in polynomial time a non-decreasing function such that

$$f(y) \leq \overline{f}(y) \leq (1 + \varepsilon) f(y). \tag{8}$$

Let $\overline{f}_2(y)$ be the upper hull of $\overline{f}(y)$. By (8), $\overline{f}_2(y)$ is smaller than or equal to the upper hull of $(1+\varepsilon)f(y)$, which in turn by Lemma 3 is smaller than $2(1+\varepsilon)f(y)$ (notice that $(1+\varepsilon)f(y)$ is subadditive). Hence we will have

$$f(y) \le \overline{f}_2(y) \le 2(1+\varepsilon)f(y). \tag{9}$$

Since $\overline{f}_2(y)$ is concave, if we replace c_{ij} with $\overline{c}_{ij}$ and $f(y)$ with $\overline{f}_2(y)$, we can provide an approximate solution to Problem 2 with approximation factor $\log(|V|\,g_{max})$. From (9) and Lemma 1 we will then have a solution to our original problem with approximation factor $2(1+\varepsilon)\log|Vg_{max}|$.

The problem that remains to be solved is the construction of the upper hull of $\overline{f}(y)$ in polynomial time. There are at most $W' = W + 2$ points in the set $\{(y, \overline{f}(y)),\ y = 0, 1, \ldots, W\} \cup \{(0,0), (W,0)\}$ and the upper hull of the points in this set can be constructed (i.e., its break points can be determined) in time $W' \log W'$ [17]. However, in our case $W = |V|\,g_{max}$ and hence the straightforward construction of the upper hull requires pseudopolynomial construction time.

To address the latter problem, we construct first a non-decreasing step function $\widehat{f}_1(y)$ with polynomial number of jump points (y is a jump point of $\widehat{f}_1(y)$ if $\widehat{f}_1(y-1) \neq \widehat{f}_1(y)$) that is a good approximation to $f(y)$, and then we construct the upper hull of $\widehat{f}_1(y)$. Since $\widehat{f}_1(y)$ has polynomial number of jump points its upper hull will also have polynomial number of break points and can be constructed in polynomial time.

We have by definition $\overline{f}(0) = 0$, $\overline{f}(1) = f(1) > 0$. Consider the sequence of integers $\widehat{f}_0 = 0$, $\widehat{f}_k$, $k = 1, \ldots, K$, generated by Algorithm 2.

Algorithm 2

Input: Algorithm for computing $\overline{f}(y)$, $\epsilon_1 > 0$.
Output: The sequence, $\widehat{f}_k$, $k = 0, 1, \ldots, K$.

1. $\widehat{f}_0 = 0$, $\widehat{f}_1 = \overline{f}(1)$, $y_1 = 1$, $k = 2$,
2. $\widehat{f}_k = (1 + \epsilon_1)\overline{f}(y_{k-1})$
3. If $\widehat{f}_k > \overline{f}(W)$, set $y_k = W$, $K = k$ and stop. Else,
4. Determine y_k such that $\overline{f}(y_k - 1) \le \widehat{f}_k \le \overline{f}(y_k)$, i.e.,
5. $k = k + 1$, go to step 2.

The sequence $\widehat{f}_k$, $k = 0, 1, \ldots, K$ can be used to construct a step function that is a good approximation to $\overline{f}(y)$. This is shown in the next lemma.

Lemma 4. *a) In Algorithm 2, $K = O\left(\frac{1}{\epsilon_1} \log \frac{\overline{f}(W)}{\widehat{f}_1}\right)$. The worst case running time of Algorithm 2 is $O\left(T \log(W) \frac{1}{\epsilon_1} \log \frac{\overline{f}(W)}{\widehat{f}_1}\right)$, where T is the worst case time (over all y, $1 \le y \le W$) needed to compute $\overline{f}(y)$.*
b) Consider the step function defined as follows: if $y_k \le y < y_{k+1}$ for some k, $1 \le k \le K - 1$, then $\widehat{f}(y) = \widehat{f}_k$, and $\widehat{f}(W) = \widehat{f}_K$. It holds,

$$\widehat{f}(y) \le \overline{f}(y) \le (1 + \epsilon_1)\widehat{f}(y) \tag{10}$$

Based on (10), we can use function $\widetilde{f}(y) = (1+\epsilon_1)\widehat{f}(y)$ to approximate $f(y)$.

Lemma 5. *Let $\epsilon_0 > 0$, $\epsilon_1 > 0$ be given. Let $f(y)$ be the optimal solution to Problem 2 and assume that we compute for a given y the approximate function $\overline{f}(y)$ so that $f(y) \le \overline{f}(y) \le (1+\epsilon_0) f(y)$.*
a) For the purposes of computing the step function $\widehat{f}(y)$ satisfying (10), $\overline{f}(y)$ may be assumed non-decreasing.
b) It holds $f(y) \le \widetilde{f}(y) \le (1 + (\epsilon_1 + \epsilon_0 + \epsilon_1\epsilon_0)) f(y)$,
c) The number of jump points of $\widehat{f}(y)$, hence of $\widetilde{f}(y)$, is $O\left(\frac{1}{\epsilon_1} \log\left(|V| g_{\max}\right)\right)$ and the running time of Algorithm 2 is $O\left(\frac{1}{\epsilon_1} T (\log\left(|V| g_{\max}\right))^2\right)$, where T is the worst case time (over all y, $1 \le y \le W$) needed to compute $\overline{f}(y)$.

From the discussion above we have polynomial time Algorithm 3 for computing the server locations. For simplicity, we pick a single ϵ for all the approximations. Algorithm 3 has a guaranteed performance ratio of $2(1+\epsilon)^2 \log\left(|V| g_{\max}\right)$ and its worst case running time is dominated by steps 1 and 2

$$O(|E||V|^3(\log\log|V| + 1/\epsilon) + |V||S|\log|S|(\log\left(|V| g_{\max}\right))^2/\epsilon). \qquad (11)$$

Algorithm 3 Polynomial Time Algorithm For Calculating Server Locations.

Input: Polynomial Algorithm for Problem 1, Algorithms 2 and 3, $\epsilon > 0$.
Output: Array of server locations.

1. For $i \in V$, $j \in H$, compute $\overline{c}_{ij}$ so that $c_{ij} \le \overline{c}_{ij} \le (1+\epsilon)c_{ij}$, $i \in V$, $j \in H$.
2. For $j \in H$, construct the step functions $\widehat{f}_j(y)$, from $\overline{f}_j(y)$ according to Algorithm 2, using as subroutine the algorithm for computing, for a given $y > 0$, $\overline{f}_j(y)$ such that $f_j(y) \le \overline{f}_j(y) \le (1+\epsilon)f_j(y)$.
3. Construct the upper hull of $\widetilde{f}_j(y) = (1+\epsilon)\widehat{f}_j(y)$. Let $\phi_j(y)$ be this upper hull.
4. Use Algorithm 1 to solve Problem 3, where c_{ij} is replaced by $\overline{c}_{ij}$ and $f_j(y)$ is replaced by $\phi_j(y)$.

---

5   Numerical Results

In this section we evaluate the total cost of using servers and utilizing the link bandwidth on random network topologies using two different routing methods.

Ten different random network topologies are generated with $|V| = 100$ nodes and $|E| = 500$ edges, using the graph generator *random_graph( )* from the LEDA package [18]. For each network the delay of a link is picked randomly with uniform distribution among the integers $[1, 100]$ and the cost is generated in such a manner that it is correlated to its delay. Thus, for each link l a parameter b_l is generated randomly among the integers $[1, 5]$. The cost of link l is then $b_l (101 - d_l)$. Hence the link cost is a decreasing function of its delay. We run the algorithm for 6 different delay constraints $D = \{100, 200, 400, 500, 600, 800\}$. We assume that the the same server types can be placed in each of the locations. For our simulations we use 4 different server types with capacities and costs equal to $\{(100, 3000)\,(150, 3500)\,(250, 4000)\,(350, 5000)\}$ respectively. We set the factors $\alpha = 0.1$ and $\beta = 0.2$ and assume the load in requests per unit of time originated by each node is randomly chosen among the integers $[1, 100]$. We also assume $H = V$, i.e., servers may be placed in any of the nodes. We run two sets of experiments which differ in the manner the request round-trip routes are selected. Specifically we consider the following algorithms:

MinDelay: In this algorithm the minimum delay round-trip routes are selected without considering the route cost. A route thus selected is rejected if its delay is larger than the specified constraint. This manner of selecting routes has been employed in [7].

MinCost: This is the algorithm proposed in the current work. That is, the round-trip routes are selected so that they are of minimum cost provided that they satisfy the delay constraint.

Table 1. Average total cost for different delay constraints.

D	100	200	400	500	600	800
MinCost	255795	179854	149608	147340	145956	145343
MinDelay	257547	198303	191054	191054	191054	191054

For the simulations we used the pseudopolynomial algorithm since it works sufficiently well for the selected instances and its implementation is considerably simpler than the polynomial algorithm. In Table 1 we present the average total cost of using the servers and utilizing the link bandwidth. We make the following observations. The cost for both algorithms decreases as the delay constraint increases and levels off after a certain value of the delay constraint. We also observe in Table 1 that the total cost of MinCost algorithm is always smaller than MinDelay, as expected, and the significance is becoming more pronounced for larger delays. This behavior is due to the manner in which routes are selected by the two algorithms for a given delay constraint. For strict delay constraints, both algorithms choose mainly the permissible minimum delay round-trip routes and hence they have similar performance. For looser constraints, the fact that MinCost picks the minimum cost round-trip routes that satisfy the delay constraint instead of simply the minimum delay route (as MinDelay does) allows it to reduce the routing cost.

6   Conclusions

In this paper we presented a pseudopolynomial approximation algorithm and a polynomial time algorithm for the NP-hard problem of replicated server placement with QoS constraints. The pseudopolynomial algorithm works well in several practical instances and is simpler than the polynomial time algorithm. The polynomial time algorithm is significant from a theoretical point of view and can be useful to employ if the problem instance renders the pseudopolynomial time algorithm very slow.

In this work we did not consider link capacities. It is an interesting open problem to incorporate the latter constraint into the problem. Another problem of interest is to consider the case where not all the database is replicated to each of the servers.

References

1. Krishnan, R., Raz, D., Shavitt, Y.: The cache location problem. IEEE/ACM Transactions on Networking **8** (2000) 568–582
2. Li, B., Golin, M., Italiano, G., Deng, X.: On the optimal placement of web proxies in the internet. In: IEEE INFOCOM. (1999)
3. Jia, X., Li, D., Hu, X., Du, D.: Optimal placement of web proxies for replicated web servers in the internet. The Computer Journal **44** (2001) 329–339
4. Qiu, L., Padmanabhan, V., Voelker, G.: On the placement of web server replicas. In: IEEE INFOCOM. (2001) 1587–1596
5. Chen, Y., Katz, R., Kubiatowicz, J.: Dynamic replica placement for scalable content delivery. In: First International Workshop on Peer-to-Peer Systems. (2002) 306–318
6. Jamin, S., Jiu, C., Kurc, A., Raz, D., Shavitt, Y.: Constrained mirror palcement on the internet. In: IEEE INFOCOM. (2001) 31–40
7. Sayal, M., Breitbart, Y., Scheuermann, P., Vingralek, R.: Selection algorithms for replicated web servers. In: Workshop on Internet Server Performance, Madison, Wisconsin. (1998)
8. Tang, X., Xu, J.: On replica placement for QoS-aware content distribution. In: IEEE INFOCOM. (2004)
9. Garey, M.R., Johnson, D.S.: Computers and Intractability. A Guide to the Theory on NP-Completeness. W. H. FREEMAN AND COMPANY (1979)
10. Kellerer, H., Pferschy, U., Pisinger, D.: Knapsack Problems. Spinger-Verlag (2004)
11. Mahdian, M., Markakis, E., Saberi, A., Varizani, V.: Greedy facility location algorithms analyzed using dual fitting with factor-revealing LP. Journal of the ACM **50** (2003) 795–824
12. Hajiaghavi, M.T., Mahdian, M., Mirrokni, V.S.: The facility location problem with general cost functions. Networks **42** (2003) 42–47
13. Siachalou, S., Georgiadis, L.: Efficient QoS routing. Computer Networks **43** (2003) 351–367
14. Mieghem, P.V., de Neve, H., Kuipers, F.: Hop-by-hop quality of service routing. Computer Networks **37** (2001) 407–423
15. Andonov, R., Rajopadhye, S.: A sparse knapsack algo-tech-cuit and its synthesis. In: International Conference on Application Specific Array Processors ASPA '94, IEEE (1994)
16. Lorenz, D.H., Raz, D.: A simple and efficient approximation scheme for the restricted shortest path problem. Operations Research Letters **28** (2001) 213–221
17. de Berg, M., Schwarzkoph, O., Kreveld, M.V., Overmars, M.: Computional Geometry: Algorithms and Applications. Springer-Verlag (2000)
18. K.Mehlhorn, Naher, S.: Leda: A platform for combinatorial and geometric computing. Cambridge University Press (2000)

Dimensioning Approaches for an Access Link Assuring Integrated QoS*

Hung Tuan Tran and Thomas Ziegler

Telecommunications Research Center Vienna (ftw.),
Donaucity Strasse 1, 1220, Vienna, Austria,
Phone: +43 1 505 28 30/50, Fax: +43 1 505 28 30/99
{tran,ziegler}@ftw.at

Abstract. This paper proposes some model-based alternatives for the solution of dimensioning an access link that conveys a mixture of realtime and elastic traffic. The novel feature of the proposed provisioning approaches is that the bandwidth they provision assures *simultaneously packet-level QoS for realtime traffic* and *flow-level QoS for elastic traffic*. This is achieved by appropriately adapting and combining packet-level and flow-level analytical models. Simulation is performed to demonstrate the operation of the proposed approaches.

Keywords: bandwidth provisioning, QoS, access links

1 Introduction

Quality of Service (QoS) in IP networks, widely known as Internet QoS, has been for the last decade a hot topic in both research and industry networking communities. In this paper we tackle one specific issue from the massive set of Internet QoS problems. Namely, we develop quantitative rules for bandwidth provisioning for the case of access links to achieve *simultaneously* QoS requirements of all live traffic classes crossing these links.

It is well known that traffic in IP networks today can be largely categorized into two classes: realtime (streaming) traffic and elastic traffic. The former one is the traffic of interactive voice and video applications (VoIP, video on demand). The latter one is traffic of digital data transmission (e.g. file transfers and downloads). Owing to the realtime application natures, streaming traffic requires strict QoS in terms of packet loss, delay and jitter for each started connection. The QoS requirement in case of elastic traffic is usually the mean transfer delay, or equivalently the mean throughput for each connection. Expressing in technical words, we can say that streaming traffic requires *packet level QoS guarantees* (packet loss, delay and jitter), while elastic traffic requires *flow level QoS guarantees* (flow blocking probability, mean throughput of flows).

* This work has been partly supported by the Austrian government's Kplus Competence Center Program, and by the European Union under the E-Next Project FP6-506869.

M. Ajmone Marsan et al. (Eds.): QoS-IP 2005, LNCS 3375, pp. 221–234, 2005.
© Springer-Verlag Berlin Heidelberg 2005

A popular belief is that in the current Internet, it is sufficient to deliver strict QoS only to the realtime traffic class and leave the elastic traffic be treated in the best effort manner, i.e. let it get the QoS that the network can instantaneously offer. Such scenario can simply be achieved, for example by

- *i)* traffic classification distinguishing 2 traffic classes, by means of, e.g., the DiffServ paradigm, and
- *ii)* traffic prioritization giving absolute high priority to the realtime traffic, and
- *iii)* forcing the load of realtime traffic to remain under a low fraction of the link capacity (e.g. 10%).

Nevertheless, from the users point of view, QoS of elastic traffic can also be of importance and the users may not necessarily always be in favor only of the streaming applications. As a straightforward consequence, for having a better profit (by being able to serve elastic, QoS-oriented applications) network operators would require appropriate dimensioning rules with regard to QoS of both realtime and elastic traffic.

From the bullet *iii)* above, it might be thought that once the average fraction of high priority traffic is kept small, it will have little effect on the QoS of the elastic traffic. Thus, dimensioning a link based on elastic QoS can largely be done without taking into account the presence of real-time traffic. However, this is not really the case. For example, [1] reveals the so called *local instability periods*, where the input elastic load is larger than the instantaneous capacity left from the real-time traffic. In such local instability periods, the QoS (e.g. per-flow throughput) of elastic traffic is significantly degraded. *Thus, we argue that the dimensioning rules should be integrated rules, meaning that they must pay attention not only to the elastic traffic itself, but also to the presence of the streaming traffic.*

The main contribution of this paper is *QoS-aware, integrated provisioning schemes*, that simultaneously and explicitly take into account packet level QoS requirements for streaming traffic and flow level QoS requirements for elastic traffic. The provisioning rules are constructed mainly by means of the model-based approaches and their reliability is validated with simulation experiments.

We note that the scope of this work is appreciated whenever economical bandwidth usage is required. Because over-provisioning is often deployed in today's backbone networks, the expected application of our work is indeed shifted to the *access environment* in which economical resource provisioning strategies still play an important role. Typically, it is useful for dimensioning an access link connecting the access multiplexer (e.g. the Digital Subscriber Line Access Multiplexer DSLAM in the context of ADSL) with the edge router of the backbone network.

The paper is organised as follows. In Section 2 we review the related work. In Section 3, we detail the technical steps of our dimensioning alternatives. We present analytical results and discussions in Section 4, validating the proposed alternatives. Finally, we conclude the paper in Section 5.

2 Overview on Related Work

Basically, we can categorize any link dimensioning scheme along two aspect angles:

- *Accommodated Traffic Nature:* this refers to the issue of what kind of traffic the dimensioning scheme deals with. The traffic accommodated by the link to be dimensioned may be only realtime streaming, or only elastic, or a mixture of both traffic kinds;
- *QoS Nature:* this refers to the target QoS criteria the dimensioning scheme adopts. The QoS criteria for dimensioning may be flow level QoS criteria (per-flow throughput, flow blocking probability), or packet level QoS criteria (packet loss, delay and jitter), or the combination of them.

As a matter of fact, dimensioning rules can be derived from a vast variety of resource management mechanisms that have been elaborated so far in the research community. However, there is often a common shortage observed in previous work. Firstly, regarding the accommodated traffic nature, many of them devote attention only to one kind of traffic (either streaming or elastic), neglecting thus the effects of the remaining traffic on the resource management. Secondly, although some other work deals with the combination of two traffic kinds (streaming and elastic), but concerning the QoS nature, the QoS guarantees are in most cases treated only at either packet level, or only at flow level. Thus, the true QoS requirements of one traffic class is hidden behind that of another class, preventing us from having a complete picture on the QoS evolution.

Among the available model-based dimensioning schemes, a close queueing network model in heavy traffic situations has been worked out to dimension the link with regard to the per-flow throughput requirement of elastic traffic in [2]. The streaming traffic is not considered, however, in this work. The $M/G/1-PS$ (Processor Sharing) or $M/G/R-PS$ model [3,4] with limited or unlimited access rate provides an alternative for bandwidth provisioning, but they again work with regards only to the elastic traffic. The integrated model proposed in [5] is suitable for provisioning with regard to the mixture of streaming and elastic traffic. However, it neglects the packet level QoS, which is primarily important for the streaming traffic, and considers only flow-level QoS. The model presented in [6] provides the way to dimension the link with respect to the packet loss of streaming traffic, but the elastic traffic is not considered. Another model in [7] does combine the packet level and flow level models, but the work therein is restricted only to their interaction for TCP, i.e. for elastic traffic. The work of [8] proposes a close form approximation for a max-min fair sharing model, enabling the capacity dimensioning for VPN access links. The traffic and criteria for dimensioning pertain again only to the elastic traffic and its mean transfer time. A summary on the previous related work is reported in Table 1.

In overall, we observe the lack of having an integrated provisioning scheme, which pays attention simultaneously to the packet level QoS for streaming traffic, and flow level QoS for elastic traffic. Motivated by this observation, our work aims to define some possible alternatives for dimensioning an access link with regards to the integrated QoS feature of the conveyed traffic. To be more specific, we consider the integrated QoS target combining

- the packet loss probability requirement for streaming traffic $p_{loss} < \epsilon$, and
- the per-flow, mean throughput requirement for elastic traffic γ_{avg}.

Table 1. An overview on previous work.

Considered QoS\Traffic nature	only streaming	only elastic	streaming & elastic
packet level	[6]		
flow level		[2–4, 8]	[5]
packet & flow level		[7]	No available work

3   Our Dimensioning Approaches

We envision in this work three model-based provisioning alternatives. The common feature of the alternatives is that they perform bandwidth provisioning by leveraging a priori chosen analytical models. Briefly, the essentials are as follows:

- **Approach 1:** Provisioning is done based on the assumption of service differentiation with a high priority granted to streaming traffic. The needed bandwidth is derived from a procedure considering both packet-level and flow-level models. Each model takes into account the presence of both streaming and elastic traffic.
- **Approach 2:** Provisioning is done based on separate consideration of streaming and elastic traffic. The bandwidth needed for (or more precisely, the maximum bandwidth fraction of) streaming traffic is derived from a packet level queueing model, whereas the bandwidth needed for elastic traffic is derived separately from a flow level queueing model. The provisioned bandwidth is simply obtained by an additive rule applied to the two obtained bandwidth values.
- **Approach 3:** This approach first leverages the TCP protocol feature, namely the inverse square root relation between a TCP connection's throughput and its packet loss. Then among further potentials to proceed, we make an attempt to exploit the traffic aggregation effect. Provisioning is done without the assumption of service differentiation, i.e. no priority scheduling is deployed. The needed bandwidth is derived from the single fluid queue model, where the input traffic is assumed to have a known correlation structure.

In the next subsections, we go into the details of the proposed alternatives.

3.1   Approach 1

This approach comprises two consecutive steps.

Step 1: Given the requirement on the streaming packet loss probability, we seek in this step the maximal allowable streaming load (normalized to the link capacity). Assuming the service differentiation concept and that a high priority is assigned to the streaming traffic, we resort to a packet-level analytical model enabling the interconnection between the streaming packet loss probability p_{loss} and its allowable load $\rho_{s,max}$. We propose to use the assimilated M/G/1/K queue with server vacation. In this model, the server representing the link delivers streaming packets as long as the buffer for streaming traffic is not empty. Only in case this buffer is empty, elastic packets can be served. During the time the link is occupied by an elastic packet, incoming streaming packets have to be queued and they see the server as if it is on vacation. The time of one server vacation is the time to deliver one elastic packet. The assumption of Poison arrival process of streaming packets is considered to keep the tractability of the model.

For this model, it is derived in [9] that the packet loss probability for streaming traffic is

$$p_{loss} = 1 - \frac{(1 - h_0)\lambda_s^{-1}}{E(V)\pi_0 + E(S)(1 - h_0)}. \tag{1}$$

Here λ_s is the mean packet arrival rate of streaming traffic, $E(S)$, $E(V)$ are the mean service time and mean vacation time of the server, h_0 is the probability that no streaming packet arrival occurs during a vacation time V, π_0 is the steady state probability that no streaming packet is left in the system at a packet departure epoch. All of these parameters are either given as input parameters (in case of $E(S)$, $E(V)$, λ_s) or calculated from the steady state distribution of the model[1] (in case of h_0, π_0), see [9].

Given the basic relation (1) and a value of p_{loss}, and assuming a fixed link capacity, we apply a binary search to find the maximum allowable input rate of streaming traffic $\lambda_{s,max}$. In turn, the maximum allowable load $\rho_{s,max} = \lambda_{s,max}/link\_capacity$ of the streaming traffic is calculable. Note that once the value of p_{loss} is fixed, the value of $\rho_{s,max}$ is insensitive to the concrete value of the chosen link capacity. It means that if we choose different values for the link capacity, it results in different values of $\lambda_{s,max}$-s, but the ratio $\lambda_{s,max}/link\_capacity$ remains the same.

Step 2: After computing $\rho_{s,max}$ in Step 1, we switch to a flow level model to define the needed bandwidth assuring the per-flow throughput requirement for elastic traffic. *The flow-level model takes into account the presence of both streaming and elastic traffic.* We resort to the integrated flow-level queueing model of [5]. Originally, the model therein is used for achieving admission control, i.e. the opposite of the dimensioning task is considered, where the link capacity C is fixed in advance and the maximum allowable input traffic load is to be decided. The model takes the following input parameters

---

[1] The steady state distribution of the model is given by the probabilities
$\lim_{t\to\infty} Pr(\text{number of streaming packets in the system at time } t \text{ is } k)$ for $0 \le k \le K$.

- ρ_e: the elastic traffic load, computed as $\lambda_e/\mu_e C$. Here, λ_e is the arrival rate of elastic flows, and $1/\mu_e$ is the average size of elastic flows;
- ρ_s: the streaming traffic load, computed as $\lambda_s d_s/\mu_s C$. Here, λ_s is the arrival rate of streaming flows, d_s is the constant rate, and $1/\mu_s$ is the average lifetime of a streaming flow.

Under the *quasi stationary* assumption, i.e. the assumption that the ratio λ_e/λ_s is large enough, the analytical model provides the expression for the mean per-flow throughput of elastic traffic as $\gamma_e = \frac{\rho_e(1-B)}{E(Q_e)}$. Here, $E(Q_e)$ is the average number of currently active elastic flows. B is the flow blocking probability[2], which is itself a function of ρ_s, ρ_e, and the link capacity.

In our dimensioning task, we are given the average per-flow throughput $\gamma_{e,req}$, and we are interested in the required bandwidth value C. This is a reverse task of the admission control task the original model addresses. *The key prerequisite in our case is that the streaming traffic load ρ_s is taken identically to $\rho_{s,max}$ computed in Step 1.* We have implemented a binary search to find the bandwidth that yields the average per-flow throughput sufficiently close to the requirement $\gamma_{e,req}$ (currently we allow the bias error of 10^{-3}).

3.2 Approach 2

The steps of this approach are as follows:

Step 1: This step is identical to the Step 1 of Approach 1, i.e. the maximal allowable load $\rho_{s,max}$ of the streaming traffic is computed based on the $M/G/1/K$ with server vacation model.

Step 2: Unlike Approach 1, this step considers only the presence of elastic traffic. Based on the offered load of elastic traffic λ_e/μ_e, and on the target per-flow throughput $\gamma_{e,req}$, we compute the bandwidth C_e to be provisioned *only for elastic traffic*. This is done by using a flow-level model, namely either the $M/G/1\text{-}PS$ or the $M/G/1\text{-}PS$ with limited access rate [3].

Let $\rho_e = \lambda_e \sigma_e/C_e$. The use of the $M/G/1\text{-}PS$ model gives a simple expression for the average per-flow throughput as

$$\gamma_e = C_e(1 - \rho_e). \tag{2}$$

The use of the model $M/G/1 - PS$ with limited access rate r brings

$$\gamma_e = \frac{1}{\frac{1}{r} + \frac{f(\rho_e)}{C_e(1-\rho_e)}\left[1 - \left(\frac{C_e}{r} - m\right)(1 - \rho_e)\right]}, \tag{3}$$

where m is the integer part of C_e/r, and $f(\rho) = \frac{(\frac{C_e\rho_e}{r})^m/m!}{(1-\rho_e)\sum_{k=0}^{m-1}(\frac{C_e\rho_e}{r})^k/k! + (\frac{C_e\rho_e}{r})^m/m!}$.

---

[2] In this model, the same blocking condition is applied to both streaming and elastic flows. Denote the current number of streaming and elastic flows by Q_s and Q_e. Let d_{min} be the minimum throughput required for elastic flows, and $d = max(d_s, d_{min})$. A new flow is blocked, if it violates the condition $Q_s d_s + Q_e d_{min} \leq C - d$.

The expression (2) or (3) coupled with a binary search enables us to find the right bandwidth value C_e, assuring the target throughput $\gamma_{e,req}$.

Step 3: The final link bandwidth C is computed as $C = C_e/(1 - \rho_{s,max})$

Approach-2 differs from Approach-1 in that the mutual effects between streaming and elastic are only considered at the packet level in Step 1. At the flow level the two traffic classes are treated separately .

3.3 Approach 3

The steps of this approach are as follows:

Step 1: In this step we leverage the TCP-specific, "inverse square root" loss-throughput expression. Given the target per-flow throughput, we compute the packet loss probability allowable for the elastic traffic. Then, we take the minimum of the target values for packet loss probabilities of streaming (which is given) and elastic traffic (which has been now computed). In this way, the stricter QoS requirement serves as the provisioning criterion, implicitly assuming conservative bandwidth provision.

There exist some formulae with different levels of sophistication, which expresses the relation between a TCP connection's throughput and the observed packet loss (see e.g. [10]). Any of them can be used for the dimensioning purpose here.

Step 2: Based on the common target loss probability derived in Step 1, we use a single fluid queue model to derive the needed bandwidth. As in this approach we do not assume the deployment of service differentiation, all packets are put into one queue at the output port of the link. This kind of all-into-one aggregation might give a reason for applying a simplified traffic model for the input traffic. Currently, we are using the Gaussian traffic model, but the principle is not restricted only to this model, i.e. other reasonable aggregate traffic model can be considered as well.

In case of the Gaussian input traffic, the loss probability formula for a finite buffer with size x is derived in [11] as

$$p_{loss} = \alpha e^{-m_x/2}, \tag{4}$$

where α and m_x are the parameters calculated based on the characteristics of the Gaussian traffic and the link bandwidth value C. More precisely, to calculate the loss probability we need to know the average rate, the autocovariance function of the aggregate input traffic, the buffer size, and the link capacity. In a reverse direction, given the loss probability and the traffic characteristics, similarly to Approach 1, we implement a binary search to find an adequate bandwidth amount, using the above loss expression (4).

Note that the root of Approach-3 is to use the relation between TCP flow throughput and its packet loss probability. Naturally, this key point enables other provisioning versions as well. For example, after getting the loss probability for the elastic traffic, one can proceed in such a way that one does not take the minimum of two loss targets as we have done in Step 1. Instead, one can directly consider such a *packet level* scheduling system consisting of two separate queues, a high priority queue for streaming and a lower priority queue for elastic traffic. The dimensioning rule is done by exactly solving this system, thus getting the quantitative relation between the link capacity and the per-class packet loss rates.

However, the main obstacle of such elaboration is that it is hard to find a good and reasonable packet-level model description for each class of the input traffic. The variation of elastic packet sizes, the complicate nature of packet arrival process in the access environment (stemming from a well-known long range dependent property, excluding the use of the amenable Poisson assumption) make such a packet level model not easy to construct and in turn a tractable solution is hardly available.

4   Analytical Results and Discussions

Having developed some provisioning alternatives, it is important to verify whether the bandwidth specified by these alternatives really yields the target QoS. We perform this investigation by means of analytical and simulation experiments. The traffic scenarios we use for testing Approach-1 and Approach-2 are as follows:

- **Streaming Traffic:** each streaming flow has the fixed rate $d_s = 16kbps$. The duration time of each flow is exponentially distributed and has a mean value $60s$. The arrival process of streaming flows is Poissonian.
- **Elastic Traffic:** the flow size distribution is either Pareto (with the shape parameter 1.2), or exponential. The mean flow size is $200kbits$. The arrival process of elastic flows is also Poissonian.

The investigation principle throughout this section is the following. First, we compute the needed bandwidth amount based on the integrated QoS criteria (i.e. the packet loss probability for streaming traffic and the average per-flow throughput for elastic traffic) with different approaches. If not stated otherwise, the average per-flow throughput criterion is $100kbps$, i.e. $\gamma_{e,req} = 100kbps$, and the packet loss probability criterion is varied from 10^{-12} to 10^{-3}. After computing the bandwidth value, we set the link capacity to the calculated bandwidth in the corresponding simulation scenario to be run[3]. Then, the packet loss probability and the average per-flow throughput obtained with simulation are compared with those serving as the original criteria for the provisioning approaches.

Although in Approach-2, either the $M/G/1-PS$ model or the $M/G/1-PS$ with limited access rate model can be used in Step 2, we observe small differences between the bandwidth values provided by these two models[4]. Therefore, in the

---

[3] We use ns [12] as the simulation platform.

[4] For the same target per-flow throughput, the $M/G/1-PS$ with limited access rate model in general suggests to deploy more bandwidth than the $M/G/1-PS$ model. However, our numerical experiments show that with the limited access rate of 1Mbps, the relative difference in bandwidth amount is in a very small range of around 0.1%.

following, we report only the results with the use of the $M/G/1-PS$ model. For each individual scenario, we repeat the simulation ten times and report here the relevant quantity (either the packet loss probability or the mean per-flow throughput) identical to the average values we get over such a series of simulation runs.

The first point to be examined is the maximum allowable traffic load $\rho_{s,max}$ of streaming traffic given a packet loss target. Remind that we use the $M/G/1/K$ vacation queue to compute this $\rho_{s,max}$ value. To remain conservative and to alleviate the under-estimation of packet loss ratio that the assimilated Poisson process might introduce, we use the worst-case vacation scenario. It means that the packet size of elastic traffic is set to the maximum value of 1500 bytes (MTU size), causing the possible maximum vacation of the server. The numerical results are reported in Table 2.

Table 2. Maximum allowable streaming traffic fraction in the link with buffer size of 15 packets.

Loss criterion for streaming traffic	ρ_{max} packet size = 60Bytes	ρ_{max} packet size = 120 Bytes	ρ_{max} packet size = 200 Bytes
1e-12	0.124550	0.237425	0.356955
1e-11	0.124557	0.237427	0.356964
1e-10	0.124634	0.237549	0.357178
1e-09	0.125000	0.238281	0.359375
1e-08	0.128906	0.250000	0.375000
1e-07	0.140625	0.281250	0.406250
1e-06	0.156250	0.312500	0.500000
1e-05	0.218750	0.375000	0.562500
1e-04	0.250000	0.500000	0.625000
1e-03	0.312500	0.625000	0.750000

We have two observations from the obtained results. On the one hand, the allowable load of streaming traffic in the link increases with the relaxation of the loss requirement. On the other hand, for the same loss requirement, the larger the streaming packet size, the higher the allowable streaming load. While the first fact is straightforward, the second one can be intuitively explained as follows. Assuming the same input streaming load and the same link capacity, the larger the packet size, the smaller the number of arriving packets. Since the buffer size is measured in packets, it follows that the larger the packet size, the lower the potential occurrence of packet loss events. Thus, looking in the reverse direction, in order to keep the packet loss ratio the same, the input load must be smaller in case of smaller packet size.

From now on, we keep the packet size of streaming traffic fixed at 60 bytes. Table 3 presents the bandwidth value calculated with the two approaches for different streaming loss criteria and elastic arrival rates. A general observation from Table 3 is that Approach-2 suggests a smaller (around 5-10% less) bandwidth amount than Approach-1.

For the simulation checking the goodness of the results in Table 3, we present here the results for 4 scenarios distinguished from each other by the input pa-

Table 3. The required bandwidth (measured in Mbps) for different streaming loss criteria and elastic loads.

Loss criterion for streaming traffic	Elastic arrival rate $\lambda_e = 12.3\ flows/s$		Elastic arrival rate $\lambda_e = 24.3\ flows/s$		Elastic arrival rate $\lambda_e = 31.7\ flows/s$	
	Approach-1	Approach-2	Approach-1	Approach-2	Approach-1	Approach-2
1e-12	3.058594	2.924210	5.879883	5.665658	7.607422	7.356217
1e-11	3.058594	2.924234	5.879883	5.665703	7.607422	7.356276
1e-10	3.058594	2.924491	5.880859	5.666201	7.607422	7.356923
1e-09	3.060547	2.925714	5.882812	5.668571	7.613281	7.360000
1e-08	3.080078	2.938833	5.915039	5.693989	7.652344	7.393002
1e-07	3.131836	2.978909	6.010742	5.771636	7.775391	7.493818
1e-06	3.203125	3.034074	6.141602	5.878519	7.941406	7.632593
1e-05	3.521484	3.276800	6.717773	6.348800	8.675788	8.243200
1e-04	3.703125	3.413333	7.042969	6.613333	9.086920	8.586667
1e-03	4.108398	3.723636	7.781250	7.214545	10.02100	9.367273

Table 4. Scenario identification.

Scenario name \ Parameters	$\lambda_e\ [flows/s]$	file size distribution
Scenario 1	12.3	Exponential
Scenario 2	12.3	Pareto
Scenario 3	24.3	Exponential
Scenario 4	24.3	Pareto

rameters related to the elastic traffic, namely the flow arrival rate and the flow size distribution. The scenario identification is shown in Table 3. Note that in the analytical model-based provisioning presented in Section 3, the flow size distribution does not play any role in the bandwidth calculation. In other words, both Approach-1 and Approach-2 are insensible to elastic flow size distribution. To examine whether this is valid in simulation, we use two kinds of flow size distribution: exponential and Pareto. The latter one reflects the real heavy-tail property of today's web-traffic, i.e. only a few downloads of very large files constitute the major part of elastic traffic.

The average per-flow throughput for elastic traffic and the packet loss probability for streaming traffic obtained with simulation of corresponding scenarios are reported in Figs. 1-4 and Table 5. Recall that if the dimensioning approaches work well, the packet loss obtained with simulation for streaming traffic should be smaller than the associated original loss criterion, and the per-flow throughput obtained with simulation for elastic traffic should be around $100kbps$.

We see in the figures that the per-flow throughput requirement $\gamma_{e,reg} = 100kbps$ is largely met over all the cases. There is an exception case with Scenario-2, where the mean throughput tends to decrease with the relaxation of the streaming loss criterion (see Fig. 2). We believe that it is partly because of the following. Relaxing the streaming loss criterion allows more streaming load into the link, making it more bursty. In such cases the assimilated Poisson arrival process used in the $M/G/1/K$ queue might be too optimistic from the aspect of estimating the streaming packet loss events. Consequently, adopting a Poisson arrival process for streaming traffic may underestimate the needed bandwidth, leading to the degradation of both the truly obtained streaming loss rate and

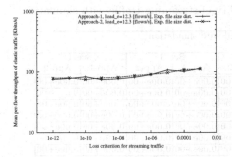

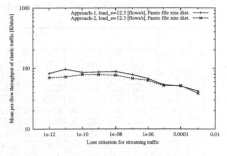

Fig. 1. Mean per-flow throughput obtained with simulation, Scenario-1.

Fig. 2. Mean per-flow throughput obtained with simulation, Scenario-2.

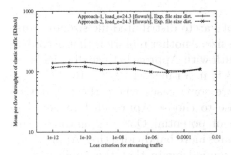

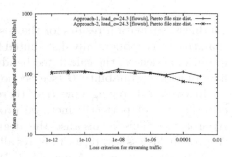

Fig. 3. Mean per-flow throughput obtained with simulation, Scenario-3.

Fig. 4. Mean per-flow throughput obtained with simulation, Scenario-4.

the elastic mean throughput. Indeed, the streaming packet loss probability is not fulfilled in some cases where the original loss criterion is in a range larger than 10^{-4} (see the last rows of Table 5). In addition, we think that the heavy-tail property of web file size distribution also plays a role in causing the observed phenomenon, but the detailed explanations would need further investigations.

Except the before mentioned case, it can be stated that for most cases, the proposed provisioning approaches work well, fulfilling both the throughput and loss requirements. We often get zero loss probability with simulation (see Table 5), meaning that during the simulation time (which has been chosen to be 2000s), there was no packet loss event.

It is also worth making some additional remarks here. In simulation we do observe that the per-flow throughput exhibits flow size dependency, i.e. large flows get higher throughput than that of short flows. When taking the average throughput value by dividing the total throughput by the number of flows, this flow size dependency, of course, is masked out. Recall that the analytical models used in Approach-1 and Approach-2 both assume a perfect rate sharing between elastic flows, which according to our revealed experiences now, does not perfectly correspond to the situation reflected in simulation. The effect of this unfair rate sharing becomes more aggravating when Pareto file size distribution is used. In fact, we can see in Figs. 1-4 and Table 5 that from the aspect of dimensioning

Table 5. Checking the packet loss probability of streaming traffic with simulation.

Loss criterion	Loss obtained with simulation							
	Sce-1		Sce-2		Sce-3		Sce-4	
	App-1	App-2	App-1	App-2	App-1	App-2	App-1	App-2
1e-12	0.00e+00	0.00e+00	0.00e+00	0.00e+00	0.00e+00	0.00e+00	0.00e+00	0.00e+00
1e-11	0.00e+00	0.00e+00	0.00e+00	0.00e+00	0.00e+00	0.00e+00	0.00e+00	0.00e+00
1e-10	0.00e+00	0.00e+00	0.00e+00	0.00e+00	0.00e+00	0.00e+00	0.00e+00	0.00e+00
1e-09	0.00e+00	0.00e+00	0.00e+00	0.00e+00	0.00e+00	0.00e+00	0.00e+00	0.00e+00
1e-08	0.00e+00	0.00e+00	0.00e+00	0.00e+00	0.00e+00	0.00e+00	0.00e+00	0.00e+00
1e-07	0.00e+00	0.00e+00	0.00e+00	0.00e+00	0.00e+00	0.00e+00	0.00e+00	0.00e+00
1e-06	0.00e+00	0.00e+00	0.00e+00	0.00e+00	0.00e+00	0.00e+00	0.00e+00	0.00e+00
1e-05	0.00e+00	0.00e+00	1.05e-06	3.08e-05	0.00e+00	0.00e+00	5.37e-06	1.15e-06
1e-04	0.00e+00	0.00e+00	4.23e-04	6.95e-03	0.00e+00	1.87e-05	4.72e-05	9.37e-05
1e-03	1.60e-04	6.67e-04	3.96e-02	7.96e-02	6.83e-04	2.79e-03	5.85e-03	2.66e-02

goodness, simulation results for the case of Pareto distributed file size are worse than those of exponentially distributed file size. Another remark is that according to our experiences, the calculated bandwidth with Approach-1 usually yields the flow blocking probability B (see Section 3.1) in a range of $10^{-5} - 10^{-4}$.

In overall, knowing now that there are some cases where the target QoS criteria are not perfectly met, we propose to choose Approach-1 rather than Approach-2. This is because the degree of potential QoS degradations with Approach-1 is less severe, as it always allocates more bandwidth than Approach-2.

Now let us turn to Approach-3. We remind again that the application of Approach-3 requires the knowledge on the aggregate traffic characteristics, namely the average rate λ_{agg}, the covariance σ^2_{agg}, and the autocovariance function $C_{\lambda_{agg}}(l)$ of the aggregate input traffic rate. Such data can either be given a priori or be derived from traffic load measurement like the way presented in [13]. We demonstrate here the case when the aggregate traffic has a priori known autocovariance function. For representative purposes, we choose two specific formulae[5], namely

- $C_\lambda(l) = \sigma^2_{agg} 0.9^{|l|}$ (we refer to this case as a *simple case*), and

- $C_\lambda(l) = \frac{\sigma^2_{agg}}{2}(|l-1|^{2H} + |l+1|^{2H} - 2|l|^{2H})$ (we refer to this case as a *fbm case*). This is the discrete version of the autocovariance of fraction Brownian motion process. H is the Hurst parameter and chosen to be 0.78, the value suggested for the traffic observed on an access link in an earlier work [14].

In order to retain somewhat the closeness to the chosen scenarios used earlier in Approach-1 and Approach-2, the incoming mean traffic rate in Step 2 of Approach-3 is set to be identical to that was obtained from the final solution of Step 2 of Approach-1, i.e. $\lambda_{agg} = (\rho_{s,max} + \rho_e)C_{App-1}$. The variance coefficient (or the index of dispersion) is chosen to be 0.2628 according to [14], consequently $\sigma^2_{agg} = 0.2628\lambda_{agg}$. Other parameters are: the buffer size is 15 packets of size

---

[5] Note that we use here the autocovariance function in discrete time domain, i.e. traffic load is measured in a slot-by-slot manner. The parameter l refers to the distance expressed in slots between two relevant load values.

Table 6. The bandwidth (measured in Mbps) assigned by Approach-3 for different streaming loss criteria and elastic loads.

Loss criterion for streaming traffic	Elastic arrival rate $\lambda_e = 12.3\ flows/s$		Elastic arrival rate $\lambda_e = 24.3\ flows/s$		Elastic arrival rate $\lambda_e = 31.7\ flows/s$	
	simple case	fbm case	simple case	fbm case	simple case	fbm case
1e-12	2.980394	2.886794	5.794899	5.688822	7.521267	7.408859
1e-11	2.970969	2.879671	5.781115	5.678462	7.505468	7.396907
1e-10	2.961196	2.872433	5.767144	5.668092	7.489236	7.384829
1e-09	2.951972	2.865964	5.754128	5.658931	7.474825	7.374976
1e-08	2.955381	2.872386	5.764962	5.674022	7.490803	7.395738
1e-07	2.987155	2.907521	5.830667	5.744309	7.577850	7.488127
1e-06	3.034414	2.960291	5.926001	5.845065	7.703135	7.619561
1e-05	3.303089	3.230026	6.433663	6.353931	8.344591	8.262391
1e-04	3.460386	3.385441	6.727323	6.647187	8.728160	8.645469
1e-03	3.822507	3.746789	7.403577	7.322581	9.599811	9.516119

$60bytes$, the RTT value is $10ms$, and the simplest expression for per-flow TCP throughput $B(p) = \frac{1}{RTT}\frac{\sqrt{2}}{\sqrt{p}}$ is used.

The bandwidth values suggested by Approach-3 are reported in Table 6. Note that extensive analysis-simulation investigations in [11] have already confirmed the accurateness of the loss probability formula (4) derived from the Gaussian traffic model. As a straightforward implication, the bandwidth value decided based on this loss formula should indeed assure the expected loss target. In other words, if we generate a traffic pattern identical to the Gaussian process characteristics (including the mean rate, the variance, the autocovariance function) that were used originally as the input parameters for Approach-3, and inject such the aggregate traffic into the link with capacity identical to the value assigned by the output of Approach-3, we will get back indeed the desired loss probability. This simulation task would only be just a confirmation of work in [11], therefore we skip achieving it.

5   Conclusions

In this paper, we have worked out and demonstrated three alternative approaches for dimensioning an access link. The essential feature distinguishing these proposed provisioning approaches from earlier work is that they decide the link bandwidth based *simultaneously on both packet level QoS criteria for realtime traffic and flow level QoS criteria for elastic traffic*. In order to achieve this aim, Approach-1 and Approach-2 leverage a novel principle, which decides first the maximum allowable streaming traffic load by means of a packet level analytical model and then using one more, flow level model to deduce the final bandwidth. The novel principle of Approach-3 is the common analytical model applied to the traffic mixture without traffic classification. Analysis and simulation investigations have indicated that the approaches work well in a major parts of examined scenarios. More extensive validations remain as our follow-on work.

References

1. F. Delcoigne, A. Proutiere, and G. Regnie. Modelling integration of streaming and data traffic. In *Proceedings of 15th ITC Specialist Seminar on IP traffic*, July 2002.
2. A. W. Berger and Y. Kogan. Dimensioning bandwidth for elastic traffic in high-speed data networks. *IEEE/ACM Transactions on Networking*, 8(5):643–654, October 2000.
3. S. B. Fredj, T. Bonald, A. Proutiere, G. Regnie, and J. W. Roberts. Statistical bandwidth sharing: A study of congestion at flow level. In *Proceedings of ACM SIGCOMM 2001*, pages 111–122, 2001.
4. Z. Fan. Dimesioning Bandwidth for Elastic Traffic. In *in LNCS 2345, Proceedings of NETWORKING 2002*, pages 826–837, 2002.
5. N. Benameur, S. B. Fredj, S. O-Boulahia, and J. W. Roberts. Quality of service and flow level admission control in the Internet. *Computer Networks*, 40:57–71, 2002.
6. B. Ahlgren, A. Andersson, O. Hagsan, and I. Marsh. Dimensioning links for IP Telephony. In *Proceedings of IPTEL*, 2001.
7. P. Lassila, H. van den Berg, M. Mandjes, and R. Kooij. An integrated packet/flow level model for TCP performance analysis. In *Proceedings of ITC 18*, pages 651–660, 2003.
8. K. Ishibashi, M. Ishizuka, M. Aida, and H. Ishii. Capacity Dimensioning of VPN Access Links for Elastic Traffic. In *Proceedings IEEE International Conference on Communications, ICC*, pages 1547–1551, 2003.
9. A. Frey and Y. Takahashi. A note on an M/GI/1/N queue with vacation time and exhaustive service discipline. *Operations Research Letter*, 21(2):95–100, 1997.
10. E. Altman, K. Avrachenko, and C. Barakat. A stochastic model of tcp/ip with stationary random losses. In *Proceedings of ACM SIGCOMM*, pages 231–242, August 2000.
11. H. S. Kim and N. B. Shroff. Loss Probability Calculations and Asymptotic Analysis for Finite Buffer Multiplexers. *IEEE/ACM Transactions on Networking*, 9(6):755–768, December 2001.
12. ns simulator homepage. http://www.isi.edu/nsnam/ns.
13. H. T. Tran and T. Ziegler. Adaptive Bandwidth Provisioning with Explicit Respect to QoS Requirements. In *LNCS 2811, Proceedings of QoFIS 2003*, pages 83–92, October 2003.
14. I. Norros. On the use of Fractional Brownian Motion in the theory of connectionless network. *IEEE Journal on Selected Areas in Communications*, 13(6):953–962, Augustus 1995.

Aggregation Network Design for Offering Multimedia Services to Fast Moving Users

Filip De Greve, Frederic Van Quickenborne, Filip De Turck,
Ingrid Moerman, and Piet Demeester

Department of Information Technology (INTEC), Ghent University,
Sint-Pietersnieuwstraat 41, B-9000 Gent, Belgium
{Filip.DeGreve,Frederic.VanQuickenborne}@intec.ugent.be

Abstract. Nowadays, passengers in fast moving vehicles can consume narrow-bandwidth services such as voice telephony or email but there is no widely deployed platform to support broadband multimedia services. In this paper, an aggregation network architecture for the delivery of multimedia services to train passengers is presented. For the topology design and capacity assignment of this network, a novel method is presented that calculates the required dynamic tunnels to meet the traffic demands of the fast moving users while achieving low congestion and optimizing the utilization of the network resources. The method enables that under conditions of train delay the Quality of Service (QoS) requirements of the traffic flows are guaranteed. The capacity assignment efficiency is demonstrated for different network scenarios and due to the time-dependent complexity of the problem, heuristic approaches are designed to solve problems of realistic size.

1 Introduction

1.1 Motivation

By 2007 the national railroad company of Belgium (NMBS) would like to build its own digital communication network "GSM-R" (GSM for Railways) in replacement of their currently out-dated network [1]. Therefore the NMBS is investing in the construction of 450 new antennas. These antennas will also be shared with some major telecom-operators in order to improve the (narrow-bandwidth) voice service of train passengers. However, current applications such as multimedia content delivery, video phoning and on-line gaming require a high level of QoS and are generally characterized by high bandwidth requirements. They cannot be offered to the passengers with this kind of network.

The lack of broadband services in fast moving vehicles such as trains, busses and vessels is stated in [2] and while trials are on the way to find out whether commuters will ultimately pay for broadband services, the aggregation part of the broadband network (see Fig. 1) is not optimally designed to cope with fast moving users; it is typically designed to cope with traffic demands of fixed users. Such a design (even with addition of an admission control system to limit the

M. Ajmone Marsan et al. (Eds.): QoS-IP 2005, LNCS 3375, pp. 235–248, 2005.

impact of unexceptional circumstances) won't be sufficient to maintain the QoS guarantees of the passenger data traffic at all times without overdimensioning the network. In this paper we present a cost effective design method that calculates the required dynamic tunnels to meet the traffic demands of fast moving users while achieving low congestion and optimizing the utilization of the network resources. It is no goal to elaborate on the operation of the enabling platform or specific service realization issues (such as handovers or roaming issues).

We can state that multimedia services are taken for granted in fixed networks and narrow-bandwidth services are being deployed for fast moving users but to our knowledge the network resource optimization and routing have never been studied for fast moving users. The environment is characterized by groups of users moving at equal speed along a (predictable) trajectory, causing dynamically changing traffic conditions. Without loss of generality we will focus on train scenarios and take into account that in many cases installed optical fibers are already present along the railroad tracks. For example, the NMBS has its own telecom division that exploits a fiber optic network covering almost the entire Belgian railroad track [1]. In many other cases train companies (such as North America's largest railroad company [3]) have given permission to telecommunication companies to install fibers along their railroads. Enabling these fibers is cheaper than digging, renting or activating additional fibers.

The considered network architecture is represented in Fig. 1. While passengers are connected to the internal train network, the train remains connected to the closest wireless station near the railroad track. These wireless stations are bundled in access networks. While trains are moving on the railroad track, the train's connection hops from access network to access network. The passenger traffic is gathered in aggregated traffic flows per train and per QoS class on the Access Gateway (AGW). The aggregation network has AGWs connecting to the access networks and Service Gateways (SGWs) connecting to the provider's domain. The connectivity in the aggregation network is achieved by setting up dynamic tunnels between the AGWs and the SGW in which the aggregated traffic flows are mapped. The SGW is constantly updated with information about the current (and future) positions of the trains and every turn the next hop gateway towards the moving users gets updated. The set-up protocol for the dynamic tunnels and the admission control mechanism for QoS-aware aggregation networks are described in [4]. Mainly due to economical reasons a network is preferred consisting of standard QoS-aware Ethernet switches (IEEE 802.1s, IEEE 802.1q&p compliant) with separated hardware queues per QoS class. We assume that the AGWs and SGW have a fixed position, successively placed along the railroad track (e.g. every 5 or 10 km). The presented method will calculate the network equipment that needs to be installed, where to place fibers between the network nodes, how to set up and adjust the AGW-SGW tunnels and how to route the traffic flows at every moment in time.

This paper is organized as follows. The introduction is completed with related work on the topics of network design for fast moving users. In the next section the theoretical model for the dimensioning problem is presented and the model

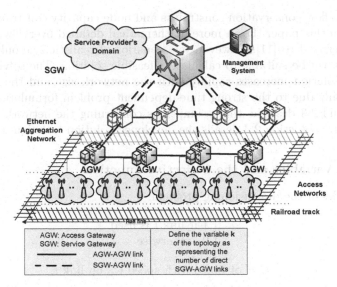

Fig. 1. Schematic representation of the network architecture.

is illustrated for linear railroad tracks. In Section 4 heuristic solution techniques are described and the performance will be evaluated by gradually increasing the railroad complexity towards ring-shaped (Section 5) and grid-like (Section 6) railroad tracks. The influence of larger scale topologies, variations on the cost parameters and different traffic patterns will be discussed.

1.2 Related Work

Today, organizations such as IEEE and 3GPP are establishing wireless specifications for new wireless technologies which have to meet the bandwidth requirements of tomorrow, e.g. the cellular UMTS, integrated Beyond 3G solutions or the future WiMAX standard for limited mobility (IEEE 802.16e). Broadband network design and dimensioning studies mainly assume traffic of fixed users [5, 6]. Unfortunately fast moving users have not been taken into account. Related solutions for moving vehicles are proposed in [7] (with focus on the wireless part) and in [8] (application-based approach for disconnection tolerant networking). In [9] the routing and dimensioning issues of time-scheduled connections in QoS networks are addressed.

2 Theoretical Model

A path flow based Integer Linear Programming (ILP [10]) formulation of the theoretical model [11] is used to calculate the exact dimensioning and tunnel path determination. This formulation is introduced by means of different variables and the objective function (subsection 2.1). The ILP constraints such as link capacity constraints, flow conservation constraints and node capacity constraints are not discussed in this paper. For a more mathematical detailed overview the reader is kindly referred to [11]. However, for scenarios of realistic size only heuristic approaches will be suitable to calculate a feasible solution. The solving time of the exact solution increases rapidly with the network size and the number of trains, mainly due to the strong time-dependent problem formulation. Subsections 2.2 and 2.3 describe the assumptions concerning the network and traffic parameters. Subsection 2.4 details the developed solution techniques.

2.1   The Variables and the Objective Function

The variables of the dimensioning problem:

$$u_l = \begin{cases} 1 \text{ if link } l \text{ is used} \\ 0, \text{ otherwise} \end{cases} . \tag{1}$$

$$x_l^v = \# \text{ of fibres with speed } C_v \text{ on link } l. \tag{2}$$

$$y_{pijk} = \begin{cases} 1 \text{ if path } p \text{ is used between AGW } i \text{ and SGW } k \quad \text{for flow } j \\ 0, \text{ otherwise} \end{cases} . \tag{3}$$

$$z_n^{vw} = \# \text{ of cards with } O_{vw} \text{ interfaces of speed } C_v \text{ on node } n. \tag{4}$$

and their vector representations:

$$U = [u_l], \ X = [x_l^v], \ Y = [y_{pijk}], \ Z = [z_n^{vw}] \tag{5}$$

Each card has a specific cost depending on the speed of the Ethernet interfaces (C_v) and the number of interfaces (O_{vw}) installed on the card. Each required link has a specific cost depending on the installation cost of the link. The cost function of the dimensioning problem is represented in a simplified form (6) depending on the parameters α, β and γ. We will focus on three different parts of the cost function: the topology cost, the node cost and the routing cost.

$$c = f_0(U) + \alpha f_1(U) + \beta f_2(Z) + \gamma f_3(Y) \tag{6}$$

The term $f_0(U)$ represents the cost to install and start using the fibers that are placed along the railroad track (links represented as full lines in Fig. 1). The term $\alpha f_1(U)$ represents the cost to install and start using the fibers that connect the AGWs to the SGWs if these fibers are not placed along the railroad track (links represented as dotted lines in Fig. 1). The topology cost is defined by their sum $f_0 U + \alpha f_1(U)$. If α equals 1, there is no cost difference in exploiting fibers that are placed along or not along the railroad. The term $\beta f_2(Z)$ represents the cost to install network equipment with network interface cards with sufficient link capacity to handle the data traffic of the fast moving users (= the node cost). We distinguish between different line card types of different speeds and with different port ranges. The parameter β enables to increase or decrease the node cost compared to the topology cost. The term $\gamma f_3(Y)$ represents the sum of the hop counts of all AGW-SGW tunnels that are required to maintain connectivity with the moving trains. We will tune γ in such a way that this term is negligible in the objective function. Because no actual cost is associated with this term, it will be used as a tiebreaker to prefer the network with the shortest average tunnel length. This γ-value can be made dependent on the specific QoS class of the flows that are transported in these tunnels. By adjusting γ according the QoS class a more expensive but QoS sensitive routing can be achieved. This can be used as online routing algorithm to optimize the real-time performance of the network.

We introduce scenarios based on 3 business models (with increasing node cost) without elaborating on the exact values. In all scenarios γ equals 0.00001.

- Scenario 1: α=100.0; β=0.01
  This corresponds to the case where the railroad company owns installed fibers along the railroad tracks (as for the Belgium railroad scenario) but they only need to be connected to the service provider.
- Scenario 2: α=1.0; β=0.01
  This corresponds to the case where a telecom operator with good connectivity along the railroad track, wants to design a network from scratch while the train company owns installed fibers along the track.
- Scenario 3: α=1.0; β=100.0
  This corresponds to the case where a railroad company wants to design a network from scratch while a network operator already has installed fibers along the track.

2.2 Network Model

The simulated networks in this paper will have increasing topology complexity: from linear and rings to grids. In our model the AGWs and SGWs have Layer 3 abilities (to connect to the access network and to the provider's network) but the data traffic is switched in Layer 2 Ethernet switches. In this paper we assume the presence of a single SGW in the network. In these circumstances the use of additional core nodes would only increase the total network cost. Therefore no extra Ethernet core switches are added and all AGWs have direct candidate links to the SGW as presented in Fig. 1. For the simulations we assumed the following range of Ethernet speeds: 100 Mb/s, 1 Gb/s and 10 Gb/s and the following port ranges: 1-port, 2-port and 4-port line cards.

2.3 Traffic Model

On the railroad track, rail lines are defined. The time schedules of these lines will be used as input to define time-dependent traffic flows in the network. In this paper the assumption of two crossing trains per line is made without loss of generality. For simplicity we assume that the traffic profile of the users itself doesn't change and we assume constant unidirectional traffic demands to focus on the rapidly moving aspect of the user. We assume traffic loads of approximately 1 Gb/s per train based on [12]. The user traffic will be gathered per train and per QoS class in basic routing units, named flows. We will define three traffic demand types: exact demands, static demands and train delay insensitive demands.

Exact Demands. Exact traffic demands imply optimization of the network resources with knowledge of the exact point (= the exact AGW) where the two trains cross each other and of the exact moment in time when the two trains cross each other. However, if one train suffers from delay and the place where the crossing of two trains occurs, moves to another AGW, the network could suffer from insufficient resources to deliver the requested demand to the crossing trains and loss of information would occur.

Static Demands. By neglecting all time-related aspects of the traffic demands, exact demands can be transformed into static demands. This is done by adding all demands that are requested for a particular AGW, and this for every AGW separately. This results in a time-independent demand from the SGW to each AGW and implies that both trains could cross in every AGW simultaneously. Static demands are required if the network is lacking a dynamic reservation mechanism and results in overdimensioning of the network resources.

Train Delay Insensitive (TDI) Demands. To tackle the problem of loss of information in case of train delays, a new approach has been developed. In this case we re-interpret the traffic demands by neglecting the exact time-position relation between multiple trains. This implies that single trains are not connected to all the AGWs at the same time but we neglect the information of when and where trains will cross each other exactly. In other words, the network is dimensioned to support that trains could cross in any AGW along their track.

2.4   Solution Techniques: Integer Linear Programming

The network topology, node- and link-related parameters are modelled by using the in-house developed TRS (Telecom Research Software) library [13]. Based on these model parameters and the constraints, the objective function (6) is minimized by using a Branch and Bound based ILP solution technique [10]. From the optimized variables z_n^{vw}, u_l, x_l^v and y_{pijk} the minimal network cost and the dynamic traffic routing can be derived.

3   Dimensioning Linear Railroad Tracks

The first test network is a linear railroad track with a single rail line of varying length: ranging from 2 up to 10 successive AGWs. The dimensioning of this rather simple network topology results in some interesting guidelines.

3.1   Evaluation Results

The three scenarios introduced in Section 2.2 are evaluated for two traffic profiles: profile A (2 crossing trains, load of 700 Mb/s and load of 800 Mb/s in the opposite direction) and profile B (doubled loads compared to profile A, respectively 1400 and 1600 Mb/s).

First we examine the network cost of the static demands with respect to the exact demands. The absolute value of the additional cost for profile B will always be larger than for profile A. Because of the self-evidence of the profile-dependent cost difference, the results are a representation of the average of both profiles (unless mentioned otherwise). The additional network cost in case of static demands is represented on Fig. 2 for the defined scenarios and profiles.

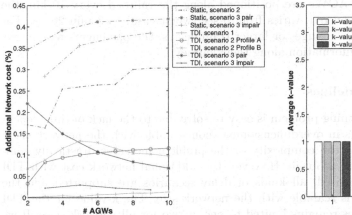 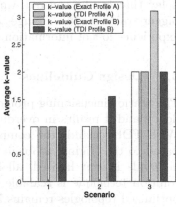

Fig. 2. Network cost for linear tracks with 2 up to 10 AGWs.

Fig. 3. Optimal k-value for linear tracks with 2 up to 10 AGWs.

Because scenario 1 (not depicted on the figure) has a dominant topology cost, the static dimensioning is not significantly bigger however the routing cost increases a lot. For the other two scenarios the static demand case creates a network overdimensioning (up to 42%) that increases with increasing influence of the routing cost and that is not scalable with the network size. This leads to increasing additional network costs with increasing number of AGWs.

Secondly, we examine the network cost of the TDI demands with respect to the exact demands. The additional network cost is also represented on Fig. 2. We define the k-value of an optimized network topology as being the amount of required links originating from the SGW towards the AGWs. This is represented in Fig. 3. For scenario 1 the additional costs are very low because the topology cost has a major influence and the optimal topology only requires a single link originating from the SGW. This leads to increasing additional network costs with increasing number of AGWs for both profiles. For scenario 2 the profiles show a different behavior: profile A is similar to scenario 1. However, for profile

B the optimal TDI topology changes by adding an additional SGW link and the additional cost starts decreasing with increasing amount of AGWs. Scenario 3 shows a similar behavior for both profiles (but significantly different in case of pair or impair number of AGWs): the optimal k is two (due to dominant topology cost) and the additional cost is decreasing with increasing number of AGWs. The optimal k increases for increasing influence of the node cost.

The additional cost is never bigger than 22% (in the rather theoretical 2 AGW case) but more important in case of dominant routing cost (scenario 2, profile B and scenario 3, both profiles) the additional cost decreases with the network size and in case of dominant topology cost (scenario 1, both profiles and scenario 2, profile A) the additional cost remains limited for realistic rail line lengths. Which train delays or train schedule changes can be supported by the exact dimensioning case? If the crossing trains are running 90 km/h on the average and if AGWs are positioned along the railroad every 10 km, the delay that can be covered, varies from 0 sec (worst case) to 3 min 20 sec. For bigger train schedule changes or larger train delays, the passengers will always experience loss of information along their journey.

3.2   Design Guidelines

The static dimensioning problem is easy to solve due to the lack of time dependency and it results in overdimensioning (non-scalable with the network size). With TDI demands the complexity of the problem increases drastically compared to the original problem. However, the additional network cost with TDI demands is kept limited, all kinds of delay scenarios are supported and the solution technique is scalable with the network size. The k-value for the TDI optimized topologies remains limited to one or two for all network sizes. If we examine this in further detail, the cheapest solutions can be found by considering only links towards edge points of the rail line. Links to intermediate AGWs are never used (except if the tiebreaker term of the objective function is decisive). The developed heuristic solutions in this paper will only consider candidate links towards the rail line edge points and will make use of TDI demands.

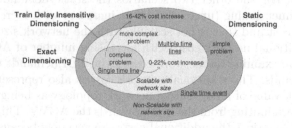

Fig. 4. Traffic models.

4 Design of the Heuristic Dimensioning Method

Two alternative solution techniques are implemented and they can be concisely described as follows:

```
1. Reduce the set of candidate SGW-AGW links
2. Redefine traffic demand profiles
3. (a) Solve the reduced ILP problem
   (b) Use heuristic approach named "K-scanning"
```

First the set of candidate AGW-SGW links is reduced by solely maintaining links connecting the SGW to rail line edge points. After this topology reduction, the traffic demand profiles can be redefined towards more simple traffic demands concentrated in the rail line edge points with addition of minimum capacity constraints for the links along the railroad lines. Subsequently, the same ILP solution technique as presented in Section 2 can be used and will obviously have lower calculation times than previously. However, for larger network problems (e.g. 30 nodes, 10 rail lines and 20 trains would take several days or even weeks on a PC with a 1GHz CPU speed) a heuristic approach that approximates the solution of this reduced ILP formulation, is still desired.

Therefore, the heuristic scans the candidate k-values and starts by calculating the network cost for the minimum and maximum k-value. First of all, a ranking of optimal sets of k AGWs is created. Secondly, the heuristic selects a range of N solutions (N$\geq$1) from the ranking. Thirdly, the minimal solution is calculated by solving the dimensioning problem in a heuristic manner for each of the N sets. The calculation times are reduced because the time dependency between the different rail lines is neglected and only a limited amount of possible paths to route the flows is explored. By increasing the N-value the amount of effort the heuristic spends in finding a suitable solution, increases. Finally, scanning of the solution space is continued by increasing or decreasing the temporary optimal k-value until no further improvement of the optimal network cost is found.

For the examples of Section 3, the "K-scanning" heuristic finds the ideal solution for all scenarios, all profiles and all network sizes. In the remainder of the paper we will further examine the heuristic solution for more complex railroad topologies. Important to notice is that both methods utilize every single link along the railroad track, not depending on traffic profiles or network parameters.

5 Evaluation Results for Ring-Shaped Railroad Tracks

In this section ring-shaped railroad tracks of different sizes (from 9 up to 24 AGWs) and with multiple rail lines (from 1 up to 6) are studied.

5.1 Performance Study of the Heuristics

In Fig. 5 the network cost calculated with the reduced TDI demands is represented in function of the ratio β/α for a ring with 9 AGWs and the traffic profile

Table 1. Traffic profiles for ring network with two crossing trains per rail line (see Fig. 6).

Rail lines (in 100 Mb/s)	Line 1	Line 2	Line 3
Profile A	4/5	6/7	8/9

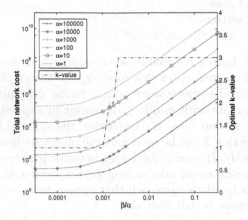

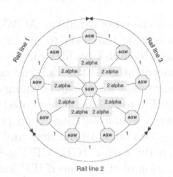

Fig. 5. Network cost and optimal k-value for ring with 9 AGWs and 3 rail lines.

Fig. 6. Test network with 9 AGWs and rail lines.

presented in Table 1. The performance is presented in Table 2 for $\alpha=1$. For high ratios, the solution for the reduced TDI demands has limited extra cost (3.4% for $\beta/\alpha=0.1$). This extra cost is caused by the fact that the heuristic utilizes every AGW-AGW link while this might not be the best choice for all traffic profiles. Take into account that with increasing rail line lengths the relative additional cost caused by this assumption will decrease. However, most important is the confirmation that only candidate SGW links to the rail line edge points are required to find the optimal solution. Based on the considered topology it can easily be proven that the additional cost for low β/α (where the topology cost becomes dominant) converges towards

$$1 - \frac{L_{st} + 2.\alpha}{L_r + 2.\alpha} \tag{7}$$

with L_r the link cost of the network along the railroad and L_{st} the link cost of its minimal spanning tree. For our test network (7) equals 9% for $\alpha=1$, 3% for $\alpha=10$ or 0.5% for $\alpha=100$. In Section 6 an alternative heuristic for low β/α ratios is presented, improving the global heuristic performance. The results of this method, named "Single-K" heuristic, are represented in the final column. Also represented in Fig. 5 is the k-value of the optimal solution. The k-curve is only depending on the ratio β/α and not on the individual α or β values. For higher ratios the k-value equals the maximum value (in this case 3) and for lower ratios it equals the minimum value (always 1). If the routing and topology

Table 2. Performance of reduced ILP, K-scanning and Single-K heuristic.

β/α	Exact solution Absolute value	Reduced ILP Add. cost (%)	K-scanning Add. cost (%)	Optimal k-value	Single-K Add. cost (%)
10	5624	1.2	1.2	3	46.5
1	567	2.7	2.7	3	45.7
0.1	69.3	3.4	3.4	3	40.4
0.0225	26.279	0	4.2	2	16.9
0.01	18.37	0	0.3	1	8.6
0.001	10.8501	7.6	7.6	1	2
0.0001	10.0895	8.9	8.9	1	0.2
0.00001	10.00895	9	9	1	0.02

cost are of similar magnitude, intermediate k-values may be found (only value 2 in this case). In any case this curve remains monotonically increasing for β/α. By applying the "K-scanning" heuristic identical solutions are found for low and high β/α ratios but near the k-slope (between 0.01 and 0.1) the heuristic approach finds an up to 4.5% more expensive solution (see Table 2).

5.2   Influence of Increasing Rail Line Lengths

If the length of the rail lines is increased without changing the traffic profiles, the slope in the k-curve moves to lower β/α ratios. This behavior is depicted in Fig. 7 for ring networks with 3 rail lines but increasing number of AGWs: from 9 to 24 AGWs. This behavior can easily be explained: for fixed α and for longer rail lines, the routing cost will become bigger if the rail lines become longer. This means that additional SGW links can be added on lower β values. The accuracy of the heuristic approach, compared to the optimal solution with reduced TDI demands, improves with the length of the rail lines: for a network with 24 AGWs the additional cost is below 2% near the k-slope. In other words, the performance of the heuristic is scalable with increasing rail line lengths.

5.3   Shape of the k-Curve

The specific shape of the k-value curve is dependent on the number of rail lines and the traffic profiles of these rail lines. We examine the influence of the traffic profiles on the k-curve for a ring of 24 AGWs and 6 rail lines. As can be seen on Fig. 8, the network costs for the 3 traffic profiles (see Table 3) are more or less the same but the k-curves are significantly different. This shape cannot be predicted and therefore the presented heuristic method scans the k-curve from bottom-to-top or top-to-bottom. The minimum k-value always equals one. However, not all intermediate k-values are present and even the maximum value is not necessary known in advance. While six might be expected, the maximum k-value remains five for traffic profile C due to light loaded network regions.

The curve for profile C is not monotonically increasing like the previously simulated curves (dotted line on Fig. 8). This is caused by using N=1 in the heuristic method. If N=3, the heuristic scans more solutions and the optimal

Table 3. Traffic profiles for ring with 24 AGWs, 6 rail lines and 2 crossing trains per rail line.

Rail lines (in 100 Mb/s)	Line 1	Line 2	Line 3	Line 4	Line 5	Line 6
Profile A	4/5	6/7	8/9	6/7	8/9	4/5
Profile B	10/10	10/10	10/10	10/10	10/10	10/10
Profile C	4/5	4/5	4/5	12/14	16/18	8/10

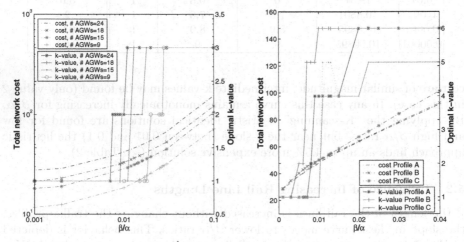

Fig. 7. Influence of increasing rail line length for ring with 9 AGWs, 3 rail lines and the traffic profiles of Table 1.

Fig. 8. Network cost and optimal k-value for ring with 24 AGWs, 6 rail lines and the traffic profiles of Table 3.

nondecreasing curve is found as depicted on the figure. The calculation time increases (from 155 sec to 598 sec on a PC with 1GHz CPU speed) but remains limited and the additional cost gain is at most 1.5% in the affected area. The ideal N-value is dependent on the desired precision and calculation time.

6   Evaluation Results for Grid-Shaped Railroad Tracks

In this final section we introduce a new heuristic and the results are presented on Fig. 9 for a grid example with 21 nodes, 8 rail lines and 16 trains. The new heuristic, named "Single-K", improves scenarios where the optimal k-value equals one or in other words for β/α ratios with dominant topology cost. The previously designed heuristics won't be optimal because they are not designed to cope with an environment where the traffic profile of the fast moving users becomes obsolete. The K-scanning heuristic will always use every AGW-AGW link while the optimal solution will try to minimize the number of utilized links. The additional network cost will increase with the available amount of loops in the network and will also converge towards (7): 14% for $\alpha=1$, 9% for $\alpha=10$ and 2% for $\alpha=100$. Therefore the Single-K heuristic is developed and it can be summarized as follows:

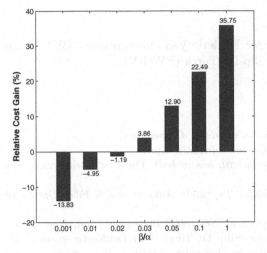

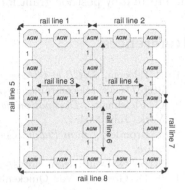

Fig. 9. Relative cost gain for Single-K heuristic with respect to the K-scanning heuristic.

Fig. 10. Test network with 21 AGWs and 8 rail lines.

1. Calculate the center node based on the traffic demands and the topology
2. Connect the center node with the SGW and remove all other AGW-SGW links (i.e. k=1)
3. Calculate the minimum spanning tree originating from the center node
4. Calculate the dimensioning for this topology

For the network in Section 5 the performance for β/α ratios lower than 0.001 is improved drastically to 2% or less. For the grid network presented in this section the Single-K heuristic outperforms the K-scanning heuristic for β/α values below 0.02: gaining 13.8% for β=0.001 and 5% for β=0.01.

7   Conclusion

In this paper we presented a design technique for aggregation networks with fast moving users: network topology design, resource optimization and path determination of the dynamic tunnels. Our approach reduces the capacity planning cost substantially. We found that traffic demands have to be defined as Train Delay Insensitive demands in order to be able to fulfill the QoS guarantees of the passengers at all times. Based on the presented results we can conclude that for an optimal network design the candidate links that need to be considered for connecting the Service Gateway, are links towards rail line end points. This gives way to a reduced Integer Linear Programming formulation which is easier to solve. Due to the time-dependent complexity of the problem, we designed heuristic methods that are able to solve problems of realistic size. The performance of the heuristics is tested and results are promising. Moreover, this technique is scalable because the heuristic's performance doesn't degrade for increasing length of the rail lines.

Acknowledgment

Research is funded by PhD grant for Frederic Van Quickenborne (IWT Vlaanderen) and by postdoc grant for Filip De Turck (FWO-V).

References

1. B-Telecom, *SNCB et télécom: plus d'un siècle d'expérience*, http://www.telecomrail.com/pdf/b-telecom_m21_fr.pdf.
2. G. Fleishman , *Destination wi-fi, by rail, bus or boat*, The New York Times, New York, July 2004.
3. J. Bromley, *Union Pacific Celebrates Twentieth Anniversary of Fiber Optic Program*, http://www.uprr.com/notes/corpcomm/2004/0628_fiberoptic.shtml, June 2004.
4. F. De Greve, F. Van Quickenborne, Filip De Turck, Piet Demeester et al. *Evaluation of a tunnel set-up mechanism in QoS-aware Ethernet access networks*, 13th IEEE Workshop on Local and Metropolitan Area Networks, San Francisco, Nov 2003.
5. B.Van Caenegem, W. Van Parys, F. De Turck, et al., *Dimensioning of survivable WDM networks*, IEEE Journal on Selected Areas in Communications 16 (7): 1146-1157, Sep 1998.
6. W.D. Grover, J. Doucette, "Topological design of survivable mesh-based transport networks," Annals of Operations Research: Topological Network Design in Telecommunication Systems, P. Kubat, J. MacGregor Smith (Editors), P.L. Hammer (Editor-in-Chief), Kluwer Academic Pub., vol. 106 (2001), pp.79-125, Sep 2001.
7. T. Van Leeuwen, I. Moerman, H. Rogier, et al., *Broadband wireless communication in vehicles*, Journal of the Communications Network 2: 77-82 Part 3, Jul-Sep 2003.
8. J. Ott, D. Kutscher, *Drive-thru Internet: IEEE 802.11b for Automobile Users*, IEEE Infocom 2004 Conference, March 2004.
9. F. De Turck, P. Demeester, H. Alaiwan, *Efficient bandwidth scheduling over switched QoS networks*, The 8th Int. Telecommunication Network Planning Symposium, pp. 249-254, Sorrento, Italy, October 18-23, 1998.
10. G. L. Nemhauser and A. L. Wolsey, *Integer and combinatorial optimization* , John Wiley & Sons, 1988.
11. F. Van Quickenborne, F. De Greve et al., *Optimization Models for Designing Aggregation Networks to Support Fast Moving Users*, Mobility and Wireless in Euro-NGI, Wadern, June 7-9, 2004.
12. B. Lannoo, P. Demeester, et al., *Radio over fiber technique for multimedia train environment*, NOC, Vienna, July 2003.
13. K. Casier and S. Verbrugge, *TRS: Telecom Research Library*, http://www.ibcn.intec.ugent.be/projects/internal/trs, 2003.

Management of Non-conformant TCP Traffic in IP DiffServ Networks

Paolo Giacomazzi, Luigi Musumeci, and Giacomo Verticale

Dipartimento di Elettronica e Informazione
Politecnico di Milano, Milano
{giacomaz,musumeci,vertical}@elet.polimi.it

Abstract. TCP traffic that enters an IP network that supports Differentiated Services, is policed packet by packet. The border router marks differently those packets that are conforming to the Commited Information Rate (CIR) and those that are in excess and, therefore, non-conforming. The latter packets have a higher probability of being dropped in the buffers both of the border router and of the core router. This behavior has an implication when TCP sources are present. The TCP congestion control mechanism adapts the source rate to a speed that depends on the packet drop probability; therefore, in case of congestion, not only the excess traffic is dropped, but the TCP source is made to work at a speed that is even lower than the contracted CIR. In this paper, we propose a mechanism capable of controlling the amount of excess traffic that is allowed to enter the DiffServ network in a way that, even in case of congestion, the TCP throughput is not lower than the CIR.

1 Introduction

The Internet Engineering Task Force (IETF) has defined the Differentiated Services (DiffServ) framework as a simple mechanism to provide Quality of Service to traffic aggregates, that is, large sets of flows with similar service requirements. The DiffServ approach does not need per-flow resource reservation and requires the traffic control functions to be performed only the border routers, while simpler functions are implemented by internal routers. In particular, border routers are responsible for ensuring that individual traffic flows conform to the traffic profile specified by the network provider. They are also in charge of the classification function that groups individual traffic flows into a small number of traffic classes according to the similarity of their Quality of Service requirements. In this way, internal routers can manage a few traffic classes as opposed to a large number of individual traffic flows. As the number of traffic flows increases, internal routers are not overloaded by the packet processing burden, which is mainly limited to packet forwarding. In turn, this approach makes the network scalable.

All traffic entering a DiffServ network is *classified* and then *conditioned* to comply with profile requirements. Traffic conditioning is performed according to shaping and/or policing techniques.

The DiffServ service class is specified in the DiffServ field of each IP packet. In particular, the 8-bit Type of Service (ToS) field in the IPv4 header is replaced by the DiffServ field. Six bits of the DiffServ field constitute the Differentiated Services Code Point (DSCP) [1], [2], which identifies the set of operations, called Per Hop Behavior (PHB) [3], [4], performed by routers on all incoming packets.

M. Ajmone Marsan et al. (Eds.): QoS-IP 2005, LNCS 3375, pp. 249–259, 2005.

The DiffServ technique requires that a Service Level Agreement [3] is established between network subscribers and service providers. The packet classification and conditioning functions are ruled by the Traffic Conditioning Agreement, which is part of the Service Level Agreement.

The packet classification function is based on information included in the packet header, such as source and destination addresses, port numbers, and protocol types. In addition, packet classification can be based on information stored inside routers. The conditioning operation includes metering, marking, shaping, and dropping (Figure 1).

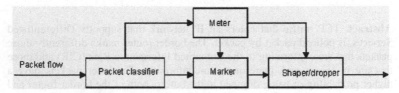

Fig. 1. Traffic conditioning operation in a DiffServ border router

After packet classification, metering functions are performed to verify that the incoming traffic satisfies profile requirements. Then, the marker module shown in Figure 1 assigns each packet a Differentiated Services Code Point, based on the results of classification and metering functions. The marker can also modify the existing value of the Differentiated Services Code Point. This can occur when packets are transmitted across two Internet domains ruled by different administrations.

When a packet violates the traffic profile, it may be dropped or forwarded with a lower service priority level (policing action). The shaping module holds the packets in a queue until they can be forwarded in compliance with the associated profile.

The Internet Engineering Task Force (IETF) has defined three types of DiffServ forwarding: Expedited Forwarding (EF) [5], Assured Forwarding (AF) [6], and Best Effort (BE). Expedited Forwarding provides minimal delay, jitter and packet loss, and it guarantees the required bandwidth. Packets that violate traffic profile requirements are dropped. The Expedited Forwarding service is suitable for delay-sensitive applications, such as voice and video. The Assured Forwarding service classifies IP packets into four traffic classes and three levels of drop precedence. In case of congestion, high-drop-precedence packets are more likely to be dropped than low-drop-precedence packets.

The implementation of the Assured Forwarding service requires an active queue management algorithm capable of solving possible long-term congestion problems within each Assured Forwarding class by dropping packets, while handling short-term congestion problems by queueing packets. Packets must be dropped gradually, based on a smooth congestion indication, in order to avoid dramatic congestion situations in the network.

In the Assured Forwarding service, the packet dropping operation can be performed according to a well known technique, called RIO [7], explained in detail in the following Section.

In a typical application based on the Assured Forwarding Per Hop Behavior [6], it is expected that IP packets are forwarded with high probability as long as they satisfy profile requirements. Packet sources are also permitted to exceed their profile requirements. In this case, excess traffic is forwarded with lower probability. Moreover, it is important to note that the network must not reorder packets of the same flow, neither in-profile nor out-of-profile.

It is likely that the bulk of the Assured Forwarding flows will be generated by applications relying on the TCP transport protocol, such as web-browsing and e-commerce. The TCP continuously increments bandwidth occupation by repeatedly increasing the data transmission rate and monitoring the behavior of the network. As soon as the network starts dropping packets, the TCP reduces its transmission rate. With the Assured Forwarding service, this feature of TCP can lead to poor performance. If a user is allowed to send packets exceeding profile requirements, these packets will be classified as out-of-profile by border routers and discarded by the network with the same probability as Best-Effort packets. In case of congestion, these packets can experience significant losses, which, in turn, trigger a dramatic reduction of the transmission rate at the TCP level. As a consequence, the performance of a TCP flow transported with the Assured Forwarding service is mainly determined by its out-of-profile component. Even if the network has sufficient bandwidth for in-profile packets, the losses experienced by out-of-profile packets downgrade the overall performance of the TCP flow.

In a previous paper [8], we proposed a new marking technique as a solution to TCP performance degradation when non-conformant packets are transmitted with high drop precedence. The proposed marking technique, named Enhanced Token Bucket (ETB), is adaptive, that is, the amount of excess bandwidth allocated by border routers to each TCP flow is variable in order to prevent TCP packet losses caused by excess low-priority traffic in the network.

This adaptive technique requires a congestion signaling procedure from internal to border routers. The paper shows that a simple congestion signaling technique, named Congestion Signaling Algorithm (CSA), can guarantee the required rate of TCP traffic flows. In particular, it is possible to make border routers aware of congestion that occurs in downstream internal routers, in order to adjust the percentage of non-conformant packets entering the network for each Assured Forwarding flow. The main limitation of the previous work was that the signaling information flowing from the internal to the border routers was supposed to travel out-of-band with no loss and no delay. This paper introduces a signaling protocol, based on the ICMP protocol, capable of making the border routers know what internal routers are congested.

The performance obtained with the proposed protocol shows that the achieved throughput is almost equal to what obtained with the ideal protocol and with only a small signaling overhead. Further increasing the signaling overhead yields a lower throughput but always no lower than the CIR.

The paper is organized as follows. Section 2 describes the models of network devices and traffic conditioning adopted in the paper. In Section 3, we explain the proposed signaling protocol. In Section 4, simulation results are presented and, finally, conclusions are drawn in Section 5.

2 Network Architecture

In this section, we describe the models of the network devices and of the traffic conditioning procedures. We also present the ETB packet marking technique and discuss the Congestion Signaling Algorithm (CSA) that we have designed to guarantee the provisioned rate of TCP traffic flows.

We assume that each Assured Forwarding Per Hop Behavior is specified by its average rate r [kbit/s]. All in-profile packets must be delivered with a low loss probability. At the access link, users can transmit at an average rate higher than r. Out-of-profile packets, that is packets exceeding r, are handled as Best-Effort traffic.

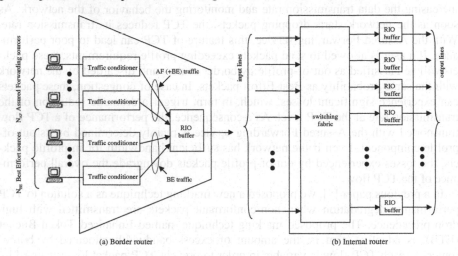

(a) Border router   (b) Internal router

Fig. 2. (a) The border router, (b) the internal router

2.1 Modeling Border and Internal Routers

Both Assured Forwarding and Best-Effort users can access border routers directly (Figure 2a). The number of Assured Forwarding and Best-Effort users is N_{AF} and N_{BE}, respectively. In border routers, traffic conditioning functions, including the packet dropping operation, are performed by the traffic conditioner (Figure 1). We have a conditioner for each input line and a common buffer (RIO buffer), controlled by a RIO packet discarding procedure.

The RIO packet discarding procedure regulates network congestion by selectively discarding in-profile and out-of-profile packets. In particular, when buffer occupancy is high, out-of-profile packets are dropped with higher probability than in-profile packets.

The traffic conditioner can modify the Per Hop Behavior of non-conformant packets by downgrading part of the Assured Forwarding traffic to Best Effort. Therefore, each traffic conditioner receiving Assured Forwarding traffic can feed the RIO buffer with both Assured Forwarding and Best-Effort traffic.

The RIO buffer discards packets based on their Per Hop Behavior. The packet dropping probability increases with the congestion level and is higher for the Best-Effort traffic. Internal routers (Figure 2b) do not perform complex traffic conditioning functions. Packets enter internal routers from input lines and are routed to the appropriate output lines on the basis of their destination. Each output line is provided with a RIO buffer performing the same selective packet discarding procedures adopted by border routers.

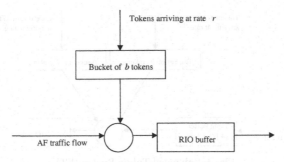

Fig. 3. Token Bucket traffic filter

The traffic conditioning function is performed by a Token Bucket traffic filter (Figure 3), provided with a bucket of b tokens. For the sake of simplicity, we assume that packets have constant length L [bytes]. The Token Bucket is filled at a $10^3 r/(8L)$ [tokens/s] rate, or, alternatively, at an r [kbit/s] rate. When an Assured Forwarding packet enters the filter, if at least one token is available in the bucket, the packet is accepted as conformant in the RIO buffer and one token is removed from the bucket. Otherwise, if there are no tokens in the bucket, the packet is accepted as non-conformant in the RIO buffer.

Therefore, the Token Bucket splits the traffic flow into a *conformant flow* (in-profile packets) at an average r [kbit/s] rate, and a *non-conformant flow* (out-of-profile packets), whose Per Hop Behavior is downgraded.

2.2 Control of the Percentage of Non-conformant Packets at the Token Bucket

A poor performance in the transport of TCP traffic over the Assured Forwarding service is observed if the percentage of downgraded Assured Forwarding packets is not kept low. In particular, when the number of Assured Forwarding traffic sources is high, a large number of non-conformant packets can be offered to border routers. This traffic is forwarded through the RIO buffer with a higher packet loss probability. This can significantly reduce the throughput of TCP connections. In fact, packet losses trigger the slow start and congestion avoidance procedures of TCP and, in turn, these procedures slow down TCP connections. Thus, even if conformant packets are forwarded with a lower loss probability, the higher loss probability of non-conformant packets has a negative effect on the overall performance of TCP flows [8]. Therefore, if the percentage of non-conformant packets is not kept low, the provisioned throughput of Assured Forwarding flows cannot be guaranteed.

To control throughput, we monitor the congestion status of the RIO buffer at border routers. We implement the following control procedures. When the RIO buffer is not congested, the Token Bucket filter can increase the percentage of non-conformant traffic in order to exploit the bandwidth available on the output line. Conversely, when the RIO buffer is congested, the Token Bucket filter must reduce the percentage of non-conformant traffic. In this way, Assured Forwarding packets, which are mostly conformant, are discarded with a lower probability and, therefore, the performance of the Assured Forwarding traffic is downgraded to a lower degree during congestion periods in the RIO buffer.

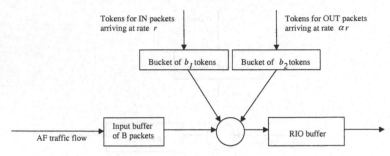

Fig. 4. Enhanced Token Bucket (ETB)

In order to implement a Token Bucket filter that can automatically regulate the percentage of non-conformant packets, we propose an Enhanced Token Bucket (ETB) (Figure 4). The traffic filter is provided with an input buffer and a second bucket of tokens to handle non-conformant packets. The bucket of conformant tokens has depth equal to b_1 and is filled at rate r, while the bucket of non-conformant tokens has depth b_2 and is filled at rate αr. By properly setting the α factor, the percentage of non-conformant packets can be differently limited for each Assured Forwarding flow in the network.

When an Assured Forwarding packet enters the filter, one of the following events can occur: if the input buffer is full, the packet is dropped; otherwise it is queued in the input buffer. Each packet in the input buffer is served according to the following three-step procedure:

1. if the conformant bucket has at least one token, it is served as conformant and a token is removed from the bucket;
2. if the conformant bucket is empty, but the out-of-profile bucket has at least one token, then the packet is served with the Best Effort Per Hop Behavior and a token is removed from the corresponding bucket;
3. if both buckets are empty, the packet is left in the input buffer.

In-profile and out-of-profile accepted packets are forwarded to the RIO buffer. Note that for TCP applications the queueing delay is not critical.

The α parameter represents the percentage of non-conformant packets transmitted by the Token Bucket. The Congestion Signaling Algorithm (CSA) adjusts the value of the α parameter when the RIO buffer is congested.

The RIO buffer can be either non-congested (when $avg\_q\_tot < min\_th\_out$) or congested (when $avg\_q\_tot > min\_th\_out$). Let us assume that the arrival of the first OUT packet during a congestion period occurs at time t_0. Two timers, T_1 and T_2 (with $T_1 < T_2$) are started and a packet counter, P, is set to 1. P is a modulo-8 counter and it returns to the 0 value at the arrival of the 8-th OUT packet. Let us assume that the 8-th arrival occurs at time t. One of the following two conditions holds:

Case 1: $(t - t_0) > T_1$, that is timer T_1 expires before the arrival of the 8-th OUT packet. In this case, the α parameter is decreased by $\Delta\alpha$, in such a way that $\alpha = \max\{\alpha\text{-}\Delta\alpha, min\_\alpha\}$. $\Delta\alpha$ is the decrease/increase step and min_α is the minimum value allowed for α. In particular, α can be decreased at most by $\Delta\alpha$ every T_1 seconds, thus, the maximum decreasing rate of α evaluates to $\Delta\alpha/T_1$. After α has been decreased, both T_1 and T_2 are restarted.

Case 2: $(t - t_0) < T_1$, that is, the 8-th packet arrives before the expiry of timer T_1. In this case, the value of α is not decreased to prevent an overly rapid reduction of α.

In any case, when T_2 expires (that is, when the α parameter has not been decreased for T_2 consecutive seconds), if the RIO buffer is not congested, the value of α is automatically increased by $\Delta\alpha$ and timer T_2 is restarted. In this way, the maximum increasing rate of the α parameter after a congestion period is set to $\Delta\alpha/T_2$.

3   The Protocol for Signalling Congestion to Internal Nodes

In non-congested border routers, Assured Forwarding flows can enter the network with a high percentage of non-conformant packets. If these traffic flows enter a congested RIO buffer of an internal router, they are likely to experience high packet loss due to their non-conformant component. This packet loss will in turn significantly downgrade TCP performance. Therefore, it is important to regulate the percentage of non-conformant packets of border routers not only with local information about the congestion status of the RIO buffer, but also with remote information about the congestion status in RIO buffers of internal routers. This allows border routers to fine tune the value of α on the basis of the congestion status of internal routers.

Let us assume that a reference packet, P, entering the network from the border router BR, is marked as non-conforming by BR. Moreover, let us assume that the internal router IR drops P. In this case, IR sends a signaling packet to BR. The address of BR can be obtained if the DiffServ network adopts MPLS or a link state routing protocol like OSPF.

As soon as BR receives the signaling packet, it assumes that IR is congested. In this way, the ETB stops increasing α and, if non-conforming packets keep on arriving, the ETB may further decrease α. Moreover, BR starts a timer, $T3$. If an additional congestion signaling packet is received from the IR, the timer is restarted; otherwise, when $T3$ expires, IR is not considered in congestion any more. From simulations (reported in the following Section) we observed that the starting value for $T3$ can be higher than the Round Trip Time. In the considered scenario we observed only a minimal throughput loss with starting values for $T3$ as high as 200 ms.

In case of high speed links, the packet loss rate in the internal routers can be very high when congestion occurs, therefore the border routers can receive several signaling packets. Considering that all those packets that arrive before time T3 expires are redundant, the internal router may send only a fraction of them. We present a simple solution where, before sending a signaling packet, the internal router makes a Bernoulli experiment with probability of success equal to p. If the experiment is successful the packet is sent, otherwise it is discarded. In simulations, we registered a negligible performance loss with success probabilities as low as 5%.

4   Simulation Results

The main objective of this section is to assess through simulation that, by adopting the Enhanced Token Bucket, the Congestion Signaling Algorithm, and the new signaling protocol described in the previous section, the provisioned rate of TCP traffic flows can be guaranteed.

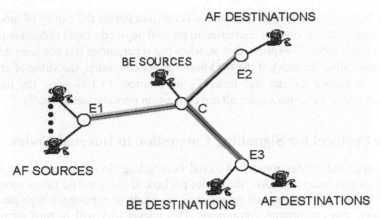

Fig. 5. Simulation Scenario

4.1 Simulation Scenario

The simulation scenario is shown in Figure 5. N_{AF} Assured Forwarding sources are connected via 2-Mbit/s links to the border router E1. Similarly, N_{BE} Best-Effort sources are connected with 2-Mbit/s links to the internal router. Half of the AF sources send TCP traffic to the AF destinations connected to the border router E3, while the other half of the AF sources sent traffic to the AF destinations connected to the border router E2. All the BE traffic goes to the BE destinations connected to the border router E3. In the DiffServ Network, all the internal links have a capacity of 30 Mbit/s.

We consider two cases: the homogenous case and the heterogeneous one. In the former case, all the AF sources have a Commited Information Rate (CIR) equal to 1 Mbit/s. In the latter case half of the sources (type 1 sources) have a CIR equal to 1.6 Mbit/s and the other half (type 2 sources) has a CIR equal to 400 kbit/s. Not that, in the heterogeneous case, half of the type 1 sources and half of the type 2 send traffic towards border router E2, while the other halves of the sources send traffic towards the border router E3.

4.2 Homogeneous Case

Figure 6 shows the throughput achieved by the AF sources that send packets along the path E1 – E2, which does not carry any Best Effort traffic.

The throughput is plotted ad a function of the total number of AF sources and of the type of Enhanced Token Bucket. The line labeled "Ideal ETB" shows the result obtained when the border routers have instantaneous knowledge of the congestion status of internal routers. The lines labeled "T3=10 ms", "T3=100 ms", and "T3=400 ms" are obtained with the proposed signaling protocol and with different starting values of timer T3. Along the considered path, congestion happens only when the number of AF sources is so large that the sum of their CIRs saturates the capacity of the internal links. Under all circumstances, the proposed signaling protocol behaves very well yielding results similar to the ideal case with all the considered values for the timer T3.

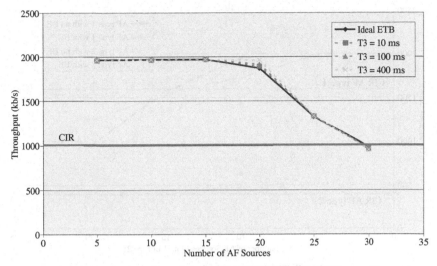

Fig. 6. Throughput along the path E1 – E2 without Best Effort traffic

Figure 7 shows the throughput achieved by the AF sources that send packets along the path E1 – E3, which is shared by AF and Best Effort traffic. In this scenario we consider a number of BE sources equal to the number of AF sources. When congestion occurs, the AF throughput quickly drops, but never falls below the CIR. As the Figure shows, the throughput goes below the CIR, only when the sum of the CIR for the various sources equals the available bandwidth. In this scenario, the proposed signal protocol yields results similar to ideal case only when the starting value for T3 is lower or equal to 100 ms. As the starting value for T3 increases, the throughput decreases but never goes below the CIR.

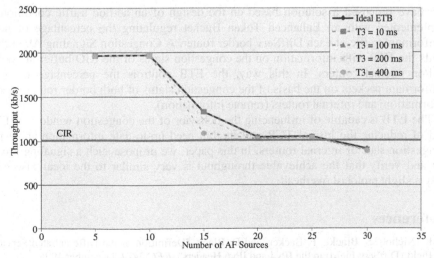

Fig. 7. Throughput along the path E1 – E3 with Best Effort traffic

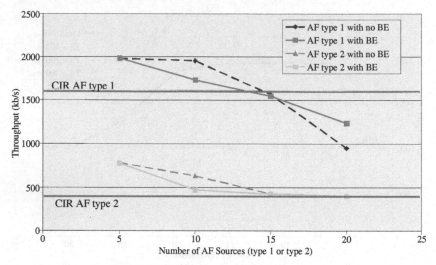

Fig. 8. Througput in the heterogeneous case

Figure 8 shows the throughput achieved in the heterogeneous case using the proposed protocol and a starting value for T3 equal to 200 ms. The proposed solution (ETB plus the signaling protocol) is effective in making several types of traffic sharing the same network while receiving a different degree of QoS. In the considered scenario, the two types of traffic always receive more bandwidth than the CIR until overbooking occurs. In this case, the traffic with the lower CIR receives a better treatment. On the other hand, the sources with a higher CIR obtain a higher share of available bandwidth when traffic is low.

5   Conclusions

We have proposed a solution based on the design of an ad-hoc traffic conditioner, implemented with an Enhanced Token Bucket regulating the percentage of nonconformant traffic at each DiffServ border router. A Congestion Signaling Algorithm feeds the ETB with information on the congestion status of the RIO buffer at border and/or internal routers. In this way, the ETB controls the percentage of nonconformant packets on the basis of the congestion status of both border routers (local information) and internal routers (remote information).

The ETB is capable of influencing the behavior of the congestion window of TCP and of reducing the loss of TCP packets, but need up-to-date information about the congestion state of internal routers. In this paper, we propose such a signaling protocol and verify that the achievable throughput is very similar to the ideal case with only a slight protocol overhead.

References

1. K. Nichols, S. Blacke, F. Backer, and D. Black, "Definition of the Differentiated Services Field (DiffServ Field) in the IPv4 and IPv6 Headers", *RFC 2474*, December 1998.
2. D. Grossman, "New Terminology and Clarifications for Diffserv", *RFC 3260*, April 2002.

3. S. Blake, D. Black, M. Carlson, E. Davies, Z. Wang, W. Weiss, "An Architecture for Differentiated Services", *RFC 2475*, December 1998.
4. S. Brim, B. Carpenter, F. Le Faucheur, "Per Hop Behavior Identification Codes", *RFC 3140*, June 2001.
5. B. Davie, A. Charny, J.C.R. Bennet, *et al.*, "An Expedited Forwarding PHB", *RFC 3246*, March 2002.
6. J. Heinanen, F. Baker, W. Weiss, J. Wroclawski, "Assured Forwarding PHB Group", *RFC 2597*, June 1999.
7. D. Clark and W. Fang, "Explicit Allocation of Best-Effort Packet Delivery Service", *IEEE/ACM Transactions on Networking*, Vol. 6, N. 4, pp. 362-373, August 1998
8. P. Giacomazzi, L. Musumeci, G. Verticale. "Transport of TCP/IP Traffic over Assured Forwarding IP Differentiated Services." *IEEE Network* 17(5), p. 18-28. IEEE, Piscataway, NJ (USA). September/October 2003.

Performance of Active Queue Management Algorithms to Be Used in Intserv Under TCP and UDP Traffic*

Ali Ucar[1] and Sema Oktug[2]

[1] Renault MAIS, I.T. Istanbul, Turkey
ali.ucar@renault.com.tr
[2] Department of Computer Engineering,
Istanbul Technical University, Maslak 34469 Istanbul, Turkey
oktug@cs.itu.edu.tr

Abstract. Active Queue Management techniques are recommended to overcome the performance limitations of TCP congestion control mechanisms over drop-tail networks. Flow Random Early Drop (FRED), GREEN, Stochastic Fair Blue (SFB), Stabilized RED (SRED) are some of the active queue management algorithms which are flow-based in nature. These algorithms can be used to implement Intserv. The main objective of this paper is to present a comparative analysis of the performance of the FRED, GREEN, SFB, and SRED algorithms using the NS-2 network simulator. In this work, the simulations are carried out for the comprehensive analysis and comparison of the algorithms. The algorithms are tested in terms of average queue size, fairness, utilization and packet loss rate by applying various number of flows under TCP, and TCP/UDP traffic.

1 Introduction

Today's Internet only provides Best Effort Service in which traffic is processed as quickly as possible, but there is no guarantee as to timeliness or actual delivery. As demands for service quality and that several service classes arise, the Internet Engineering Task Force (IETF) has proposed many service models and mechanisms to satisfy the required Quality of Service (QoS). The most common of them are the Integrated Services/RSVP (Intserv) model [1, 2], the Differentiated Services (DS) model [3, 4], MPLS [5], Traffic Engineering [6] and Constraint Based Routing [7]. The Integrated Services requires flow-specific state in the routers [8].

IETF has also recommended active queue management mechanisms to overcome the performance limitations of TCP over drop-tail networks. Active queue management algorithms are built based on the idea to convey congestion notification early to the TCP endpoints so that they can reduce their transmission rates before queue overflow. Flow Random Early Drop (FRED)[9], GREEN[11], Stabilized RED (SRED)[13], Stochastic Fair Blue (SFB)[14] are active queue management algorithms which are flow-based in nature.

The main objective of this paper is to present a performance comparison of the FRED, GREEN, SFB, and SRED algorithms utilizing the NS-2[15] Network Simulator. The NS-2 simulation tool is utilized to conduct comprehensive analysis on the performance of the algorithms in terms of average queue size, fairness, utilization and

---

* This work is supported by the Istanbul Technical University Research Fund under Grant 1872

M. Ajmone Marsan et al. (Eds.): QoS-IP 2005, LNCS 3375, pp. 260–270, 2005.

packet loss rate under pure TCP traffic, and a heterogeneous traffic of TCP and UDP flows. In these simulations, round trip times of TCP sources are varied. In the simulations employing heterogeneous traffic, 10 percent of the traffic is UDP based. It is observed that the performance of the algorithms change according to the metric considered.

The algorithms discussed here have been presented by the original authors under some scenarios. However, in this work, we try to compare the performance of the algorithms considering a wide range of parameters which may put them in trouble. We believe that a comparative study of this kind can provide a better understanding of these flow-based active queue management techniques proposed for TCP/IP congestion control and makes the life easier in deciding the appropriate algorithm to deploy.

This paper is organized as follows: Section 2 gives a brief description of the algorithms. The topology used, the performance metrics considered, and the comparison of the employed active queue management algorithms is presented in Section 3. Section 4 concludes the paper by giving future directions.

2   Brief Description of the Algorithms

This section gives a brief description of the FRED, GREEN, SRED, and SFB techniques.

2.1   FRED, Flow Random Early Drop

RED gateways attempt to mark packets sufficiently frequently to control the average queue size [10]. The mechanism works as follows. The gateway detects incipient congestion by computing the average queue size; when this exceeds a preset threshold, arriving packets are marked (or dropped) with a certain probability that is a function of the average queue size. The average queue size is kept low, but occasional bursts can pass through unharmed. During congestion the probability of marking a packet from a particular flow is roughly proportional to the bandwidth share of that flow.

The aim of Flow Random Early Drop (FRED) is to reduce the unfairness observed in RED. FRED generates selective feedback to a filtered set of connections which have a large number of packets queued, instead of indicating congestion to randomly chosen connections by dropping packets proportionally as done by RED.

FRED behaves very similar to RED, but has the following additions:

- introduces the parameters min_q and max_q, to control the minimum and maximum number of packets each flow allowed to buffer.
- introduces the global variable avg_{cq}, an estimate of the average per-flow buffer count; flows with less than avg_{cq} packets queued are favored over flows with more.
- maintains a count of buffered packets q_{len} for each flow that currently has any packets buffered.
- maintains a variable strike for each flow, which counts the number of times the flow has failed to respond to congestion notification; FRED penalizes flows with high strike values.

In FRED, a connection is allowed to buffer min_q packets without loss. All additional packets undergo RED's random drop. Min_q is used as the threshold to decide whether to deterministically accept a packet from a low bandwidth connection. An incoming packet is always accepted if the connection has fewer than min_q packets buffered and the average buffer size is less than max_{th}. When the number of active connections is small ($N \ll min_{th} /min_q$), FRED allows each connection to buffer min_q number of packets without dropping. Some flows, however, may have substantially more than min_q packets buffered. If the queue averages more than min_{th} packets, FRED will drop randomly selected packets. FRED never lets a flow buffer more than max_q packets, and counts the number of times each flow tries to exceed max_q in the per-flow strike variable. Flows with high strike values are not allowed to queue more than avg_{cq} packets; that is, they are not allowed to use more packets than the average flow. This allows adaptive flows to send bursts of packets, but prevents non-adaptive flows from consistently monopolizing the buffer space. The detailed FRED algorithm is given in [9].

2.2 GREEN

GREEN [11] is presented to prevent congestion from ever occurring and ensure a high degree of fairness between the flows. This goal is achieved by applying the knowledge of the steady state behavior of TCP connections to packet drop policy. Besides, GREEN maintains high link utilization, low packet loss, and short queues. The aim of GREEN is to give each connection its fair share of bandwidth while keeping the queue length low. Mathis et al. [12] show that the throughput of a connection satisfies the following equation under certain circumstances:

$$BW = (\frac{MSS.c}{RTT.\sqrt{p}})^2 \qquad (2.1)$$

Where BW is throughput / bandwidth of the connection, MSS is maximum segment size, c is a constant, RTT is round trip time, p is packet loss probability.

In a scenario, where there are N active flows at a router at a particular outgoing link of capacity L, a flow is said to be active only if it has sent at least one packet in a certain window of time. The fair share of a flow is L/N, putting L/N for BW in Equation 2.1; the following equation is obtained for loss probability p.

$$p = (\frac{N.MSS.c}{L.RTT})^2 \qquad (2.2)$$

By keeping p at this value, GREEN makes flows send at their fair rate. The constant c is taken as 0.93 as recommended in [12].

2.3 SRED, Stabilized RED

Stabilized RED[13] (SRED) is an other flow based active queue management scheme, which aims to improve throughput and fairness. Besides discarding packets preemptively with a load-dependent probability, SRED tries to stabilize its buffer occupancy at a level independent of the number of active connections. It uses a statistical mechanism to collect state information of the misbehaving flows and to analyze the information.

The SRED technique is solely based on the idea of checking whether an incoming packet belongs to the same flow with a randomly chosen zombie from the zombie list.

Zombie List is a list of M recent flows, each having extra information of count and timestamp. The list is initially empty. Upon packet arrival, the new coming packet is compared with a randomly chosen zombie. A hit is said to occur if the test is successful. Active number of flows can easily be estimated using the hit rate.

In hit case: if the new packet belongs to the same flow with the zombie then a hit is said to occur. The count of the zombie is increased by one and the timestamp is set to packet arrival time.

In no hit case: with probability p the flow identifier of the zombie is overwritten by the new packet's flow identifier. The count and timestamp are reset to 0 and packet arrival time, respectively. With probability $1-p$ nothing is done.

2.4 SFB, Stochastic Fair Blue

Stochastic Fair Blue (SFB) [14], is introduced to protect TCP flows against non-responsive flows using the BLUE algorithm[16]. SFB maintains $N \times L$ accounting bins. The bins are organized in L levels with N bins in each level. In addition, SFB maintains L independent hash functions. Using the hash functions, SFB maps a flow into one of the N accounting bins in that level. An accounting bin keeps track of queue occupancy statistics of a flow belonging to a particular bin. Every bin stores a marking/dropping probability p_m that is updated according to buffer occupancy. Upon a packet arrival, for each of the levels, it is hashed to a bin. If the number of packets mapped to a bin exceeds a certain threshold, p_m is increased. If the number of packets drops to zero, p_m is decreased. A non-responsive flow quickly rises p_m to 1 in all the bins it is associated to in all the levels. It is always possible for a responsive flow to share a bin or bins with a non-responsive flow. Unless the number of non-responsive flows is very large than responsive ones, a responsive flow is hashed into at least one bin that is not occupied by a non-responsive flow. The marking or dropping decision is based upon p_{min}, the minimum p_m value of all bins the flow is mapped into. If p_m is 1, the packet is declared as belonging to a non-responsive flow and the flow's rate is limited.

3   Comparison of Active Queue Management Algorithms

Two sets of experiments are performed. The first set of experiments is carried out by injecting TCP flows with varying RTTs. Where as in the second, a heterogeneous network is established by injecting both TCP and UDP flows where 10% of all flows are UDP.

3.1   TCP Flows with Varying RTTs

N sources and N sinks are connected to the routers over 10Mbs links with propagation delays varying from 500/Nms to 500ms. The simulations are run with 50,100, 250, and 500 flows. The butterfly topology in Fig.1 is used in the simulations.

The bottleneck has a bandwidth of 155Mbs and a delay of 30ms. Within the first second of the simulation, FTP connections are started, and the simulation is run for

180 seconds. FRED, GREEN, SFB, SRED are implemented at the gateway and the link utilization, fairness, packet loss, and average queue size are measured at this gateway. DropTail queue is also used to provide a baseline when assessing the performance of the other techniques.

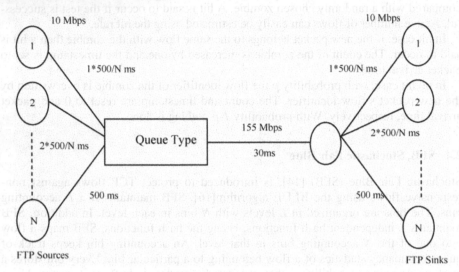

Fig. 1. Network Topology

The performance of the algorithms is studied under four different metrics which are *fairness*, *link utilization*, *packet loss rate*, and *average queue size*.

Fairness
To asses the algorithms performance, *Jain's Fairness Index* [2] is used. This fairness index quantifies the degree of similarity between all the flow bandwidths. Given the set of throughputs $(x_1, x_2, ..., x_n)$, the fairness index is calculated as follows:

$$f(x_1, x_2, ..., x_n) = \frac{(\sum_{i=0}^{n} x_i)^2}{n \sum_{i=0}^{n} x_i^2}$$

(3.1)

The fairness index always lies between 0 and 1. Hence, a higher fairness index indicates better fairness between flows. The fairness index is 1 when all the throughputs are equal. When all the throughputs are not equal, the fairness index drops below 1.

Fig. 2 shows the fairness index values of the algorithms for 50, 100, 250, and 500 flows.

As seen in Fig. 2 GREEN outperforms other queue management schemes in terms of fairness. GREEN's fairness increases as the number of flows increases while it remains nearly constant in other algorithms. GREEN slows down the flows with short RTTs by dropping packets, it does not speed up the flows with long RTTs.

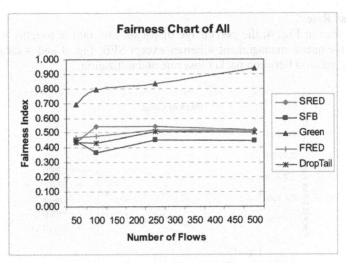

Fig. 2. Jain's Fairness Index vs. Number of Flows, for All Queue Types

SRED and FRED have better fairness compared to DropTail queue, but the difference is very small. SFB performs worst in giving each flow its fair share of bandwidth. The possible reason for this is the misclassification problem. Flows with small RTTs make the bins polluted, which degrades the performance of SFB.

Link Utilization

Fig.3 shows the utilization of the algorithms for 50,100,250, and 500 flows.

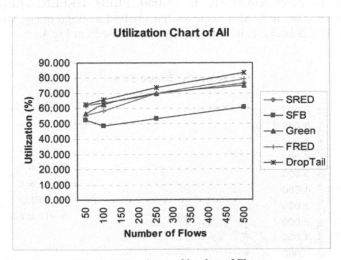

Fig. 3. Utilization vs. Number of Flows

DropTail achieves the highest utilization for aggregate flow but it sacrifices fairness. FRED, GREEN and SRED show similar characteristics in terms of utilization and they all perform well compared to SFB.

Packet Loss Rate

As clearly seen in Fig. 4, the percentage of packet loss rate is roughly same for all studied active queue management schemes except SFB. Fig. 3 and 4 clearly present the opposite relation between packet loss rate and utilization.

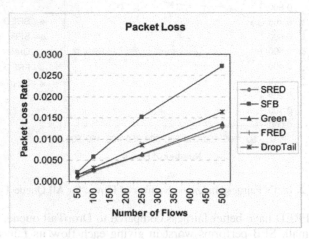

Fig. 4. Packet Loss vs. Number of Flows

Average Queue Size

Fig. 5 shows the average queue size measurements of the algorithms. Looking at the results, it can be said that, as the number of flows increases, the average queue size for DropTail increases remarkably. In contrast, FRED, GREEN, SRED, and SFB keep the queue sizes low. SFB achieves the smallest average queue size in all the simulations. This is because, it drops more packets as seen in Fig.4.

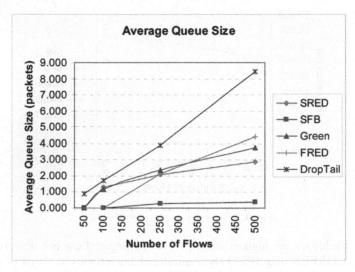

Fig. 5. Average Queue Size vs. Number of Flows

3.2 Heterogeneous Network with TCP and UDP Flows

Internet relies heavily on the cooperative nature of TCP congestion control in order to limit packet loss and fair share of network resources. However, new applications which are not responsive to congestion signals given by the network are also deployed. A lot of research is being conducted to protect routers against these non-responsive flows because they increase the packet loss in the network and impact the performance of responsive flows.

In the second set of simulations, a heterogeneous flow environment is simulated with responsive and non-responsive flows competing for the network resources. The behavior of each technique toward the responsive and non-responsive flows is studied. The throughput of every individual flow is recorded. In our simulations, it is taken that the UDP sources form 10 percent of the overall traffic sources.

The simulations are run with 400 connections using the topology in Fig. 6. The UDP flows transmit at the rate of 0.5Mbs. The simulations are run for 100 seconds. The buffer size of the queue is configured as 200KB.

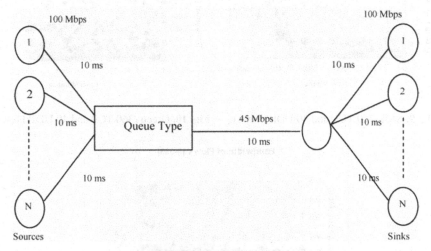

Fig. 6. Network Topology

As Fig.7 – Fig. 11 show, Drop Tail, SFB, FRED and SRED fail in limiting the shares of the UDP flows and TCP flows suffer a lot. GREEN limits the shares of the UDP flows, but it also fails in achieving fair share among the TCP flows. In both cases, SFB suffer from misclassification since UDP flows polluted its bins. Thus, SFB can not protect the responsive flows. Since FRED keeps state for flows which have packets queued, it requires large buffers to work well. Otherwise, FRED can not detect non-responsive flows which may not have enough number of packets queued. This probably shows why FRED also fails under this environment.

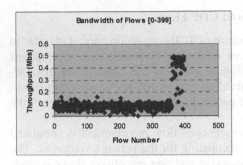

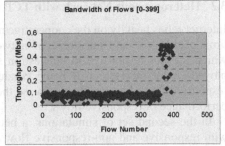

Fig. 7. Droptail (360 TCP and 40 UDP Flows)   **Fig. 8.** FRED (360 TCP and 40 UDP Flows)

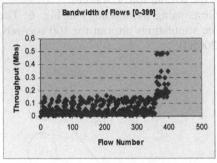

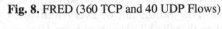
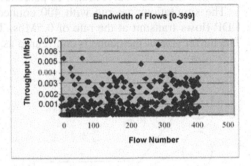

Fig. 9. SFB (360 TCP and 40 UDP Flows)   **Fig. 10.** Green (360 TCP and 40 UDP Flows)

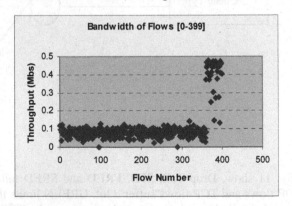

Fig. 11. SRED (360 TCP and 40 UDP Flows)

4 Conclusions and Future Work

In this paper, a comparative analysis of the performance of the flow based active queue management algorithms, which are FRED, GREEN, SFB, and SRED, is studied. First, an environment of TCP flows with varying RTTs is created and each algorithm is examined in terms of fairness, average queue size, packet loss rate, and link utilization. It is observed that, GREEN outperforms other queue management schemes in terms of fairness. Surprisingly, SRED, although not primarily designed with fairness concern, achieves better fairness compared to FRED and SFB whose actual business is to provide fairness among flows. FRED has better fairness compared to DropTail queue, but the difference is rather small. SFB performs worst in giving each flow fair bandwidth share. DropTail achieves the highest utilization. FRED, GREEN, and SRED show similar characteristics in terms of utilization and they all perform well compared to SFB. SFB also performs poor when packet loss rate is considered. The other techniques produce similar values for packet loss rate. The average queue size measurements of the algorithms show that, as the number of flows increases, FRED, GREEN, SRED, and SFB keep their queue sizes low. SFB achieves the smallest average queue size in all the experiments.

In the heterogeneous network case, where TCP flows compete with the non-responsive UDP flows, Drop Tail, SFB, FRED, and SRED fail in limiting the shares of the UDP flows and the TCP flows suffer a lot. GREEN limits the shares of the UDP flows, but it also fails in sharing bandwidth fairly among the TCP flows. In both cases, SFB suffers from misclassification. UDP flows pollute its bins. Thus, SFB can not protect the responsive flows. FRED also can not detect non-responsive flows. It needs large buffers to operate well, 200KB of buffer size is not sufficient in our scenario.

As future work, trace files obtained for multimedia traffic, and web traffic, will be used in the simulations. The performance of SFB can be investigated by varying the number of bins. FRED tests can be repeated with larger buffer size values.

References

1. J. Postel, 1981, Transmission Control Protocol, RFC 793, IETF
2. D. Chiu and R. Jain, 1989, Analysis of the Increase and Decrease Algorithms for Congestion Avoidance in Computer Networks, Comp. Networks and ISDN Sys., vol. 17, no. 1
3. S. Floyd, 1994, TCP and Explicit Congestion Notification, ACM Comp. Commun. Rev., vol. 24, no. 5
4. A. Charny, D. Clark, and R. Jain, 1995, Congestion Control with Explicit Rate Indication, Proc. ICC
5. S. Floyd and V. Jacobson, 1991, Traffic Phase Effects in Packet-switched Gateways, ACM Comp. Commun. Rev., vol. 21, no. 2
6. B. Braden, 1998, Recommendations on Queue Management and Congestion Avoidance in the Internet, RFC 2309, IETF
7. P. Ferguson and G. Huston, 1998, Quality of Service: Delivering QoS on the Internet and in Corporate Networks, Wiley
8. V. Jacobson, 1988, Congestion Avoidance and Control, ACM Comp. Commun. Rev., vol. 18, no. 4
9. D. Lin and R. Morris, 1997, Dynamics of Random Early Detection, in Proc. of ACM SIGCOMM
10. S. Floyd and V. Jacobson, 1993, Random Early Detection Gateways for Congestion Avoidance, IEEE/ACM Trans. Net., vol. 1, no. 4
11. W. Feng, A. Kapadia, S, Thulasidasan, 2002, GREEN: Proactive Queue Management over a Best-Effort Network, In Proceedings of IEEE Globecom, Taipei, Taiwan
12. M. Mathis, J. Semke, J. Mahdavi, and T. Ott, 1997, The Macroscopic Behavior of the TCP Congestion Avoidance Algorithm," Computer Communication Review, vol. 27, no. 3

13. Teunis J. Ott, T. V. Lakshman, and Larry H. Wong, 1999, SRED: Stabilized RED, in Proceedings of INFOCOM, vol. 3
14. W. Feng, D. Kandlur, D. Saha, and K. Shin, 2001, Stochastic Fair Blue: A Queue Management Algorithm for Enforcing Fairness, in Proc. of IEEE INFOCOM
15. NS, Network Simulator, http://www.isi.edu/nsnam/ns
16. W. Feng, D. Kandlur, D. Saha, and K. Shin. Blue: A New Class of Active Queue Management Algorithms. In UM CSE-TR-387-99, April 1999.

An Analytical Framework to Design
a DiffServ Network Supporting EF-, AF- and BE-PHBs*

Mario Barbera, Alfio Lombardo, Giovanni Schembra, and Andrea Trecarichi

D.I.I.T. – University of Catania, Viale A.Doria 6 – 95125 Catania – Italy
{mbarbera,lombardo,schembra,ctreca}@diit.unict.it

Abstract. Differentiated Services (DiffServ) architecture enables IP networks to offer different QoS levels to different users and applications. In this context, the definition of analytical tools to dimension network resources and buffer parameters has become a challenging problem for telecommunications research. The target of this paper is to define a new model of a DiffServ domain, loaded by TCP sources requiring different PHB's. In particular three different PHBs are considered: the AF-PHB, used for TCP data traffic with throughput requirements, the EF-PHB, chosen by all aggregates of TCP traffic with both throughput and delay requirements, and the BE-PHB, assigned to TCP traffic with no QoS requirements. The analytical framework is applied to a case study to evaluate how a different setting of the bandwidth reserved for EF traffic can impact the performance offered by the Diffserv network, not only to EF traffic but also to AF and BE traffic.

1 Introduction

The development of the Internet in both applications and infrastructures has determined the need to provide users with quality of service (QoS) guarantees. The most relevant solution proposed in the literature is the Differentiated Services (DiffServ) architecture, proposed in [1] This architecture enables IP networks to offer different QoS levels to different users and applications, locating traffic classification and conditioning functions at the edge routers, thus relieving core routers from complex tasks. Core routers discriminate between packets exclusively on the basis of the information impressed at the network edge. Here we will consider the two standardized and most widely used PHBs, Assured Forwarding (AF) [2] and Expedited Forwarding (EF) [3]. In addition, we will consider another PHB, Best-Effort (BE), which was defined to manage network traffic with no QoS requirements.

The AF-PHB was proposed to provide individual or aggregate flows with guarantees in terms of throughput and burstiness, according to negotiated profiles. When packets arrive at the edge routers in a DiffServ domain, a profile meter performs measurements and assigns packets a drop precedence (DP) according to the measurement results, and stores the DP in the DiffServ Code Point (DSCP) of the packets [4]. Congested DiffServ nodes try to protect packets with a lower DP value from being lost by preferably discarding packets with a higher DP value. As suggested in [1], this can be achieved by employing Active Queue Management (AQM) mechanisms in the

---

* This work was funded by the TANGO project of the FIRB program of the Italian Ministry for Education, University and Research.

M. Ajmone Marsan et al. (Eds.): QoS-IP 2005, LNCS 3375, pp. 271–285, 2005.

core routers such as WRED (Weighted RED), proposed in [5] as an extension of the classical Random Early Detection (RED) [6].

The EF-PHB is used to build a low loss, low latency, low jitter, assured bandwidth, end-to-end service through differentiated service domains.

In order to exploit the DiffServ facilities in a real environment, the definition of an analytical tool to dimension network resources and buffer parameters is today a challenging problem [7-10]. The target of this paper is to define a new model of a Diff-Serv domain where traffic is generated by data sources requiring different PHBs. In order to maintain a low level of model complexity for any buffer queue dimension or network topology, the fluid-flow approach is adopted [11]. All the three different PHBs described so far are considered in the paper: the AF-PHB, used for TCP data traffic with throughput requirements, the EF-PHB, chosen by all aggregates of TCP with both throughput and delay requirements, and the BE-PHB, assigned to TCP traffic with no QoS requirements.

In this paper we assume that all network nodes implement the above-mentioned PHBs using three different buffers, one for each PHB, and sharing the same link capacity through a scheduler. As suggested in [3], priority queue scheduling is widely considered as the canonical example for EF PHB implementation. Thus the scheduler considered in the proposed model gives strict priority to EF traffic, also assuring minimum throughput guarantees to AF traffic and sharing out excess bandwidth equally among AF and BE traffic aggregates. In order to prevent EF flows from causing starvation of other PHBs, each EF traffic aggregate is policed by a shaper.

The analytical framework is applied to a case study to evaluate how a different setting of the bandwidth reserved for EF traffic can impact the performance offered by the DiffServ network, not only to EF traffic but also to AF and EF traffic.

The rest of the paper is organized as follows. In Section 2, an analytical model of the system is presented. Section 3 applies the model to a case study and gives numerical results. Finally, Section 4 presents our conclusions.

2 System and Model

In this section we assume that all applications adopt the TCP protocol and we limit our study to a steady-state analysis of the system. For this reason, we only consider the greedy source model; however, sources that only transmit during an activity period, or sources that have to transmit finite-size files can easily be modeled as in [12]

In the rest of this section we will first introduce the TCP behavior model (Section 2.1), and then describe the DiffServ router network model (Section 2.2).

2.1 TCP Behavior Modeling

Let us consider a network workload of $N^{(F)}$ flows denoted as $f_1, f_2, ..., f_{N^{(F)}}$, each emitted by $N^{(F)}$ different sources. Since organizations negotiate a traffic profile for the aggregate of all their flows with a DS network provider, in the proposed model the traffic flows are assumed to be grouped together in such a way as to form $N^{(A)}$ traffic aggregates (of course $N^{(A)} \leq N^{(F)}$), denoted as $a_1, a_2, ..., a_{N^{(A)}}$. So, each of the $N^{(F)}$ flows belongs to one of the $N^{(A)}$ traffic aggregates. For the sake of simplicity, all the flows belonging to the same traffic aggregate are assumed to follow the same path within the DiffServ network. In this way all the flows belonging to a generic traffic aggregate will have the same average behavior. It is important to stress the fact that the complex case of different flows belonging to the same traffic aggregate but following different paths within the network can easily be taken into account by slightly modifying some aspects of the proposed model, but this will be omitted here.

Among the numerous versions proposed for the TCP congestion control algorithm, we chose TCP-New-Reno [13] because it is one of the most common in the present Internet. In order to characterize the source model completely, we need to derive the average time variation of:

- the congestion window $W_n(t)$ of the generic flow f_n. In our framework we will not consider the end-to-end flow control algorithm, assuming TCP throughput to be bounded by the congestion control algorithm alone. For this reason $W_n(t)$ also represents the transmission window of the n-th TCP flow.
- the threshold $T_n(t)$ separating the Slow Start range and the Congestion Avoidance range for the congestion window of the generic flow f_n.

The time variation of the window size $W_n(t)$ is the sum of two contributions: the first term, $AI_n(t)$, corresponds to the additive-increase part, the second, $MD_n(t)$, corresponds to the multiplicative-decrease part:

$$\frac{dW_n(t)}{dt} = AI_n(t) + MD_n(t) \qquad n = 1, 2, ..., N^{(F)} \qquad (1)$$

For each ACK packet reaching a generic TCP source the congestion window size $W_n(t)$ increases by one packet during the Slow Start phase, while it increases by $1/W_n(t)$ packets during the Congestion Avoidance phase. So, if we indicate the arrival rate of ACKs for the flow f_n as $r_n^{(ACK)}(t)$, we can write

$$AI_n(t) = \frac{r_n^{(ACK)}(t)}{[W_n(t)]^{u(W_n(t) - T_n(t))}} \qquad n = 1, 2, ..., N^{(F)} \qquad (2)$$

where $u(x)$ is the step function (its value is 1 for $x \geq 0$ and 0 elsewhere), while $r_n^{(ACK)}(t)$ will be derived in Section 2.2.

In order to derive the multiplicative-decrease term, we observe that a New-Reno TCP source behaves differently according to whether it detects a packet loss by receiving a triple duplicate ACK (TD loss) or because a timeout (TO loss) expires. For this reason we need to distinguish between the rate of TD losses, $\gamma_n^{(TD)}(t)$, and the rate of TO losses, $\gamma_n^{(TO)}(t)$, detected by the source of the flow f_n at the instant t.

According to the New-Reno TCP congestion control algorithm, the source halves its congestion window when a TD loss occurs, while it sets its congestion window to one when a timeout expires. Consequently, the variation of the congestion window $W_n(t)$ is equal to $-(W_n(t)/2)$ when a TD loss is detected, while it is equal to $(1 - W_n(t))$ when a TO loss is detected. From these considerations we derive the relationship for $MD_n(t)$:

$$MD_n(t) = -\frac{W_n(t)}{2} \cdot \gamma_n^{(TD)}(t) + (1 - W_n(t)) \cdot \gamma_n^{(TO)}(t) \qquad n = 1, 2, ..., N^{(F)} \qquad (3)$$

In order to derive $\gamma_n^{(TD)}(t)$ and $\gamma_n^{(TO)}(t)$ we assume that information about losses travels through the network with the packets sent out by the generic TCP source along the same path. So in Section 2.2 we will consider the network as also being passed through by $N^{(F)}(t)$ ghost flows, each carrying loss indications relating to a TCP source; let us denote the rate of loss indications for the generic flow f_n at the time instant t as $r_n^{(LOSS)}(t)$. It is important to note that the rate of loss indication $r_n^{(LOSS)}(t)$ and the total packet loss rate detected by the source $\gamma_n(t) = \gamma_n^{(TO)}(t) + \gamma_n^{(TD)}(t)$, are quite different. Loss indications are not actual information sent by the network routers to the sources, but only a modeling artifice. Nevertheless $\gamma_n^{(TD)}(t)$ and $\gamma_n^{(TO)}(t)$ can be derived from $r_n^{(LOSS)}(t)$ considering the way a generic TCP New-Reno source operates in detecting losses. To this end we need to consider the number of ACKs, $N_n^{(ACK)}(t - \tau, t)$, received by the source of the generic flow f_n during a time interval τ that ends at the time instant t:

$$N_n^{(ACK)}(t - \tau, t) = \int_{t-\tau}^{t} r_n^{(ACK)}(v)dv \qquad n = 1, 2, ..., N^{(F)} \qquad (t > \tau) \qquad (4)$$

At the generic instant t, if the number of ACKs received in the last time interval equal to the retransmission timeout (RTO) is less than 3, i.e. $N_n^{(ACK)}(t - RTO) < 3$, any losses are detected at the instant t as TO losses, and therefore the rate of loss indications at the instant $t - RTO$, $r_n^{(LOSS)}(t - RTO)$, becomes a TO loss rate at instant t, $\gamma_n^{(TO)}(t)$. If, on the contrary, there exists a time interval $[t - \tau, t]$ with a duration τ less than RTO, where the number of ACKs received is equal to 3, i.e. $N_n^{(ACK)}(t - RTO) = 3$, then losses are detected as TD losses, and therefore the loss rate at the instant $t - \tau$, $r_n^{(LOSS)}(t - \tau)$, becomes a TD loss rate at the instant t, $\gamma_n^{(TD)}(t)$.

So, assuming that the retransmission timeout can be approximated by four times the average round trip time, $RTT_n(t)$ [14], we have:

$$\gamma_n^{(TO)}(t) = \begin{cases} r_n^{(LOSS)}(t - 4 \cdot RTT_n(t)) & \text{if } N_n^{(ACK)}(t - 4 \cdot RTT_n(t), t) < 3 \\ 0 & \text{elsewhere} \end{cases} \qquad (5)$$

$$\gamma_n^{(TD)}(t) = \begin{cases} r_n^{(LOSS)}(t - \tau) & \text{with } \tau : N_n^{(ACK)}(t - \tau, t) = 3 \text{ if } \tau < 4 \cdot RTT_n(t) \\ 0 & \text{elsewhere} \end{cases} \qquad (6)$$

The relationship to compute $RTT_n(t)$ will be derived in Section 2.2.

The second differential equation we need in order to complete the source model is the one relating to the threshold $T_n(t)$ separating the slow start range from the Congestion Avoidance range for the congestion window of the generic flow f_n. This threshold is set to half of the congestion window every time a loss is detected (both TD losses and TO losses); so its variation is equal to zero when no losses are detected, and to $\frac{1}{2}W_n(t) - T_n(t)$ otherwise:

$$\frac{dT_n(t)}{dt} = \left(\frac{W_n(t)}{2} - T_n(t) \right) \cdot \gamma_n(t) \qquad n = 1, 2, ..., N^{(F)} \qquad (7)$$

Up to now we have derived $2 \cdot N_n^{(F)}$ differential equations, but some of them are duplicated because all the flows belonging to the same aggregate have the same average behavior. Consequently $2 \cdot N_n^{(A)}$ differential equations are sufficient (two equations per aggregate).

In order to simplify the notation we indicate the arrival rate of ACKs for the aggregate a_m, and the rate of loss indications for the aggregate a_m, as $R_m^{(ACK)}(t)$ and $R_m^{(LOSS)}(t)$, respectively. Obviously we have:

$$r_n^{(ACK)}(t) = \frac{R_m^{(ACK)}(t)}{N_{a_m}^{(F)}} \qquad n = 1, 2, ..., N^{(F)}, \text{ and } m : f_n \in \mathfrak{I}_{a_m}^{(F)} \tag{8}$$

$$r_n^{(LOSS)}(t) = \frac{R_m^{(LOSS)}(t)}{N_{a_m}^{(F)}} \qquad n = 1, 2, ..., N^{(F)}, \text{ and } m : f_n \in \mathfrak{I}_{a_m}^{(F)} \tag{9}$$

where $\mathfrak{I}_{a_m}^{(F)}$ is the set of flows belonging to the traffic aggregate a_m, and $N_{a_m}^{(F)}$ is the number of flows in the aggregate.

For the next derivations it is necessary to calculate the total emission rate $\Lambda_m(t)$ at the generic instant t for each traffic aggregate a_m:

$$\Lambda_m(t) = N_{a_m}^{(F)} \cdot \frac{W_n(t)}{RTT_n(t)} \qquad m = 1, 2, ..., N^{(A)}, \forall n : f_n \in \mathfrak{I}_{a_m}^{(F)} \tag{10}$$

2.2 DiffServ Network Modeling

The proposed model is able to describe the behavior of a DiffServ domain in which three types of PHB can be defined: Expedited Forwarding (EF), Assured Forwarding (AF) and Best-Effort (BE).

Let $\mathfrak{I}_{EF}^{(A)}$, $\mathfrak{I}_{AF}^{(A)}$ and $\mathfrak{I}_{BE}^{(A)}$ be the sets of EF, AF and BE traffic aggregates, respectively. Each traffic aggregate belonging to $\mathfrak{I}_{EF}^{(A)}$ or $\mathfrak{I}_{AF}^{(A)}$ is characterized by a service profile. We will assume that the service profile of an EF traffic aggregate is defined by its Average Transmission Rate (ATR) and its Maximum Transmission Rate (MTR), while the service profile of an AF traffic aggregate is completely defined by both its Committed Information Rate (CIR) and its Committed Burst Size (CBS).

In order to provide aggregate flows with QoS guarantees, according to the negotiated profiles, we assume that routers in the DiffServ domain have the internal structure shown in Fig. 1. More exactly, Fig. 1 presents the generic ingress node architecture; the generic interior node structure is the same without traffic conditioning elements.

Before continuing with the description of the network model we need to introduce some further notation:

- K_m: number of buffers passed through by packets (and corresponding ACKs) belonging to the traffic aggregate a_m;
- $\mathfrak{I}_{a_m}^{(B)} = \{b_1, b_2, ..., b_{K_m}\}$: ordered set of AQM buffers in the path followed by packets (and corresponding ACKs) belonging to the traffic aggregate a_m;
- d_j: constant propagation delay (expressed in seconds) of the generic unidirectional output link associated to the AQM buffer b_j;
- $\lambda_{j,m}(t)$: arrival rate (expressed in packets per second) at the input of the AQM buffer $b_j^{(m)} \in \mathfrak{I}_{a_m}^{(B)}$ of packets belonging to the traffic aggregate a_m;

- $\mu_{j,m}(t)$: output rate (expressed in packets per second) from the AQM buffer $b_j^{(m)} \in \mathfrak{I}_{a_m}^{(B)}$ of packets belonging to the traffic aggregate a_m;
- $v_{j,m}(t)$: loss rate (expressed in packets per second) for sources belonging to the traffic aggregate a_m, at the output of the AQM buffer $b_j \in \mathfrak{I}_{a_m}^{(B)}$.

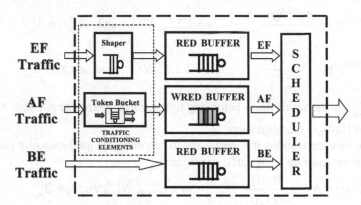

Fig. 1. Architecture of a generic ingress node

To better understand how we model the interactions between network elements and traffic aggregates we will discuss the equations related to the three different types of PHB separately. A further subsection will be devoted to describing the interactions between traffic aggregates adopting different PHBs within the scheduler in each node.

2.2.1 EF-PHB Modeling

The shaper prevents the transmission rate of EF traffic aggregates from exceeding its declared maximum transmission rate (MTR), R_{MAX}. To this end it is implemented with a buffer having a service capacity equal to R_{MAX}. Usually, the shaper buffer size $L^{(SH)}$ is small, in order to reduce delay and jitter in packet delivery.

Let $q_m^{(SH)}(t)$ be the shaper queue size associated with the generic EF traffic aggregate a_m. The differential equation that describes the time variation of $q_m^{(SH)}(t)$ is:

$$\frac{dq_m^{(SH)}(t)}{dt} = \begin{cases} \left[\Lambda_m(t) - R_{MAX_m}\right]^+ & \text{if } q_m^{(SH)}(t) = 0 \\ \Lambda_m(t) - R_{MAX_m} & \text{if } 0 < q_m^{(SH)}(t) < L_m^{(SH)} \\ \left[\Lambda_m(t) - R_{MAX_m}\right]^- & \text{if } q_m^{(SH)}(t) = L_m^{(SH)} \end{cases} \quad \forall a_m \in \mathfrak{I}_{EF}^{(A)} \quad (11)$$

where $[f(t)]^+$ is equal to $f(t)$ when it is positive, and equal to zero otherwise, while $[f(t)]^-$ is equal to $f(t)$ when it is negative, and equal to zero otherwise.

The shaper output rate $\Lambda_m^{(SH)}(t)$ can be obtained through the following relationship:

$$\Lambda_m^{(SH)}(t) = \begin{cases} R_{MAX_m} & \text{if } q_m^{(SH)}(t) > 0 \\ \min\left(\Lambda_m(t), R_{MAX_m}\right) & \text{if } q_m^{(SH)}(t) = 0 \end{cases} \quad (12)$$

Finally we can observe that the packet loss rate $\gamma_m^{(SH)}(t)$ at the ingress of the shaper is equal to the total arrival rate minus the output rate and the shaper queue variation:

$$\gamma_m^{(SH)}(t) = \Lambda_m(t) - \Lambda_m^{(SH)}(t) - \frac{dq_m^{(SH)}(t)}{dt} \tag{13}$$

Substituting (11) and (12) in (13) we obtain:

$$\gamma_m^{(SH)}(t) = \begin{cases} \left[\Lambda_m(t) - R_{MAX_m}\right]^+ & \text{if } q_m^{(SH)} = L_m^{(SH)} \\ 0 & \text{otherwise} \end{cases} \tag{14}$$

After shaping in the edge router, packets belonging to EF traffic aggregates pass through a set of RED buffers.

In this case the arrival rate $\lambda_{j,m}(t)$ at the generic RED buffer $b_j \in \mathfrak{S}_{a_m}^{(B)}$ is:

$$\lambda_{j,m}(t) = \begin{cases} \Lambda_m^{(SH)}(t) & \text{if } j = 1 \\ \mu_{j-1,m}(t - d_{j-1}) & \text{if } 1 < j \le K_m \end{cases} \qquad \forall m: \ a_m \in \mathfrak{S}_{EF}^{(A)} \tag{15}$$

The set of differential equations to obtain the average queue length $q_j(t)$ and the loss probability $p_j(t)$ for a generic RED buffer was derived in [12] and will be omitted here.

Let $\tau_j(t)$ be the queuing time in the buffer $b_j \in \mathfrak{S}_{a_m}^{(B)}$ at the instant t. If we assume that the service rate $c_j^{(EF)}(t)$ assigned to buffer $b_j \in \mathfrak{S}_{a_m}^{(B)}$ by the corresponding scheduler changes slowly, we can approximate $\tau_j(t)$ as:

$$\tau_j(t) \cong \frac{q_j(t)}{c_j^{(EF)}(t)} \qquad \forall j: \ b_j \in \mathfrak{S}_{a_m}^{(B)} \quad \text{and} \quad a_m \in \mathfrak{S}_{EF}^{(A)} \tag{16}$$

In order to derive the output rate $\mu_{j,m}(t)$ we observe that the total number of packets belonging to the traffic aggregate a_m that will be served by the buffer $b_j \in \mathfrak{S}_{a_m}^{(B)}$ up to the time instant $t + \tau_j(t)$ is equal to the total number of packets arriving in the buffer up to the instant t, minus the total number of packets lost in the same buffer, that is:

$$\int_0^{t+\tau_j(t)} \mu_{j,m}(v)dv = \int_0^t \lambda_{j,m}(v)dv - \int_0^t \lambda_{j,m}(v)p_j(v)dv \tag{17}$$

Calculating the derivative of both members of (17) we obtain:

$$\mu_{j,m}(t + \tau_j(t)) = \frac{\lambda_{j,m}(t)(1 - p_j(t))}{1 + d\tau_j(t)/dt} \qquad \forall j: b_j \in \mathfrak{S}_{a_m}^{(B)} \quad \text{and} \quad a_m \in \mathfrak{S}_{EF}^{(A)} \tag{18}$$

Now we are interested in deriving the rate of packet losses, $v_{j,m}(t)$, suffered by the traffic aggregate a_m at the output of the generic buffer $b_j \in \mathfrak{S}_{a_m}^{(B)}$. We assume that the packet loss rate at the output of the buffer b_j at the time instant $t + \tau_j(t)$ is equal to the packet loss rate at the input of the same buffer at the time instant t, that is:

$$v_{j,m}(t + \tau_j(t)) = \begin{cases} \gamma_m^{(SH)}(t) + \lambda_{1,m}(t) \cdot p_1(t) & \text{if } j = 1 \\ v_{j-1,m}(t - d_{j-1}) + \lambda_{j,m}(t) \cdot p_j(t) & \text{if } 1 < j \le K_m \end{cases} \tag{19}$$

$$\forall j: \ b_j \in \mathfrak{S}_{a_m}^{(B)} \quad \text{and} \quad a_m \in \mathfrak{S}_{EF}^{(A)}$$

From the above derivations it is easy to express the arrival rate of ACKs, $R_m^{(ACK)}(t)$, and the rate of loss indications, $R_m^{(LOSS)}(t)$, relating to the sources belonging to the aggregate a_m, in the following way:

$$R_m^{(ACK)}(t) = \mu_{K_m,m}\left(t - d_{K_m}\right) \qquad \forall a_m \in \mathfrak{I}_{EF}^{(A)} \tag{20}$$

$$R_m^{(LOSS)}(t) = \nu_{K_m,m}\left(t - d_{K_m}\right) \qquad \forall a_m \in \mathfrak{I}_{EF}^{(A)} \tag{21}$$

To complete the set of equations related to EF-PHB we need to derive the round-trip time $RTT_n(t)$ relating to the generic flow f_n belonging to the traffic aggregate a_m. It can be approximated by the following relationship:

$$RTT_n(t) = \frac{q_m^{(SH)}(t)}{R_{MAX_m}} + \sum_{b_j \in \mathfrak{I}_{a_m}^{(B)}} \left(\frac{q_j(t)}{c_j^{(EF)}(t)} + d_j \right) \quad n = 1,2,...,N^{(F)} \text{ and } m : f_n \in \mathfrak{I}_{a_m}^{(F)} \tag{22}$$

To evaluate network performance for EF traffic it is useful to calculate the mean delay D_n and jitter J_n suffered by packets belonging to the EF traffic aggregate a_m.

Let $\mathfrak{I}_{a_m}^{(B/s-d)}$ be the set of buffers passed through by packets belonging to the EF traffic aggregate a_m from source hosts to destination hosts. The delay $D_m(t)$ at the generic time instant t can be calculated as follows:

$$D_m(t) = \frac{q_m^{(SH)}(t)}{R_{MAX_m}} + \sum_{b_j \in \mathfrak{I}_{a_m}^{(B/s-d)}} \left(\frac{q_j(t)}{c_j^{(EF)}(t)} + d_j \right) \qquad m = 1,2,...,N^{(A)} \tag{23}$$

The mean delay D_m can be calculated as the temporal mean value of (23), that is:

$$D_m = \overline{D_m(t)} \qquad m = 1,2,...,N^{(A)} \tag{24}$$

while, as known, the mean jitter J_m suffered by packets belonging to the traffic aggregate a_m can be derived as the difference between the mean square value of (23) and the square of (24), that is

$$J_m = \overline{D_m^2(t)} - \overline{D_m(t)}^2 \qquad m = 1,2,...,N^{(A)} \tag{25}$$

2.2.2 AF-PHB Modeling

At the edge routers a token bucket for each AF traffic aggregate marks packets as IN if they are "in-profile", or OUT if they are "out-of-profile". As known, a token bucket can be seen as a virtual buffer of size CBS filled with a rate of CIR tokens per second.

Let $Tc_m(t)$ be the instantaneous token counter related to the token bucket associated with the traffic aggregate a_m, and let CIR_m and CBS_m respectively be its Committed Information Rate and Committed Burst Size. Furthermore, we indicate the output rate of IN and OUT packets at the output of the token bucket as $\Lambda_m^{(IN)}(t)$ and $\Lambda_m^{(OUT)}(t)$, respectively. When the token counter is not zero all the packets are marked as IN and the rate of IN packets will therefore be equal to the rate of incoming packets; when, on the other hand, the token counter is zero, at most CIR_m packets per second are marked as IN, while the remaining packets are marked as OUT.

In other words, the rate of IN packets at the output of the token bucket is given by:

$$\Lambda_m^{(IN)}(t) = \begin{cases} \Lambda_m(t) & \text{if } Tc_m(t) > 0 \\ \min(CIR_m, \Lambda_m(t)) & \text{if } Tc_m(t) = 0 \end{cases} \tag{26}$$

The output rate of OUT packets will be the difference between the rate of incoming packets and the output rate of IN packets:

$$\Lambda_m^{(OUT)}(t) = \Lambda_m(t) - \Lambda_m^{(IN)}(t) \tag{27}$$

Therefore $Tc_m(t)$ can be calculated as follows:

$$\frac{dTc_m(t)}{dt} = CIR_m - \Lambda_m^{(IN)}(t) \quad \text{with the constraint} \quad 0 \le Tc_m(t) \le CBS_m \tag{28}$$

After being marked in the edge router, packets belonging to AF traffic aggregates pass through a set of WRED buffers. In this case we have to distinguish between the arrival rate of IN packets, $\lambda_{j,m}^{(IN)}(t)$, and the arrival rate of OUT packets, $\lambda_{j,m}^{(OUT)}(t)$, at the generic WRED buffer $b_j \in \mathfrak{S}_{a_m}^{(B)}$ (with $a_m \in \mathfrak{S}_{AF}^{(A)}$). The relationships that we will derive in this section have similar expressions for IN packets and OUT packets, so we will often substitute the IN (or OUT) superscript with X. This means that the reader can replace the (X) superscript with either (IN) or (OUT). According to this notation:

$$\lambda_{j,m}^{(X)}(t) = \begin{cases} \Lambda_m^{(X)}(t) & \text{if } j = 1 \\ \mu_{j-1,m}^{(X)}(t - d_{j-1}) & \text{if } 1 < j \le K_m \end{cases} \qquad \forall m: \ a_m \in \mathfrak{S}_{AF}^{(A)} \tag{29}$$

Let $\mathfrak{S}_{b_j}^{(A)}$ be the set of traffic aggregates a_m that pass through the buffer b_j. The total arrival rate $\Phi_j^{(X)}(t)$ of X packets at the buffer b_j will be:

$$\Phi_j^{(X)}(t) = \sum_{a_m \in \mathfrak{S}_{b_j}^{(A)}} \lambda_{j,m}^{(X)}(t) \tag{30}$$

The equation that regulates variations in the queue length $q_j(t)$ of the generic WRED buffer $q_l^{(AF)}(t)$, derives from the Lindley equation, and is:

$$\frac{dq_j(t)}{dt} = \begin{cases} -c_j^{(AF)}(t) + (1 - p_j(t)) \cdot \left(\Phi_j^{(IN)}(t) + \Phi_j^{(OUT)}(t) \right) & \text{if } q_j(t) > 0 \\ \left[-c_j^{(AF)}(t) + (1 - p_j(t)) \cdot \left(\Phi_j^{(IN)}(t) + \Phi_j^{(OUT)}(t) \right) \right]^+ & \text{if } q_j(t) = 0 \end{cases} \tag{31}$$

where $c_j^{(AF)}(t)$ is the service rate assigned to buffer $b_j \in \mathfrak{S}_{a_m}^{(B)}$ by the scheduler.

The drop probability $p_j(t)$ in the generic buffer $b_j \in \mathfrak{S}_{a_m}^{(B)}$ at time $t \ge 0$, can be calculated as follows:

$$p_j(t) = p_j^{(IN)}(t) \cdot \frac{\Phi_j^{(IN)}(t)}{\Phi_j^{(IN)}(t) + \Phi_j^{(OUT)}(t)} + p_j^{(OUT)}(t) \cdot \frac{\Phi_j^{(OUT)}(t)}{\Phi_j^{(IN)}(t) + \Phi_j^{(OUT)}(t)} \tag{32}$$

According to the WRED mechanism [5], $p_j^{(X)}(t)$ is

$$p_j^{(X)}(t) = \begin{cases} 0 & \text{if } m_j(t) < T_{MIN_j}^{(X)} \\ \dfrac{m_j(t) - T_{MIN_j}^{(X)}}{T_{MAX_j}^{(X)} - T_{MIN_j}^{(X)}} \, p_{MAX_j}^{(X)} & \text{if } T_{MIN_j}^{(X)} \le m_j(t) \le T_{MAX_j}^{(X)} \\ 1 & \text{if } m_j(t) > T_{MAX_j}^{(X)} \end{cases} \qquad (33)$$

where $T_{MIN}^{(IN)}$, $T_{MAX}^{(IN)}$, $p_{MAX}^{(IN)}$, $t_{MIN}^{(OUT)}$, $t_{MAX}^{(OUT)}$, $p_{MAX}^{(OUT)}$ are the WRED parameters, and $m_j(t)$ is the average queue length estimated by the WRED algorithm. The differential equation that describes the temporal behavior of $m_j(t)$ is derived in [12].

Following considerations similar to those made in Section 2.2.1 we can complete the model for AF-PHB traffic aggregates, calculating $\mu_{j,m}^{(X)}(t)$, $\lambda_{j,m}^{(X)}(t)$, $R_m^{(ACK)}(t)$, $R_m^{(LOSS)}(t)$ and $RTT_n(t)$ with the following equations:

$$\mu_{j,m}^{(X)}\!\left(t+\tau_j(t)\right) = \frac{\lambda_{j,m}^{(X)}(t)\left(1-p_j(t)\right)}{1+d\tau_j(t)/dt} \qquad \forall j : b_j \in \mathfrak{S}_{a_m}^{(B)} \text{ and } a_m \in \mathfrak{S}_{AF}^{(A)} \qquad (34)$$

$$v_{j,m}\!\left(t+\tau_j(t)\right) = \begin{cases} \lambda_{1,m}^{(IN)}(t)\cdot p_1^{(IN)}(t) + \lambda_{1,m}^{(OUT)}(t)\cdot p_1^{(OUT)}(t) & \text{if } j=1 \\ v_{j-1,m}\!\left(t-d_{j-1}\right) + \lambda_{j,m}^{(IN)}(t)\cdot p_j^{(IN)}(t) + \\ + \lambda_{j,m}^{(OUT)}(t)\cdot p_j^{(OUT)}(t) & \text{if } 1 < j \le K_m \end{cases} \qquad (35)$$

$$\forall j: \ b_j \in \mathfrak{S}_{a_m}^{(B)} \text{ and } a_m \in \mathfrak{S}_{AF}^{(A)}$$

$$R_m^{(ACK)}(t) = \mu_{K_m,m}\!\left(t-d_{K_m}\right) \qquad \forall a_m \in \mathfrak{S}_{EF}^{(A)} \qquad (36)$$

$$R_m^{(LOSS)}(t) = v_{K_m,m}\!\left(t-d_{K_m}\right) \qquad \forall a_m \in \mathfrak{S}_{EF}^{(A)} \qquad (37)$$

$$RTT_n(t) = \sum_{b_j \in \mathfrak{S}_{a_m}^{(B)}} \left(\frac{q_j(t)}{c_j^{(AF)}(t)} + d_j \right) \qquad n = 1, 2, \ldots, N^{(F)} \text{ and } m : f_n \in \mathfrak{S}_{a_m}^{(F)} \qquad (38)$$

2.2.3 BE-PHB Modeling

The set of equations relating to BE-PHB can be obtained simply by considering the set of equations relating to EF-PHB (Section 2.2.1), cutting out all the terms that contain the superscript (SH), and substituting $c_j^{(EF)}(t)$ with $c_j^{(BE)}(t)$.

2.2.4 Scheduler Modeling

Let $\mathfrak{S}^{(L)}$ be the set of all the unidirectional output links present in the network. Let C_l be the transmission capacity (in packets per second) of the link $l \in \mathfrak{S}^{(L)}$.

As shown in Fig. 1, we assume that the generic scheduler in each node works on three buffers, respectively devoted to EF, AF and BE traffic. Let $c_l^{(EF)}(t)$, $c_l^{(AF)}(t)$ and $c_l^{(BE)}(t)$ be the service rates assigned by the scheduler connected to the generic link l, to the EF, AF and BE buffers.

In order to guarantee the QoS requirements of traffic aggregates belonging to EF-PHB and AF-PHB we assume that two fixed portions $\hat{C}_l^{(EF)}$ and $\hat{C}_l^{(AF)}$ of the total available bandwidth C_l in each link are reserved for EF traffic and AF traffic, respectively.

The portion $\hat{C}_l^{(EF)}$ has to be chosen in such a way that the actual bandwidth assigned to each EF traffic aggregate $a_m \in \mathfrak{I}_{EF}^{(A)}$ is greater than a minimum value R_{MIN_m} (corresponding to its Average Transmission Rate) and less than or equal to a maximum value R_{MAX_m} (corresponding to its Maximum Transmission Rate). A possible choice for $\hat{C}_l^{(EF)}$ is:

$$\hat{C}_l^{(EF)} = \sum_{a_m \in \mathfrak{I}_l^{(EF)}} \left[R_{MIN_m} + \beta_l \cdot \left(R_{MAX_m} - R_{MIN_m} \right) \right] \tag{39}$$

where $\mathfrak{I}_l^{(EF)}$ is the set of all the EF traffic aggregates whose packets pass through the link l, and β_l is a design parameter ($0 \le \beta_l \le 1$).

As regards AF traffic, on the other hand, we assume that the reserved fixed portion of the output link capacity of the generic link l is:

$$\hat{C}_l^{(AF)} = \sum_{a_m \in \mathfrak{I}_l^{(AF)}} CIR_m \tag{40}$$

where $\mathfrak{I}_l^{(AF)}$ is the set of all the AF traffic aggregates passing through the link l.

We assume that a call admission control technique requires the condition $C_l > \hat{C}_l^{(EF)} + \hat{C}_l^{(AF)}$ to be met, for each $l \in \mathfrak{I}^{(L)}$.

The unreserved bandwidth UB_l in the generic link $l \in \mathfrak{I}^{(L)}$ will be:

$$UB_l = C_l - \hat{C}_l^{(EF)} - \hat{C}_l^{(AF)} \tag{41}$$

In this framework we assume that the generic scheduler equally shares the unreserved bandwidth between AF and BE traffic, while the bandwidth assigned to EF traffic is maintained constant and equal to $\hat{C}_l^{(EF)}$. If the total arrival rate of AF traffic in the corresponding buffer, $\lambda_l^{(AF)}(t)$, is temporarily less than its available bandwidth, $\left(\hat{C}_l^{(AF)} + UB_l/2 \right)$, the free bandwidth will be assigned to BE traffic, and vice versa.

From these considerations it is easy to derive the following relationships:

$$c_l^{(EF)}(t) = \hat{C}_l^{(EF)} \tag{42}$$

$$c_l^{(AF)}(t) = \begin{cases} \hat{C}_l^{(AF)} + \dfrac{UB_l}{2} + \left[\dfrac{UB_l}{2} - \lambda_l^{(BE)}(t) \right]^+ & \text{if } q_l^{(AF)}(t) > 0 \text{ and } q_l^{(BE)}(t) = 0 \\[2mm] \min\left(\lambda_l^{(AF)}(t), \ \hat{C}_l^{(AF)} + \dfrac{UB_l}{2} \right) & \text{if } q_l^{(AF)}(t) = 0 \text{ and } q_l^{(BE)}(t) > 0 \\[2mm] \hat{C}_l^{(AF)} + \dfrac{UB_l}{2} & \text{otherwise} \end{cases} \tag{43}$$

$$c_l^{(BE)} = C_l - c_l^{(EF)}(t) - c_l^{(AF)}(t) \tag{44}$$

where $q_l^{(AF)}(t)$ and $q_l^{(BE)}(t)$ are the queue lengths of the AF and BE buffers connected, through the scheduler, with the output link l.

3 Case Study

In this section the proposed model is applied to evaluate how a different setting of the bandwidth reserved for EF traffic can impact the performance offered by the Diffserv network. To this end, let us consider the network topology presented in Fig. 1, made up of three ingress nodes (named N_1, N_2, N_3), one interior node (N_4) and one egress node (N_5). Each ingress node is directly connected to three LANs through high-speed links with negligible propagation delay. In Fig. 2 each of these LANs is labeled with the symbols L_{EFx}, L_{AFx} or L_{BEx}, according to whether it emits an Expedited Forwarding (EF), Assured Forwarding (AF) or Best Effort (BE) traffic aggregate, x being a number identifying the ingress node N_x to which the LAN is connected.

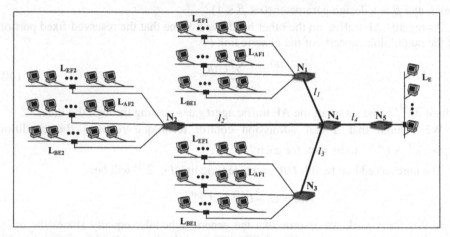

Fig. 2. Network topology

The network is then loaded by 9 traffic aggregates, $a_1 - a_9$, following the paths listed in Table 1. In the same table the PHB types and the traffic profile parameters for each traffic aggregate are listed.

Table 1. Description of the traffic aggregates considered

Traffic Aggregate Identifier	Path Followed (Source-Routers-Destination)	PHB Type	Traffic Profile
a_1	$L_{A1} - A - D - E - L_E$	EF	$(R_{MIN}, R_{MAX}) = (750, 1250)$
a_2	$L_{A2} - A - D - E - L_E$	AF	$(CIR, CBS) = (500, 10)$
a_3	$L_{A3} - A - D - E - L_E$	BE	/
a_4	$L_{B1} - B - D - E - L_E$	EF	$(R_{MIN}, R_{MAX}) = (750, 1250)$
a_5	$L_{B2} - B - D - E - L_E$	AF	$(CIR, CBS) = (500, 10)$
a_6	$L_{B3} - B - D - E - L_E$	BE	/
a_7	$L_{C1} - C - D - E - L_E$	EF	$(R_{MIN}, R_{MAX}) = (750, 1250)$
a_8	$L_{C2} - C - D - E - L_E$	AF	$(CIR, CBS) = (500, 10)$
a_9	$L_{C3} - C - D - E - L_E$	BE	/

As can be noted, the traffic profiles are expressed in terms of R_{MIN} and R_{MAX} for EF aggregates, and of CIR and CBS for AF aggregates, while no traffic profile is required by BE aggregates. The sizes of all the shapers, located at the ingress nodes in order to limit the EF traffic entering the network, are assumed to be equal to 3 packets. Each traffic aggregate is assumed to be emitted by 10 TCP greedy sources.

Packet size is assumed to be fixed and equal to 500 bytes. The unidirectional links connecting the nodes with each other have the same propagation delay $d_j = 50$ ms and same output capacity $C_j = 2500$ packtes/sec. Each of these links is associated with three buffers, denoted as EF, AF and BE buffers according to the PHB type for the traffic aggregates that pass through them, respectively. The buffers of the same type are assumed to adopt the same AQM algorithm. Table 2 shows for each link the configurations of all the types of buffer. For all buffers it is also assumed that: $\alpha = 0.0001$.

As seen in Section 2.2.4, for each link l_i the portion of the output capacity that is reserved for the EF buffer connected to it is determined by the value of the parameter β_i, by means of (39)The aim is to study how the variations in this parameter can influence the performance offered to the EF, AF and BE traffic aggregates by the Diffserv network.

The results given by the proposed model are shown in Figs. 3, 4, and 5, obtained assuming, for the sake of simplicity, that $\beta_1 = \beta_2 = \beta_3$. It is easy to show that choosing different values of β_4 always results in an excessive link underutilization, and then we will assume that: $\beta_1 = \beta_2 = \beta_3 = \beta_4$. The common value of $\beta_1, \beta_2, \beta_3, \beta_4$ will conventionally be denoted as β. As β increases from 0 to 1, the bandwidth reserved for the EF traffic by the scheduler associated with the links l_1, l_2, l_3 increases from R_{MIN} to R_{MAX}, that is from 750 to 1250 packets per second (see Table 1).

Table 2. Buffer configurations

Output Link	Type of Buffer	AQM Algorithm	Configuration Parameters
l_1, l_2, l_3	EF	RED	$\left(T_{MIN}, T_{MAX}, p_{MAX}\right) = (5, 30, 0.1)$
	AF	WRED	$\left(T_{MIN}^{(OUT)}, T_{MAX}^{(OUT)}, p_{MAX}^{(OUT)}\right) = (50, 100, 0.5)$ $\left(T_{MIN}^{(IN)}, T_{MAX}^{(IN)}, p_{MAX}^{(IN)}\right) = (100, 150, 0.1)$
	BE	RED	$\left(T_{MIN}, T_{MAX}, p_{MAX}\right) = (50, 100, 0.5)$
l_4	EF	RED	$\left(T_{MIN}, T_{MAX}, p_{MAX}\right) = (30, 90, 0.1)$
	AF	WRED	$\left(T_{MIN}^{(OUT)}, T_{MAX}^{(OUT)}, p_{MAX}^{(OUT)}\right) = (50, 150, 0.5)$ $\left(T_{MIN}^{(IN)}, T_{MAX}^{(IN)}, p_{MAX}^{(IN)}\right) = (150, 300, 0.1)$
	BE	RED	$\left(T_{MIN}, T_{MAX}, p_{MAX}\right) = (50, 150, 0.5)$

Fig. 3 and 4 show the average delay and jitter suffered by the packets belonging to each EF traffic aggregate versus β. As expected, delay and jitter decrease as β increases. When the bandwidth reserved for EF traffic is set to its maximum value, that

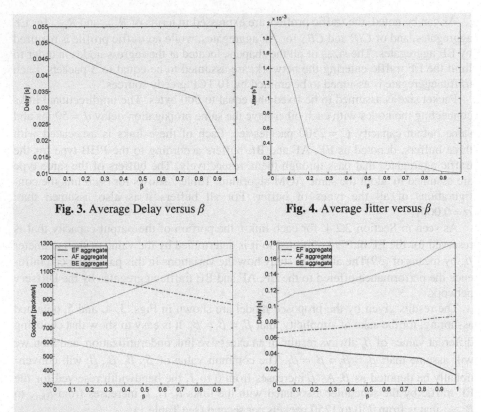

Fig. 3. Average Delay versus β **Fig. 4.** Average Jitter versus β

Fig. 5. Average Goodput of EF, AF and BE traffic aggregates versus β **Fig. 6.** Average Delay of EF, AF and BE traffic aggregates versus β

is when β is equal to unity, all the EF buffers are always empty while the shapers are almost always full, and then the average delay is approximately equal to the sum of the propagation delay of all the links associated with the same buffers, while the average jitter is approximately equal to zero. This is the ideal condition for the EF traffic aggregates, but is obtained at a cost of excessive bandwidth reservation. These plots can be used as design tools to choose the minimum value of β, and then the minimum EF reserved bandwidth, in order to meet the real-time application minimum requirements in terms of delay and jitter.

Fig. 5 and 6 show the average goodput and delay of the EF, AF and BE traffic aggregates versus β. It is useful to evaluate from these plots how the performance offered to AF and BE traffic aggregates, mainly expressed in terms of goodput and sometimes also in terms of delay, can be influenced by a different choice of β. Since the proposed fluid-flow model allows the average behavior of the network to be derived in much less time than per-packet simulation, it can be very helpful to designers to properly dimension available network resources.

4 Conclusions

In this paper a fluid-flow model of a complex DiffServ domain loaded with both real-time applications, data transfer applications with QoS requirements, and data transfer applications with no QoS requirements has been proposed. The proposed model models all the network elements needed to support three types of standard PHB's: EF, AF and BE. EF traffic aggregates are assumed to be conditioned while AF traffic aggregates are policed by profile meters based on token buckets. A fluid-flow approach has been used because it allows to derive the average network behavior in much less time than per-packet simulation. Finally, the proposed analytical framework has been applied to a case study to evaluate how a different setting of the bandwidth reserved to EF traffic can impact the performance offered by the Diffserv network not only to the same EF traffic, but also to the AF and the EF traffic. Besides, this case study has shown how the proposed model could be very helpful to designers to properly dimension available network resources.

References

1. S. Blake, D. Black, M. Carlson, E. Davies, Z. Wang and W. Weiss, "An architecture for differentiated services". *RFC 2475*, December 1998.
2. J. Heinanen, F. Baker, W. Weiss and J. Wrocklawski, "Assured Forwarding PHB Group", RFC 2597, June 1999.
3. B. Davie, A. Charny, F. Baker, J.C.R. Bennett, K. Benson, J. Le Boudec, A. Chiu, W. Courtney, S. Cavari, V. Firoiu, C. Kalmanek, K. Ramakrishnam and D. Stiliadis, "An Expedited Forwarding PHB (Per-Hop Behavior)", RFC 3246, March 2002.
4. K. Nichols, S. Blake, F. Baker and D. Black, "Definition of the Differentiated Services Field (DS Field) in the IPv4 and IPv6 Headers", RFC 2474, December 1998.
5. http://www.cisco.com/univercd/cc/td/doc/product/software/ios120/12cgcr/qos_c/qcpart3/qcwred.htm
6. S. Floyd, V. Jacobson. "Random Early Detection Gateways for Congestion Avoidance". IEEE/ACM Transactions on Networking (1993).
7. A. Abouzeid, S. Roy. "Modeling random early detection in a differentiated services network". Computer Networks (Elsevier), 40 (4): 537-556, November 2002
8. N. Malouch, Z. Liu. "On steady state analysis of TCP in networks with Differentiated Services". Proceeding of Seventeenth International Teletraffic Congress, ITC'17, December 2001
9. I. Yeom, A. Reddy. "Modeling TCP behavior in a differentiated-services network". IEEE/ACM Transaction on Networking, Vol.9, Issue 1, 31-46, February 2001
10. Y. Chait, C. Hollot, V. Misra, D. Towsley, and H. Zhang, "Providing throughput differentiation for TCP flows using adaptive two color marking and multi-level AQM" IEEE INFOCOM 2002, New York, NY, 23-27 June 2002.
11. V. Misra, W. Gong, D. Towsley, "Fluid-based analysis of a network of AQM routers supporting TCP flows with an application to RED". SIGCOMM '00 (August 2000).
12. M.Barbera, A. Lombardo, G. Schembra, "A Fluid-Based Model of Time-Limited TCP Flows", Computer Networks (Elsevier), 44(3): 275-288, February 2004
13. S. Floyd, T. Henderson "The NewReno Modification to TCP's Fast Recovery Algorithm". RFC 2582, April 1999.
14. M. Handley, S. Floyd, J. Padhye, J. Widmer, "TCP Friendly Rate Control (TFRC): Protocol Specification," RFC3448, January 2003.

A Performance Model for Multimedia Services Provisioning on Network Interfaces*

Paola Laface[1], Damiano Carra[2], and Renato Lo Cigno[2]

[1] Dipartimento di Elettronica – Politecnico di Torino,
Corso Duca degli Abruzzi, 24 – 10129 Torino, Italy
paola.laface@polito.it
[2] Dipartimento di Informatica e Telecomunicazioni – Università di Trento,
Via Sommarive, 14 – 38050 Povo, Trento, Italy
{carra,locigno}@dit.unitn.it

Abstract. This paper presents a method for the performance evaluation of multimedia streaming applications on IP network interfaces with differentiated scheduling. Streaming applications are characterized by the emission of data packets at constant intervals, changing during the connection lifetime, hence a single source can be effectively modeled by an MMDP.
We propose an MMDP/D/1/K model to represent the aggregate arrival process at the network interface, assuming that multimedia packet dimensions are approximately constant. A method for solving the above queuing system providing upper and lower bounds to the packet loss rate is presented and results for realistic VoIP applications are discussed and validated against accurate event-driven simulations showing the efficiency and accuracy of the method.

Keywords: Multimedia Traffic, QoS Network Planning, Markov Modeling, Approximate Solutions

1 Introduction

Multimedia and Quality of Service (QoS) are probably the most repeated words in networking research during the past fifteen years or so. Spawned by research on ATM (Asynchronous Transfer Mode) in the late '80s and early '90s, topics related to offering the appropriate QoS in integrated packet networks received even more attention when the application context moved to IP-based networking.

One of the key aspects of heterogeneous service provisioning is the guarantee of the QoS the service will receive during its lifetime. Enforcing QoS encompasses a number of different aspects, ranging from service architecture, to network dimensioning, to protocol design and many others. All the design aspects pivot around the performance evaluation of the provisioned service: without a means to evaluate the performance, it is not possible to select an appropriate Service Level Agreement (SLA) between the network and the user.

In this paper we explore an analytical approach based on the solution of DTMCs (Discrete Time Markov Chains) embedded in a more general CTC (Continuous Time

---

* This work was supported in Torino by the Italian Ministry for University and Research (MIUR) under the FIRB project TANGO.

M. Ajmone Marsan et al. (Eds.): QoS-IP 2005, LNCS 3375, pp. 286–299, 2005.

Chain) to evaluate the performance of several classes of multimedia services, namely those, as voice and video, that are characterized by intermittent or variable bit rate, and piecewise constant inter-packet emission times. After describing the general framework, we focus our attention on voice services comparing the analytical solution with simulations. We consider IP Telephony, or VoIP for short, which is by far the most diffused (and widespreading) multimedia application on IP networks.

The main contribution of this work is providing a simple and efficient analytical framework to predict the performance of multimedia services in several possible scenarios, like for instance an access link to the Internet or DiffServ [1] interfaces with *Expedited Forwarding Per Hop Behavior* [2]. Though the mathematical modeling is not entirely novel (as discussed in Sect. 2), since similar problems were tackled studying ATM networks (see [3, 4] and the references therein, or [5] for a review), the solution technique we propose yields the derivation of exact solutions for upper and lower bounds on the packet loss performance, which enables service planning and SLA definition. The bounds are shown to be tight. We generalize the solution technique to the case of n-state sources transmitting at different rates in each of the n states.

To conclude with, we note that the approximations we introduce ensure exact bounding of the solution and do not suffer from numerical instabilities. Furthermore, to the best of our knowledge, previous works treated only On-Off sources, while our approach is amenable to application on multi-state sources with arbitrary state rates.

2 Problem Formulation

Multimedia services are related to audio and video, i.e., to connection oriented, streaming applications, whose emission characteristics are studied fairly well. Indeed, although voice and video can potentially generate packets at variable intervals, all existing applications encode blocks of information at fixed time intervals. Moreover, the packet size is often fixed, either by the encoder, the protocol or the application program. In other words, multimedia streaming or conversational applications can be efficiently modeled as sources with a piecewise constant emission rate of fixed size packets.

In queuing theory notation, these are Modulated Deterministic Processes (MDP). If the time between state transitions is exponentially distributed and the source can transmit with n different speeds, then the source is an n-state Markov Modulated Deterministic Process (MMDP), whose states s_k are characterized by the emission of packets with rate δ_k. On-Off sources are a subset with two states and $\delta_0 = 0$ and $\delta_1 = c \neq 0$.

A superposition of MMDP is not an MMDP, unless the sources are appropriately synchronized. Source synchronization is not an unrealistic scenario (VoIP calls generated by the same media conversion gateway are either synchronized or can be synchronized easily). Approximating the superposition of MMDPs with a single MMDP is equivalent to neglect short term congestion, due to the arrival process higher variability. The impact of this approximation is discussed in Sect. 5, where event driven simulations are used to validate both the modeling assumptions and the bounding approximations.

We consider a single network interface (can be at the access or in any section of the network) assuming it is the only point of potential congestion, and model the system as an MMDP/D/1/K queuing system. Figure 1 shows an example of the system with four

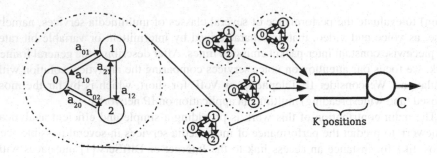

Fig. 1. Arrival process: superposition of three-state sources.

three-state sources. The states of each source are labeled as 0,1,2, with transmission rates δ_0^i, δ_1^i, δ_2^i, not necessarily equal one another.

The number of states m of the arrival modulating process is, in the most general case, $m = \prod_{i=1}^{M} n_i$, where M is the number of considered sources and n_i is the number of possible transmission rates of source i. No constraints are posed on δ_k^i, but all packets are of the same length for deterministic service times.

This same modeling approach was taken in [3] limiting the analysis to homogeneous On-Off sources. When possible we use the same notation used in that work, to help readers familiar with it.

For homogeneous sources, the arrival modulating process is an $(n-1)$-dimensional quasi-birth-death CTMC. Figure 2 shows the modulating CTMC for $n = 3$, where the first index i is the number of sources in state 1, the second index j is the number of sources in state 2 and $M - i - j$ sources are in state 0. Evolution along rows and columns follows a simple birth-and-death process; evolution along diagonals means that a source can switch from state 1 to state 2 and vice versa.

Although the arrival process is modulated by a CTMC, the overall queuing system is not Markovian, since phase transitions (i.e., transitions of the above CTMC) can occur at any time between deterministic arrivals and departures.

3   Model Analysis

Define:

- $\{X(t), t \geq 0\}$: the finite, irreducible CTMC representing the arrival process;
- $S = \{\underline{f}\}$: the state space of $X(t)$; the vector $\underline{f} = [f_0, f_1, \ldots, f_F]$ represents the arrival process phase; $f_k = 0, 1, \ldots, M_k$ is the number of sources sending at rate δ_k; if sources are all homogeneous $F = n - 1$ and $M_k = M$, $\forall k$; if sources are all different $F = \prod_{i=1}^{M} n_i$ and $M_k = 1$, $\forall k$;
- $R = [r_{\underline{f}, \underline{f}'}]$: the $m \times m$ infinitesimal generator of $X(t)$, $m = ||S||$;
- $1/\gamma_{\underline{f}}$ ($\underline{f} \in S$): the mean sojourn time in state $\underline{f}$.

Two examples help understanding the system.

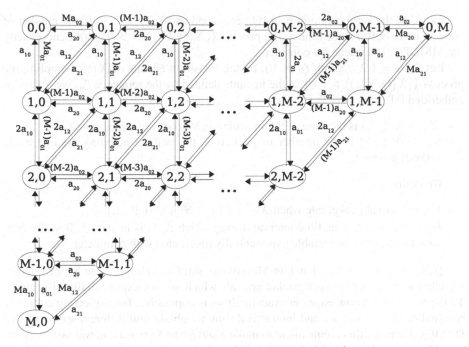

Fig. 2. Markov chain describing the evolution of the MMDP deriving from the superposition of M sources with three different emission rates.

Example 1: Heterogeneous On-Off Sources. We have F types of On-Off sources, M_0 with emission rate δ_0, M_1 with emission rate δ_1, $\cdots$, M_F with emission rate δ_F, $M = \sum_{i=0}^{F} M_i$; the components f_k of vector $\underline{f}$ represent the number of active sources with rate δ_k. The cardinality of S is $m = (M_0 + 1) \cdot (M_1 + 1) \cdot \ldots \cdot (M_F + 1)$, and the CTMC is a combination of birth-death processes, i.e., only transitions of the type

$$[f_0, \ldots, f_i, \ldots, f_F] \rightarrow [f_0, \ldots, f_i + 1, \ldots, f_F];$$
$$[f_0, \ldots, f_i, \ldots, f_F] \rightarrow [f_0, \ldots, f_i - 1, \ldots, f_F]$$

are allowed.

Example 2: Homogeneous Multirate Sources. We have M sources that can transmit n different rates δ_k, $k = 0, 1, \ldots, n$; the components f_k of vector $\underline{f}$ still represent the number of active sources with rate δ_k; we have $F = n + 1$ and $m = \binom{M + n - 2}{M}$, since we don't have to explicitly represent the number f_0 of sources with rate δ_0 because the relation $M = \sum_{k=0}^{F} f_k$ holds. The CTMC is no more the product of birth-death chains and 'diagonal' transitions (as shown in Fig. 1) are admitted, thus we have three possible transition types:

$$[f_1, \ldots, f_i, \ldots, f_F] \rightarrow [f_1, \ldots, f_i + 1, \ldots, f_F];$$
$$[f_1, \ldots, f_i, \ldots, f_F] \rightarrow [f_1, \ldots, f_i - 1, \ldots, f_F];$$
$$[f_1, \ldots, f_i - 1, \ldots, f_F] \rightarrow [f_1, \ldots, f_j + 1, \ldots, f_F].$$

Let C be the service rate in packets per second and $Y(t) = k \leq K$ the number of packets in the buffer at the time t. The process $\{(X(t), Y(t)), t > 0\}$ represents exactly the MMDP/D/1/K queue we consider.

Let ξ_n (n = 1,2, ... with $\xi_0 = 0$), be the transition epochs of $X(t)$. Sampling the process $\{(X(t), Y(t)), t > 0\}$ in the instants defined by the sequence ξ_n, we obtain an embedded DT process $\{(X_n, Y_n), n = 0, 1, ...\}$, where

- $X_n = X(\xi_n^+)$ is the state of the modulating Markov process at time ξ_n^+;
- $Y_n = Y(\xi_n^+)$ is the number of packets in the buffer (including the one being served) at time ξ_n^+.

We define

- V_f: the arrival packet rate when $X_n = \underline{f}$ ($\underline{f} \in S; n = 0, 1, ...$);
- $U_{\underline{f}} = \xi_{n+1} - \xi_n$: the time interval during which $X(t)$ is in state $\underline{f}$. It is (by construction) a random variable exponentially distributed with parameter $\gamma_{\underline{f}}$.

$\{(X_n, Y_n), n = 0, 1, ...\}$ is non-Markovian, since sampling ξ_n can happen during a packet service and between packet arrivals which are not exponentially distributed. To the best of our knowledge, an exact analysis is impossible, but neglecting either the residual or elapsed service and interarrival time we obtain four different approximated DTMCs. Formally this is equivalent to make a *service* and *arrival* renewal assumptions.

Service Renewal Assumption. The elapsed service time T_n^s after the n-th transition is equal either to 0 or to $1/C$. $T_n^s = 0$ (the packet is not yet served) overestimates the system load; $T_n^s = 1/C$ (the packet is served entirely) underestimates the system load.

Arrival Renewal Assumption. The elapsed inter-arrival time T_n^a after the n-th transition is equal either to 0 or to $1/V_{\underline{f}}$. $T_n^a = 0$ (the first packet of the new phase arrives following the new rate and the last packet of the previous phase is neglected) underestimates the system load; $T_n^a = 1/V_{\underline{f}}$ (the last packet of the old phase arrives in any case on phase transition) overestimates the system load.

Under these assumptions we can get four cases:

$$\text{LL: if } T_n^s = 1/C \text{ and } T_n^a = 0 \qquad \text{UU: if } T_n^s = 0 \text{ and } T_n^a = 1/V_{\underline{f}}$$
$$\text{LU: if } T_n^s = 1/C \text{ and } T_n^a = 1/V_{\underline{f}} \quad \text{UL: if } T_n^s = 0 \text{ and } T_n^a = 0$$

The interesting cases are LL and UU, that yield lower and upper bounds to the system load and hence on the loss probability, while the UL and LU cases are approximations, but it is not easy to tell whether they are upper or lower bounds. It must be noted at this point that the authors in [3], besides considering only homogeneous On-Off sources ($F = 1$ and the state identifier $\underline{f}$ is a scalar and not a vector), approximate the $\{(X_n, Y_n), n = 0, 1, ...\}$ process with a limiting passage, assuming that $E[U_{\underline{f}}] \gg E[1/V_{\underline{f}}], 1/C$, which is, for $E[1/V_{\underline{f}}] \to 0$ and $1/C \to 0$, equivalent to the UL approximation.

The renewal assumptions they introduced refer only the buffer evolution (the Y_n component of the (X_n, Y_n) process). They approximate the real evolution of the buffer with incremental geometric random variables. In formulas:

$$Y_{n+1} = \begin{cases} \min\{K, Y_n + I_{\underline{f}}\}, & \text{if } V_{\underline{f}} > C \\ \max\{0, Y_n - O_{\underline{f}}\}, & \text{if } V_{\underline{f}} < C \end{cases} \tag{1}$$

where

$$I_{\underline{f}} = \begin{cases} \lfloor (V_{\underline{f}} - C)U_{\underline{f}} \rfloor & \text{in UL and LL} \\ \lceil (V_{\underline{f}} - C)U_{\underline{f}} \rceil & \text{in LU and UU} \end{cases} \tag{2}$$

and

$$O_{\underline{f}} = \begin{cases} \lfloor (C - V_{\underline{f}})U_{\underline{f}} \rfloor & \text{in } UL \text{ and } UU \\ \lceil (C - V_{\underline{f}})U_{\underline{f}} \rceil & \text{in } LL \text{ and } LU \end{cases} \tag{3}$$

are the geometric increments with parameter $\rho_{\underline{f}} = exp\{-\gamma_{\underline{f}}/|V_{\underline{f}} - C|\}$, where $\gamma_{\underline{f}}$ is the average holding time of state $\underline{f}$.

3.1 Solution of the $\{(X_n, Y_n), n = 0, 1, ...\}$ Embedded Markov Chain

Let

$$a_{k,h}^{\underline{f}} = P\{Y_{n+1} = h | X_n = \underline{f}, Y_n = k\} \tag{4}$$

be the probability that the number of packets in the buffer passes from k to h while the arrival process is in phase $\underline{f}$, and $\mathbf{A}_{\underline{f}} = [a_{k,h}^{\underline{f}}]$ of dimension $(K + 1) \times (K + 1)$ the transition probability matrix (notice that it is a stochastic matrix by construction). Then the transition probability from state $(\underline{f}, k)$ to state $(\underline{f}', h)$ of (X_n, Y_n) is:

$$q_{(\underline{f},k),(\underline{f}',h)} = a_{k,h}^{\underline{f}} p_{\underline{f},\underline{f}'}$$

and the transition probability matrix of (X_n, Y_n) is

$$\mathbf{Q} = [q_{(\underline{f},k),(\underline{f}',h)}] .$$

The renewal assumptions defined in Sect. 3 affects only the $a_{k,h}^{\underline{f}}$ distribution, while the structure and the solution of the chain are unaffected. The computation of this distribution is straightforward, though it can be a bit cumbersome. In [17] the detailed computation for the UU and LL assumptions we use in this paper are reported. The resulting DTMC is finite and ergodic by construction.

$\mathbf{Q}$ is highly structured and can be recursively partitioned into blocks changing one of the $\underline{f}$ components at a time. Recall that in general we have at most M_k sources that can transmit with rate δ_k which corresponds to the state component f_k, then we have

$$\mathbf{Q} = \begin{bmatrix} \mathbf{Q}_{0,0} & \mathbf{Q}_{0,1} & \mathbf{Q}_{0,2} & \cdots & \mathbf{Q}_{0,M_0} \\ \mathbf{Q}_{1,0} & \mathbf{Q}_{1,1} & \mathbf{Q}_{1,2} & \cdots & \mathbf{Q}_{1,M_0} \\ \mathbf{Q}_{2,0} & \mathbf{Q}_{2,1} & \mathbf{Q}_{2,2} & \cdots & \mathbf{Q}_{2,M_0} \\ \cdots & \cdots & \cdots & \cdots & \cdots \\ \mathbf{Q}_{M_0-1,0} & \mathbf{Q}_{M_0-1,1} & \mathbf{Q}_{M_0-1,2} & \cdots & \mathbf{Q}_{M_0-1,D_0} \\ \mathbf{Q}_{M_0,0} & \mathbf{Q}_{M_0,1} & \mathbf{Q}_{M_0,2} & \cdots & \mathbf{Q}_{M_0,D_0} \end{bmatrix} = [\mathbf{Q}_{f_0,f_0'}] . \tag{5}$$

The block $\mathbf{Q}_{f_0,f_0'}$ refers to transitions from the states where the first component of $\underline{f}$ is equal to f_0 to the states where the first component amounts to f_0'.

The block decomposition can be iterated and the general form of a block is

$$\mathbf{Q}_{f_{k-1},f'_{k-1}}^{f_0,f_1,\ldots f_{k-2}} = [\mathbf{Q}_{f_k,f'_k}^{f_0,f_1,\ldots f_{k-1}}] \qquad k = 1, 2, \ldots F \tag{6}$$

that refers to transitions from the states where the number of sources transmitting at rate $\delta_0, \delta_1, \ldots, \delta_{k-1}$ is fixed to $f_0, f_1, \ldots f_{k-1}$ and the number of sources transmitting at rate δ_k passes from f_k to f'_k.

Finally, the last block partitioning is

$$\mathbf{Q}_{f_F,f'_F}^{f_0,f_1,\ldots f_{F-1}} = [p_{\underline{f},\underline{f}'}\mathbf{A}_{\underline{f}}] . \tag{7}$$

The main performance index we're interested in is the packet loss probability P_l. Given the system structure, losses can occur only in states $\underline{f}$ for which $V_{\underline{f}} \geq C$, with the equality holding for the UU approximation and not for the LL one. Since services and arrivals are deterministic within a single phase of the arrival process, P_l can be computed starting from the excess arrivals within phases:

$$P_l = \frac{\sum_{\underline{f}:V_{\underline{f}} \geq C} \sum_{j=0}^{K} E[R_{\underline{f},j}]\pi_{\underline{f},j}}{\sum_{\underline{f}=0}^{m-1} \sum_{j=0}^{K} E[N_{\underline{f},j}]\pi_{\underline{f},j}} \tag{8}$$

where

- $\underline{\pi}_{\underline{f}} = (\pi_{\underline{f},0}, \pi_{\underline{f},1}, \ldots, \pi_{\underline{f},K})$ is the steady state distribution of the embedded DTMC defined by $\mathbf{Q}$;
- $E[N_{\underline{f},j}]$ is the average number of packets arriving in phase $\underline{f}$ given that the number of packets in the buffer at the beginning of the phase is j;
- $E[R_{\underline{f},j}]$ is the average number of packets rejected in the above conditions.

$E[N_{\underline{f},j}]$ and $E[R_{\underline{f},j}]$ depend on the distribution of $a_{k,h}^{f}$. Their computation is reported in [17].

Solution Method: Ad-Hoc Block Reduction. The numerical solution of the system may pose problems as the dimension of the matrix $\mathbf{Q}$ increases. Recall that $\mathbf{Q}$ has dimension $[m \cdot (K+1)] \times [m \cdot (K+1)]$, so that as soon as the number of sources and the buffer dimension increase above a few tens the dimension of $\mathbf{Q}$ grows to thousands.

If we restrict the analysis to homogeneous sources or to a limited number of source classes (which is the problem we're interested in), $\mathbf{Q}$ has a banded structure, so that efficient Block Reduction techniques [6] can be used to solve the linear system. Unfortunately the block and band structure depends on the transition structure of the arrival modulating process, so that a general description is cumbersome, and a case-by-case analysis is required to obtain the best solution. In [17] the structure for 3-state sources is reported, while in the following we concentrate on 2-state sources for VoIP applications.

3.2   Application to Packetized Voice

Packetized voice applications are characterized by the presence of VAD (Voice Activity Detector) devices, that suppress the voice encoding when the speaker is silent and transmit *silence descriptors* for comfort noise instead of voice, at a much lower transmission rate. Voice packets have a constant dimension that depends on the encoder and silence descriptors have a constant dimension that in the general case can be different from the one of voice packets. Our model however dictates constant service times regardless of the source state, so we assume that voice and silence packets are equal in size.

A voice source can be described as a two-state source: when in state 1 (say High transmission rate) it generates voice packets equally spaced at fixed rate δ_1; when the source is in state 0 (say Low transmission rate), it generates silence description packets equally spaced at fixed rate δ_0. High and Low holding times are $1/\lambda$ and $1/\mu$ respectively. Given M of these sources we obtain an (M+1)-state MMDP modulating the arrivals. The state f is monodimensional with one component $f_1 = i$ to simplify the notation. The embedded chain transition probabilities are:

$$p_{\underline{f},\underline{f}'} = p_{i,j} = \begin{cases} (M-i)\lambda/\gamma_i & i = 0, ..., M-1; \; j = i+1 \\ i\mu/\gamma_i & i = 1, ...M; \; j = i-1 \\ 0 & otherwise \end{cases} \tag{9}$$

where $\gamma_i = (M-i)\lambda + i\mu$.

When the number of active voice sources is i, we have an aggregate arrival rate $V_i = i\delta_1 + (M-i)\delta_0$.

The probability transition matrix $\mathbf{Q}$ has the following banded structure,

$$\mathbf{Q} = \begin{bmatrix} \mathbf{0} & \alpha_0\mathbf{A}_0 & \cdots & & \\ \beta_1\mathbf{A}_1 & \mathbf{0} & \alpha_1\mathbf{A}_1 & \cdots & \\ & \beta_2\mathbf{A}_2 & \mathbf{0} & \cdots & \\ & & & \cdots & \alpha_{M-1}\mathbf{A}_{M-1} \\ & & & \beta_M\mathbf{A}_M & \mathbf{0} \end{bmatrix} \tag{10}$$

where

- $\mathbf{A}_i$ and $\mathbf{0}$ are $(K+1) \times (K+1)$ matrices;
- $\mathbf{A}_i = [a_{k,h}^i]; \; a_{k,h}^i = P\{Y_{n+1} = h | X_n = i, Y_n = k\}$;
- $\alpha_i = (M-i)\lambda/\gamma_i$ for $i = 0, 1, ... M-1$, and $\beta_i = i\mu/\gamma_i$ for $i = 1, ... M$.

The main diagonal is zero because sources can only move between the High and Low states. In this particular case the Block Reduction algorithm used is the following:

$$\underline{\pi}_i = (\pi_{i,0}, \pi_{i,1}, ..., \pi_{i,K})$$

$$\underline{\pi}_i = \underline{\pi}_{i+1}\mathbf{U}_i \quad i = 0, ... M-1$$

where

$$\begin{aligned} \mathbf{U}_0 &= -\beta_1\mathbf{A}_1 & i = 0 \\ \mathbf{U}_i &= -\beta_{i+1}\mathbf{A}_{i+1}(\mathbf{I} - \alpha_{i-1}\mathbf{U}_{i-1}\mathbf{A}_{i-1})^{-1} & i = 1, ... M-1 \end{aligned} \tag{11}$$

and

$$\underline{\pi}_M(\mathbf{I} - \alpha_{M-1}\mathbf{U}_{M-1}\mathbf{A}_{M-1}) = \mathbf{0};$$

$\underline{\pi}_i$ are normalized during the iteration.

We solved the system with the Open Source application Octave [7] on standard PC hardware for any value of M and K of practical interest.

4   System Simulation

As a numerical example we examine IP Telephony. We consider standard applications like NetMeeting[8], Open H323[9], or any other application using either H.323[10] or SIP[11] standards for signaling. All applications use RTP[12, 13] upon UDP/IP as transport protocol.

We implemented an ad-hoc simulator[15] because the system is simple enough to discourage the use of a general purpose network simulator as *ns-2*[16], and, most of all, because we want to control all details of the implementation and its efficiency, so as to be able to estimate accurately loss probabilities as low as 10^{-7}. To obtain such low estimates with the accuracy indicated in Sect. 5 some simulations were run for up to 10^{11} packets.

Among the different encoders we consider G.729[14] encoder with VAD[1]. Voice packets contain 40 bytes of data, that, with RTP/UDP/IP headers make 80 bytes packets. In case of header compression, the total packet dimension becomes 44 bytes. We assume that also silence packets contain 40 bytes of data to preserve deterministic service.

Sources are homogeneous corresponding to the case of Sect. 3.2; $\delta_{\mathrm{High}} = \delta_1 = (1/20)\,\mathrm{ms}^{-1}$, $\delta_{\mathrm{Low}} = \delta_0 = (1/160)\,\mathrm{ms}^{-1}$. The mean High and Low periods are equal, and we consider two different situations: intra-word silence detection, with $T_o = 1/\mu = 1/\lambda = 0.5\,\mathrm{s}$, and macro silence detection, with $T_o = 1/\mu = 1/\lambda = 5\,\mathrm{s}$.

Voice is a delay (and delay jitter) sensitive application: a single interface must not introduce excessive delay, thus limiting the buffer requirements. We consider two cases of maximum allowed delay: $d_{\max} = 5\,\mathrm{ms}$ and $d_{\max} = 10\,\mathrm{ms}$. This last constraint defines the dimension of the buffer dedicated to VoIP applications: $B = \left\lfloor \dfrac{d_{\max}C}{S_P} \right\rfloor$ where S_P is the packet size in bits. For instance, dedicating 1 Mbit/s to VoIP services with $d_{\max} = 10\,\mathrm{ms}$ yields a 15 packets buffer.

Figure 3 shows the two possibilities we're faced with, when multiplexing sources with deterministic arrivals. The lower part of the figure refers to the case when sources can be synchronized, as, for instance, when all sources belong to a same packetizing gateway. This case maps exactly to the MMDP/D/1/K queuing system. The upper part refers to a case where sources cannot be synchronized and within a single phase pack-

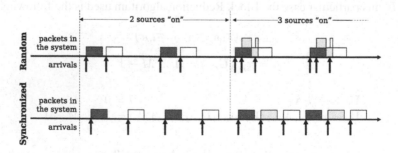

Fig. 3. Comparison between synchronized and random arrivals.

---

[1] Any other standard, like GSM, G.723, G.711, etc. would only change the packet size or packet interarrival time.

ets are not equally spaced and can overlap, leading to short term congestion and queue buildup. Our simulator handle both cases and in Sect. 5 we discuss the impact on performance.

5   Numerical Examples

We focus our attention on three different link capacities, assuming that the whole capacity is reserved for VoIP services: 512 kbit/s, 2 Mbit/s and 10 Mbit/s. In the case of 512 Kbit/s we assume that header compression is present. For each capacity, we evaluate the performances with the two different buffer sizes and the two different mean High/Low periods. All simulations are run till a 99% confidence level is reached on a $\pm 5\%$ interval of the point estimate. For high packet loss rates the confidence interval is much lower than 5%, often below 1%. The load is varied changing the number of sources M.

Figure 4 reports the model upper and lower bound for P_l and the simulation results assuming or not synchronization. Left hand plots refer to $T_o = 5$ s, right hand ones to $T_o = 0.5$ s. Different rows refer to different capacity C, and the buffer is for the case $d_{\max} = 10$ ms. As ordinates we report both the number of sources (bottom axis) and the average offered load (top axis) for easy comparison among different scenarios. Simulation results always fall between the model estimated upper and lower bounds, also when non-synchronized sources are simulated and the model is thus an approximation. As expected, the upper and lower bounds are tighter for long High-Low periods, since the renewal assumptions have a smaller relative impact on the performance; on the other hand, the bounds are looser for $T_o = 0.5$ s, since the dimension of a packet is comparable with the phase duration and considering or not fractions of packets in the system yields to distinguishable differences. Also the short term congestion induced by non-synchronized sources is more evident if the High-Low periods are very short, and the relative simulation curve approaches the upper bound.

The role of short term congestion is greater reducing the buffer size. Fig. 5 reports the results for $d_{\max} = 5$ ms. Again left hand plots refer to $T_o = 5$ s, right hand ones to $T_o = 0.5$ s. We only report results for $C = 2$ Mbit/s and $C = 512$ kbit/s since those for $C = 10$ Mbit/s are qualitatively equal to those with $C = 2$ Mbit/s (a complete set of results can be found in [17]). As expected reducing the buffer size can dramatically change the quality of the approximation, but, most interestingly, it is only the absolute value of the buffer size and not the maximum delay introduced by the buffer or the average High-Low period that predominates, as it is clear comparing the four plots. If sources are synchronized, the model upper and lower bounds hold also for very low buffer sizes. The difference between the results in case of synchronized and not synchronized sources was observed in [18] for simple On-Off sources: our results extend those considerations to the general case of High-Low sources. We can conclude that the model fails to catch the system behavior only if sources are not synchronized (e.g., in backbone routers) and the buffer dedicated to VoIP services is extremely small.

Concluding it is interesting to compare the results yielded by our model with those obtained with simple On-Off approximations[2]. Figure 6 reports two possible compar-

---

[2] For simplicity we use the upper bounds of the two curves; the conclusions hold also for lower bounds and for simulation results.

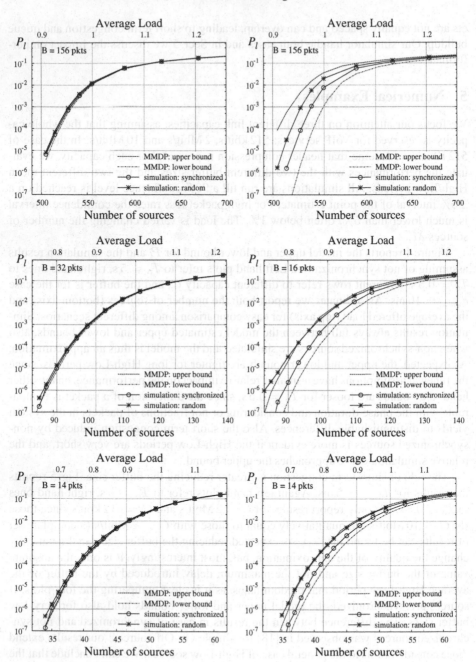

Fig. 4. Packet loss: Model vs. Simulations, $d_{\max} = 10\,\mathrm{ms}$; top row $C = 10\,\mathrm{Mbit/s}$, middle row $C = 2\,\mathrm{Mbit/s}$, bottom row $C = 512\,\mathrm{kbit/s}$; left column $T_o = 5\,\mathrm{s}$, right column $T_o = 0.5\,\mathrm{s}$.

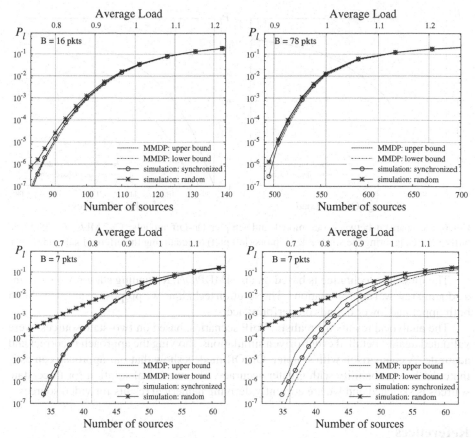

Fig. 5. Packet loss: Model vs. Simulations, $d_{max} = 5\,\text{ms}$; top row $C = 2\,\text{Mbit/s}$, bottom row $C = 512\,\text{kbit/s}$; left column High and Low mean time 5 s, right column High and Low mean time 0.5 s.

ison scenarios. In the left hand plot, where results are compared for an equal number of sources, the On-Off model underestimates P_l since during the Off phase the silence descriptors are not present and the average offered load is smaller than with High-Low sources. A more interesting perspective is offered by the right hand side plot, where results are compared for an equal average offered load. In this case the simpler On-Off model overestimates P_l. For loads of practical interest (< 0.80) the gap can be larger than an order of magnitude. The reason is that for the same offered load On-Off sources are burstier than sources with a High and a Low (with a rate different from zero) state.

6   Conclusions

This paper describes a novel analytical framework to evaluate the performance of a class of multimedia (namely those that can be described with a piecewise constant emission rate) services on a single network interface, with applications for VoIP and video services.

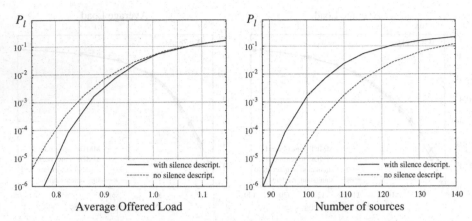

Fig. 6. Comparison between our model and simpler On-Off models; C=2 Mb/s, t_{on}=500 ms, buff=32 pkts; ignoring the silence descriptors (left plot) or equalizing the offered load (right plot).

The modeling technique is based on an MMDP/D/1/K queuing station we solve introducing renewal approximations. The solution technique we propose enables to obtain both upper and lower bounds on performance.

The analytical results for realistic VoIP scenarios, based on two-state sources, were validated against detailed event-driven simulations, showing the approach is correct and accurate. A comparison with simpler, On-Off models show that our approach can estimate the loss probability with greater accuracy. The numerical solution for three-state sources is sketched and we are currently deriving numerical results for video sources.

References

1. S. Blake et al. *An Architecture for Differentiated Services*, RFC 2475, IETF, Dec. 1998.
2. B. Davie et al., *An Expedited Forwarding PHB (Per-Hop Behavior)*, RFC3246, IETF, Mar. 2002.
3. T. Yang, D.H.K. Tsang, "A Novel Approach to Estimating the Cell Loss probability in an ATM Multiplexer Loaded with Homogeneous On-Off Sources," *IEEE Trans. on Communications*, 43(1):117–126, Jan. 1995.
4. Sang H. Kang, Yong Han Kim, Dan K. Sung, Bong D. Choi, "An Application of Markovian Arrival Process (MAP) to Modeling Superposed ATM Cell Streams," *IEEE Trans. on Communications*, 50(4):633–642, Apr. 2002.
5. H. Saito, *Teletraffic Technologies in ATM Network*, Boston, MA, USA, Artech House, 1994.
6. M. F. Neuts, *Structured Stochastic Matrices of M/G/1 Type and Their Applications*, M. Dekker, New York, 1989.
7. The Octave Web Page. http://www.octave.org/
8. Microsoft NetMeeting. http://www.microsoft.com/windows/netmeeting/
9. The OpenH323 Project Homepage. http://www.openh323.org/
10. ITU Standard H.323 Version 5, *Packet-based multimedia communications systems*, ITU, Geneva, CH, July 2003.
11. J. Rosenberg et al., *SIP: Session Initiation Protocol*, RFC 3261, IETF, June 2002,
12. H. Schulzrinne et al., *RTP: A Transport Protocol for Real-Time Applications*, RFC 3550, IETF, July 2003.

13. H. Schulzrinne et al., *RTP Profile for Audio and Video Conferences with Minimal Control*, RFC 3551, IETF, July 2003.
14. ITU Standard G.729, *Coding of speech at 8 kbit/s using conjugate-structure algebraic-code-excited linear-prediction (CS-ACELP)*, and subsequent modifications, ITU, Geneva, CH, March 2003 – Oct. 2002.
15. D. Carra, MMDP Multimedia simulator: The home page.
    http://netmob.unitn.it/mmdpms/
16. ns, network simulator (ver.2), Lawrence Berkeley Laboratory,
    http://www-mash.cs.berkeley.edu/ns
17. D. Carra, P. Laface, R. Lo Cigno, "A Performance Model for Multimedia Services Provisioning on Network Interfaces (Extended Version)," Tecnical Report DIT-04-092, Università di Trento. http://dit.unitn.it/locigno/preprints/CaLaLo04-92.pdf.
18. R. J. Gibbens, F. P. Kelly, "Distributed Connection Acceptance Control for a Connectionless Network," *Teletraffic Engineering in a Competitive World* (Eds. P. Key and D. Smith), ITC16. Elsevier, Amsterdam, 1999. 941-952.

An Integrated Multi-service Software Simulation Platform: SIMPSONS Architecture*

Laura Vellante[1], Luigi Alcuri[2], Michele L. Fasciana[2],
Francesco Saitta[2], Paola Iovanna[3], and Roberto Sabella[3]

[1] CoRiTeL – Consorzio di Ricerca sulle Telecomunicazioni, Rome, Italy
vellante@coritel.it
[2] DIE- Dipartimento di Ingegneria Elettrica,
Università degli studi di Palermo, Palermo, Italy
{luigi.alcuri,michele.fasciana,franesco.saitta}@tti.unipa.it
[3] ERI- Ericsson Lab Italy, Rome, Italy
{paola.iovanna,roberto.sabella}@ericsson.com

Abstract. This paper describes SIMPSONS (SIp and MPls Simulations On NS-2), which is a software platform for simulating multiple network scenarios, such as Telephony Over IP (TOIP), and multi-services networks extending the functionalities of Network Simulator NS-2. The innovative aspects of SIMPSONS is the complete integration of control protocols and traffic engineering mechanisms inside the same simulation tool. In fact, next generation networks must meet basically two fundamental requirements: support of Quality of Service (QoS) and Traffic Engineering (TE) functionalities. SIMPSONS is able to simulate DiffServ, MPLS, OSPF and SIP and their interaction in everything TOIP scenario. So with this powerful tool is possible to explore next generation networks and study in advance any interactions problems due to utilization of QoS management and multi-service applications. Even if both SIP and MPLS technologies have been the subjects of several works, the design and the implementation of an integrated simulation platform for SIP and MPLS interactions, still presents unsolved nuts. This work attempts to highlight the open issues of the integrated solution, giving an overview of the involved technologies and the solutions that can be implemented in such scenarios.

1 Introduction

Next Generation Networks (NGN) will be required to fulfill two main requirements: i) the ability to support different QoS in an "all-IP" context, so as to realize real multi-service networks; and ii) the capacity to optimize the use of the network resources in order to save costs. Moreover, Internet traffic is highly variable in time compared to traditional voice traffic and is not easy to forecast. This means that networks have to be flexible enough to react promptly to traffic changes while meet different QoS requirements for different service classes. The employment of Differentiated Services (DiffServ) and Multi Protocol Label Switching (MPLS) techniques on the same network infrastructure seems to be the most promising solution to achieve such requirements. DiffServ is basically a strategic solution for differentiating packets

---

* This work was funded by the TANGO project of the FIRB program of the Italian Ministry for Education, University and Research.

M. Ajmone Marsan et al. (Eds.): QoS-IP 2005, LNCS 3375, pp. 300–312, 2005.

streams that belong to the same class of service for which a network behavior better than best effort is requested. It is possible to do an admission control and apply different routing strategies for each class of service, in order to satisfy QoS requirements 0. The differentiated routing procedure can be easily implemented with the MPLS paradigm. MPLS is able to control different traffic trunks and moreover to provide the means of achieving transmission resources optimization among the classes.

At the same time, the Session Initiation Protocol (SIP) was developed in order to control and manage NGN scenarios in a simple and efficient way. SIP provides the required instruments to manage communication sessions and data transmissions. This means that SIP protocol acquires a relevant role in the management of QoS, bringing it to interact with all the mechanism introduced at lower level for the management of multi-service QoS aware networks 0. The interaction of all these elements is one of main unknown aspects of Multi-Service Network development. The main challenge to address this issue is the complexity to provide a simulation tool that is able to integrated different technologies. The amount of variables and features that an accurate study requires to consider such integrated scenario, is very high. On the other hand, it is well known the potentiality that open platform, such as NS, is able to provide to deal with such topics. As a consequence, the extension of NS modules to simulate the proposed scenario seems to be a reasonable approach to Next Generation Network features. This work aims at analyzing and preferably solving some open issues that rise from the implementation of SIP and DiffServ-over-MPLS solution. In particular, in this paper some guidelines in the analysis, study and implementation of multi Diffserv-MPLS solutions, accompanied by a complete SIP session control management, are provided. At first it gives an overview about QoS and TE mechanisms provided by DiffServ and MPLS. Then it is introduced a performance evaluation tool, that has been developed. At the end of the paper some simulation results in a sample multi service network are provided.

2 Background

The Internet is growing in size and structural complexity. This means that many structural changes have been studied and introduced into the actual infrastructure, such as:

- *Increasing connectivity:* Over the last years a steady improvement in connectivity has been observed between Autonomous Systems (AS). The exponential increase of last past years causes that the original backbone has turned into a backbone mesh, introducing new management aspects that have to be investigated.
- *Basic structural changes:* More and more backbone providers enter the scene and inter-connect themselves. They also expand their geographical scope, causing in this way the need to study interaction between providers management domains.
- *Increasing number of application types and protocols*: Many new applications and services require new protocols or use standard protocols in a different way than originally intended (e.g., SIP, MPLS, DiffServ). The convergence between telephony network and IP transport infrastructure, sponsored by telephone operators in order to reduce costs and management, has introduced the needs to offer services with different performance and Quality of Service, QoS. They influence the picture when modelling traffic distributions, and new tools of analyses are requested during the project of networks.

The main building blocks for Next Generation Network are SIP, for management of multimedia applications, DiffServ for Quality of Service, and MPLS, for TE functionalities. In this context, the reference scenario is showed in Fig. 1

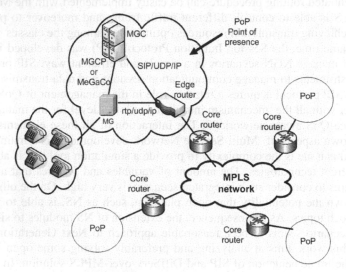

Fig. 1. ToIP Reference Scenario

The main features of such building blocks are reported in the following.

2.1 SIP

The dialogue between the elements (MGC, MG and SG) is performed by using specific signaling protocols: Session Initiation Protocols between MGCs; Media Gateway Control Protocol between MGC and MG; and QSIG/ISUP over Stream Control Transmission Protocol.

Session Initiation Protocol (SIP) is an application-layer control protocol that can establish, modify and terminate multimedia sessions or calls. These multimedia sessions include services like multimedia conferences, Internet telephony and related applications. SIP is one of the key protocols used to implement VoIP. SIP-T (SIP for telephones0) refers to a set of mechanisms for interfacing traditional telephone signaling with SIP. The purpose of SIP-T is to provide protocol translation and feature in a transparent way across points of QSIG-SIP interconnection. The SIP-T effort provides a framework for the integration of legacy telephony signaling into SIP messages.

Media Gateway Control Protocol (MGCP or MeGaCo) is a master/slave protocol used to perform the control and management of MGs. By using MGCP commands, MGC can: i) monitor the state of the network consulting MG interfaces; ii) activate QoS mechanisms (marker and policier) on a connection on the basis of management policy or specific conditions on the network; iii) reject new access requests and monitor the active connections. Both SIP and MGCP use Session Description Protocol (SDP) to describe the features of the incoming session (supported codec, RTP/UDP ports, etc.).

2.2   Differentiated Services and QoS Control

Diffserv based solutions focus on the core network and assume aggregated traffic flowing to and from the access networks. RFC 2475 gives a technical description of DiffServ 0. DiffServ can be considered as a powerful tool that ISP can use in order to offer services with the benefits of QoS control and the respect of Service Level Agreements (SLA). This is accomplished by mapping SLAs into the DiffServ Code-Points (DSCP). Diffserv offers four classes of Assured Forwarding (AF) services with three different level of precedence, each of them can be used for marking services that requires better performance than best-effort behavior. Moreover there is a single class of Expediting Forwarding (EF) reserved to streams for which low latency is an essential specification. A QoS control is applied at the ingress point of the DiffServ Domain in order to mark input packets and associate them at the appropriate flow maintaining under control the prerequisites of each different service. A policer applied at the edge router of any DiffServ domains limits the number of packets marked into a same class of service.

2.3   MPLS and TE Solutions

In the panorama of technology solutions provided to face the aforementioned NGN requirements, a key role is played by MPLS and TE. They basically offer the capability of dynamically routing the traffic over the network in order to minimize congestions and to optimize the use of network resources, while at the same time guaranteeing a certain grade of service, handling traffic fluctuations and offering multi-service capabilities 0. The MPLS scheme is based on the encapsulation of IP packets into labeled packets, that are forwarded in a MPLS domain along a virtual connection named Label Switch Path (LSP). The packet forwarding is achieved by label switching carried out by Label Switching Router (LSR). The strength of TE within MPLS is the ability to control virtual connections (i.e. LSP) in a dynamic way by means of IP based protocols. This allows introducing a high level of flexibility that is proper of IP world. One of the key functions that MPLS uses for TE is the constraint based routing (CBR); CBR is the capability to perform route calculation taking into account different constraints (i.e. the link state, resource availability besides network topology, administrative groups, priority, etc.) instead of the number of hops as in the case of plain IP routing. Another key capability of MPLS is the possibility to perform the bandwidth reservation by means of signaling protocols, such as RSVP, that are suitable extended to support such functions.

2.4   MPLS Support of DiffServ

As mentioned before, in multi-services IP networks, where QoS control and resources optimization are required, Diffserv architectures may be employed over MPLS TE mechanisms. In such a framework TE mechanisms are thought to operate on an aggregate base across all Diffserv classes of services.

In 0 a solution for Diffserv support in MPLS networks is provided. Basically, the solution gives the means to choose the mapping between Diffserv classes and MPLS LSPs, in order to perform TE strategies according to traffic QoS requirements.

2.5   DiffServ-Aware MPLS Traffic Engineering

Performing per-class TE, rather than on per-aggregate basis across all classes, is the foundation which the Diffserv-aware MPLS TE (DS-TE) 0 is built on.

DS-TE is based on the possibility of splitting the traffic according to classes of service, into multiple traffic trunks which are transported over separate LSPs. The LSP path selection algorithm and the MPLS admission control procedure can take into account the specific requirements of the traffic trunk transported on each LSP (e.g., bandwidth requirement, preemption priority). Such requirements are translated in engineering constraints applied to the routed traffic trunks. DS-TE solution basically has two goals: i) limit traffic trunk transporting a particular class to a relative percentage on core links, ii) more efficiently use network capacity by exploiting MPLS TE peculiar functionalities (e.g. preemption). For link bandwidth allocation, constraint based routing and admission control, DS-TE makes use of the concept of Class Types (CTs) previously defined in 0. A CT is a set of traffic trunk crossing a link ruled by a specific set of bandwidth constraints. DS-TE must support up to 8 CTs. The Diffserv classes can be mapped to CTs without any particular limitation. Since preemption priority associated to an LSP results independent from CT definition, it is possible to characterize an LSP by its preemption level and the CT it belongs to. In DS-TE framework, a traffic class identified by a given CT and with a certain preemption priority is called "TE-Class". In order to support such new TE-Classes, extensions to MPLS routing and signaling protocols have been already proposed 0.

3   A Tool for Investigation: SIMPSONS Description

In such a wide context, the interworking between described protocols is a delicate aspect. A simulation tool is nedeed to make us able to analyze the interaction between the various elements and to evaluate network performace. We choose to use Network Simulator 0 and we have realized SIMPSONS (SIp and MPls Simulations On NS-2). The SIMPSONS scope is to give a modular simulation tool to realize telephony applications over IP. To make it efficient it was been necessary introduce new modules for NS2 or extend the existing ones, placing particular attention to the inter-working between them. The NS2 modules which have been involved in SIMPSONS are: OSPF, MPLS, DiffServ, RSVP, and SIP.

3.1   MPLS Module and TE Extensions

A module implementing MPLS functionalities is available embedded in the NS2 distribution. It is derived from a contribution code, called MNS (MPLS Network Simulator) 0, and provides a certain set of TE functionalities. Herein the employed signaled protocol is LDP. From such implementation, we reutilize the label switching functionalities for the forwarding plane and the label distribution procedure for the control. The functions of admission control and resource reservation have been implemented separately. We have chosen not to exploit the already implemented procedures because we want to implement different strategies (different PHB-to-CT mapping, different algorithms, different booking schemes).

3.2 DiffServ Module and Its Interaction with Application and MPLS Layers

This tool makes use of NS2 DiffServ module in order to differentiate the input streams on the basis of the service, which belong to, and matching to relative service level specifications. After having established the kind of service which have to be managed inside the DiffServ domain it is necessary to assign them the corresponding DSCP, which will mark all the packets which belong to a specific service and influence the Per Hop Behavior (PHB) of DiffServ capable routers. We can differentiate the service on the basis of their performance requests described by SLA and SLS in terms of latency, packet loss rate and bandwidth. For each class of service individuated there is an assignment inside a forwarding class of DiffServ. AF classes are utilized by real-time application, for example we have configured different voice rate services each with different voice compression codec. The DiffServ module is used on each router at the edge of DiffServ Domain for marking ingress packets to each corresponding PHB by means of DiffServ CP. Each forwarding class is managed on the basis of precedence level associated to router physical and virtual queue for simulating the difference precedence level implemented in DiffServ definitions.

3.3 OSPF and TE Extensions

The starting module is the QOSPF module for NS implementing functionalities described in [15]. We have extended this module in order to simulate the TE LSA, as specified in [16] and to add a personalized constraint based routing algorithm. The first step has been the extension to simulate TE LSAs. They are opaque LSA carrying information about traffic engineering topology, including bandwidth and administrative group. The implemented constraint-based routing algorithm has the objective of balancing as well as possible the load in the network. It works in this way: given a couple source-destination and a bandwidth constraint, it chooses the shortest path between source-destination, so that every link on the path satisfies the specificied bandwidth requirement. In particular, it has been assigned at each link a cost that is inversely propotional to the available bandawidth on that link. The updating of the available bandwidth value is made by TE LSA. In a second time, we implemented TE extensions for the support of multiple service classess descibed in [16].

For each of the eight Class Type the opaque LSA carries these values: available bandwidth on link, maximum reservable bandwidth value on link, non reservable bandwidth value on link and administrative group. At each CT it was assigned a priority level. At the moment we are working to extend the constraint based routing algorithm so that it calculates the path for each CT traffic and not only for the aggregate traffic, like now. In this way, it will be possible to create bandwidth constraints also for data traffic. This will lead a scenario where all the traffic flows entering the network are goverened by engineered constraints. Moreover, the four Assured Forwarding classes may be mapped into different TE-Classes, and thus they may get different services from the connection priority point of view.

3.4 RSVP-TE

Regarding the signaling protocol, we have implemented some of the extensions of RSVP described in 0. The starting module is the RSVP module for NS2. The extensions realized regards the definitions of the following objects inserted in the protocol message: LABEL REQUEST; ERO; LABEL; SENDER TEMPLATE; SETUP; HOLDING PRIORITY. The meaning of these object is the one described in 0. At the moment only the first four objects are used, the lastest two are defined but not used. RSVP-TE module interact with three elements: admission control module, MPLS application layer and OSPF. The admission control, which we refer to is the one embedded in NS2. Regarding the interaction between RSVP-TE and OSPF, the OSPF calculates the requested path with the constraint-based routing algorithm, described in the previous section, and passes it to the RSVP-TE module that makes the ERO object with the path information received by OSPF. Along the path specified in the ERO object, RSVP-TE and MPLS together realize the label binding. Another hinge point is the interworking between MPLS and RSVP-TE modules.

The implementation of MPLS TE functionalities in the simulator regards the development of LSP set-up, bandwith reservation, constraint-based routing and interoperability with the DiffServ implementation part. A module implementing MPLS funcionalities is available embedded in the NS2 distribution. It is derived from a distribution code called MNS (MPLS Network Simulator) and provides a certain set of TE functionalities. Herein the employed signaling protocol is LDP. The interworking between MPLS and RSVP-TE has some problem by an implementation point of view. Infact, as we said, MPLS has been developed to specifically use LDP as signaling protocol; MPLS module does not support another signaling protocol, like RSVP-TE. We have implemented a transitory solution in order to make RSVP-TE able to exchange label with MPLS, even though MPLS does not have the perception of using RSVP-TE as signaling protocol.

3.5 SIP and MGC

For the management of signalling in next multi-service network we have developed an innovative module for NS2, which is able to simulate MGC behavior in environments like those described in Fig. 2

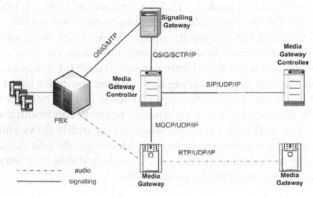

Fig. 2. Module: reference scenario

This MGC module is able to simulate a bidirectional dialogue between two SIP agents that send INVITE requests, according the specifics of SIP protocol we have simulated the complete messages exchange in the SIP RFC 0. Only when call setup is correctly performed, a new bidirectional voice session starts between two nodes.

Obviously the overall voice traffic depends on the number of call setup correctly performed. The inter-arrival process between two consecutive sent INVITE messages is poissonian. According to SIP call flow, the TRYING message is sent instantly as soon as an INVITE message is received; timers manage the forwarding of SIP messages properly: RINGING timer (provides uniform delay) and 200OK timer (provides exponential delay). Finally BYE timer controls the stop of the single call.

Between one SIP message and the following one, MGC sends a series of command to MG using MGCP protocol. One of the main problems that the simulation of a SIP-T scenery introduces is that to synchronize the departure of the vocal calls with the instant in which MGC agents receives the message 200OK; in fact, it is just in this moment that all the fit procedures for the call setup (the information exchanged through SDP, opening of RTP interfaces and UDP ports in the Media Gateway, choice of the codecs, etc.) are dispatched. NS2 provides the possibility to use particular procedures able to modify the execution of the simulation directly from C++ code. Several Tcl procedures (Instproc) were introduced for managing voice sources during the simulations. In this way MGCs can manage autonomously set up, start and stop of the source voices. In particular, Instproc CONFIGURESESSION refers to CRCX, Instproc STARTCALL refers to MDCX, Instproc STOPCALL refers to DLCX command.

By using these procedures in the simulation script it is possible for the MGC module to manage autonomously set up, start and stop of the voice sources. In real implementation, SIP protocol can benefit of a transport layer both reliable and unreliable, so we have foreseen the possibility to choose between UDP (connectionless service) and SCTP (reliable service). In the following simulation however we have adopted a connectionless service leaving to the SIP protocol the management of communication.

3.6 Module Integration

One of the main difficulties in the development of NGN architecture is the correlation between different modules for the management of QoS. Actually we have defined procedures and interaction functions between SIP, MPLS, DiffServ and OSPF modules. We have collected these interaction aspects in a collection of NS tcl libraries, so to extend scalability and scenario developments.

In the future the OSPF-TE and RSVP-TE modules will be integrated in such libraries to integrate flow-bandwidth reservation in simulation scenario, instead of implementing it separately.

4   Performance Evaluation

In our modular realization of NGN infrastructure through NS, we have developed several functional objects that could be connected between them in order to get any possible scenario. In the following we will describe the realization and the performance evaluation of a sample network scenario.

4.1 Network Topology

The first step is the choice of the core networks and the configuration of core routers with support to DiffServ and MPLS routing. Each router is constituted by a NS node in which it is activated the MPLS routing support and the OSPFTE module, in fact MPLS and OSPFTE functionalities were developed as optional modules attached to the simple node object. In this way we have maintained compatibility with previous NS applications. Next in the description of network topology is the definition of the links between core routers and their properties. In this case we have adopted RSVP duplex-link with a parametric local admission control and a null estimator, the queue scheduler adopted is a WFQ with priority enabled.

With this link we are able to make reservation and manage RSVPTE module functions through essential linking objects. For each link it is possible to configure capacity, propagation time, percent of reserved bandwidth, reserved bandwidth for signaling, local admission control, and local bandwidth estimator.

The routers are able to simulate MPLS and LDP protocol with OSPFTE extension and the links are able to support RSVPTE extensions and DiffServ Priority scheduling.

4.2 POP Description

Next step in the realization of a simulation scenario is the definition of traffic sources, in order to make SIMPSONS scalable and simple to utilize we have created a completely functional object, called POP, which includes all the elements which are characteristic of a DiffServ aware Multi-Service traffic source. In fact inside each object POP, we have put an edge router, an MGC-SIP messages generator, an RTP voice sources node, an FTP data source node, and as many MG-Monitor node as are the class of services defined inside the DiffServ domain.

The POP object could be easily connected to any core router trough the "create_pop" function while a "configure_edge" function could be used to configure specific traffic parameters for each POP. It is possible to configure for each POP, the capacity of the link with the core network, the number of Physical and Virtual queues, the DiffServ policer of each queue and the RED thresholds. The complete network topology with all nodes and links is showed in Fig. 3

4.3 Networks Dynamic Events

The main generator of simulation events is the SIP module. In fact this module simulate call requests by random generated new call incoming events. Obviously it is possible to configure the randomness of call requests by setting specific parameters, the most important are: Burst time, idle time, activity rate, minimum SS7 delay, maximum SS7 delay, default packet size and call length distribution of probability.

The simulation where conducted considering poissonian voice traffic generation and the configuration parameters where set according the Brady Voice Model.

Moreover we have introduced 4 class of service each with a different cbr and it is possible for each POP to regulate the random probability of new incoming calls to belong at each specific class. For the simulation example we have considered the following call generation parameters:

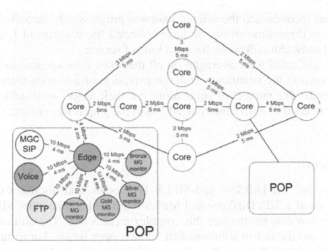

Fig. 3. NS description of network scenario, in evidence the core network and POP structure

- *Premium*: probability = 0.042; rate = 64Kbps
- *Gold*: probability = 0.084; rate = 32Kbps
- *Silver*: probability = 0.336; rate = 8Kbps
- *Bronze*: probability = 0.538, rate = 5Kbps

With these particular values we obtained approximately the same offered bit-rate for each kind of service. Whenever a new call request event occurs, the SIP module sends an INVITE message towards the SIP node of destination POP with the information about kind of call and codec supported. The SIP module of destination POP respond with a TRYING message and interrogate MGC-CAC module for network status and QoS parameters. CAC module estimates network utilization by interrogating MG-monitor modules for the specific kind of services and answers with a network status GREEN, YELLOW or RED.

- In GREEN network status, the call is accepted with the requested type of service, and the SIP module sends a 200OK message
- In YELLOW status, the call is accepted with a lower call of service and the SIP module sends a 200OK message with the new class information
- In RED status, the call is suspended and the control is passed to the Edge MPLS module that will search for an alternative path with available bandwidth for the class of service. In the case of success the call is accepted and a 200OK message is sent back, while in case of failure the call is refused and a 406ERROR is sent to signalize the network congestion status at the originating POP.

4.4 Simulation Description

Using the network scenario described above, we have conducted several simulations in order to test the interaction and the functionality of each module and to give an example of potentiality of SIMPSONS. We have launched a campaign of simulation with MPLS and CBR flow reservation and the results where compared which the results of a campaign of simulations with MPLS and OSPF. Each simulation was long

3600 simulated seconds, and the offered load was progressively increased from 70 % to 300 %. From these simulations we have collected the statistics of latency, packet loss rate and bandwidth utilization for each kind of service.

The results collected were averaged on all the voice flow streams received by all the POP involved in the simulation with the exclusion of a starting transitory period long 100 seconds. The number of simulations and their length were sufficient to limit 95-percentile interval to a maximum of 10^{-4} so in the graphs the variance could not be appreciated.

5 Conclusion

This paper reviews the DiffServ and MPLS-TE models, and discusses the architectural framework of a SIP, DiffServ and MPLS integrated architecture. SIMPSONS is a NS-20 extension able to simulate this complex architecture. The aim of such a tool is the analysis and the test of solutions that address open issues. For instance, regarding the DS-TE context, 0 shows how bandwidth constraint models present the tradeoff between bandwidth sharing, to achieve better efficiency under normal condition, and class isolation under overload network condition. Thus, besides the investigation of new models, it is possible to study such a trade-off employing different routing and admission control algorithms. Moreover, in all the Diffserv-over-MPLS solutions, there is the optional possibility to dynamic adjust Diffserv PHBs parameters (e.g. queue bandwidth) during the establishment of a new LSP. This option may involve a number of configurable parameters: an investigation about the trade-off between complexity and performance gain could be the matter of a research study.

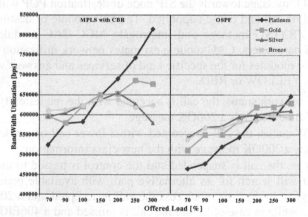

Fig. 4. Bandwidth Utilization Vs. Network Offered Load for both MPLS/CBR and OSPF

Another open issue is how to map PHBs in CTs, and how to associate preemption priorities to form TE-Classes in the DS-TE framework. In fact, there are TE mechanisms, such as soft-preemption, where reservation contention may not reflect forwarding plane congestion. SIMPSONS allows all the different QoS aspects, either relating to the control or the data plane, to be taking into account simultaneously in the same simulation context. Moreover the modular structure of NS provides the means for adding new modules to implement new functionalities. With NS it is also possible to

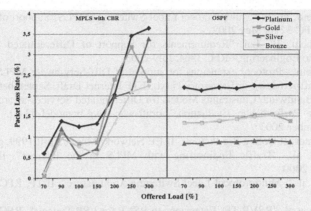

Fig. 5. Packet Loss Rate Vs. Network Offered Load for both MPLS/CBR method and OSPF

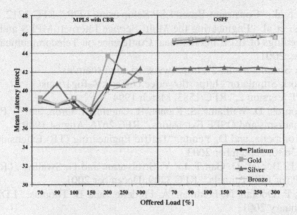

Fig. 6. Mean Transmission Latency Vs. Network Offered Load for both MPLS/CBR and OSPF

model different traffic sources from simple real-time voice applications to more sophisticated video conference and video streaming applications.

This aspect belong with the possibility of simulating different clients at the edges of the network makes this tools useful for the study of a large number of different network scenarios.

Figures from 4 to 6 depict the results of a simulation conducted by means of SIMPSONS where a Constraint Based Routing is tested versus shortest path algorithm in a Diffserv scenario with four real-time voice classes, where SIP is used to manage telephony session with an end-to-end QoS-based call admission control applied. A tool such this, which is able to simulate from the simplest TE scenario to the most complex DS-TE solution, is surely very useful for similar investigation research task. Moreover, the prospective to provide trends to design VPNs in a Diffserv-MPLS environment makes the employment of a simulating tool very attractive.

References

1. D. Awduche, A. Chiu, A. Elwalid, I. Widjaja, and X. Xiao, "Overview and Principles of Internet Traffic Engineering," RFC 3272, May 2002

312   Laura Vellante et al.

2. F. Le Faucheur, et al., "Multi-Protocol Label Switching (MPLS) Support of Differentiated Services," RFC 3270, May 2002
3. F. Le Faucheur, W. Lai, "Requirements for Support of Differentiated Services-aware MPLS Traffic Engineering," RFC 3564, July 2003
4. F. Le Faucheur, "Protocol extensions for support of Diff-Serv-aware MPLS Traffic Engineering," draft-ietf-tewg-diff-te-proto-05.txt, IETF Internet Draft, September 2003
5. W. S. Lai, "Bandwidth Constraints Models for Differentiated Services-aware MPLS Traffic Engineering: Performance Evaluation," draft-wlai-tewg-bcmodel-03.txt, IETF Internet Draft, September 2003
6. Xiao, Ni, "Internet QoS: a Big Picture," IEEE Network, March/April 1999, pp. 8-18
7. X. Xiao, et al., "Traffic Engineering with MPLS in the Internet", IEEE Network, March/April 2000.
8. E. Rosen, et al. " Multiprotocol Label Switching Architecture", IETF, RFC 3031, January 2001
9. D. Awduche et al, "RSVP-TE: Extensions to RSVP for LSP Tunnels", RFC 3209, December 2001.
10. B. Jamoussi, et. al., "Constraint-Based LSP Setup using LDP," RFC 3212, January 2002
11. P. Trimintzios, et al. "Engineering the Multi-Service Internet: MPLS and IP-Based Techniques" Proceedings of IEEE International Conference on Telecommunications, Bucharest, Romania, June, 2001.
12. "The Network Simulator - ns-2", http://www.isi.edu/nsnam/ns/
13. "MPLS Network Simulator", http://flower.ce.cnu.ac.kr/~fog1/mns/
14. J. Moy, "OSPF Version 2," RFC 1583, March 1994
15. G. Apostolopoulos, D. Williams, S. Kamat, R. Guerin, A. Orda, and T. Przygienda, "QoS Routing Mechanisms and OSPF Extensions," RFC 2676, August 1999
16. D. Katz, K. Kompella, and D. Yeung, "Traffic Engineering (TE) Extensions to OSPF Version 2," RFC 3630, September 2003
17. D. Awduche, L. Berger, D. Gan,T. Li, V. Srinivasan, and G. Swallow, "RSVP-TE: Extensions to RSVP for LSP Tunnels," RFC 3209, December 2001
18. L. Andersson, P. Doolan, N. Feldman, A. Fredette, and B. Thomas, "LDP Specification," RFC 3036, January 2001
19. T. Kimura, S. Kamei, T Okamoto, "Evaluation of DiffServ-aware constraint-based routing schemes for multiprotocol label switching networks," Networks, 2002. ICON 2002. 10th IEEE International Conference on, 27-30 Aug. 2002 Page(s): 181 –185
20. V. Fineberg, C. Chen, X. Xiao, "An end-to-end QoS architecture with the MPLS-based core," IP Operations and Management, 2002 IEEE Workshop on, 2002 Page(s): 26 –30
21. E. Rosen, A. Viswanathan, R. Callon, "Multiprotocol Label Switching Architecture," RFC 3031
22. Blake, S., Black, D., Carlson, M.,Davies, E., Wang, Z., And Weiss, W. "An Architecture for Differentiated Services". RFC 2475. December 1998.
23. S. Floyd. "Comments on Measurement-based AdmissionsControl for Controlled-Load Service". Computer Communication Review, 1996.
24. L.Alcuri, F. Saitta, "Telephony over IP: A QoS Measurement-Based End to End Control Algorithm", CCCT 2003 Orlando (FL), August 2003
25. Handley M., Schulzrinne H., Schooler E., Rosenberg J., "SIP: Session Initiation Protocol", RFC 2543. IETF. March 1999.
26. A. Vemuri, J. Peterson, "SIP for Telephones (SIP-T): Context and Architectures", draft-ietf-sipping-sipt-04.IETF. June 2002. Best Current Pactice
27. Best Current Practice for Telephony Interworking", draft-zimmerer-sip-bcp-t-00.txt . IETF. April 2000. Informational
28. M. Arango, A. Dugan, I. Elliott, C. Huitema, S. Pickett, "Media Gateway Control Protocol (MGCP Request for Comments: 2705.IETF. October 1999

Optimal Load Balancing
in Insensitive Data Networks

Juha Leino* and Jorma Virtamo

Networking Laboratory, Helsinki University of Technology,
P.O.Box 3000, FIN-02015 HUT, Finland
{Juha.Leino,Jorma.Virtamo}@hut.fi

Abstract. Bonald et al. have recently characterized a set of insensitive dynamic load balancing policies by modelling the system as a Whittle network. In particular, they derived optimal "decentralized" strategies based on limited state information and evaluated their performance in simple example networks. In this paper, we consider the specific case of a data network where each flow can be routed on one of a set of alternative routes. By using the linear programming formulation of MDP theory we are able to analyze optimal routing policies that utilize the full global state information. In the ordinary LP formulation of MDP theory, the global balance condition appears as a linear constraint on the decision variables. In order to retain insensitivity, we impose stricter detailed balance conditions as constraints. As a further extension, the MDP-LP approach allows joint optimization of the routing and resource sharing, in contrast to the earlier work where the resource sharing policy was required to be separately balanced and fixed in advance. The various schemes are compared numerically in a toy network. The advantage given by global state information is in this case negligible, whereas the joint routing and resource sharing gives a clear improvement. The requirement of insensitivity still implies some performance penalty in comparison with the best sensitive policy.

1 Introduction

Load balancing has important applications in computer and communication systems. The ability to route the service demands to different available service resources of the system can have a significant effect on the performance. Load balancing policies can be categorized by the information available at the time of a decision. Static load balancing policies are the simplest ones. The customers are divided between the servers probabilistically regardless of the system state. The optimal static load balancing can be solved straightforwardly as an optimization problem [1]. A more complex problem arises if the load balancing decisions depend on the system state. These dynamic or adaptive policies are more efficient than the static ones as the customers can be routed to less utilized servers.

---

* Corresponding author.

M. Ajmone Marsan et al. (Eds.): QoS-IP 2005, LNCS 3375, pp. 313–324, 2005.
© Springer-Verlag Berlin Heidelberg 2005

However, the problem of optimal dynamic load balancing is difficult; even the simplest load balancing schemes are hard to solve.

In this paper, we study load balancing in data networks. We assume that there are several routes the flows can utilize. The balancing is realized at the flow level. Each arriving flow is routed to one of the routes and the same route is utilized until the flow is finished. It is also assumed that every time the number of flows in the system changes, the bandwidth resources of the network are instantaneously reallocated among the flows. Thus, between epochs of a flow arrival or departure each flow receives a constant bandwidth. The load balancing problem amounts to dividing the traffic into different routes in a way that maximizes the performance of the network. While we specifically discuss load balancing in telecommunications, the methods used can be applied to other load balancing applications as well.

The most basic load balancing scenario consists of a single source of flows and parallel identical links. The optimal routing policy depends on the arrival process and the flow size distribution. The policy of joining the shortest queue is the best in many cases [2, 3]. However, it is not always the optimal policy [4]. All in all, the sensitivity to detailed traffic parameters makes solving of optimal load balancing difficult.

In this paper, we consider routing policies that are insensitive, i.e. the performance does not depend on the flow size distribution. The analysis of insensitive flow-level dynamic systems has been pioneered by Bonald and Proutière [5, 6], who apply Whittle networks as a model for such systems. Notably, they have introduced the concept of balanced fairness (BF) as the most efficient insensitive bandwidth allocation when static routing is used. Bonald et al. [7] have also recently studied insensitive dynamic routing in Whittle networks and characterized a set of insensitive dynamic load balancing policies. In particular, they discuss optimal "decentralized" strategies based only on local knowledge.

We present a method for determining the optimal insensitive load balancing policy utilizing global knowledge. Our work is based on the theory of Markov decision processes (MDP). Markov decision process is a Markovian process that can be controlled in some way and the theory of MDP's provides a mean to determine the optimal control policy. MDP theory has been applied to routing problems [8] as well as to other telecommunication problems [9]. In this paper, we apply the linear programming (LP) formulation of the MDP theory to the insensitive load balancing problem.

In the ordinary LP formulation of the MDP theory, global balance condition appears as a linear constraint on the decision variables. In order to retain insensitivity, we impose stricter detailed balance conditions as constraints. As a further extension, the MDP-LP approach allows joint optimization of the routing and resource sharing, in contrast to the earlier work where the resource sharing policy was required to be separately balanced and fixed in advance. The idea of obtaining insensitivity by jointly balancing the routing and resource sharing appears already in [5] but, to our knowledge, has not been applied previously to optimal load balancing. The various schemes are compared numerically in a toy network. The advantage given by global state information is in this case negligible, whereas the joint routing and resource sharing gives a clear improvement. The requirement of insensitivity still implies some performance penalty in comparison with the best sensitive policy.

This paper is organized as follows. The next section introduces Whittle queueing networks and Markov decision processes. In Sect. 3, the queueing network model is used to model a communication network. In Sect. 4, we present different routing schemes using the network model and MDP theory. Section 5 provides a numerical example. Finally, we conclude the paper in Sect. 6.

2 Theoretical Framework

2.1 Insensitivity in Processor Sharing Networks

We consider an open queueing network of N processor sharing (PS) nodes. The state of the network is $x = (x_1, \ldots, x_N)$, where x_i is the number of customers at node i. The capacity $\phi_i(x)$ of a node i may depend on the state of the network. The capacity is divided equally between the customers at the node. For the purposes of this paper, it is sufficient to consider only simple networks, where a customer upon completion of the service at any node leaves the network. The arrivals at node i are Poissonian with rate λ_i. The customers at node i require i.i.d. exponentially distributed service with mean σ_i. We define q_{xy} as the transition rate from state x to state y.

The stochastic process describing the state of the system is a Markov process. An invariant measure of the system $\pi(x)$ satisfies the global balance condition

$$\sum_y q_{xy}\pi(x) = \sum_y q_{yx}\pi(y) \quad \forall x. \tag{1}$$

A queueing network is insensitive if the steady-state distribution depends only on the traffic intensities $\rho_i = \lambda_i \sigma_i$ of the nodes and not on any other characteristics such as the arrival rates or the distribution of the service times.

The presented queueing network without internal routing is insensitive if and only if the the detailed balance conditions

$$\pi(x)q_{xy} = \pi(y)q_{yx} \quad \forall x, y \tag{2}$$

are satisfied. These constitute a stricter requirement than the global balance condition (1). More generally, when internal routing is allowed these detailed balance conditions should be replaced with partial balance conditions [5].

The service rates are said to be balanced if

$$\phi_i(x)\phi_j(x - e_i) = \phi_j(x)\phi_i(x - e_j) \quad \forall x_i > 0, x_j > 0, \tag{3}$$

where e_i is a vector with 1 in component i and 0 elsewhere. The balance condition is equivalent to the detailed balance condition if the arrival rates λ_i do not depend on the state. All balanced service rates can be expressed with a unique balance function Φ so that $\Phi(0) = 1$ and

$$\phi_i(x) = \frac{\Phi(x - e_i)}{\Phi(x)} \qquad \forall x : x_i > 0. \tag{4}$$

If the service rates are balanced, the stationary distribution depends only on the traffic loads at the nodes and can be expressed as

$$\pi(x) = \pi(0)\Phi(x)\prod_{i=1}^{N} \rho_i^{x_i}, \tag{5}$$

where $\pi(0)$ is the normalization constant.

2.2   Linear Programming Formulation of MDP

In many stochastic processes, the state transitions can be controlled by taking a sequence of actions. Markov processes that can be affected in this way are called Markov decision processes, see e.g. [10, 11]. In each state, an action is selected from a set of feasible actions. The aim is to choose the actions that either minimize the system costs or maximize the system value often defined on infinite time horizon. In a communication network, a typical objective is to maximize throughput or to minimize blocking probability. The optimal controls can be determined using policy-iteration or value-iteration algorithms or linear programming approach. The LP formulation makes it possible to use probabilistic actions. Routing is stochastic in an insensitive network, hence the MDP-LP approach is suited for the optimal routing problem discussed in this paper.

In the LP approach, the decision problem is formulated as a linear optimization problem. Let A_x be the finite set of feasible actions in state x. The $\pi(x, a) \geq 0$ are the decision variables of the problem. They can be interpreted as the probability that the system is in state x and action $a \in A_x$ is selected and they satisfy the normalization constraint $\sum_x \sum_{a \in A_x} \pi(x, a) = 1$. The object is to minimize a cost function $\sum_x \sum_{a \in A_x} c(x, a)\pi(x, a)$, where $c(x, a)$ is the cost rate when action a is used in state x. In an ordinary MDP problem, the decision variables have to satisfy the global balance condition

$$\sum_y \sum_{a \in A_x} q_{xy}(a)\pi(x, a) = \sum_y \sum_{a \in A_y} q_{yx}(a)\pi(y, a) \quad \forall x, \tag{6}$$

where $q_{xy}(a)$ is the transition intensity when action a is selected. If insensitivity is required, the global balance equation is replaced with stricter partial balance equations, which in our case of no internal routing between the nodes are equivalent to the detailed balance equations

$$\sum_{a \in A_x} q_{xy}(a)\pi(x, a) = \sum_{a \in A_y} q_{yx}(a)\pi(y, a) \quad \forall x, y. \tag{7}$$

The stricter condition might lead to a worse solution, hence the problem with the detailed balance condition gives a lower bound for the performance of the system. The choice of optimal actions is formulated as a linear optimization problem:

$$\min_{\pi(x,a)} \sum_x \sum_{a \in A_x} c(x,a)\pi(x,a), \tag{8}$$

$$\text{s.t.} \sum_{a \in A_x} q_{xy}(a)\pi(x,a) = \sum_{a \in A_y} q_{yx}(a)\pi(y,a) \quad \forall x, y, \tag{9}$$

$$\sum_x \sum_{a \in A_x} \pi(x,a) = 1, \tag{10}$$

$$\pi(x,a) \geq 0 \quad \forall x, a. \tag{11}$$

The problem can be solved using a generic linear programming algorithm.

3   Application to Data Networks

The queueing network described in the previous section can be used to model a communication network [6]. The flows of a traffic class on a route are the customers of a PS node in the queueing network. The capacity of the node is the allocated bit rate of the traffic class. The network is insensitive if and only if the detailed balance condition is satisfied. The steady state distribution depends only on the traffic loads on different routes as long as the flow arrivals are Poissonian.

The network consists of links $l \in \mathcal{L}$ with capacities C_l. We assume that there are K traffic classes. Class-k flows arrive at rate ν_k. The mean flow size of class k is σ_k. Class-k flows can be routed to one of the routes R_k. Each route r is a subset of the links $r \subset \mathcal{L}$. x is the system state vector, where the element $x_{k,r}$ is the number of class-k flows on route r. The arrival rate of class-k traffic on route r is $\lambda_{k,r}(x)$. The rates need to satisfy the traffic constraints

$$\sum_{r \in R_k} \lambda_{k,r}(x) \leq \nu_k \quad \forall x, k. \tag{12}$$

The bitrate allocated for class-k flows on route r is denoted $\phi_{k,r}(x)$ and assumed to be equally shared between these flows. The allocations have to satisfy the capacity constraint

$$\sum_k \sum_{r \in R_k : l \in r} \phi_{k,r}(x) \leq C_l \quad \forall x, l. \tag{13}$$

If the arrival rates of the flows are fixed, i.e. $\lambda_{r,k}(x) = \lambda_{r,k}$, the detailed balance condition is satisfied when the capacity allocation is balanced. Every balanced capacity allocation can be defined by a unique balance function $\Phi(x)$ so that $\Phi(0) = 1$ and

$$\phi_{k,r}(x) = \frac{\Phi(x - e_{k,r})}{\Phi(x)} \quad \forall x_{k,r} > 0. \tag{14}$$

Balanced fairness is the balanced capacity allocation that is the most efficient [5, 12] in the sense that at least one link is saturated in each state. Balanced
fairness can be defined recursively by $\Phi_{bf}(0) = 1$ and

$$\Phi_{bf}(x) = \max_l \frac{1}{C_l} \sum_k \sum_{r:l \subset r, x_{k,r} > 0} \Phi_{bf}(x - e_{k,r}). \tag{15}$$

If the capacity allocation is balanced, a network with dynamic routing is
insensitive if and only if the routing is balanced [5]. Similarly to capacity allocation, routing is balanced if there exists a balance function Λ such that $\Lambda(0) = 1$
and

$$\lambda_{k,r}(x) = \frac{\Lambda(x + e_{k,r})}{\Lambda(x)} \quad \forall x. \tag{16}$$

A network can be insensitive even if the capacity allocation is not balanced [5]. While balanced fairness is the most efficient balanced allocation more
efficient solutions may exist if both allocation and routing are considered simultaneously. The detailed balance condition is satisfied with any balance functions
Ψ such that

$$\frac{\lambda_{k,r}(x - e_{k,r})}{\phi_{k,r}(x)} = \frac{\Psi(x - e_{k,r})}{\Psi(x)} \quad \forall x : x_{k,r} > 0. \tag{17}$$

4   Optimal Insensitive Routing Policies

In this section, we present several insensitive routing methods using the presented
network model. With each method, our aim is to find the policy that minimizes
the flow blocking probability of the system. Because we consider insensitive
routing, the blocking probabilities do not depend on the flow size distribution.
More efficient routing policies may be found if the requirement for insensitivity
is omitted hence the best insensitive policy is a lower bound for the performance
of the sensitive network.

We assume that the flows have a lower bound for a useful bandwidth, see
e.g. [13]. An arriving class-k flow is blocked if the minimum bit rate $\phi_k^{\min}$ cannot
be provided. The state space S of the system is

$$S = \left\{ x : \frac{\phi_{k,r}(x)}{x_{k,r}} \geq \phi_k^{\min}, \forall k, r \in R_k \right\}. \tag{18}$$

The criterion we use to determine the optimal policy is the flow blocking
probability. The blocking probability of traffic class k is

$$\sum_x \pi(x) \left(1 - \frac{\sum_{r \in R_k} \lambda_{k,r}(x)}{\nu_k} \right) \tag{19}$$

and the overall blocking probability is

$$\sum_x \pi(x) \left(1 - \frac{\sum_k \sum_{r \in R_k} \lambda_{k,r}(x)}{\nu} \right), \tag{20}$$

where $\nu = \sum_k \nu_k$ is the overall offered traffic.

4.1 Static Routing

The simplest routing policy is the static routing. The routing does not depend on the network state but the traffic classes are routed in some fixed ratios among the different routes. Determining the best static routing is straightforward, see e.g. [1].

4.2 Dynamic Routing with Local Information

When a routing decision is made the best result is achieved if the state of the whole network is known at the time of the decision. In practice, this information is not always available. If a traffic class knows only the number of flows belonging to that class on each route it is said to have local knowledge. The routing decision depends only on the local state $\lambda_{k,r}(x) = \lambda_{k,r}(x_k)$, where $x_k = \{x_{k,r}\}_{r \in R_k}$. The routing is decentralized as the routing decisions of the different classes are made independently.

In [7], the authors present a method to determine optimal insensitive routing with local knowledge. The details can be found in the article. Finding the policy is very straightforward and fast. The optimal policy is *simple*, i.e. there is only one local state where traffic is rejected.

In a sense, the assumption of only local knowledge is too restrictive. If flows on a route are blocked the source knows that the route is congested at some link. The capacity allocated for the flows depends on the network state hence the source can get some information on the global network state by observing the bandwidth available on each route.

4.3 Dynamic Routing with Balanced Fairness Allocation

When the state of the whole network is known at the time of the routing decision the routing is more efficient than with only local knowledge. If a link is congested the traffic can be routed to another link with less traffic. We assume that the capacity is allocated according to balanced fairness as it is the most efficient balanced allocation.

The routing problem can be formulated as a Markov decision process. In each state, the set of actions are the feasible routing alternatives. The actions that lead to the lowest blocking probabilities while satisfying the constraints constitute the optimal routing policy.

In the MDP problem, the state of the process is the network state and the actions are the routing decisions. The decision vector is $d = (d_1, \ldots, d_K)$, where $d_k = r$ if class-k traffic is directed to route $r \in R_k$ and $d_k = 0$ if the class is blocked. $\pi(x, d)$ is a decision variable and corresponds to the probability that the system is in state x and routing d is used. The time the system is blocking class-k traffic in state x is $\sum_{d:d_k=0} \pi(x, d)$. Using this notation the Markov decision problem can be formulated as a linear programming problem as follows:

$$\min_{\pi(x,d)} \sum_k \frac{\nu_k}{\nu} \sum_x \sum_{d:d_k=0} \pi(x, d), \tag{21}$$

$$\text{s.t. } \nu_k \sum_{d:d_k=r} \pi(x,d) = \frac{\phi_{k,r}(x+e_{k,r})}{\sigma_k} \sum_d \pi(x+e_{k,r},d) \quad \forall x, k, r \in R_k, \quad (22)$$

$$\sum_x \sum_d \pi(x,d) = 1, \quad (23)$$

$$\pi(x,d) \geq 0 \quad \forall x,d. \quad (24)$$

If the flow sizes are exponentially distributed, the stochastic process describing the state of the network is Markovian and the best nonbalanced routing policy can be determined using MDP-theory. The detailed balance constraints (22) are replaced with the global balance constraints

$$\sum_k \nu_k \sum_{d:d_k \neq 0} \pi(x,d) + \sum_k \sum_{r \in R_k} \frac{\phi_{k,r}(x)}{\sigma_k} \sum_d \pi(x,d)$$

$$= \sum_k \sum_{r \in R_k} \nu_k \sum_{d:d_k=r} \pi(x-e_{k,r},d) +$$

$$+ \sum_k \sum_{r \in R_k} \frac{\phi_{k,r}(x+e_{k,r})}{\sigma_k} \sum_d \pi(x+e_{k,r},d) \quad \forall x \quad (25)$$

and the LP problem is solved.

4.4 Joint Optimal Routing and Capacity Allocation

The previous policy assumed the capacity allocation to be separately balanced and fixed in advance (balanced fairness) and only the routing was considered. Better results can be obtained if routing and capacity allocation are considered simultaneously. The actions in each state consist of the routing decision and the capacity allocation decision.

Let $C_{k,r} = \min_{l \in r}(C_l)$ be the maximum bandwidth available for class-k traffic on route r. The allocation actions are modeled with an allocation vector $b = (b_{1,1}, \ldots, b_{1,r}, b_{2,1}, \ldots, b_{K,1}, \ldots, b_{K,r})$, where $b_{k,r} = 1$ if bandwidth $C_{k,r}$ is allocated to class k on route r and 0 if no capacity is allocated.

When routing is considered, the accepted traffic may not exceed the offered traffic. The problem formulation takes this constraint into account implicitly. When capacity allocation is considered, the allocated capacity on any link may not exceed the capacity of the link. In addition to the detailed balance constraints, the capacity constraints need to be added explicitly to the problem. Additional constraints are also needed to guarantee the minimum bit rate $\phi_k^{\min}$ for the accepted flows. The MDP-LP formulation of the problem reads

$$\min_{\pi(x,d,b)} \sum_k \frac{\nu_k}{\nu} \sum_x \sum_{d:d_k=0} \sum_b \pi(x,d,b), \quad (26)$$

$$\text{s.t. } \nu_k \sum_{d:d_k=r} \sum_b \pi(x,d,b) = \frac{C_{k,r}}{\sigma_k} \sum_d \sum_{b:b_{d,r}=1} \pi(x+e_{k,r},d,b) \quad \forall x, \quad (27)$$

$$\sum_k \sum_{r \in R_k : l \in r} C_{k,r} \sum_d \sum_{b : b_{k,r}=1} \pi(x,d,b) \leq C_l \sum_d \sum_b \pi(x,d,b) \quad \forall x, l, \quad (28)$$

$$x_{k,r} \phi_k^{\min} \sum_d \sum_b \pi(x,d,b) \leq C_{k,r} \sum_d \sum_{b : b_{k,r}=1} \pi(x,d,b) \quad \forall x, k, r \in R_k, (29)$$

$$\sum_x \sum_d \sum_b \pi(x,d,b) = 1, \quad (30)$$

$$\pi(x,d,b) \geq 0 \quad \forall x, d, b, \quad (31)$$

where (28) represents the link capacity constraint and (29) is the minimum bit rate constraint. The inner double summation of the left hand side of (28) gives the proportion of time the system is in state x and capacity $C_{k,r}$ is allocated to class k on route r. Division by the double sum of the right hand side yields the capacity allocation to class k on route r in state x.

A better performance may be achieved if the detailed balance equations (27) are replaced by the global balance equations, as is done in the usual MDP-LP formulation,

$$\sum_k \nu_k \sum_{d : d_k \neq 0} \sum_b \pi(x,d,b) + \sum_k \sum_{r \in R_k} \frac{C_{k,r}}{\sigma_k} \sum_d \sum_{b : b_{k,r}=1} \pi(x,d,b)$$

$$= \sum_k \sum_{r \in R_k} \nu_k \sum_{d : d_k = r} \sum_b \pi(x - e_{k,r}, d, b) +$$

$$+ \sum_k \sum_{r \in R_k} \frac{C_{k,r}}{\sigma_k} \sum_d \sum_{b : b_{k,r}=1} \pi(x + e_{k,r}, d, b) \quad \forall x. \quad (32)$$

Note, however, that in this case the system is not insensitive and the MDP assumptions require the flow size distribution to be exponential.

4.5 Extensions to the Model

The LP formulation makes it possible to easily introduce additional constraints to the problem. The traffic classes can have service requirements that need to be satisfied. Any constraint that can be expressed as a linear equation in the decision variables can be included in the problem.

The following examples are given using the formulation with free capacity allocation but the same constraints can be written using the formulation with fixed capacity allocation. For instance, we may require that the blocking probability of class k is below α

$$\sum_x \sum_{d : d_k = 0} \sum_b \pi(x,d,b) \leq \alpha. \quad (33)$$

Another possible requirement is that the average throughput of accepted class-k flows is at least c_k

$$\sum_{r \in R_k} C_{k,r} \sum_{x : x_{k,r} > 0} \sum_d \sum_{b : b_{k,r}=1} \pi(x,d,b) \geq c_k. \quad (34)$$

5   Numerical Results

We study the different policies considered above in a simple network illustrated in Fig. 1. There are two links with unit capacities $C_i = 1, i = 1, 2$. The minimum bit rate $\phi_k^{\min}$ required for a flow is taken to be 0.2.

There are three traffic classes. Flows in the adaptive class with arrival intensity ν_0 can be routed to either of the links, while both links receive also dedicated background traffic. The arrival rate of the background traffic on link i is ν_i. The number of background flows on link i is x_i and and the number of adaptive flows is $x_{0,i}, i = 1, 2$. We assume that each background class makes up 10% of the total load and that the mean flow sizes σ_0, σ_1 and σ_2 are identical. The allocated bit rates on link i are $\phi_{0,i}(x)$ and $\phi_i(x)$. If balanced fairness is used the allocation is $\phi_{0,i}(x) = \frac{x_{0,i}}{x_{0,i}+x_i}$ and $\phi_i = \frac{x_i}{x_{0,i}+x_i}$.

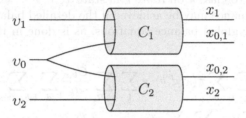

Fig. 1. Example network.

We compare the policies presented in Sect. 4 in the example network with different traffic loads. The network is insensitive with all the presented policies, i.e. the stationary distribution does not depend on the traffic characteristics. Figure 2 illustrates the blocking probabilities. The static routing has the poorest performance as expected. The dynamic policies utilizing balanced fairness perform almost identically with the policy based on global information being slightly better than the one restricted to the use of local knowledge, but the difference is hardly noticeable. If also the capacity allocation is considered jointly with the routing the results are clearly better. For comparison, the optimal sensitive result obtained with exponentially distributed flow sizes is also depicted and demonstrates the performance penalty that has to be paid for the requirement of insensitivity.

An interesting point is that the best decentralized policy is very close to the optimum with global information. If this is the case also in more complex networks the result has useful consequences. Determining the decentralized policy is significantly easier than solving the LP problem corresponding to the MDP formulation. The requirement of real time global knowledge of the network state is hard to satisfy, hence a policy based on it is hard to implement. On the other hand, the local knowledge is easily attainable.

While it seems counter-intuitive, the optimal capacity allocation does not utilize all the link capacities in every state. There is a trade-off between rejecting traffic and wasting capacity. When the load of the network is low it is better to waste part of the capacity instead of blocking traffic.

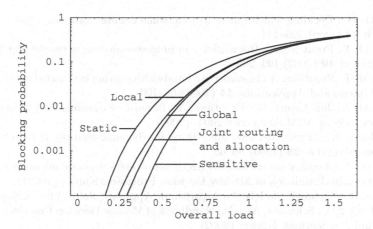

Fig. 2. Blocking probabilities.

6   Conclusions

Load balancing is an important topic on communication networks. Optimal load balancing is sensitive to detailed traffic parameters, hence finding the optimal policy is usually feasible only with the assumption of an exponential flow size distribution. In this paper, we presented a method to determine the best insensitive routing policy by utilizing the linear programming formulation of the Markov decision theory. In contrast to the global balance conditions of the ordinary MDP-LP formulation, detailed balance conditions were imposed as constraints. The performance of the system utilizing the insensitive policy does not depend on the flow size distribution.

We illustrated the performance in an example network. The performance with global knowledge is only slightly better than with local knowledge. Better results are obtained if both the capacity allocation and the routing are considered simultaneously. Still, even in this case, some performance penalty has to be paid for the insensitivity.

Acknowledgements

This work was financially supported by the Academy of Finland (grant n:o 74524).

References

1. Combe, M.B., Boxma, O.J.: Optimization of static traffic allocation policies. Theoretical Computer Science **125** (1994) 17–43
2. Ephremides, A., Varaiya, P., Walrand, J.: A simple dynamic routing problem. IEEE Transactions on Automatic Control **25** (1980) 690–693
3. Towsley, D., Panayotis, D., Sparaggis, P., Cassandras, C.: Optimal routing and buffer allocation for a class of finite capacity queueing systems. IEEE Transactions on Automatic Control **37** (1992) 1446–1451

4. Whitt, W.: Deciding which queue to join: some counterexamples. Operations Research **34** (1986) 226–244
5. Bonald, T., Proutière, A.: Insensitivity in processor-sharing networks. Performance Evaluation **49** (2002) 193–209
6. Bonald, T., Proutière, A.: Insensitivite bandwidth sharing in data networks. Queueing Systems and Applications **44** (2003) 69–100
7. Bonald, T., Jonckheere, M., Proutière, A.: Insensitive dynamic load balancing. In: Proceedings of ACM Sigmetrics 2004. (2004)
8. Krishnan, K.: Markov decision algorithms for dynamic routing. IEEE Communications Magazine **28** (1990) 66–69
9. Altman, E.: Applications of Markov decision processes in communication networks: A survey. In: Handbook of Markov Decision Processes. Kluwer (2002)
10. Tijms, H.: Stochastic Models: An Algorithmic Approach. John Wiley & Sons (1994)
11. Feinberg, E.A., Schwartz, A., eds.: Handbook of Markov Decision Processes – Methods and Applications. Kluwer (2002)
12. Bonald, T., Proutière, A.: On performance bounds for balanced fairness. Performance Evaluation **55** (2004) 25–50
13. Massoulié, L., Roberts, J.W.: Arguments in favour of admission control for TCP flows. In: Proceedings of ITC 16. (1999)

Hybrid IGP+MPLS Routing
in Next Generation IP Networks:
An Online Traffic Engineering Model

Antoine B. Bagula

Department of Computer Science,
University of Stellenbosch, 7600 Stellenbosch, South Africa
Phone: +2721 8084070, Fax: +2721 8084416
bagula@cs.sun.ac.za

Abstract. Hybrid routing approaches which either combine the best of traditional IGP and MPLS routing or allow a smooth migration from traditional IGP routing to the newly proposed MPLS standard have scarcely been addressed by the IP community. This paper presents an on-line traffic engineering (TE) model which uses a hybrid routing approach to achieve efficient routing of flows in IP networks. The model assumes a network-design process where the IP flows are classified at the ingress of the network and handled differently into the core using a multiple metric routing mechanism leading to the logical separation of a physical network into two virtual networks: An *IGP network* carrying low bandwidth demanding (LBD) flows and an *MPLS network* where bandwidth-guaranteed tunnels are setup to route high bandwidth demanding (HBD) flows. The hybrid routing approach uses a route optimization model where (1) link weight optimization (LWO) is implemented in the *IGP network* to move the IP flows away from links that provide a high probability to become bottleneck for traffic engineering and (2) the *MPLS network* is engineered to minimize the interference among competing flows and route the traffic away from heavily loaded links. We show that a cost-based optimization framework can be used by an ISP to design simple and flexible routing approaches where different metrics reflecting the ISP's view of its TE objectives are deployed to improve the use and efficiency of a network. This is achieved through extensions to a mixed cost metric derived from the route optimization model. The IGP+MPLS routing approach is applied to compute paths for the traffic offered to a 15- and 50-node network. Preliminary simulation reveals performance improvements compared to both IGP and MPLS routing in terms of the routing optimality and the network reliability.

1 Introduction

The Internet has developed beyond the best effort service delivered by traditional IP protocols into a universal communication platform where (1) better IP services delivery is attempted using enhancements to traditional Interior Gateway Protocol (IGP) protocols and (2) new protocols are standardized to support Traffic Engineering (TE): a network management technique allowing the traffic to be efficiently routed through a routed or switched network by effecting QoS agreements between the offered traffic and the available resources.

M. Ajmone Marsan et al. (Eds.): QoS-IP 2005, LNCS 3375, pp. 325–338, 2005.
© Springer-Verlag Berlin Heidelberg 2005

1.1   The Dualism IGP/MPLS TE

Traffic engineering has raised debates splitting the Internet Engineering Task Force (IETF) into divergent groups with different views concerning how the future Internet will be engineered. On one hand, there are the advocates of the destination-based TE model who point to (1) the ability of the Internet to support substantial increases of traffic load without the need for sophisticated TE mechanisms and (2) the need for routing mechanisms which keep the Internet's backbone simple while moving the complexity and intelligence at the edges of the network and at customer sites. The *link weight optimization (LWO)* model proposed in [1] implements this approach using traditional IGP routing with appropriate adjustments to the IGP link weights. On the other hand, proponents of the source- or flow-based routing approach have adopted a TE model closer to traditional data communications technologies such as frame relay and ATM where a smarter IP backbone is implemented using Multiprotocol Label Switching (MPLS) [2]. MPLS is based on a routing model borrowed from the ATM *virtual connection* paradigm where the traffic is routed over bandwidth-guaranteed tunnels referred to as Label Switched Paths (LSPs). These LSPs are set up and torn down through signaling.

It can be credited to the *LWO* model the advantages of (1) simplicity (2) capability of using diverse performance constraints and (3) compatibility with traditional IGPs. However there are drawbacks associated to this model. These include (1) the need to change routing metrics (patterns) leading to routing inconsistencies during the transient behavior from one routing pattern to another (2) the NP-hardness of the problem of finding link metrics which minimize the maximum utilization and (3) the inability to achieve optimal routing patterns for some networks even with link weight changing.

MPLS has recently experienced a wide deployment on the ISP backbone as a protocol that merges various technologies such as ATM and Frame relay over one backbone running IP (see [3] for listings of MPLS vendors and users). MPLS is based on a TE model where packet forwarding and routing are decoupled to support efficient traffic engineering and VPN services. However, a cautionary note was sounded on MPLS [4] as a protocol that presents security and privacy concerns and increases the network management burden resulting from the signaling operations required to setup and tear down the LSPs.

1.2   Contributions and Outline

Hybrid routing approaches which either combine the best of IGP and MPLS routing or allow a smooth migration from traditional IGP routing toward the newly standardized MPLS protocol have scarcely been addressed by the IP community. There have been a few number of papers proposing hybrid IGP+MPLS routing in offline traffic engineering settings where the network topology and traffic matrix are known a priori. These proposals include [5–8]. We proposed a hybrid IGP+MPLS routing approach referred to as *Hybrid* in [9] and showed through simulation its relative efficiency compared to pure IGP and MPLS routing. *Hybrid* is based on an online TE model where (1) the IP flows are classified at the ingress of the network based on a cutoff parameter expressing the limit between low bandwidth demanding (LBD) and high bandwidth demanding

(HBD) flows and (2) handled differently in the core of the network using the IGP model to route LBD flows while HBD flows are carried over MPLS tunnels (LSPs).

This paper builds upon the work done in [9] to improve the performance of *Hybrid* with the objective of finding an optimal network configuration minimizing loss, maximizing bandwidth and reducing the network complexity. We adopt a routing approach which is based on an online TE model similar to [9] where the IP flows are classified at the ingress of the network into LBD and HBD traffic classes and handled differently in the core leading to the logical separation of a physical network into two virtual networks: (1) an *IGP network* carrying LBD flows and (2) a *MPLS network* where bandwidth-guaranteed tunnels are setup to route HBD flows. The main contributions of this paper are

- **Hybrid IGP+MPLS Route Optimization.** We present an online traffic engineering model using ingress flow classification as in [9] but deploying a different route optimization model in the core where (1) link weight optimization (*LWO*) is implemented in the *IGP network* to move the IP flows away from links that provide a high potential to become bottleneck for traffic engineering and (2) the *MPLS network* is engineered to minimize the interference among competing flows and route the flow requests away from heavily loaded links.
- **Cost-Based Optimization Framework.** We show through logarithmic transform of a multiplicative link metric derived from the route optimization model that a cost-based optimization framework can be used by an Internet Service Provider (ISP) to design simple and flexible routing approaches where different metrics reflecting the ISP's view of its TE objectives (reliability, optimality, delay, etc) are deployed to improve the use and efficiency of a network.
- **Network Performance Improvements.** We applied the routing approach to a 15- and 50-node network to evaluate the performance of the IGP+MPLS routing approach compared to MPLS, IGP routing and the *Hybrid* algorithm recently proposed in [9]. Simulation reveals that while achieving the same optimality as the MPLS approach, the new IGP+MPLS routing approach results in further performance improvements in terms of network reliability and quality of the paths.

The rest of this paper is structured as follows. Section 2 presents the IGP+MPLS routing approach. An application of the IGP+MPLS routing approach to compute paths for the flows offered to a 15- and 50-node network is presented in section 3. Our conclusions are presented in section 4.

2   The IGP+MPLS Routing Approach

Consider a network represented by a directed graph $(\mathcal{N}, \mathcal{L})$ where $\mathcal{N}$ is a set of nodes, $\mathcal{L}$ is a set of links, $L = |\mathcal{L}|$ is the number of links and $N = |\mathcal{N}|$ is the number of nodes of the network. Assume that the network carries IP flows that belong to a set of service classes $\mathcal{S} = \{S_{LBD}, S_{HBD}\}$ where S_{LBD} and S_{HBD} define the set of low bandwidth demanding (LBD) and high bandwidth demanding (HBD) flows respectively. Let C_ℓ denote the capacity of link ℓ and let $\mathcal{P}_{i,e}$ denote the set of paths connecting the ingress-egress pair (i, e). Assume a flow-differentiated services where a request to route a class service $s \in S$ flow of $d_{i,e}$ bandwidth units between an ingress-egress pair (i, e) is received and that future demands concerning IP flow routing requests are not known.

Let $L_p = \sum_{\ell \in p} L_\ell(n_\ell, r_\ell, \alpha, \beta)$ denote the cost of path p where $L_\ell(n_\ell, r_\ell, \alpha, \beta)$ is the cost of link ℓ when carrying n_ℓ flows, r_ℓ is the total bandwidth reserved by the IP flows traversing link ℓ and (α, β) is a pair of network calibration parameters.

The flow routing problem consists of finding the best feasible path $p_s \in \mathcal{P}_{i,e}$ where

$$L_{p_s} = \min_{p \in \mathcal{P}_{i,e}} L_p \qquad (1)$$

$$d_{i,e} < \min_{\ell \in p_s}(C_\ell - r_\ell) \qquad (2)$$

Equations (1) and (2) express respectively the optimality of the routing process and the feasibility of the flows.

2.1 The Cost-Based Optimization Framework

We proposed in [9] a route optimization model minimizing the interference among competing flows through link loss minimization (reliability) and link bandwidth usage maximization (quantifying the network optimality). This combined reliability and optimality objective was achieved by using a routing metric multiplying the link loss probability w_ℓ by power values of the link interference n_ℓ and the link congestion distance $D_\ell = C_\ell - \beta(r_\ell + d_{i,e})$ under an *"equal probability"* assumption where the link loss probability is set to a constant value $w_\ell = 1/L$. The resulting mixed cost metric is expressed by

$$L_\ell(n_\ell, r_\ell, \alpha(s), \beta(s)) = w_\ell n_\ell^{\alpha(s)}/(C_\ell - \beta(s)(r_\ell + d_{i,e}))^{(1-\alpha(s))} \qquad (3)$$

The Link Weight Optimization

The *"equal probability"* of the link loss probability ($w_\ell = 1/L$) assumed in [9] may become a limitative factor in network conditions where some links have a higher probability to become bottleneck for traffic engineering than others or where some links are more unreliable than others. This is illustrated by the *symmetric* network of Figure 1 (a) which depicts a *route-multiplexed* network configuration where the two flows (s_1, d_1) and (s_3, d_3) interfere with the flow (s_2, d_2) on link $(9, 10)$. Under heavy traffic profile, this network configuration may lead to overloading the link $(9, 10)$ which thus becomes a bottleneck for traffic engineering. Under failure of link $(9, 10)$, the *route-multiplexed* network configuration leads to the loss of all the three flows (s_1, d_1), (s_2, d_2) and (s_3, d_3). The *route-separated* network configuration of Figure 1 (b) leads to a balanced network where each of the three flows is routed over its own path. The *path separation* paradigm illustrated above may be implemented by assigning a higher weight to the links which have a higher probability to attract more traffic such as the link $(9, 10)$ while allocating lower weights to the other links $\ell \neq (9, 10)$.

We consider a flow optimization model where the link weights are expressing the probability for a link to attract more traffic flows. This link weight also expresses the

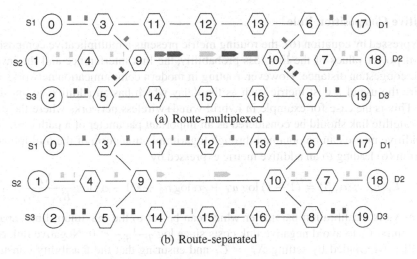

(a) Route-multiplexed

(b) Route-separated

Fig. 1. The "symmetric" network.

potential for the network to reroute the IP flows under failure of the link. The link weight w_ℓ also referred to as the *link loss probability* is defined by

$$w_\ell = \sum_{k \in \mathcal{K}} \delta_{\ell,k}/L \qquad (4)$$

where $\mathcal{K} = \cup_{i,e} \mathcal{K}_{i,e}$ is the set of disjoint paths on all ingress-egress pairs of the network and $\mathcal{K}_{i,e}$ is the set of disjoint paths from node i to node e and δ_ℓ is defined by

$$\delta_{\ell,k} = \begin{cases} 1 & \text{path } k \text{ traverses link } \ell \\ 0 & \text{otherwise} \end{cases} \qquad (5)$$

The New Mixed Cost Metric

We adopt the same *reliability-related optimality* paradigm as [9] using an IGP cost metric to route LBD flows while finding the paths to carry HBD flows based on a TE metric. We extend the link cost model (3) to consider the impact of the probability of a link to attract more or less traffic than another ($w_\ell \neq 1/L$) on the network performance as illustrated by the *path separation paradigm*. The new mixed cost metric is expressed by

$$L_\ell(n_\ell, r_\ell, \alpha, \beta) = w_\ell^{(1-s)} n_\ell^{s\alpha}/(R_\ell(i,e) - \beta r_\ell)^{s(1-\alpha)} \qquad (6)$$

where $R_\ell(i,e) = C_\ell - \beta d_{i,e}$ is the subscribed bandwidth of the flow offered to the I-E pair (i,e) on link ℓ, $\beta \in \{0,1\}$ is a parameter expressing the link residual bandwidth model implemented and $0 \leq \alpha \leq 1$ is a calibration parameter expressing the trade-off between reliability and optimality. Note that the link metric (6) yields (1) Link weight optimization ($L_\ell = w_\ell$) for $s = 0$ and (2) Least Interference Optimization ($L_\ell = n_\ell^\alpha/(R_\ell(i,e) - \beta r_\ell)^{(1-\alpha)}$) for $s = 1$.

Additive Composition Rule

As expressed by equation (6), the routing metric presents a multiplicative composition rule of power values of the link loss probability, the link interference and the inverse of the congestion distance. However, routing in modern communication networks may require the use of a cost metric such as the delay which has an additive composition rule. This is the case for example in hybrid wired+wireless networks where the delay on a satellite link should be considered as an important parameter of a path cost. Such an additive metric can be obtained through logarithmic transform of the multiplicative function (6) leading to an additive metric expressed by

$$L_\ell(n_\ell, r_\ell, \alpha, \beta) = (1 - s) \log w_\ell + s\alpha \log n_\ell + s(1 - \alpha) \log \frac{X_\ell}{R_\ell - \beta r_\ell} \qquad (7)$$

where X_ℓ is a calibration parameter which can be set to the maximum link capacity $X_\ell = \max_\ell C_\ell$ to avoid negative link costs since $\log \frac{1}{R_\ell - \beta r_\ell} < 0$. Negative link costs may also be avoided by setting $X_\ell = C_\ell$ and ensuring that the feasibility constraint expressed by equation (2) is respected.

A generalization of this cost model to the case where the delay (d_ℓ) or some other additive metrics are considered is as follows

$$L_\ell(n_\ell, r_\ell, \eta_i) = \eta_0 \log w_\ell + \eta_1 \log n_\ell + \eta_2 \log \frac{X_\ell}{R_\ell - \beta r_\ell} + \eta_3 \log d_\ell \qquad (8)$$

where $\sum_{k=0}^{3} \eta_k = 1$, η_k is the weight allocated to the k th parameter on the link cost, d_ℓ is an additive routing metric such as the queuing or propagation delay. Note that the parameters η_K are weights which are used to express the relative importance of each of the four different parameters of the additive cost metric. In the rest of this paper we will refer to routing models using the multiplicative cost metrics (3) and (6) as *Hybrid* and $Hybrid_{lwo}$ respectively. The routing models using the additive link metrics (7) and (8) will be referred to as $Hybrid_{log}$ and $Hybrid_{gen}$ respectively. The delay cost metric (9) referred to as *Delay* will be compared to the additive metrics $Hybrid_{log}$ and $Hybrid_{gen}$ in section 3.

$$L_\ell = \frac{1}{C_\ell - r_\ell} + d_\ell \qquad (9)$$

2.2   The Path Selection Model

The basic idea behind our path selection model is to differentiate flows into classes based on their bandwidth requirements and route LBD flows using *LWO* and a mix of interference minimization and congestion distance maximization for HBD flows.

Key Features

The path selection algorithm proposed is based on the following key features

- The link loss probability estimation is based on a counting process where (1) the KSP algorithm proposed in [11] is implemented to find disjoint paths for each ingress-egress pair and (2) the loss probability on a link is set to a value expressing the number of disjoint paths traversing that link. The core of this counting process has been embedded in the path selection algorithm below.

- We consider a flow classification model where for flow routing requests with bandwidth demands uniformly distributed in the range $[1, M]$, flows are classified into S_{LBD} and S_{HBD} classes based on a cut-off parameter $1 \leq \tau \leq M$ used to define the limit between LBD and HBD flows. This parameter will be used as a network calibration parameter defining the cost model to be deployed for routing the different flows.
- The *IGP network* uses Dijkstra's algorithm to route the IP flows while a modified constraint-based routing approach is deployed in the *MPLS network* where the network is pruned to meet the bandwidth demand requirements before running Dijkstra's algorithm for finding the least cost paths.

The Path Selection Algorithm

Consider a request to route a class s flow of $d_{i,e}$ bandwidth units between two nodes i and e. Let $d_{i,e}$ be uniformly distributed in the range $[1, M]$ and $1 \leq \tau \leq M$ is a cut-off parameter defining the limit between LBD and HBD flows. The algorithm proposed executes the following steps to route this flow

1. **Link Loss Probability Estimation.** Compute new link loss probabilities upon topology change
   (a) **Initialization.** Set $n = 0$ and $w_\ell = 0$ for each link $\ell \in \mathcal{L}$.
   (b) **Iteration.** For each I-E pair (i,e)
       For each node ν neighbor to node i
       - find the shortest path $p \in \mathcal{P}_{i,e}$ from i to e.
       - for each link $\ell \in p$ set $w_\ell = w_\ell + 1$ and $L_\ell = \infty$.
       - $n := n + 1$.
   (c) **Termination.** For each link $\ell \in \mathcal{L}$ set $w_\ell = w_\ell/n$.
2. **Network Calibration.** Set $s = 0$ if $d_{i,e} < \tau$ or $s = 1$ if $d_{i,e} \geq \tau$.
3. **Path Selection.**
   (a) **Prune the Network.** Set $L_\ell(n_\ell, r_\ell, \alpha, \beta) = \infty$ for each link ℓ whose link slack $C_\ell - r_\ell \leq d_{i,e}$.
   (b) **Find a Least Cost Path.**
       - **Traffic Aggregation.** If $s == 1$ and there is an existing path p with sufficient bandwidth to carry the HBD flow then (1) set $p_s = p$ and (2) goto step 4.
       - **New Path Finding.** Apply Dijkstra's algorithm to find a new least cost path $p_s \in \mathcal{P}_{i,e}$.
4. **Route the Request.**
   - Assign the traffic demand $d_{i,e}$ to path p_s.
   - Update the link occupancy and interference. For each link $\ell \in p_s$ set $r_\ell := r_\ell + d_{i,e}$ and $n_\ell := n_\ell + 1$.

Note that the path selection algorithm has the same complexity as Dijkstra's algorithm: $O(N^2)$.

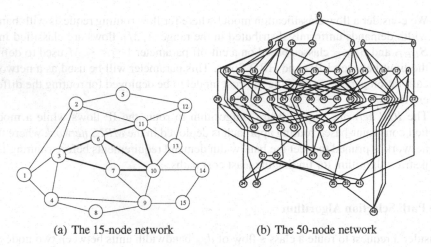

(a) The 15-node network   (b) The 50-node network

Fig. 2. The test networks.

3   An Implementation

This section presents simulation experiments conducted using a 15-node test network
obtained from [13] and a 50-node test network taken from [14] to compare the performance of (1) IGP routing using the OSPF model (2) MPLS routing using the LIOA
[15] algorithm (3) hybrid IGP+MPLS routing using the two multiplicative metrics
$Hybrid$ and $Hybrid_{lwo}$ (4) hybrid routing using the two additive metrics $Hybrid_{log}$
and $Hybrid_{gen}$ and (5) the widely known delay metric $Delay$ combining queuing
and propagation delays. The 50-node network used in our experiments includes 2450
ingress-egress pairs and 202 links capacitated with 38,519,241 units of bandwidth. Figure 2 (b) presents a graphical representation of the 50-node network. The 15-node network illustrated by Figure 2 (a) is the same as the test network used in [13] but in
contrast to [13] which consider only four traffic flows we assume that each node of the
15-node network is a potential source (ingress) and destination (egress) of traffic.

We consider two types of traffic: (1) *uniform* where the demands $d_{i,e}$ are uniformly
distributed in the interval $[1, M]$ as illustrated by Figure 3(a)and *bursty* where the demands $d_{i,e}$ present periods of bursts over the simulation period as illustrated by Figure 3(b).

3.1   Performance Parameters

The relevant performance parameters used in the simulation experiments are (1) the
network optimality expressed by the percentage flow acceptance referred to as *ACC*
and the average link utilization referred to as *UTIL* (2) the network reliability expressed
by the average link interference *AV* and the maximum link interference *MAX* and (3)
the quality of the paths expressed by the *path set optimality* and *path set disjointness*.
The impact of the hybrid IGP+MPLS routing approach on the network simplicity (expressed by the reduction in number of LSPs required to route HBD flows) has not been

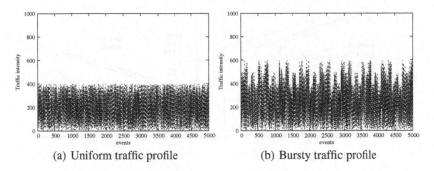

(a) Uniform traffic profile   (b) Bursty traffic profile

Fig. 3. The traffic profiles.

considered in this paper since the static flow differentiation model considered should lead to the same performance pattern as [9] in terms of network simplicity.

ACC is the percentage of flows which have been successfully routed by the network. *UTIL* defines the average link load expressing how far the links are from their congestion region. This parameter determines the potential for the network to support traffic growth: a lower utilized network offers a higher potential to support an increase of traffic load than a highly utilized network. *AV* is the average number of flows carried by the network links. It expresses the average number of flows which must be re-routed upon failure: an algorithm which achieves lower average interference is more reliable since it leads to re-routing fewer flows upon failure. *MAX* is the maximum number of flows carried by a link. It is another reliability parameter expressing the maximum number of flows which will be re-routed upon failure of the most interfering link.

It is widely recognized that offline TE algorithms designed based on a centralized routing model can optimize routing globally. We proposed a preplanned flow optimization algorithm referred to as *FLD* in [11] for setting up LSPs in a network using offline TE. We consider the *path set optimality* as an optimality parameter expressing how close the set of paths computed by an algorithm are to the set of paths computed using the FLD algorithm. The K-shortest path (KSP) algorithm has been widely deployed for facility restoration in several types of networks. Similarly, we consider the *path set disjointness* as a reliability parameter expressing how close the set of paths computed using a defined algorithm are to the set of disjoint paths computed using a version of the K-shortest path algorithm proposed in [11].

3.2 Simulation Experiments

Four simulation experiments were conducted to analyze (1) the impact of the demand range $[1, M]$ on the network optimality expressed by the flow acceptance and the link utilization (2) the impact of the demand range $[1, M]$ on the network reliability expressed by the average and maximum link interference (3) the impact of the traffic profile (bursty or uniform) on the network efficiency and (4) the quality of the paths carrying the IP flows.

The parameter values for the simulation experiments are presented in in terms of the offered flow routing requests, the flow request rates λ, the flow holding time $1/\mu$,

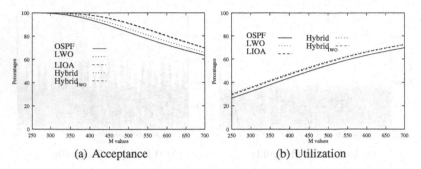

(a) Acceptance   (b) Utilization

Fig. 4. Optimality: 50-node network.

the cut-off parameter τ, the maximum bandwidth demand M, the number of simulation trials $T = 50$ and the number of flow requests per trial $N = 50000$. The flow arrival and services processes are Poisson. Each flow bandwidth demand $d_{i,e}$ is uniformly distributed in the range $[1, M]$.

The results of the experiments are presented in Figures 4 and 5 and Tables 1 and 2. Each entry in the tables presents the point estimates (averages) of each of the performance parameters described above computed at 95% confidence interval using $\alpha = 0.5$. These averages have been computed for different demand ranges and different traffic profiles. Table 2 compares the quality of the paths in terms of *path set optimality* (a measure of the network optimality) and *path set disjointness* (a measure of the network reliability).

Experiment 1. *The Impact of the Demand Range* $[1, M]$ *on the Network Optimality*

For flow routing requests whose demand are uniformly distributed in the range $[1, M]$, different traffic conditions (profiles) may be obtained through variation of the upper bound of the demand range M by setting its value higher to simulate heavy load conditions and lower to express light load conditions. The curves depicted by Figure 4(a) reveal (1) the same flow acceptance for the five algorithms under light traffic profile ($M \leq 300$) (2) the same and higher flow acceptance for LIOA , Hybrid and $Hybrid_{lwo}$ compared to both LWO and OSPF under heavier traffic conditions ($M > 300$) and (3) better flow acceptance for LWO compared to OSPF under heavier traffic conditions ($M > 300$). The curves depicted by Figure 4(b) reveal lower average link utilization for OSPF compared to the other four algorithms. This performance is achieved at the expense of reduced flow acceptance as illustrated by Figure 4(a).

Experiment 2. *The Impact of the Demand Range* $[1, M]$ *on the Network Reliability*

The average and maximum link interference are performance measures reflecting the average and maximum number of flows which should be rerouted upon link failure. Figures 5 (a) and (b) illustrating how the average and maximum interference respectively vary with the load conditions present shapes which show that the reliability of the network decreases with M increases. Figure 5 (a) depicting the average interference for the five algorithms reveals that OSPF achieves the highest value compared to the other

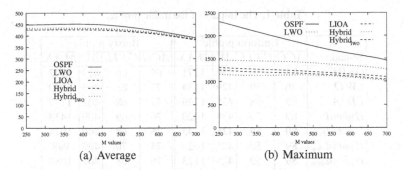

(a) Average   (b) Maximum

Fig. 5. Reliability: 50-nodes network.

four other algorithms. The results illustrated by Figure 5 (b) show that LWO achieves the lowest maximum interference while OSPF achieves the highest values of the maximum interference and presents a shape which reveals a higher degradation of the network reliability with the variation of M (traffic conditions) compared to the other four algorithms. The LWO performance is in agreement with the use of the link loss probability as a weight set to avoid the concentration of traffic flows on some links while the OSPF inefficiency results from the destination-based routing paradigm implemented by OSPF where no attention is paid to potential bottlenecks for traffic engineering.

Experiment 3. *The Impact of the Traffic Profile on the Network Performance*

The results presented in Table 1 show that (1) for the same network parameters, the network performance degrades under bursty traffic profile and (2) in general $Hybrid_{lwo}$ performs better than the other algorithms in terms of reliability and optimality under both uniform and bursty traffic profiles. These results also reveal that (1) the hybrid routing approaches using additive ($Hybrid_{log}$) and multiplicative ($Hybrid_{lwo}$) composition rules achieve the same performance and (2) using hybrid the additive metrics ($Hybrid_{log}$ and $Hybrid_{gen}$) leads to better network reliability results compared to the *Delay* model. These results confirm the relative efficiency of $Hybrid_{lwo}$ compared to $Hybrid$ and the improvements of LWO on OSPF routing. These relative efficiency results from the implementation of the link weight optimization.

Experiment 4. *The Quality of Paths*

We conducted simulation experiments to compare the commonality between the path sets computed using the FLD and KSP algorithms and the path sets found by each of the eight routing algorithms ($OSPF$, $LIOA$, LWO, $Hybrid$, $Hybrid_{lwo}$, $Hybrid_{log}$, $Hybrid_{gen}$ and $Delay$) to find how close these path sets are to the FLD and KSP computed path sets under uniform and bursty traffic profile. The results of these experiment are defined in terms of commonalty between the path sets defined as follows:

- $\mathcal{P}_x$ and $\mathcal{P}_y$ denote the sets of paths found by the algorithms x and y respectively.
- $\mathcal{P}_x \cap \mathcal{P}_y$(strong) : paths common to both $\mathcal{P}_x$ and $\mathcal{P}_y$ whose x and y values differ by less than 5%.

Table 1. The impact of the traffic profile.

50-node	Uniform profile				Bursty profile			
	ACC	UTIL	AV	MAX	ACC	UTIL	AV	MAX
OSPF	83	56	450	1771	69	66	417	1542
LWO	86	59	429	1134	72	68	403	1070
LIOA	92	58	425	1229	77	68	402	1137
$Hybrid$	89	59	438	1620	76	69	409	1435
$Hybrid_{lwo}$	**90**	**59**	**425**	**1133**	**76**	**68**	**400**	**1074**
$Hybrid_{log}$	89	59	425	1023	75	70	400	998
$Hybrid_{gen}$	89	59	425	1123	76	68	401	1080
$Delay$	90	56	381	1419	76	66	400	1189

15-node	Uniform profile				Bursty profile			
	ACC	UTIL	AV	MAX	ACC	UTIL	AV	MAX
OSPF	82	79	38	105	75	80	35	99
LWO	81	80	37	78	74	81	34	73
LIOA	82	80	38	76	75	81	35	71
$Hybrid$	83	79	38	86	76	80	35	79
$Hybrid_{lwo}$	**82**	**79**	**37**	**65**	**76**	**80**	**34**	**62**
$Hybrid_{log}$	81	79	37	66	75	80	34	62
$Hybrid_{gen}$	81	79	37	66	75	80	34	62
$Delay$	81	79	39	76	75	80	36	72

- $\mathcal{P}_x \cap \mathcal{P}_y$ (weak) represents the paths common to both $\mathcal{P}_x$ and $\mathcal{P}_y$ whose x and y values differ by more than 5%.
- $\mathcal{P}_x \setminus \mathcal{P}_y$: paths discovered only by x.
- $\mathcal{P}_y \setminus \mathcal{P}_x$: paths discovered only by y.

Simulation revealed the same performance pattern (quality of paths) for both traffic profiles where the newly proposed $Hybrid_{lwo}$ algorithm achieves the best *path set optimality* and *path set disjointness*. A summary of these results is presented in Table 2 where (1) the first column indicates the algorithm whose path set is compared to the FLD or KSP path set (2) *STRONG* denotes the strong correlation while *WEAK* denote the weak correlation (3) FLD denote the set of paths found by FLD only while KSP

Table 2. The quality of paths: 50-node network.

X	Path set optimality				Path set disjointness			
	STRONG	WEAK	FLD	X	STRONG	WEAK	KSP	X
OSPF	50	21	8	21	49	20	11	20
LIOA	62	18	8	12	61	17	10	12
LWO	63	16	8	13	61	16	10	13
$Hybrid$	53	22	8	17	52	22	10	16
$Hybrid_{lwo}$	**65**	**17**	**8**	**10**	**63**	**16**	**10**	**11**
$Hybrid_{log}$	64	17	8	11	62	17	10	11
$Hybrid_{gen}$	64	17	8	11	62	17	10	11
$Delay$	63	17	8	12	62	17	10	11

denote the set of paths found by KSP only and (4) X denote the set of paths found by an algorithm X only. Assuming that an algorithm whose path sets correspond more strongly to the FLD path sets achieves better optimality, the results presented in Table 2 show that $Hybrid_{lwo}$ and its additive version $Hybrid_{log}$ achieve the best optimality while OSPF performs worse. Similarly if we assume that an algorithm whose path sets correspond more strongly to the KSP path sets achieve better reliability, similar performance patterns are revealed where $Hybrid_{lwo}$ and its additive version $Hybrid_{log}$ achieve the best reliability results while OSPF routing achieves the worse.

4   Conclusion and Future Research

This paper presents an online TE model combining link weight optimization and interference minimization to achieve efficient routing of flows in IP networks. The model is based on a hybrid IGP+MPLS routing approach where flows are classified into low bandwidth demanding (LBD) and high bandwidth demanding (HBD) flows at the ingress of the network and routed differently in the core where an *IGP network* is used to route LBD flows while bandwidth-guaranteed tunnels are setup in an *MPLS network* to carry HBD flows. The hybrid routing approach uses a route optimization model where (1) link weight optimization (LWO) is implemented in the *IGP network* to move the IP flows away from links that provide a high probability to become bottleneck for traffic engineering and (2) the *MPLS network* is engineered to minimize the interference among competing flows and route the traffic away from heavily loaded links. Through logarithmic transform, a multiplicative cost metric is transformed into an additive cost metric which may be used in routing situations where additive metrics such as the delay on a satellite link are considered as important parameters of a path cost. This is typical of hybrid wireless/wired routing environments. Preliminary simulation using a 15- and 50-node network models reveal that the hybrid newly proposed IGP+MPLS routing approach using link weight optimization (1) performs better than both MPLS routing and IGP routing in terms of network reliability and optimality (2) finds better paths than both MPLS routing and IGP routing and (3) may be deployed using a cost-based optimization framework where a multiplicative or an additive composition rule of the link metric are used with the same efficiency.

This paper has presented a significant suggestion on how differentiated metrics handling can be used in a cost-based framework to improve the efficiency of a network. However, further simulation experiments using different topologies with different mesh degrees and various traffic profiles are required to validate the multiple metric routing approach proposed in this paper as an efficient tool to be deployed by ISPs. The deployment of the routing approach proposed in this paper to achieve inter-domain traffic engineering in the emerging Internet panorama is another direction for future research.

References

1. B. Fortz, M. Thorup, "Internet Traffic Engineering by Optimizing OSPF Weights", *Proceedings of IEEE INFOCOM*, March 2000.
2. E.C. Rosen, A. Viswanathan, and R. Callon. "Multiprotocol Label Switching Architecture", *Request for Comments, http://www.ietf.org/rfc/rfc3031.txt*, 2001.

3. The MPLS Resource Center, *http://www.mplsrc.com/*.
4. R. Bush, "Complexity – The Internet and the Telco Philosophies, A Somewhat Heretical View", *NANOG/EUGENE http://psg.com/ randy/021028.nanog-complex.pdf*, October 2002.
5. W. Ben-Ameur et all. "Routing Strategies for IP-Networks", *Telekntronikk Magazine*, 2 March 2001.
6. E. Mulyana, U. Killat. "An Offline Hybrid IGP/MPLS Traffic Engineering Approach under LSP constraints", *Proceedings of the 1st International Network Optimization Conference INOC 2003, Evry/Paris France*, October 2003.
7. S. Koehler, A. Binzenhoefer, "MPLS Traffic Engineering in OSPF Networks- A Combined Approach", *18th ITC Specialist Seminar on Internet Traffic Engineering and Traffic Management*, August-September 2003.
8. A. Riedl, "Optimized Routing Adaptation in IP Networks Utilizing OSPF and MPLS", *Proceedings of the ICC2003 Conference*, May 2004.
9. A.B Bagula, "Online Traffic Engineering: A Hybrid IGP+MPLS Routing Approach", *Proceeding of the QofIS2004, Barcelona, Spain*, September 2004.
10. F. Faucheur et al., "Use of IGP Metric as a second TE Metric", IETF draft draft-lefaucheur-te-metric-igp-00.txt, july 2001.
11. A. Bagula, A.E. Krzesinski. "Traffic Engineering Label Switched Paths in IP Networks using A Pre-Planned Flow Optimization Model", *Proceedings of the Ninth International Symposium on Modeling, Analysis and Simulation of Computer and Telecommunication Systems* p70-77, August 2001.
12. J. Moy. "OSPF Version 2", *Request For Comments, http://www.ietf.org/rfc/rfc1583.txt*, March 1994.
13. K. Kar, M. Kodialam, T.V. Lakshman. "Minimum Interference Routing with Application to MPLS Traffic Engineering", *Proc. IEEE INFOCOM, Vol 2, pp 884–893, Tel-Aviv, Israel*, March 2000.
14. Optimized multipath. *http://www.fictitous.org/omp/random-test-cases.html*
15. A. Bagula, M. Botha, and A.E Krzesinski, "Online Traffic Engineering: The Least Interference Optimization Algorithm", *Proceedings of the ICC2004 Conference, Paris*, June 2004.

Leveraging Network Performances with IPv6 Multihoming and Multiple Provider-Dependent Aggregatable Prefixes

Cédric de Launois*, Bruno Quoitin**, and Olivier Bonaventure

Université catholique de Louvain,
Department of Computing Science and Engineering
{delaunois,quoitin,bonaventure}@info.ucl.ac.be
http://www.info.ucl.ac.be

Abstract. Multihoming, the practice of connecting to multiple providers, is becoming highly popular. Due to the growth of the BGP routing tables in the Internet, IPv6 multihoming is required to preserve the scalability of the interdomain routing system. A proposed method is to assign multiple provider-dependent aggregatable (PA) IPv6 prefixes to each site, instead of a single provider-independent (PI) prefix. This paper shows that the use of multiple PA prefixes per sites not only allows route aggregation but also can be used to reduce end-to-end delay by leveraging the Internet path diversity. We also quantify the gain in path diversity, and show that a dual-homed stub AS that uses multiple PA prefixes has already a better Internet path diversity than any multihomed stub AS that uses a single PI prefix, whatever its number of providers. We claim that the benefits provided by the use of IPv6 multihoming with multiple PA prefixes is an opportunity to develop the support for quality of service and traffic engineering.

Keywords: BGP, IPv6 Multihoming, Path Diversity.

1 Introduction

Today, the Internet connects more than 17000 *Autonomous Systems* (AS) [1], operated by many different technical administrations. The large majority of ASes are *stub* ASes, i.e. autonomous systems that do not allow external domains to use their infrastructure. Only about 20% of autonomous systems provide transit services to other ASes [2]. They are called *transit* ASes. The Border Gateway Protocol (BGP) [3] is used to distribute routing announcements among routers that interconnect ASes.

The size of the BGP routing tables in the Internet has been growing dramatically during the last years. The current size of those tables creates operational issues for some Internet Service Providers and several experts [4] are

---

* Supported by a grant from FRIA (Fonds pour la Formation à la recherche dans l'Industrie et dans l'Agriculture, rue d'Egmont 5 – 1000 Bruxelles, Belgium).
** Supported by the Walloon Government within the WIST TOTEM project. http://totem.info.ucl.ac.be

M. Ajmone Marsan et al. (Eds.): QoS-IP 2005, LNCS 3375, pp. 339–352, 2005.
© Springer-Verlag Berlin Heidelberg 2005

concerned about the increasing risk of instability of BGP. Part of the growth of the BGP routing tables [5] is due to the fact that, for economical and technical reasons, many ISPs and corporate networks wish to be connected via at least two providers to the Internet. For more and more companies, Internet connectivity takes a strategic importance. Nowadays, at least 60% of those domains are multihomed to two or more providers [1,2]. Therefore, it can be expected that IPv6 sites will continue to be multihomed, primarily to enhance their reliability in the event of a failure in a provider network, but also to increase their network performances such as network latency. In order to preserve the scalability of the interdomain routing system, every IPv6 multihoming solution is required to allow route aggregation at the level of their providers [4]. Among the several IPv6 multihoming methods proposed at the IETF [6], a popular solution is to assign multiple provider-dependent aggregatable (PA) IPv6 prefixes to each site, instead of a single provider-independent (PI) prefix. Both IPv4 and IPv6 multihoming methods are described in section 2.

We show in this paper that the use of multiple PA prefixes introduces other benefits than simply allowing route aggregation. We explain in section 4 how stub ASes that use multiple PA prefixes can exploit paths that are otherwise unavailable. In other words, we explain how the use of PA prefixes increases the number of concurrent paths available. Next, we show that lower delays can often be found among the new paths. Our simulations suggest that a delay improvement is observed for approximately 60% of the stub-stub pairs, and that the delay improvement could be higher in the actual Internet.

In section 5, we quantify the gain in terms of Internet path diversity. We show that a dual-homed stub AS that uses multiple PA prefixes has already a better Internet path diversity than any multihomed stub AS that uses a single PI prefix, whatever its number of providers.

2   IPv4 and IPv6 Multihoming

This section provides some background on traditional IPv4 multihoming and IPv6 multihoming.

In the current IPv4 Internet, the traditional way to multihome is to announce, using BGP, a single prefix to each provider, see fig. 1 and 2. In fig. 1, AS 123 uses provider-aggregatable addresses. It announces prefix 10.0.123.0/24 to its providers AS 10 and AS 20. AS 10 aggregates this prefix with its 10.0.0.0/8 prefix and announces the aggregate to the Internet. In fig. 2, AS 123 announces a provider-independent prefix to its providers. This prefix is then propagated by BGP routers over the Internet. Throughout this paper, we will refer to this technique as *traditional IPv4 multihoming*, or simply *IPv4 multihoming*.

The way stub ASes multihome in IPv6 is expected to be quite different from the way it is done currently in IPv4. Most IPv6 multihoming mechanisms proposed at the IETF rely on the utilization of several IPv6 provider-aggregatable prefixes per site, instead of a single provider-independent prefix, see [6,7] and the references therein. Figure 3 illustrates a standard IPv6 multihomed site.

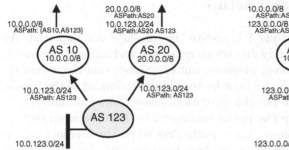

Fig. 1. IPv4 Multihoming using a provider-aggregatable prefix.

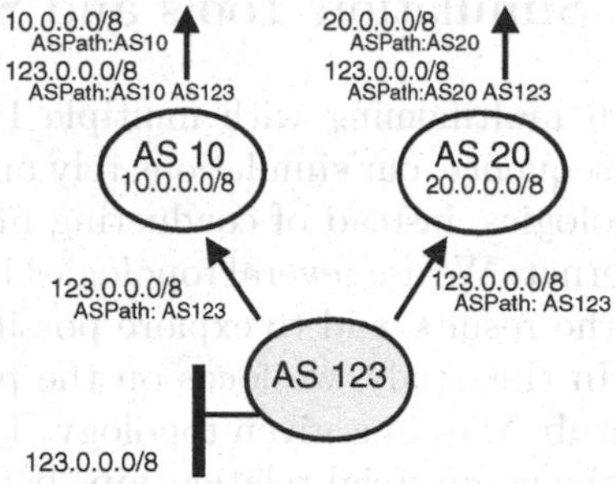

Fig. 2. IPv4 Multihoming using a provider-independent prefix.

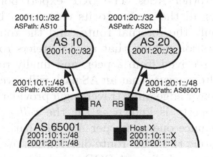

Fig. 3. IPv6 Multihoming.

In fig. 3, AS 10 and AS 20 provide connectivity to the multihomed site AS 65001. Each provider assigns to AS 65001 a site prefix, respectively 2001:10:1::/48 and 2001:20:1::/48. The two prefixes are advertised by the site exit routers RA and RB to every host inside AS 65001. Finally, these prefixes are used to derive one IPv6 address per provider for each host interface. In this architecture, AS 65001 advertises prefix 2001:10:1::/48 only to AS 10, and AS 10 only announces its own IPv6 aggregate 2001:10::/32 to the global Internet. This new solution is expected to be used only by stub ASes. Transit ASes are not concerned by these solutions since they will receive provider-independent IPv6 prefixes. Consequently, in this study, we focus only on stub ASes.

The use of multiple PA prefixes is natural in an IPv6 multihoming environment. However, it is not impossible to use the same multihoming technique in IPv4, i.e. to delegate two IPv4 prefixes to a site. Unfortunately, due to the current lack of IPv4 addresses, the need to delegate several IPv4 prefixes to a multihomed site makes this solution less attractive. Thus, throughout this document, we will call the new multihoming technique presented here for IPv6 simply as *IPv6 multihoming*; although the same concept could also be applied to IPv4 multihomed sites, and although other IPv6 multihoming techniques exist.

3   Simulation Tools and Setup

IPv6 multihoming with multiple PA prefixes is currently not deployed. As a consequence, our simulations rely on various generated and inferred Internet-like topologies, instead of conducting measurement experiments on the actual IPv4 Internet. We use several topologies in order to delimit the impact of the topology on the results, and to explore possible evolution scenarios for the Internet.

In this study, we focus on the paths announced by BGP between each pair of stub ASes in a given topology. These paths depend on the topology but also on the commercial relationships between ASes together with their BGP routing policies. The commercial agreements between two ASes are usually classified as customer-provider or peer-to-peer relationships [8,9]. These relationships are either inferred [8,9], or directly provided by the topology description. We then compute, for each AS, the BGP configuration that corresponds to its commercial relationships with the other ASes. The BGP export policies basically define that an AS announces all the routes to its customers, but announces to its peers and providers only the internal routes and the routes of its customers. Moreover, the configuration defines that an AS prefers routes received from a customer, then routes received from a peer, and finally routes received from a provider [8,9]. These filters ensure that an AS path will never contain a customer-to-provider or peer-to-peer edge after traversing a provider-to-customer or peer-to-peer edge. This property is known as the the *valley-free* property, and is defined in [8]. We announce one prefix per AS. The paths for a given topology are obtained by simulating the BGP route distribution over the whole topology. For this purpose, we use a dedicated BGP simulator, named C-BGP [10]. C-BGP supports import and export filters, and uses the full BGP decision process. In the absence of intradomain structures, the tie-breaking rule used by C-BGP for choosing between two equivalent routes is to prefer the route learned from the router with the lowest router address. As soon as all the routes have been distributed and BGP has converged, we perform traceroute measurements on the simulated topology and deduce the paths.

4   Improving Delays with Multiple Prefixes per Site

We will show in this section how the use of multiple PA prefixes can reduce the end-to-end delay by leveraging the Internet path diversity. Section 4.1 explains how stub ASes that use PA prefixes can exploit paths that are unavailable using a single PI prefix. Among the newly available paths, some offer lower delays. In section 4.3, we evaluate how often this improvement in network latency occurs. The topology used for the simulation is presented in section 4.2.

4.1   Impact of PI and PA Prefixes on Available AS Paths

Fig. 4 shows an AS-level interdomain topology with shared-cost peerings and customer-provider relationships. An arrow labelled with "$" from AS x to AS

y means that x is a customer of y. A link labelled with "=" means that the ASes have a shared-cost peering relationship [8]. In this figure, both S and D are dual-homed ASes.

In IPv4, D traditionally announces a single provider-independent prefix to each of its providers. This PI prefix is propagated by BGP routers all over the Internet. In particular, if AS S is single-homed, it will receive a single route from its provider to reach the dual-homed AS D. This route is the best route known by the provider to join AS D. If AS S is also dual-homed, as illustrated in figure 4, S will receive two BGP routes $ECAD$ and $FCAD$ towards D one from each of its providers, as shown in fig. 5.

When stub ASes use IPv6 multihoming with multiple PA prefixes, additional routes exist.

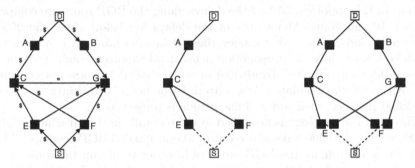

Fig. 4. Topology.   Fig. 5. IPv4 path tree.   Fig. 6. IPv6 path tree.

Suppose that both ASes S and D use IPv6 multihoming with multiple PA prefixes. Each host in S has two IPv6 addresses. One is derived from the prefix allocated by E, while the other one is derived from the prefix allocated by F. Similarly, each host in D has two IPv6 addresses. When selecting the source address of a packet to be sent, the host in S could in theory pick any of its two addresses. However, for security reasons, IPv6 providers must refuse to convey packets with source addresses outside their address range [6, 7]. For example, E refuses to forward a packet with a source address belonging to F. Using traditional IPv4 multihoming, two BGP routes towards D (e.g. $SECAD$ and $SFCAD$) are advertised by E and F to S, fig. 5. In an IPv6 multihoming scenario, since both S and D have two prefixes, S can reach D via A or B depending on which destination prefix is used, and via E or F depending on which source prefix is used. So, S has a total of four paths to reach D: $SECAD$, $SEGBD$, $SFCAD$ and $SFGBD$, see fig. 6.

4.2 A Two-Level Topology with Delays

In order to simulate delays along paths, we cannot rely on topologies provided by Brite [11], Inet [12], or GT-ITM [13] since they either do not model business relationships or do not provide delays along links.

A topology that contains both delays and commercial relationships is available at [14]. In this topology, the interdomain links and the business relationships are given by a topology inferred from multiple routing tables [9]. For each peering relationship found between two domains in this topology, interdomain links are added. The different points of presence of each domain are geographically determined by relying on a database that maps blocks of IP addresses and locations worldwide. The intradomain topology is generated by first grouping routers that are close to each other in clusters, and next by interconnecting these clusters with backbone links. The delays along the links is the propagation delay computed from the distance between the routers. The IGP weights used are the delays for links shorter than 1000 km, twice the delay for links longer than 1000 km but shorter than 5000 km and 5 times the delay for links longer than 5000 km. This is used to penalize the long intradomain links and favor hot-potato routing. In this topology, 55% of the delays along the BGP route are comprised between 10 and 50ms. About 20% of the delays are below 10ms and 25% sit between 50 and 100ms. We consider these delays as minimal bounds for the real delays, since only the propagation delay is taken into account. Factors that increase delays like limited bandwiths or congestion delays are not considered here. Although the simulated delays are inferior bounds to delays observed in the global Internet, their order of magnitude is preserved.

The resulting topology is described in more details in [14]. It contains about 40,000 routers, 100,000 links and requires about 400,000 BGP sessions. C-BGP [10] is used to simulate the BGP protocol in order to obtain the router-router paths and their delays between multihomed stub ASes. For computation time reasons, we conduct the simulation for a subset of 2086 multihomed stub ASes randomly chosen among the 8026 multihomed stub ASes.

4.3   Simulation Results

Figure 7 plots the lowest delay obtained when stub ASes use traditional IPv4 multihoming (x-axis), against the lowest delay obtained when stub ASes use IPv6 multihoming with multiple PA prefixes (y-axis). The gray-scale indicates the number of stub-stub paths, on a logarithmic scale. The diagonal line that appears represents stub-stub pairs for which both multihoming techniques yield to the same lowest delay.

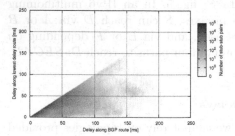
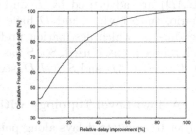

Fig. 7. Delay along the BGP route versus delay along the lowest delay route.

Fig. 8. Distribution of the relative delay improvement.

As explained in section 4.1, the use of multiple PA prefixes provides additional paths, beside traditional paths that are still available. As a consequence, delays can only improve, and no dot can appear above the diagonal line. A dot under this diagonal line indicates that the use of multiple PA prefixes introduces a new path with a delay lower than the delay along the best BGP path obtained when a single PI prefix is used. We can see that a lot of dots are located under this line. Sometimes, the improvement can even reach 150ms in this topology.

Figure 8 shows the cumulative distribution of the relative delay improvement. It shows that no improvement is observed for approximately 40% of stub-stub pairs. However, the relative improvement is more than 20% for 30% of stub-stub paths. Delays are cut by half for about 8% of stub-stub pairs.

As said in section 4.2, the delays observed in this topology are expected to be minimal bounds to those seen in the real Internet. Thus, we can reasonably assume that the absolute delay improvements presented in figure 7 will not be lower in the actual Internet.

These simulation results show that improving delays is a benefit of IPv6 multihoming with multiple PA prefixes, without increasing the BGP routing tables.

5   Leveraging Internet Path Diversity with Multiple Prefixes

Section 4.1 has shown that stub ASes using multiple PA prefixes can exploit paths that are otherwise unavailable. In other words, the use of multiple PA prefixes increases the number of paths available, i.e. the Internet path diversity. We have shown that better delays can often be found among the new paths. An increase in path diversity can also yields other benefits, like better possibilities of load balancing. In this section, we propose to quantify the Internet path diversity that exists when a multihomed stub AS uses either multiple PA prefixes or a single PI prefix.

We first detail the inferred and generated topologies that are used in the simulations. Next, we present and discuss the results of simulations made on an inferred AS-level Internet topology. Finally, we evaluate the impact of the topology on the path diversity.

5.1   Internet Topology

In section 4, we used a large router-level Internet topology that models delays. In this section, we use AS-level topologies instead, for two reasons. A first reason is the computation time. The topology used in section 4 is unnecessarily complex for an AS-level simulation since it models routers and delays. A second reason is that we want to consider other topologies in order to estimate the variability of our results with respect to the topology.

We first use an AS-level Internet topology inferred from several BGP routing tables using the method developed by Subramanian et al. [9].

We next generate several AS-level Internet-like topologies, using the GHITLE topology generator [15]. A topology is generated level by level, from the dense core to the customer level. Four levels are usually created : a fully-meshed dense core, a level of large transit ASes, a level of local transit ASes, and a level of stub ASes. Additional levels can be created if needed. We developed GHITLE for this analysis because no existing generator could produce a topology that provides details about customer-provider and peer-to-peer relationships, and where the number of Internet hierarchy levels and nodes in each level could be specified. In particular, Inet [12] does not provide commercial relationships between ASes. Brite [11] and GT-ITM [13] do not produce a hierarchical topology with more levels than just transit and stub levels.

5.2   A New Path Diversity Metric

In order to measure the path diversity for a given destination AS, we first build the tree of paths from all source AS towards the destination AS. As explained in section 4.1, this path tree depends on the multihoming technique used. Next, we use a new, fine-grain, path diversity metric to evaluate the diversity of this tree. This metric takes into account the lengths of the paths and how much they overlap. We define this new path diversity metric, from a source AS S to a destination AS D, as follows.

Let $P_1, P_2, ..., P_n$ be the n providers of S. We first build the tree of all paths starting from providers P_i of S to destination D, for $i = 1, ..., n$. This tree represents all the BGP paths for D that are advertised by the providers P_i to S. Our path diversity metric is computed recursively link by link, from the leaves to the root. It returns a number between 0 and 1. We first assign an initial diversity of 0.5 to each link in the tree. This number is chosen in order to best distribute the values of the path diversity metric in the range $[0, 1]$. At each computation step, we consider two cases, to which all other cases can be reduced. Either two links are in sequence, or the links join in parallel at the same node.

In the first case, two links with diversity d_1 and d_2 in sequence can be merged into a single link with a combined diversity $d_{1,2} = d_1 \cdot d_2$. The combined diversity $d_{1,2}$ is a number in $[0, 1]$ lower than both d_1 and d_2, so that the metric favors short paths over longer one. This computation step also implicitly gives higher

Alg. 1. Computing Diversity Metric _____

```
Diversity(root)
{
    d = 0 ;
    if ( Children(root) == ∅ )
        return 1 ;

    foreach child ∈ Children(root) {
        d_child = 0.5 · Diversity(child) ;
        d = d + d_child − d · d_child ;
    }

    return d ;
}
```

importance to the path diversity that exists near the root. This property ensures that the metric prefers trees where paths join lately near the destination node over trees where paths merge near the source node.

In the second case, when a link with a diversity d_1 and another link with a diversity d_2 join in parallel, we merge the two links into a single link with a combined diversity $d_{1,2}$, computed as $d_{1,2} = d_1 + d_2 - d_1 \cdot d_2$. The resulting diversity is greater than both d_1 and d_2, which corresponds adequately to an improvement in terms of path diversity.

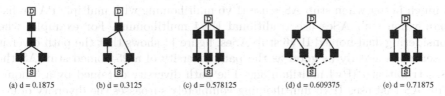

(a) d = 0.1875   (b) d = 0.3125   (c) d = 0.578125   (d) d = 0.609375   (e) d = 0.71875

Fig. 9. Path diversity metric examples.

A recursive algorithm to compute this metric is presented in Alg. 1. A detailed description of this metric together with a comparison with other path diversity metrics is available at [16]. Examples of values for this metric are illustrated in figure 9.

5.3 Simulation Results

Figure 10 presents the path diversity available to stub ASes that use traditional IPv4 multihoming in the inferred AS-level Internet topology. Figure 11 shows the path diversity when all stub ASes use IPv6 multihoming with multiple PA prefixes, in the same inferred topology.

The figures show $p(x)$: the percentage of couples *(source AS, destination AS)* having a path diversity greater than x. The results are classified according to the number of providers of the destination stub AS. The number of providers is indicated beside each curve. Figure 10 shows for example that only 12% of

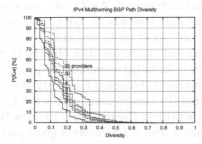

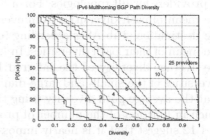

Fig. 10. AS-level path diversity for the inferred Internet topology, using traditional IPv4 multihoming.

Fig. 11. AS-level path diversity for the inferred Internet topology, using IPv6 multihoming.

single-homed stub ASes using traditional IPv4 multihoming have a diversity better than 0.2. This percentage raises to 22% for dual-homed stub ASes. Fig. 11 shows that about 50% dual-homed IPv6 stub ASes have a path diversity better than 0.2.

We can observe that the diversity remains the same when considering only single-homed destinations. Indeed, only one prefix is announced by a single-homed stub AS, using either IPv4 or IPv6 multihoming technique. The use of IPv6 multihoming does not introduce any benefit in this case.

When comparing figures 10 and 11, it appears that the AS-level path diversity is much better when stub ASes use IPv6 multihoming with multiple PA prefixes than when stub ASes use traditional IPv4 multihoming. For example, when considering dual-homed IPv6 stub ASes, figure 11 shows that the path diversity observed is already as good as the path diversity of a 25-homed stub AS that uses traditional IPv4 multihoming. The path diversity obtained by a 3-homed stub AS that uses IPv6 multihoming completely surpasses the diversity of even a 25-homed stub AS that uses traditional IPv4 multihoming.

5.4   Impact of Topology on Path Diversity

The way Internet will evolve in the future remains essentially unknown. In order to delimit the range of variation for our results, we perform simulations with three distinct topologies.

The first is a topology that tries to resemble the current Internet.

The second is a small-diameter Internet topology, consisting of stub ASes directly connected to a fully meshed dense core. This topology simulates a scenario where ASes in the core and large transit ASes concentrates for commercial reasons. At the extreme, the Internet could consist in a small core of large transit providers, together with a large number of stub ASes directly connected to the transit core. This could lead to an Internet topology with a small diameter.

The third is a large-diameter topology, generated using eight levels of ASes. This topology simulates a scenario where the Internet continues to grow, with more and more core, continental, national and metropolitan transit providers. In this case, the Internet might evolve towards a network with a large diameter.

Figures 12 and 13 show the average path diversity in function of the number of providers for all topologies. For a given destination stub AS D, we compute the mean of path diversities from every source stub towards D. We then group the destination stub ASes according to their number of providers, and compute the mean of their path diversities. In figures 12 and 13, we can first observe that the average diversity of the inferred Internet is included between the average diversities of the small- and large-diameter Internet topologies. Figure 12 shows that the average path diversity using traditional IPv4 multihoming does not rise much in function of the number of providers, whatever the topology. Figures 10 and 12 suggest that it is nearly impossible that a stub AS achieves a good path diversity using traditional IPv4 multihoming, whatever the number of providers. In contrast, the path diversity that is obtained using IPv6 multihoming with multiple PA prefixes is much better, see figure 13. Figures 12 and 13 show that

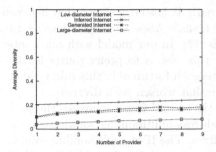
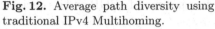

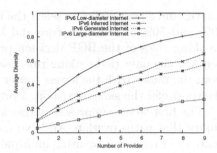

Fig. 12. Average path diversity using traditional IPv4 Multihoming.

Fig. 13. Average path diversity using IPv6 multihoming.

a dual-homed stub AS using IPv6 multihoming already gets a better diversity than any multihomed stub AS that uses traditional IPv4 multihoming, whatever its number of provider and for all considered topologies. In a small-diameter Internet, this diversity rises fast in function of the number of providers, but also shows a marginal gain that diminishes quickly. In a large-diameter Internet, the diversity rises more slowly.

So far, we have analyzed the AS-level path diversity considering one router per AS. However, a factor that can impact the path from a source to a destination is the intradomain routing policy used inside transit ASes. We have also evaluated the path diversity that exists when ISP routing policies in the Internet conform to hot-potato routing. In hot-potato routing, an ISP hands off traffic to a downstream ISP as quickly as possible. Results show that hot-potato routing has no significant impact on the AS-level path diversity, but due to space limitations, we cannot report this evaluation here.

5.5   Impact of BGP on Path Diversity

We discuss in this section how the path diversity is affected by the BGP protocol.

Multihoming is assumed to increase the number of alternative paths. However, the AS-level path diversity offered by multihoming depends on how much the interdomain routes, as distributed by BGP, overlap.

The results presented in the previous section suggest that BGP heavily reduces the path diversity, at the level of autonomous systems. Two factors explain why the diversity is so much reduced.

The primary factor is that BGP only announces one route, its best one. This behavior heavily reduces the number of paths. Unfortunately, BGP is designed as a single path routing protocol. It is thus difficult to do better with BGP.

A second factor exists that further reduces the path diversity. The tie-breaking rule used by BGP to decide between two equivalent routes often prefers the same next-hops. Let us consider a BGP router that receives two routes from its provider towards a destination D. According to the BGP decision process, the shortest AS path is selected. However the diameter of the current Internet is small, more or less 4 hops [1]. As a consequence, paths are often of the same length, and do not suffice to select the best path. It has been shown that between 40% and 50% of routes in core and large transit ASes are selected using tie-breaking rules of the BGP decision process [17]. In our model with one router per AS, the only tie-breaking rule used in this case is to prefer routes learned from the router with the lowest router address. Unfortunately this rule yields to always prefer the same next-hop, a practice that worsen path diversity.

The first factor suppresses paths, while the second factor increases the probability that paths overlap. An IPv6 multiaddress multihoming solution circumvents the first factor by using multiple prefixes. The IPv6 multihoming solution has no impact on the second factor, since it does not modify BGP and its decision process in particular.

6   Related Work

A work about IPv4 multihoming path diversity appears in [18], where the authors define two path diversity metrics to quantify the reliability benefits of multihoming for high-volume Internet servers and receivers. They notice however that their metrics have an undesirable bias in favor of long paths. Their study draws empirical observations from measurement data sets collected at servers and monitoring nodes, whereas our work is based on inferred and generated global-scale AS-level topologies.

A comparison of Overlay Routing and Multihoming Route Control appears in [19]. In this study, the authors demonstrate that an intelligent control of BGP routes, coupled with ISP multihoming, can provide competitive end-to-end performance and reliability compared to overlay routing. Our results agree with this finding. In addition, our work explicits the impact of the path diversity on performances, and shows that IPv6 multihoming with multiple PA prefixes is able to provide these benefits.

It is well known that the use of provider-dependent aggregatable prefixes preserves the scalability of the interdomain routing system [20]. To our knowledge, this is the first study that shows that the use of multiple PA prefixes also increases network performances by leveraging the Internet path diversity, compared to the use of traditional multihoming with a single PI prefix.

7   Conclusion

A proposed way to improve network performances of the interdomain is to enhance BGP. We revealed that another way is to use multiple provider-dependent aggregatable (PA) prefixes per sites, in an IPv6 Internet in particular. In this paper, we have shown that stub ASes that use multiple PA prefixes can exploit paths that are otherwise unavailable. Thus, the use of multiple PA prefixes increases the number of paths available, i.e. the Internet path diversity. Our simulations suggest that about 60% of the paths can benefit from a lower delay. We also used simulations on various topologies to quantify the path diversity benefits offered by the use of multiple PA prefixes. We have shown that a dual-homed stub AS that uses multiple PA prefixes has already a better Internet path diversity than any multihomed stub AS that uses a single provider-independent (PI) prefix, whatever its number of providers.

Our observations show that, from a performance point of view, IPv6 multi-homed stub ASes get benefits from the use of multiple PA prefixes and should use them instead of a single PI prefix as in IPv4 today. This study thus strongly encourages the IETF to pursue the development of IPv6 multihoming solutions relying on the use of multiple PA prefixes. The use of such prefixes reduces the size of the BGP routing tables, but also provides lower delays, more diverse Internet paths, which in turn yields to better possiblities to balance the traffic load and to support quality of service.

Acknowledgments

We thank Steve Uhlig and Marc Lobelle for their useful comments and support. We also thank the authors of [9] for providing the inferred Internet topology.

References

1. Huston, G.: BGP Routing Table Analysis Reports.
   http://bgp.potaroo.net (2004)
2. Agarwal, S., Chuah, C.N., Katz, R.H.: OPCA: Robust interdomain policy routing and traffic control. In: Proceedings OPENARCH. (2003)
3. Stewart, J.W.: BGP4: Inter-Domain Routing in the Internet. Addison-Wesley (1999)
4. Atkinson, R., Floyd, S.: IAB concerns & recommendations regarding internet research & evolution. Internet Draft, IAB (2004) <draft-iab-research-funding-03.txt>, work in progress.
5. Bu, T., Gao, L., Towsley, D.: On routing table growth. In: Prceedings IEEE Global Internet Symposium. (2002)
6. Huitema, C., Draves, R., Bagnulo, M.: Host-centric IPv6 multihoming. Internet Draft (2004) <draft-huitema-multi6-hosts-03.txt>, work in progress.
7. Huston, G.: Architectural approaches to multi-homing for IPv6. Internet Draft, IETF (2004) <draft-ietf-multi6-architecture-02.txt>, work in progress.
8. Gao, L.: On inferring autonomous system relationships in the internet. IEEE/ACM Transactions on Networking **vol. 9, no. 6** (2001)
9. Subramanian, L., Agarwal, S., Rexford, J., Katz, R.H.: Characterizing the internet hierarchy from multiple vantage points. In: Proceedings IEEE Infocom. (2002)
10. Quoitin, B.: C-BGP - An efficient BGP simulator. http://cbgp.info.ucl.ac.be/ (March 2004)
11. Medina, A., Lakhina, A., Matta, I., Byers, J.: BRITE: An Approach to Universal Topology Generation. In: Proceedings MASCOTS '01. (2001)
12. Jin, C., Chen, Q., Jamin, S.: Inet: Internet topology generator. Technical Report CSE-TR-433-00 (2000)
13. Calvert, K., Doar, M., Zegura, E.: Modeling internet topology. IEEE Communications Magazine (1997)

14. Quoitin, B.: Towards a POP-level Internet topology.
    http://cbgp.info.ucl.ac.be/itopo/ (2004)
15. de Launois, C.: GHITLE.
    http://openresources.info.ucl.ac.be/ghitle/ (April 2004)
16. de Launois, C.: Leveraging internet path diversity and network performances with
    IPv6 multihoming. http://www.info.ucl.ac.be/people/delaunoi/diversity/
    (2004)
17. Bonaventure, O., et al.: Internet traffic engineering. In: COST 263 Final Report,
    LNCS 2856, pp.118-179. (2003)
18. Akella, A., et al.: A measurement-based analysis of multihoming. In: Proceedings
    ACM SIGCOMM'03. (2003)
19. Akella, A., et al.: A comparison of overlay routing and multihoming route control.
    In: Proceedings ACM SIGCOMM'04. (2004)
20. Fuller, V., Li, T., Yu, J., Varadhan, K.: Classless inter-domain routing (CIDR):
    an address assignment and aggregation strategy. RFC 1519, IETF (1993)

Open-Source PC-Based Software Routers: A Viable Approach to High-Performance Packet Switching*

Andrea Bianco[1], Jorge M. Finochietto[1], Giulio Galante[2],
Marco Mellia[1], and Fabio Neri[1]

[1] Dipartimento di Elettronica, Politecnico di Torino,
Corso Duca degli Abruzzi 24, 10129 Torino, Italy
{bianco,finochietto,mellia,neri}@polito.it
[2] Networking Lab, Istituto Superiore Mario Boella,
via Pier Carlo Boggio 61, 10138 Torino, Italy
galante@ismb.it

Abstract. We consider IP routers based on off-the-shelf personal computer (PC) hardware running the Linux open-source operating system. The choice of building IP routers with off-the-shelf hardware stems from the wide availability of documentation, the low cost associated with large-scale production, and the continuous evolution driven by the market. On the other hand, open-source software provides the opportunity to easily modify the router operation so as to suit every need. The main contribution of the paper is the analysis of the performance bottlenecks of PC-based open-source software routers and the evaluation of the solutions currently available to overcome them.

1 Introduction

Routers are the key components of IP packet networks. The call for high-performance switching and transmission equipment in the Internet keeps growing due to the increasing diffusion of information and communication technologies, and the deployment of new bandwidth-hungry applications and services such as audio and video streaming. So far, routers have been able to support the traffic growth by offering an ever increasing switching speed, mostly thanks to the technological advances of microelectronics granted by Moore's Law.

Contrary to what happened for PC architectures, where, at least for hardware components, de-facto standards were defined, allowing the development of an open multi-vendor market, networking equipment in general, and routers in particular, have always seen custom developments. Proprietary architectures are affected by incompatibilities in configuration and management procedures, scarce programmability, lack of flexibility, and the cost is often much higher than the actual equipment value.

Appealing alternatives to proprietary network devices are the implementations of software routers based on off-the-shelf PC hardware, which have been recently made available by the open-source software community, such as Linux [1], Click [2] and FreeBSD [3] for the data plane, as well as Xorp [4] and Zebra [5] for the control plane,

---

* This work has been carried out in the framework of EURO, a project partly funded by the Italian Ministry of University, Education, and Research (MIUR).

M. Ajmone Marsan et al. (Eds.): QoS-IP 2005, LNCS 3375, pp. 353–366, 2005.
© Springer-Verlag Berlin Heidelberg 2005

just to name a few. Their main benefits are: wide availability of multi-vendor hardware and documentation on their architecture and operations, low cost and continuous evolution driven by the PC market's economy of scale.

Criticisms to software routers are focused on limited performance, instability of software, lack of system support, scalability problems, lack of functionalities. Performance limitations can be compensated by the natural evolution of the performance of the PC architecture. Current PC-based routers and switches have the potentiality for switching up to a few Gbit/s of traffic, which is more than enough for a large number of applications. Today, the maturity of open-source software overcomes most problems related to stability and availability of software functionalities. It is therefore important to explore the intrinsic limitations of software routers.

In this paper we focus only on the data plane, ignoring all the (fundamental) issues related to management functions and to the control plane. Our aim is to assess the routing performance and the hardware limitations of high-end PCs equipped with several Gigabit Ethernet network interface cards (NICs) running at both 1 Gbit/s and 100 Mbit/s under the Linux operating system.

The remainder of the paper is organized as follows. Provided that the Linux IP stack is implemented partly in hardware and partly in software in the operating system kernel, in Sect. 2 we give an overview of the architecture and the operation of hardware commonly available on high-end PCs, whereas in Sect. 3 we describe the current implementation of the IP stack in the Linux kernel. In Sect. 4 we describe our experimental setup, the tests performed, and the results obtained. Finally, in Sect. 5 we conclude and give some directions for future work.

2   PC Architectural Overview

A bare bones PC consists of three main building blocks: the central processing unit (CPU), random access memory (RAM), and peripherals, glued together by the *chipset*, which provides complex interconnection and control functions.

As sketched in Fig. 1, the CPU communicates with the chipset through the *system bus*, also known as front side bus (FSB) in Intel's jargon. The RAM provides temporary data storage for the CPU as long as the system is on, and can be accessed by the memory controller integrated on the chipset through the memory bus (MB). The peripherals are connected to the chipset by the peripheral component interconnect (PCI) shared bus, which allows to expand the system with a huge number of devices, including, but not limited to, permanent data storage units, additional expansion buses, video adapter cards, sound cards, and NICs. All interconnections are bidirectional, but, unfortunately, use different parallelisms, protocols, and clock speeds, requiring the implementation of translation and adaption functions on the chipset.

The following sections detail the operation of the CPU, the RAM the PCI bus, NICs and explain how the these components can be used to implement a software router.

2.1   CPU

State-of-the-art CPU cores run at frequencies up to 3.8 GHz, whereas next-generation CPUs run at 4 GHz. The front side bus is 64-bit wide and runs at either 100 MHz or

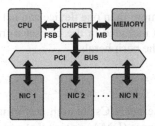

Fig. 1. PC architectural overview.

133 MHz with *quad pumped* transfers, meaning that data are transferred at a speed four times higher than the nominal clock speed. Therefore, the peak transfer bandwidth achievable ranges between 3.2 Gbyte/s and 4.2 Gbyte/s. Note that in Intel's commercial jargon, systems with FSBs clocked at 100 MHz and 133 MHz are marketed as running at 400 MHz and 533 MHz, because of quad pumping. High-end PCs are equipped with chipsets supporting multiple CPUs connected in a symmetric multiprocessing (SMP) architecture. Typical configurations comprise 2, 4, 8 or even 16 identical CPUs.

2.2 RAM

The memory bus is usually 64-bit wide and runs at either 100 MHz or 133 MHz with *double pumped* transfers, meaning that data are transferred on both rising and falling clock edges. Thus, the peak transfer bandwidth available ranges between 1.6 Gbyte/s and 2.1 Gbyte/s, and, in Intel's jargon, the two solutions are named *PC1600* and *PC2100* double data rate (DDR) RAM. In high-end PCs the memory bandwidth is doubled bringing the bus width to 128 bits by installing memory banks in pairs. Note that this allows to match the memory bus peak bandwidth to that of the front side bus.

2.3 PCI Bus

Depending on the PCI protocol version implemented on the chipset and the number of electrical paths connecting the components, the bandwidth available on the bus ranges from 1 Gbit/s for PCI 1.0, when operating at 33 MHz with 32-bit parallelism, to 2 Gbyte/s for PCI-X 266, when transferring 64 bits on both rising and falling edges of a double pumped 133 MHz clock.

The PCI protocol is designed to efficiently transfer the contents of large blocks of contiguous memory locations between the peripherals and the RAM, without requiring CPU intervention. Data and address lines are time multiplexed, therefore each *transaction* starts with an *addressing cycle*, continues with the actual data transfer which may take several *data cycles*, and ends with a *turnaround cycle*, where all signal drivers are three-stated waiting for the next transaction to begin. Some more cycles may be wasted if the *target* device addressed by the transaction has a high initial latency and introduces several *wait states*. This implies that the throughput experienced by the actual data transfer increases as the burst length gets larger, because the almost constant-size protocol overhead becomes more and more negligible with respect to useful data.

As the bus is shared, no more than one device can act as a *bus-master* at any given time; therefore, an *arbiter* is included in the chipset to regulate the access and fairly share the bandwidth among the peripherals.

2.4 NICs

Gigabit Ethernet and Fast Ethernet NICs are high-performance PCI cards equipped with at least one direct memory access (DMA) engine that can operate as bus-masters to of-fload the CPU from performing back-and-forth bulk data transfers between their inter-nal memory and the RAM. An adequate amount of on-card transmission (TX)/reception (RX) first-in first-out (FIFO) buffer memory is still needed to provide storage for data directed to (received from) the RAM.

The most common operation mode for streaming data transfers is *scatter-gather* DMA, which allows to spread small buffers all over the available RAM rather then allocating a single large contiguous RAM region which may be difficult, if not im-possible, to find because of memory fragmentation. During RAM-to-card transfers, the card fetches data *gathering* them from sparse RAM buffers; conversely, in the opposite direction, data originating from the card are *scattered* over available RAM buffers.

The simplest way to implement scatter-gather DMA is with two linked lists, one for transmission which requires RAM-to-card transfers and the other for reception which triggers card-to-RAM transfers. Each list element, dubbed *descriptor*, contains a pointer to the first location of the RAM buffer allocated by the operating system, the buffer size, a command to be executed by the card on the data, a status field detailing the result of such operation, and a pointer to the next descriptor in the list, possibly null if this is the last element. Alternatively, descriptors can be organized in two fixed-size arrays, managed by the card as circular buffers, named *rings*. The performance improvement granted by this solution is twofold: first, descriptors become smaller; second, descrip-tors on the same ring are contiguous and can be fetched in one burst, allowing for a higher PCI bus efficiency.

During normal operation, the card *i*) fetches descriptors from both rings to deter-mine which RAM buffers are available for reading/writing data and how the related data must be processed, *ii*) transfers the contents of the buffers, *iii*) performs the re-quired operations, and *iv*) updates the status in the corresponding descriptors.

Outgoing packets are read from buffers on the *transmission ring*, whereas incom-ing packets are written to buffers on the *reception ring*. Buffers pointed by descriptors are usually sized to fit both incoming and outgoing packets, even though, it is often possible to split outgoing packets among multiple buffers. Transmission stops when-ever the transmission ring empties, whereas incoming packets are dropped by the card either when the reception ring fills up or when the on-card FIFO overruns because of prolonged PCI bus unavailability due to congestion.

Each NIC is connected to one (possibly shared) hardware interrupt request (IRQs) line and collects in a *status register* information on TX/RX descriptor availability and on events needing attention from the operating system. It is then possible to selectively enable the generation of hardware IRQs to notify the CPU when a given bit in the status register is cleared/set to indicate an event, so that the operating system can take the appropriate action. Interrupts are usually generated when the reception ring is either full or almost full, when the transmission ring is either empty or almost empty, and after every packet transmission/reception.

In addition, it is usually possible to turn IRQ generation off altogether, leaving to the operating system the burden of periodically *polling* the hardware status register and react accordingly. More details on these packet reception schemes are provided later in the paper.

2.5 Putting All the Pieces Together

The hardware available on a PC allows to implement a shared bus, shared memory router, where NICs receive and store packets in the main RAM, the CPU routes them to the correct output interface, and NICs fetch packets from the RAM and transmit them on the wire. Therefore, each packet travels twice on the PCI bus, halving the bandwidth effectively available for NIC-to-NIC packet flows.

3   Linux Network Stack Implementation

The networking code in the Linux kernel is highly modular: the hardware-independent IP stack has a well defined application programming interface (API) toward the hardware-dependent device driver, which is the glue making the IP layer operate with the most diverse networking hardware.

When a NIC is receiving traffic, the device driver pre-allocates packet buffers on the reception ring and, after they have been filled by the NIC with received packets, hands them to the IP layer. The IP layer examines each packet's destination address, determines the output interface, and invokes the device driver to enqueue the packet buffer on the transmission ring. Finally, after the NIC has sent a packet, the device driver unlinks the packet buffer from the transmission ring. The following sections discuss briefly the operations performed by the memory management subsystem, the IP layer, and detail how the network stack is invoked upon packet reception.

3.1   Memory Management

In the standard Linux network stack implementation, buffer management is performed resorting to the operating system general-purpose memory management system, which requires CPU expensive operations. Some time can be saved if the buffer deallocation function is modified so as to store unused packet buffers on a recycling list in order to speed up subsequent allocations, allowing the device driver to turn to the slower general-purpose memory allocator only when the recycling list is empty.

This has been implemented in a patch [6] for 2.6 kernels, referred to as *buffer recycling* patch in the reminder of the paper, which adds buffer recycling functionalities to the e1000 driver for Intel Gigabit Ethernet NICs. As of today, the buffer recycling patch has not been officially included in the kernel source, but it could be easily integrated in the core networking system code, making it available for all network device drivers without any modification to their source code.

3.2   IP Layer

The Linux kernel networking code implements a standard RFC 1812 [7] router. After a few sanity checks such as IP header checksum verification, packets that are not addressed to the router are processed by the routing function which determines the IP address of the next router to which they must be forwarded, and the output interface on which they must be enqueued for transmission.

The kernel implements an efficient routing cache based on a hash table with collision lists; the number of hash entries is determined as a function of the RAM available when the networking code is initialized at boot time. The route for outgoing packets is first looked up in the routing cache by a fast hash algorithm, and, in case of miss, the whole routing table stored in the forwarding information base (FIB) is searched by a (slower) longest prefix matching algorithm.

Next, the time-to-live (TTL) is decremented, the header checksum is updated and the packet is enqueued for transmission in the RAM on the drop-tail TX queue associated with the correct output interface. Then, whenever new free descriptors become available, packets are transferred from the output TX queue to the transmission ring of the corresponding NIC. The maximum number of packets that can be stored in the output TX queue is limited and can be easily modified at runtime; the default maximum length for current Linux implementations is 1000 packets.

Note that, for efficiency's sake, whenever it is possible, packet transfers inside the kernel networking code are performed by moving pointers to packet buffers, rather than actually copying buffers' contents. For instance, packet forwarding is implemented unlinking the pointer to a buffer containing received packets from the reception ring, handing it to the IP layer for routing, linking it to the output TX queue, and moving it to the NIC transmission ring before processing the next pointer. This is commonly referred to as *zero-copy* operation.

3.3   Stack Activation Methods

Interrupt. NICs notify the operating system of packet reception events by generating hardware IRQs for the CPU. As shown in Fig. 2, the CPU invokes the driver hardware IRQ handler which acknowledges the NIC request, transfers all packets currently available on the reception ring to the kernel *backlog queue* and schedules the network software-IRQ (softIRQ) handler for later execution.

SoftIRQs are commonly used by many Unix flavors for deferring to a more appropriate time the execution of complex operations that have a lower priority than hardware IRQs and that cannot be safely carried out by the IRQ handler, because they might originate race conditions, rendering kernel data structures inconsistent.

The network softIRQ handler extracts packets from the backlog queue, and hands them to the IP layer for processing as described in Sect. 3.2. There is a limit on the maximum number of packets that can be stored in the backlog queue by the IRQ handler so as to upper bound the time the CPU spends for processing packets. When the backlog queue is full, the hardware IRQ handler just removes incoming packets from the reception ring and drops them. In current Linux implementations, the default size of the backlog queue is 300 packets.

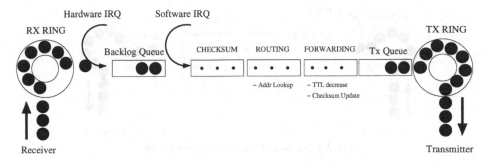

Fig. 2. Operation of the interrupt-driven network stack.

Under heavy reception load, a lot of hardware IRQs are generated from NICs. The backlog queue fills quickly and all CPU cycles are wasted extracting packets from the reception ring just for dropping them after realizing that the backlog queue is full. As a consequence, the softIRQ handler, having lower priority than the hardware IRQ handler, never gets a chance of draining packets from the backlog queue, practically zeroing the forwarding throughput. This phenomenon was first described in [8] and dubbed *receive livelock*.

Hardware IRQs are also used to notify the operating system of the transmission of packets enqueued on the transmission ring so that the driver can free the corresponding buffers and move new packets from the TX queue to the transmission ring.

Interrupt Moderation

Most modern NICs provide *interrupt moderation* or *interrupt coalescing* mechanisms (see, for example, [9]) to reduce the number of IRQs generated when receiving packets. In this case, interrupts are generated only after a batch of packets has been transmitted/received, or after a timeout from the last IRQ generated has expired, whichever comes first. This allows to relieve the CPU from IRQ storms generated during high traffic load, improving the forwarding rate.

NAPI. Receive livelock can be easily avoided by disabling IRQ generation on all NICs and letting the operating system decide when to poll the NIC hardware status register to determine whether new packets have been received. The NIC polling frequency is determined by the operating system and, as a consequence, a polling-driven stack may increase the packet forwarding latency under light traffic load.

The key idea introduced in [8] and implemented in the Linux network stack in [10] with the name new API (NAPI), is to combine the robustness at high load of a polling-driven stack with the responsiveness at low load of an interrupt-driven stack.

This can be easily achieved by enabling IRQ status notification on all NICs as in an interrupt-activated stack. The driver IRQ handler is modified so that, when invoked after a packet reception event, it enables polling mode for the originating NIC by switching IRQ generation off and by adding that NIC to the NAPI *polling list*. It then schedules the network softIRQ for execution as usual.

As shown in Fig. 3, a new function `poll` is added to the NIC driver to *i*) remove packets from the reception ring, *ii*) hand them to the IP layer for processing as described in Sect. 3.2, *iii*) refill the reception ring with empty packet buffers, *iv*) detach from the

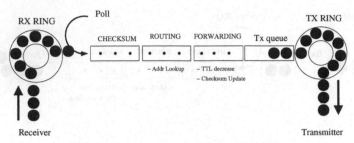

Fig. 3. Operation of the NAPI network stack.

transmission ring and free packets buffers after their content has been sent on the wire. The `poll` function never removes more than a *quota Q* of packets per invocation from the NIC reception ring.

The network softIRQ is modified so as to run `poll` on all interfaces on the polling list in a round-robin fashion to enforce fairness. No more than a *budget B* of packets can be extracted from NIC reception rings in a single invocation of the network softIRQ, in order to limit the time the CPU spends for processing packets. This algorithm produces a max-min fair share of packet rates among the NICs on the polling list.

Whenever `poll` extracts less than Q packets from a NIC reception ring, it reverts such NIC to interrupt mode by removing it from the polling list and re-enabling IRQ notification. The default values for B and Q in current Linux implementations are, respectively, 300 and 64.

Currently, not all NIC drivers are NAPI aware, but it is easy to write the NAPI `poll` handler by just borrowing code from the hardware IRQ handler.

4   Performance Evaluation

In Sect. 4.1, we describe the testbed setup we used and report some preliminary results from experiments on the NIC maximum reception and transmission rate, which motivated the adoption of a commercial router tester for subsequent tests.

Many metrics can be considered when evaluating the forwarding performance of a router: for instance, the RFC 2544 [11] defines both steady-state indices such as the maximum lossless forwarding rate, the packet loss rate, the packet delay and the packet jitter, as well as transient quantities such as the maximum-length packet burst that can be forwarded at full speed by the router without losses. In this paper, however, the different configurations were compared in terms of the *steady-state saturation forwarding throughput* obtained when all router ports are offered the maximum possible load. We considered both *unidirectional* flows, where each router port, at any given time, only receives or transmits data, as well as the *bidirectional* case, where all ports send and receive packets at the same time. All the results reported are the average of five 30-second runs of each test.

The IP routing table used in all tests is minimal and only contains routes to the class-C subnetworks reachable from each port. As a consequence, the number of IP destination addresses to which the router-tester sends packets on each subnetwork is always less than 255, so that the routing cache never overflows and the routing overhead is marginal.

4.1   Testbed Setup

The router tested is based on a high-end PC with a SuperMicro X5DPE-G2 mainboard equipped with one 2.8 GHz Intel Xeon processor and 1 Gbyte of PC1600 DDR RAM consisting of two interleaved banks, so as to bring the memory bus transfer rate to 3.2 Gbyte/s.

Almost all experiments were performed running Linux 2.4.21, except for buffer recycling tests, which were run on a patched 2.6.1 kernel. No major changes occurred in the networking code between kernel version 2.4 and 2.6. The only modification needed to make a fair performance comparison between 2.6 and 2.4 kernels is to lower the clock interrupt frequency from 1000 Hz (default for 2.6 kernels) to 100 Hz (default for 2.4 kernels).

Although Fast and Gigabit Ethernet NICs offer a raw rate of 100 Mbit/s and 1 Gbit/s, respectively, the data throughput at IP layer actually achievable may be much lower because of physical and data-link layer overhead, due to Ethernet overhead. Indeed, besides MAC addresses, protocol type and CRC, also the initial 8-bytes trailer and the final 12-byte minimum inter-packet gap, must be taken into account. Carried payload ranges from a 46-bytes minimum size up to 1500-bytes of maximum size. As a consequence, Gigabit (Fast) Ethernet NICs running at full speed must handle a packet rate ranging from 81 274 (8 127) to 1 488 095 (148 809) packets/s as the payload size decreases from 1500 byte to 46 byte.

Most of the results presented in this section were obtained for minimum-size Ethernet frames because they expose the effect of the per-packet processing overhead. However, we also ran a few tests for frame sizes up to the maximum, in order to check that both the PCI bus and the memory subsystem could withstand the increased bandwidth demand.

A number of tests were performed on Gigabit Ethernet NICs produced by Intel, 3Com (equipped with a Broadcom chipset), D-Link and SysKonnect, using open-source software generators (see [12] for a good survey) operating either in user- or in kernel-space. We compared rude [13], which operates in user-space, with udpgen [14] and packetgen [15], that, instead, live in kernel-space. The aim was to assess the maximum transmission/reception rate of each NIC-driver pair, when only one NIC was active and no packet forwarding was taking place. In order to allow for a fair comparison and to avoid a state-space explosion we left all driver parameters to their default values. The highest generation rate of 650 000 minimum-size Ethernet frames per second (corresponding to almost 500 Mbit/s) was achieved by packetgen on Intel PRO 1000 Gigabit Ethernet NICs running with the Intel e1000 driver [16] version 5.2.52.

As it was not possible to generate 64-byte Ethernet frames at full speed on any of the off-the-shelf NIC we considered, we ran all subsequent tests on an Agilent N2X RouterTester 900 [17], equipped with 2 Gigabit Ethernet and 16 Fast Ethernet ports, that can transmit and receive at full rate even 64-byte Ethernet frames.

The maximum reception rate of the different NICs was also evaluated, when only one NIC was active and packet forwarding was disabled, by generating 64-byte Ethernet frames with the RouterTester, and counting the packets received by the router with udpcount [14]. Again, the best results were obtained with the e1000 driver and Intel PRO 1000 NICs, which were able to receive a little bit more than 1 000 000 packets/s on an IRQ stack and about 1 100 000 packets/s on a NAPI stack. Given the superior performance of Intel NICs, all tests in the next two sections were performed equipping the router with seven Intel PRO 1000 MT dual port Gigabit Ethernet NICs running at either 1 Gbit/s or 100 Mbit/s. All driver parameters were left to their default values, except for interrupt moderation which was disabled when NAPI was used.

From the packet reception and transmission rates measured when forwarding was disabled, it is possible to extrapolate that, very likely, in a bidirectional two-port scenario, a router will not be able to forward more than 650 000 packets/s each way, because of the NIC's transmission rate limit.

4.2 Two-Port Configuration

We first considered the simplest unidirectional case, where packets are generated from the tester, received by the router from one NIC, and sent back to the tester through the other NIC.

In Fig. 4 we plot the forwarding rate achieved for minimum-size Ethernet frames by IRQ, IRQ moderated, and NAPI activated stacks under unidirectional traffic, as a function of the offered load. Both IRQ and IRQ moderated stacks suffer from receive livelock: when the offered load becomes greater than 230 000 packets/s, or 500 000 packets/s, respectively, the throughput quickly drops to zero. Only NAPI is able to sustain a constant forwarding rate of 510 000 packets/s (corresponding to 340 Mbit/s) under high load. A more accurate analysis of the latter case shows that all packet drops occur on the reception ring rather than from the RX queue. Such forwarding rate is more than reasonable, because a back-of-the-envelope calculation indicates that the per-packet overhead is around 2 μs or about 5500 clock cycles. On the other hand, a 2.8 GHz CPU for forwarding 1 488 095 packets/s would have to take less than 2000 clock cycles (corresponding to approximatively 700 ns) per packet. Given the superior performance of NAPI with respect to other packet forwarding schemes, in the remainder of the paper, we will consider only results obtained with NAPI.

It is now interesting to repeat the last test for different Ethernet payload sizes to assess whether the forwarding rate limitation observed in Fig. 4 depends on the per-packet processing overhead or on a more unlikely PCI bandwidth bottleneck.

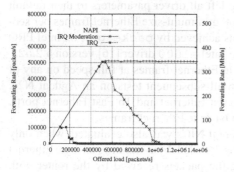

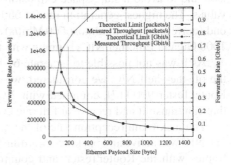

Fig. 4. Comparison of the router forwarding rate under unidirectional traffic for an IRQ, an IRQ moderated and a NAPI stack.

Fig. 5. Comparison of the router saturation forwarding rate for a NAPI stack under unidirectional traffic and different IP packet sizes.

In Fig. 5, we report in both packets/s and Mbit/s the forwarding rate we measured (empty markers) and the value the forwarding rate would have taken in absence of packet drops (filled markers) versus the Ethernet payload size, under NAPI. For 64- and 128-byte payloads the forwarding rate is most likely limited to about 500 000 packets/s by the per-packet CPU processing overhead. Conversely, for payloads of 512 byte or more, it is possible to reach 1 Gbit/s, because the corresponding packet rate is low enough and the PCI bus is not a bottleneck. The result for 256-byte frames is difficult to explain and may be related to some PCI-X performance impairment. Indeed, in [18], the authors, using a PCI protocol analyzer, show that the bus efficiency for bursts of 256 byte or less is pretty low.

Since the PCI bus seems not to be a bottleneck for minimum-size packet transfers, in Fig. 6 we compare the aggregated forwarding rate of unidirectional and bidirectional flows plotted versus the aggregated offered load. Notice that, for bidirectional traffic, the forwarding rate improves from 510 000 packets/s to 620 000 packets/s, corresponding to approximatively 400 Mbit/s. This happens because, at high offered load under bidirectional traffic, the NAPI `poll` function, when invoked, services both the reception and the transmission ring of each NIC it processes, greatly reducing the number of IRQs generated with respect to the unidirectional case, where only one NIC receives packets in polling mode and the other one sends them in IRQ mode. Indeed, measurements performed when the aggregated offered load is around 1 000 000 packets/s show that the transmitting NIC generates, under unidirectional traffic, 27 times as many interrupts than in the bidirectional case.

The curves in Fig. 7 show that the buffer recycling optimization improves the forwarding rate of a bidirectional flow of 64-byte Ethernet frames to 730 000 packets/s, roughly corresponding to 500 Mbit/s. Measurements on the router reveal that, after an initial transient phase when all buffers are allocated from the general-purpose memory allocator, in stationary conditions all buffers are allocated from the recycling list, which contains around 600 buffers, occupying approximatively 1.2 Mbyte of RAM.

All tests in the following section were performed on a Linux 2.4.21 kernel and, therefore, could not take advantage of the buffering recycling patch.

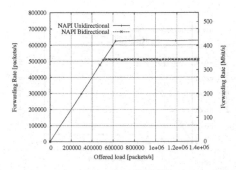

Fig. 6. Comparison of the router forwarding rate achieved by unidirectional and bidirectional traffic flows for a NAPI stack.

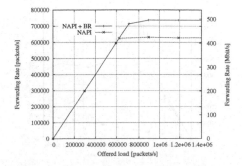

Fig. 7. Comparison of the effect of the buffer recycling (BR) patch on the forwarding rate under bidirectional traffic for a NAPI stack.

4.3 Multi-port Configurations

In this section we consider both a *one-to-one* traffic scenario, where all packets arriving at one input port are addressed to a different output port, so that no output port gets packets from more than one input port, as well as a *uniform* traffic pattern where packets received by one NIC are spread over all the remaining ports uniformly.

We first consider a router with two ports running at 1 Gbit/s and two ports running at 100 Mbit/s, under one-to-one unidirectional traffic. The offered load is distributed according to a 10:1 ratio between the ports running at 1 Gbit/s and the ports running at 100 Mbit/s,and varied from 0 to 100% of the full line rate. The quota Q and the budget B are set to their default values of 64 and 300. Figure 8 shows the total and per-flow forwarding rate versus the aggregated load offered to the router. The forwarding rate for the 100 Mbit/s flow increases steadily, stealing resources from the 1 Gbit/s flow according to a max-min fair policy, when the router bandwidth is saturated, and the aggregated forwarding rate keeps constant.

This behavior can be easily altered to implement other fairness models just changing the quota assigned to each port. Figure 9 shows the results obtained in the same scenario when the quota Q was set to 270 for ports running at 1 Gbit/s and to 27 for ports running at 100 Mbit/s. In this way, `poll` can extract from the former 10 times as many packets than from the latter and, assigning more resources to more loaded ports, aggregated performance is improved.

In Fig. 10 we plot the aggregated forwarding rate versus the aggregated offered load for 6, 10 and 14 ports running at 100 Mbit/s under bidirectional one-to-one traffic. There is a slight decrease in the saturation forwarding rate as the number of ports receiving traffic increases, which may be due to the greater overhead incurred by the NAPI `poll` handler when switching among different ports.

In Fig. 11 we compare the forwarding rate for the one-to-one and the uniform traffic pattern when 14 ports are running at 100 Mbit/s. Again, we observe a small difference in the forwarding rate achieved in the two scenarios, which is probably due to a major processing overhead incurred by the latter.

This may depend on the fact that, under uniform traffic, packets consecutively extracted from the reception ring of one port are headed for different IP subnetworks and

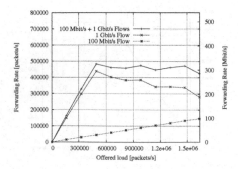

Fig. 8. Max-min fair behavior of two unidirectional traffic flows coming from ports running at different speed under NAPI.

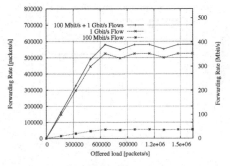

Fig. 9. Impact of different per-port quota setting on the fair share of two unidirectional traffic flows.

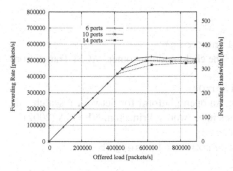

Fig. 10. Forwarding performance for one-to-one bidirectional traffic versus number of 100 Mbit/s ports.

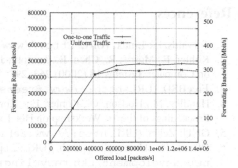

Fig. 11. Forwarding performance comparison between one-to-one and uniform bidirectional traffic with 14 100 Mbit/s ports.

must be spread among different TX queues, contrarily to what happens under one-to-one traffic, where all packets are forwarded to the same TX queue.

5   Conclusions and Future Work

In this paper we evaluated the viability of building a high-performance IP router out of common PC hardware and the Linux open-source operating system. We ran a number of experiments to assess the saturation forwarding rate in different scenarios, completely ignoring all issues related to the control plane.

We showed that a software router based on a high-end off-the-shelf PC is able to forward up to 600 000 packets/s, when considering minimum-size Ethernet frames, and to reach 1 Gbit/s when larger frame sizes are considered. Configurations with up to 14 ports can be easily and inexpensively built at the price of a small decrease in the forwarding rate. A number of tricks, such as packet buffer recycling, and NAPI quota tuning are also available to improve the throughput and alter the fairness among different traffic flows.

In the future we plan to compare the routing performance of Click and FreeBSD with that of Linux. Provided that the major bottleneck in the systems seems to be the per-packet processing overhead introduced by the CPU, we are also profiling the Linux kernel networking code so as to identify the most CPU intensive operations and implement them on custom NICs enhanced with field programmable gate arrays (FPGAs).

Acknowledgment

We would like to thank M.L.N.P.P. Prashant for running the tests on different packet sizes and the experiments for router configurations with more than 6 ports, and Robert Birke for modifying udpcount to run on top of NAPI and providing the results on the NIC reception rates. We would also like to thank the partners of the Euro project for the useful discussions, and the anonymous reviewers for helping to improve the paper.

References

1. Torvalds, L.: Linux. (URL: http://www.linux.org)
2. Kohler, E., Morris, R., Chen, B., Jannotti, J.: The click modular router. ACM Transactions on Computer Systems **18** (2000) 263–297
3. FreeBSD. (URL: http://www.freebsd.org)
4. Handley, M., Hodson, O., Kohler, E.: Xorp: An open platform for network research. In: Proceedings of the 1st Workshop on Hot Topics in Networks, Princeton, NJ, USA (2002)
5. GNU: Zebra. (URL: http://www.zebra.org)
6. Olsson, R.: skb recycling patch. (URL: ftp://robur.slu.se/pub/Linux/net-development/skb_recycling)
7. Baker, F.: RFC 1812, requirements for IP version 4 routers. URL: ftp://ftp.rfc-editor.org/in-notes/rfc1812.txt (June 1995)
8. Mogul, J.C., Ramakrishnan, K.K.: Eliminating receive livelock in an interrupt-driven kernel. ACM Transactions on Computer Systems **15** (1997) 217–252
9. Intel: Interrupt moderation using Intel Gigabit Ethernet controllers (Application Note 450). URL: http://www.intel.com/design/network/applnots/ap450.htm
10. Salim, J.H., Olsson, R., Kuznetsov, A.: Beyond softnet. In: Proceedings of the 5th Annual Linux Showcase & Conference (ALS 2001), Oakland, CA, USA (2001)
11. Bradner, S., McQuaid, J.: RFC 2544, benchmarking methodology for network interconnect devices. URL: ftp://ftp.rfc-editor.org/in-notes/rfc2544.txt (March 1999)
12. Zander, S.: Traffic generator overview. (URL: http://www.fokus.gmd.de/research/cc/glone/employees/sebastian.zander/private/trafficgen.html)
13. Laine, J.: Rude/Crude. (URL: http://www.atm.tut.fi/rude)
14. Zander, S.: UDPgen. (URL: http://www.fokus.fhg.de/usr/sebastian.zander/private/udpgen)
15. Olsson, R.: Linux kernel packet generator for performance evaluation. (URL: /usr/src/linux-2.4/net/core/pktgen.c)
16. Intel: Intel PRO/10/100/1000/10GbE linux driver. (URL: http://sourceforge.net/projects/e1000)
17. Agilent: N2X routertester 900. (URL: http://advanced.comms.agilent.com/n2x)
18. Brink, P., Castelino, M., Meng, D., Rawal, C., Tadepalli, H.: Network processing performance metrics for IA- and IXP-based systems. Intel Technology Journal **7** (2003)

Performance of a Software Router Using AltQ/CBQ – A Measurement-Based Analysis

Daniel Breest[1] and Jean-Alexander Müller[2]

[1] Leipzig University, Department of Computer Science,
Augustusplatz 10-11,
D-04109 Leipzig, Germany
mai96itt@informatik.uni-leipzig.de

[2] Hochschule für Technik und Wirtschaft Dresden (FH) – University of Applied Sciences,
Department of Information Technology and Mathematics,
Friedrich-List-Platz 1,
D-01069 Dresden, Germany
jeanm@informatik.htw-dresden.de

Abstract. An AltQ-based software router, using CBQ as key building block, offers a flexible platform to provide service differentiation in the future Internet. This paper presents an experimental analysis of the actual performance and scheduling accuracy of an AltQ/CBQ-based software router, using a very precise hardware-based measuring technique. Our results show, that even though several jitter effects can be estimated very well in advance, they can become considerably large with only a small number of concurrent flows. Moreover, the forwarding performance of the test system is still not sufficient to handle a single Fast Ethernet link, which will lead to an unpredictable delay jitter.

Keywords: Policing, Scheduling, CBQ, AltQ

1 Introduction

The vision of a network providing service differentiation for almost every application becomes more and more reality. IP telephony for example, that has been mainly used within enterprise networks for a fairly long time, seems to be ready for the broad market now. An increasing number of end customers switching to this technology requires a sophisticated resource management within and in between individual provider networks.

However, the deployment of Quality-of-Service mechanisms that are both scalable and predictable in terms of transmission properties is still a non-trivial task.

A common and widely accepted approach to maintain scalability is proposed in [1, 2]. Traffic management within the core of the network is based on a few traffic aggregates and static resource assignment. Service guarantees can then be achieved by a strict admission control at configuration time and by continuously monitoring and shaping incoming traffic at the network boundary. The scheduling of the ingress traffic needs to be as accurately as possible thereby, to precisely estimate the behaviour of the resulting traffic and to exploit the allocated resources of a given traffic aggregate as efficiently as possible (see [3]). The traffic conditioning at the network boundary additionally requires packet classification and resource management based on single micro flows which is not or only partially supported by currently available hardware routers.

M. Ajmone Marsan et al. (Eds.): QoS-IP 2005, LNCS 3375, pp. 367–378, 2005.

The deployment of a software-based router at the network boundary, using an advanced scheduler like Class Based Queueing (CBQ) as key building block, offers an interesting alternative. CBQ supports both real-time and link sharing services in an integrated fashion. See [4] for an introduction to CBQ concepts and mechanisms. However, the flexibility of a software-based solution is limited, because the packet processing is almost completely realized in software.

In addition, a complex packet scheduling discipline like CBQ introduces a much higher workload compared to that of simple FIFO scheduling. Thus, accurate scheduling requires not only an efficient implementation of the scheduling algorithm but also a machine that can cope with the workload easily.

This paper gives an overview of the general performance of a FreeBSD/AltQ-based router, using CBQ as packet scheduler, with the focus on the accuracy of the packet forwarding especially. AltQ is a queueing framework that allows flexible resource management on BSD-based operation systems. It is described in detail in [5].

The experiments, described in this paper, were carried out using a very accurate hardware-based measuring technique. The results contradict statements given for example in [5], that an FreeBSD/AltQ-based router, using commodity hardware, will have processing power enough to handle multiple Fast Ethernet connections in parallel. Although, a more powerful system was used, the packet forwarding performance was still not sufficient to handle a single Fast Ethernet link when a typical packet size distribution is taken into consideration. As it is stated in [5] too, the CBQ scheduler itself is still not as efficient as it can be, even though a refined implementation was used (see [6]). Both, the insufficient overall performance and the inefficient CBQ implementation, negatively affect the accuracy of the packet forwarding. This makes it difficult to achieve real traffic isolation and to precisely estimate the behaviour of the resulting traffic. While the packet forwarding accuracy is affected by a number of other factors too, whose influence can be derived very well, optimisation is required, as the introduced packet delay jitter will become too large for certain applications.

Section 2 describes the test-bed and measuring technique used in the experiments. Section 3 gives an overview of the packet forwarding performance of the PC-system used as software router. Section 4 analyses several issues concerning the CBQ scheduling accuracy. Section 5 finally discusses the ability of the PC-system to provide traffic isolation. Section 6 presents a short summary of the major results and gives an outlook for future work.

2   Experimental Test-Bed and Methodology

The experiments were carried out with the core test bed shown in Figure 1. All machines were equipped with Intel Fast Ethernet devices (Intel 82558/82562 chip-set) and had Linux or FreeBSD installed. Machine A was used to generate the test traffic, that was sent through the router and received by machine B. In addition, background traffic was generated by several machines from within the local network and sent to machine B via the router. The router was a Pentium III 866 MHz machine with FreeBSD 4.7 installed and AltQ 3.1 compiled into the kernel. It was started in single user mode via `boot -s` to prevent background processes from negatively affecting the packet processing of the router. The network was configured manually.

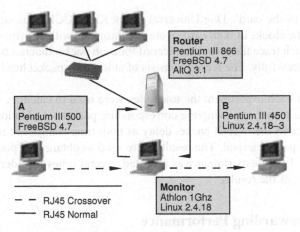

Fig. 1. Test-bed topology.

mgen [7] was used to generate UDP traffic with packets of 46, 512, 1024 and 1500 Bytes length respectively and to simulate the G.726 VoIP codec [8]. Based on the packet statistics of a hardware router of the campus network (see Table 1), Best-Effort UDP traffic, as it is typically observed in a LAN/Intranet network, was simulated using tg [9].

While the results of the experiments in a number of other papers are based on software tools like tcpdump (see [10] for example), a very accurate hardware based measuring technique was used in our experiments to collect the packet traces. Even though it will not be discussed in detail, two drawbacks shall be pointed out when using a software based solution. The accuracy of the time-stamp, recorded for a packet, then typically depends on the system clock and the frequency of the kernel timer respectively. With a timer frequency of 1000 cps for example, the resolution of the time-stamp is 1 ms at best. Given the fact, that with Fast Ethernet packets can typically arrive at the router within about 6 to 122 μs, a much higher time-stamp resolution is required. While someone could point out, that the use of the time-stamp counter of the routers processor would provide the required resolution, this would implicate the execution of the trace recording software on the router itself. This will introduce additional load and affect the router's packet forwarding ability that should be measured actually.

Therefore a dedicated monitoring machine was used to collect packet traces before and after the router. It was equipped with two Endace DAG 3.5E network measuring devices [11], each capable of monitoring the packet stream on the physical layer and of capturing a configured number of Bytes of each packet together with a high resolution time-stamp. The time-stamp is recorded at the beginning of the packet arrival and the

Table 1. Distribution of the packet length of IP packets as typically observed in a LAN/Intranet.

IP packet size (Bytes)	0-46	47-109	110-237	238-1005	1006-1500
(%)	27	24	17	5	27

clock resolution of the card's Dag Universal Clock Kit (DUCK) time-stamp engine is 60 ns [12, 13]. The clocks of both cards were synchronised with an error of ±120 ns on average. Thus, each trace file contains a record for each packet that has been forwarded by the router successfully. The record consists of at least the packet header and a 64 bit time-stamp.

At the end of each experiment the trace files were used to calculate the packet delay caused by the router, removing the corresponding packet serialisation delay which is comprised once in the overall packet delay as both time-stamps are recorded at the beginning of the packet arrival. The results were used to obtain the "packet delay distribution" graph. Every experiment was repeated several times in order to ensure the statistical validity of the results.

3   Packet Forwarding Performance

To get an overview of the system's behaviour, test traffic with packets of constant and variable length was generated (see Section 2), gradually increasing the ingress rate up to the maximum achievable rate. Each experiment was carried out using both, the standard FIFO queue and a simple CBQ configuration with packets directed to the default class.

3.1   General Performance

As shown in Table 3, the basic packet delay, which is the minimum possible processing delay that has been introduced on a significant number of packets at a very low rate of the test traffic, is almost independent of the packet length. Therefore, frame reception and transmission and the corresponding PCI transfers of the packet data have to take place in parallel and not strictly consecutive. Otherwise, it would take up to 23 μs longer to forward a packet of 1500 Bytes length using a 32 bit/33 MHz PCI bus (see Table 2 for the one time PCI transfer delay).

During frame reception, the network device performs DMA write transactions from the device buffer to the system memory, so that the PCI transfer delay is actually covered by the much larger packet serialisation delay that has been already removed from the overall packet delay (see Section 2). The same applies to frame transmission, where the network device starts serialising the packet while the packet data is actually transfered from the system memory to the device buffer using DMA read transactions (see [14]). So, the packet delay is mainly caused by the router's software layer and does

Table 2. PCI transfer delay.

IP packet size (bytes)	PCI transfer delay (μs)
46	0.35
512	3.88
1024	7.76
1500	11.36

Table 3. Basic and worst-case packet delay.

IP packet size	basic delay (μs)		worst-case delay (μs)	
(bytes)	FIFO	CBQ	FIFO	CBQ
46	24	30	124	129
512	26	33	130	137
1024	26	33	125	131
1500	26	33	118	138

comprises no or only a small part of the processing delay caused by the PCI bus and network device respectively. If it takes at least 24 μs to forward a single packet however, the packet forwarding rate of the system will be limited to about 41000 pps. This is important, as the packet arrival rate could become larger than 100000 pps, considering a typical packet size distribution (see Table 1). The consequences are discussed in detail in Section 3.2.

While almost all packets are forwarded with nearly the basic packet delay, a small number already experiences higher delays under low ingress rates. In the experiments with a packet length of 512 Bytes for example, some packets will be typically delayed by 36 μs and 56 μs respectively. This deviation from the basic packet delay could become a serious problem when the ingress rate of the test traffic increases. The delay of a large number of packets can then be affected too, due to the fact that the traffic burstiness increases as well. Although, the burstiness of the generated traffic is mainly a problem of the inaccuracy of the mgen traffic generator, results are not worthless but have to be interpreted accordingly as real network traffic is bursty too. Given this fact, the delay deviation of a single packet is inevitably passed to each packet that follows within a burst. The larger the burst the more packets are potentially affected. This burstiness effect is the main reason, that leads to the visible steps at 36 μs and 56 μs respectively and to the increasing overall packet delay, that can be observed in Figure 2.

While the resulting jitter is comparably small, i.e. the worst case delay, occurred under FIFO scheduling, is about 130 μs (see Table 3), the sensitivity of packet processing and of the resulting packet delay becomes already observable without any complex traffic management configured and with only a single packet flow active.

3.2 Impact of Variable Length Packets

A different behaviour can be noticed in the experiments with variable length packets. Besides the fact, that a significant number of packets experienced a delay of about 150 μs, a broader distribution of the packet delay can be observed in general and especially under CBQ scheduling, when the ingress rate is increased (see Figure 3).

The visible step in the range of 130 μs to 150 μs is caused by the packets of variable length. The further discussion will be restricted to the worst case behaviour observed under FIFO scheduling when a packet of maximum length (1500 Bytes) and a very small packet (46 Bytes) arrive at the router within a burst. In this case both, the inter arrival time and the processing delay of the second packet is smaller then the processing and serialisation delay of the first packet respectively. The small packet will then have

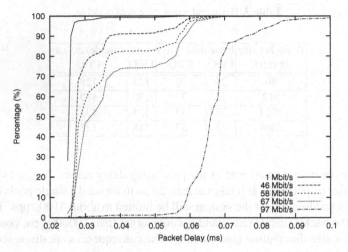

Fig. 2. Distribution of the packet delay of test traffic (512 Bytes length packets) with FIFO scheduling.

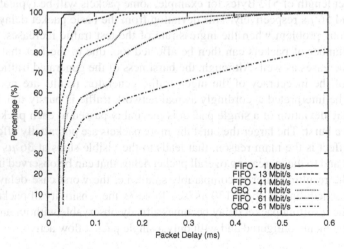

Fig. 3. Distribution of the packet delay of test traffic (variable length packets) with FIFO/CBQ scheduling.

to wait at least 117 μs in addition to its own processing delay which results in an overall delay of at least 141 μs. If another burst of small packets arrive immediately thereafter, each packet will experience at least the same delay due to the burstiness effect already mentioned in Section 3.1.

The broader delay distribution can be mainly attributed to the already described limited packet forwarding performance of the system. The router cannot forward a burst of small packets (<110 Bytes) with the required forwarding rate. This in turn leads to a congestion at the incoming interface and to an increasing packet delay within the burst.

It must be questioned then, if traffic isolation, as it should be provisioned by the CBQ scheduler, is actually possible. This aspect will be finally discussed in Section 5.

4   Accuracy of the CBQ Packet Scheduler

CBQ is aimed to support both, real-time and link-sharing services in an integrated fashion. The accuracy of the packet scheduling and the resulting delay jitter for a certain traffic entity then depends on several issues concerning both, the system hardware and the CBQ implementation itself. It has to be analysed, how background traffic influences the delay jitter of some test traffic.

4.1   Priority Scheduling

Priority scheduling can be used to achieve some kind of coarse-grained traffic differentiation. The CBQ implementation of AltQ supports up to eight priorities by default. In general, a class at a certain priority is allowed to send packets as long as there is no other class on a higher priority that is able to send packets. Only two priorities will be considered in the discussion furthermore. The test traffic is generated at a rate of 3 Mbit/s and is served with the highest priority. The background traffic is served with the lowest priority and its ingress rate is gradually increased during the overall measurement. The background traffic consists of packets of maximum length and is assumed to be always active.

The results are presented in Figure 4. It shows the distribution of the packet delay of the test traffic in dependence on the ingress rate of the background traffic. The ideal curve represents the packet delay distribution of the test traffic under unloaded conditions, i.e. with no background traffic active. The worst case jitter theoretically occurs, when a high priority packet arrives at the outgoing interface of the router shortly after the scheduler has started serving a maximum length low priority packet. It has to wait up to 122 μs on Fast Ethernet until the low priority packet has been serialised completely. In fact, packets will be typically moved to the network device buffer however, which leads to an even higher delay deviation than expected. As the device buffer is of 3 KByte size, up to two low priority packets fit into the buffer. Therefore, a significant number of packets of the test traffic will be delayed by more than 270 μs, which is twice the serialisation delay of a packet of 1500 Bytes length and includes the processing delay of the test packet.

To reduce the jitter effects caused by the device buffer, the transfer rate between system memory and network device buffer must be throttled actually. This could be done in theory using a Token-Bucket-Regulator (TBR) [5] as it is implemented in AltQ. Experiments, that were carried out using this mechanism have not lead to the desired behaviour so far, which is supposed to be caused by the deficit counter used by the TBR implementation. Further research is required here to optimise the implementation.

4.2   Weighted-Round-Robin Scheduling

It is interesting furthermore, how accurate CBQ schedules flows that share bandwidth at the same priority level. A Weighted-Round-Robin (WRR) scheduler is used to schedule

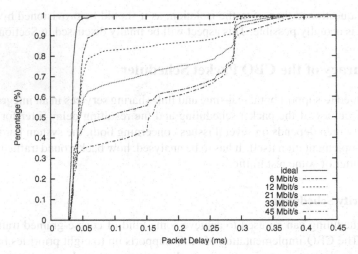

Fig. 4. Distribution of the packet delay of test traffic with low priority background traffic (1500 Bytes length packets).

those classes and to take care of each class receiving its reserved share of bandwidth. A weight (byte allocation) is calculated for each class according to its reserved bandwidth share, which is the number of Bytes a class may send in a WRR round. The jitter, that may occur to the packets of a certain traffic class, not only depends on the number of classes on the same priority but also on their bandwidth allocations. In the worst case a packet arrives at a certain class shortly after the WRR scheduler could have served this class. The packet will have to wait then, until each class with the same priority will have sent at least their byte allocation and a packet of maximum length as the AltQ implementation of CBQ uses a deficit counter.

A first experiment was carried out to measure the influence of the bandwidth allocation of a single class to the delay jitter of the packets of the test traffic. The configuration of the priority experiment was slightly modified so that both, the test traffic and the background traffic are served with the same priority. The background traffic consists of packets of maximum length. Its ingress rate is gradually increased during the experiment. The bandwidth allocation of the corresponding class was adapted accordingly, i.e. the byte allocation was increased as well. As long as its byte allocation is smaller than 1500 Bytes, the worst case jitter of the test packets should correspond to the serialisation delay of a packet of maximum length. If the byte allocation becomes larger than 1500 Bytes but not larger than 3000 Bytes, up to two maximum length packets can be sent in a round because of the deficit counter implementation. Including the processing delay of the test packet, this should result in an overall delay of about 150 μs and 270 μs respectively.

The results given in Figure 5 do not confirm this assumptions however. It shows the packet delay distribution of the test traffic in dependence on the ingress rate and the byte allocation of the background traffic respectively. The ideal curve again represents the packet delay distribution of the test traffic under unloaded conditions. With byte allocations smaller than 1500 Bytes, a significant number of packets will be delayed by more then 270 μs already, which indicates a serialisation of two packets of maximum

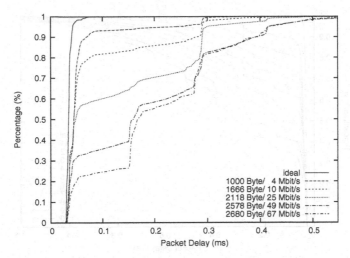

Fig. 5. Distribution of the packet delay of test traffic with background traffic (1500 Bytes length packets) at the same priority.

length. With larger allocations, up to three packets of maximum length can be sent, before the test packet will be served.

Again, the network device buffer may be the reason for this behaviour. In case the byte allocation is smaller than 1500 Bytes, it can be assumed, that the scheduler starts serving another maximum length packet, while already having one in the device buffer. With an allocation larger than 1500 Bytes, the device buffer already contains two packets of maximum length. While the scheduler cannot serve another background packet immediately, the background class is still selected to be served next by the scheduler. Up to three maximum length packets have to be serialised then, before the test packet can be served by the scheduler.

The number of active background flows and the number of the corresponding CBQ classes was modified in a second experiment with the ingress rate of each flow left constant. The byte allocation of each background class is smaller than 1500 Bytes. In contrast to the previous experiment, each background class may send a maximum length packet within a round, because of the deficit counter implementation of the scheduler. The worst-case jitter should simply be the serialisation delay of a packet of maximum length times the number of background classes.

The results given in Figure 6, that show the packet delay distribution of the test traffic in dependence on the number of background flows, do not completely confirm this assumptions. As before, the ideal curve represents the packet delay distribution of the test traffic under unloaded conditions. With ten background classes, up to eight typical steps can be observed only, each indicating the serialisation of a packet of maximum length. The suspension mechanism of CBQ may be the reason for this behaviour, i.e. not all background classes are allowed to send a packet in each round. On the other hand, the network device buffer may be involved again, which is safe to say only for the measurement with a single load class. A significant number of test packets will be delayed by more than 270 μs, which corresponds to the serialisation delay of two packets of maximum length.

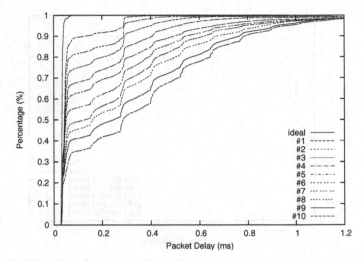

Fig. 6. Distribution of the packet delay of test traffic with a number of background flows (1500 Bytes length packets) at the same priority.

5   Ability to Isolate Traffic

A traffic management that strictly avoids interaction of certain traffic entities cannot completely achieved with CBQ for several reasons described in the previous section. While the jitter, resulting from the WRR scheduling especially, could become considerably large with only a small number of traffic flows active, it can be estimated rather precisely, as demonstrated. Nevertheless, it could become rather difficult to precisely adapt traffic parameters given the results in Section 3.

To stress this point, the priority measurement was slightly modified. VoIP traffic (G.726 codec) was generated as test traffic and traffic with packets of variable length as background traffic (see Table 1). Given the results from the priority experiments, only the jitter caused by the device buffer should be observed ideally. Including the basic processing delay, this should result in an overall delay for the VoIP packets of at most 270 μs roughly.

The results given in Figure 7 show, that up to an ingress rate of the background traffic of 10 Mbit/s, the overall delay of the VoIP packets is mainly within the expected range of 270 μs. With the background traffic becoming more intense a significant influence of the VoIP traffic can be observed however. At ingress rates of the background traffic of 20, 40, 60 and 80 Mbit/s the number of packets accurately delivered within 270 μs is shortened to about 96, 93, 88 and 70 % respectively. As it is unpredictable how the system behaves when packet processing becomes more complex, i.e. when classification and scheduling of a larger number of flows is required, it is impossible to precisely estimate the resulting jitter in advance.

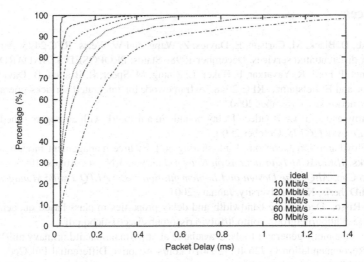

Fig. 7. Distribution of the packet delay of VoIP test traffic with background traffic (variable length packets) at the same priority.

6   Conclusion and Outlook

This paper gives a discussion of some selected aspects concerning the performance of an AltQ/CBQ-based software router using commodity hardware with the focus on the accuracy of the packet scheduling. The results are twofold.

First, it has been shown, that the performance of the test system is still not sufficient to be used in production and research environments. The system is unable to handle a half-duplex Fast Ethernet link when a typical distribution in the packet sizes is taken into consideration. This can lead to congestion in the policing/shaping router and result in unpredictable packet delay furthermore.

Jitter effects, caused by the AltQ implementation of the CBQ scheduling mechanism, have been identified as a second problem. While those effects can be estimated very well in advance, they can become considerably large with only a small number of traffic flows active. Especially the WRR scheduler can cause a large delay jitter, due to the deficit counter-based implementation.

Apart from the limited overall system performance it must be stated, that policing and shaping is not achievable with a desirable accuracy using the AltQ/CBQ combination. Both the implementation weaknesses and the per packet processing delay needs to be addressed first.

The limited packet forwarding performance for example, that has been identified as the main problem, can be increased on the one side by using a more powerful hardware, i.e. a better processor and a faster memory subsystem. As demonstrated in [15] a much larger performance improvement can probably be achieved by using a polling-based packet forwarding. The software router evaluated here uses FreeBSD 4.7 as software plattform, which is interrupt-driven. This issues will be addressed in ongoing research.

References

1. S. Blake, D. Black, M. Carlson, E. Davies, Z. Wang, and W. Weiss. RFC 2475: An architecture for differentiated services, December 1998. Status: PROPOSED STANDARD.
2. Y. Bernet, P. Ford, R. Yavatkar, F. Baker, L. Zhang, M. Speer, R. Braden, B. Davie, J. Wroclawski, and E. Felstaine. RFC 2998: A framework for integrated services operation over diffserv networks, November 2000.
3. A. Charny and J.-Y. Le Boudec. Delay bounds in a network with aggregate scheduling. In *Proceedings of QOFIS*, October 2000.
4. Sally Floyd and Van Jacobson. Link-sharing and resource management models for packet networks. *IEEE/ACM Transactions on Networking*, Vol. 3(No. 4), August 1995.
5. Kenjiro Cho. *AltQ: The Design and Implementation of the ALTQ Traffic Management System*. PhD thesis, Keio University, January 2001.
6. Fulvio Risso. Decoupling bandwidth and delay properties in class based queueing, 2000. available at http://netgroup.polito.it/fulvio.risso/pubs/iscc01-dcbq.pdf.
7. MGEN – the multi-generator toolset. available at http://manimac.itd.nrl.navy.mil/MGEN/.
8. ITU. Recommendation G.726,40,32,24,16 kbit/s Adaptive Differential PulsCode Modulation (ADPCM), December 1990.
9. Traffic generator (TG). available at http://www.postel.org/tg/tg.htm.
10. Fulvio Risso and Panos Gevros. Operational and performance issues of a CBQ router. *ACM Communication Review*, 29(5):47–58, October 1999.
11. Endace. Endace measurement systems. http://www.endace.com.
12. Jörg Micheel, Stephen Donnelly, and Ian Graham. Precision timestamping of network packets. In *Proceedings of the First ACM SIGCOMM Workshop on Internet Measurement*, pages 273–277. ACM Press, 2001.
13. Endace Measurement Systems. ENDACE DAG TIME-STAMPING WHITEPAPER. available at http://www.endace.com/Brochures/Endace_DAGTimestamping_Rev_A.pdf; 09.11.2004.
14. Intel. *82558 Fast Ethernet PCI Bus Controller with Integrated PHY*. Intel, January 1998.
15. Robert Morris, Eddie Kohler, John Jannotti, and M. Frans Kaashoek. The click modular router. In *Symposium on Operating Systems Principles*, pages 217–231, 1999.

Comparative Analysis
of SMP Click Scheduling Techniques*

Giorgio Calarco[1], Carla Raffaelli[1], Giovanni Schembra[2], and Giovanni Tusa[2]

[1] D.E.I.S., University of Bologna, Viale Risorgimento 2, 40136 Bologna, Italy
Phone: +39 51 2093776, Fax: +39 51 2093053
{gcalarco,craffaelli}@deis.unibo.it
[2] D.I.I.T., University of Catania, Viale A.Doria 6, 95125 Catania, Italy
Phone: +39 95 7382375, Fax: +39 95 338280
{schembra,gtusa}@diit.unict.it

Abstract. The interest of the scientific and commercial telecommunications community for the use of software routers running in general purpose (PC) hardware, as an alternative to the traditional special purpose hardware routers, is risen quickly in the last few years. This is due to the high level of flexibility and extensibility of this solution: the support for new protocols and network architectures and services, in fact, is easily obtained by re-programming the router itself. In addition, the diffusion of multiprocessor systems due to the progress in the semiconductor technologies allows software routers to obtain high performance if supported by multiprocessor PC hardware. Of course, in order to achieve a good use of the potentiality offered by multiprocessor architectures, the distribution of the tasks among the CPUs, and the parallel execution of the different operations, requires to be performed with some care. This paper demonstrates the benefits given by the hardware technological improvements, mainly concerning the use of multiple CPUs systems with respect to single processor ones, and shows how excellent forwarding performance can be achieved by Click software routers running on high powered PC hardware. Moreover, through the comparative analysis of different CPU scheduling approaches available in SMP Click, the paper discusses how different CPU scheduling techniques, that is, different approaches in the assignment of the tasks to the different CPUs, affect the router performance.

1 Introduction

In the last few years many modifications and improvements have been proposed for telecommunications networks by the scientific community. These proposals mainly concern new protocols and new services. However, only few of them have been really deployed in the Internet, and this deployment has happened very slowly over the time. The main reason is that most improvements need expensive hardware equipments to be replaced (e.g. network routers), and therefore service providers prefer to deploy these modifications slowly. As more and more changes and services are proposed, the need of solutions which provide network flexibility increases. To deal with the above problems, the idea to perform packet processing, routing and forwarding in software

* This work was funded by the EURO project of the COFIN programme of the Italian Ministry for Education, University and Research.

M. Ajmone Marsan et al. (Eds.): QoS-IP 2005, LNCS 3375, pp. 379–389, 2005.
© Springer-Verlag Berlin Heidelberg 2005

seems their solution. It provides the easy extensibility of network services, and the capacity to support new protocols can be ensured by simply re-programming the system, rather than substituting hardware equipment. Furthermore, it is easy to programming a software router with the ability of not only forwarding packets through the network, but also processing them. The traditional router paradigm "store-and-forward" is then extended to "store-process-and-forward" [2]. The main advantages coming from the use of software routers, which are also capturing the interest of the commercial telecommunications area in the last years, can be summarized as follows:

1) a greater flexibility, that is, the ability to adapt the network to the continuous Internet evolution;
2) a longer life-time for the equipments;
3) an easier and faster deployment of new services, which allows network providers and equipment vendors to quickly react to the user demand.

On the other hand, a critical drawback of the software processing can be a lower performance as compared to traditional hardware solutions, developed, customized and optimized for faster processing of specific protocols. Of course, an important challenge in software packet processing is to achieve performance comparable to the traditional hardware solutions. To this end there is a wide research area for the deployment of high-performance software routers. In this context, the Click modular router ([1], [3], [4]), a software toolkit for building routers and other packet processors, was developed at the MIT (Massachusetts Institute of Technology) with the aim of increasing the flexibility and the extensibility of network routers, according to what discussed so far. An important feature, which makes Click configurations easy to write, is the modularity, that is, the user can create modular routers design thanks to the concept of element, which is the component abstraction of the Click architecture. Anyone can extend the Click architecture in order to increase router functionalities, by implementing and adding new element classes, writing C++ code. In order to build a router configuration, the user has to connect a chosen set of elements appropriately. To this purpose, a configuration script has to be written in the Click language, and to create the right connection between elements, two different packet transfer mechanisms are supported by Click, the *push* and the *pull* one. In a *push* connection the packet transfer is initiated by the source element, which passes it to the downstream element. The *pull* connection, on the other hand, works in the dual way: in this case it is the downstream element that initiates the packet transfer, asking the upstream element to send it a packet.

Furthermore, in order to improve the performance provided by software-based routers, high-performance processing infrastructures, which use multiprocessor systems, can be used. Multiprocessor systems can achieve high processing power thanks to a high level of parallelism between a certain number of CPUs. Given the great availability of multiprocessor systems, SMP Click [5] was developed with the aim of improving the Click software routers performance. Obviously, to really take advantage of a multiprocessor system, making a good use of the parallelism, great attention to the CPU task assignment techniques is necessary.

This paper deals with performance analysis and comparison of Click software routers with different PC hardware configurations. For each of the considered case studies, performance of a single processor solution are compared with performance of the dual processor cases with the adoption of the different CPU scheduling approaches available in SMP Click: the adaptive and the static approach. This allows us to understand the key factors which mainly affect the performance, and therefore leads the users to the choice of the best solution. The paper is organized as follows. Section 2 describes the SMP Click basic principles and CPU scheduling management, while Section 3 refers to performance analysis. Finally, Section 4 concludes the paper.

2 Click Multiprocessor

In this section we illustrate the main Click multiprocessor principles. Specifically, in Section 2.1 we introduce its basic functionalities, while Section 2.2 refers to the different CPU scheduling techniques available in SMP Click. For an exhaustive description of SMP Click, see the related paper [5].

2.1 SMP Click Schedulable Elements

The main aim of SMP Click is to run Click configurations on multiprocessor PC architectures, through the parallel execution of packet processing tasks among different CPUs. SMP Click creates a different thread for each CPU available in the system. A CPU thread can schedule only a few elements, the so-called schedulable elements, of a Click configuration. All the schedulable elements in the configuration are placed in a task array as tasks. In addition, each CPU has its private worklist. Each task is then assigned to the worklist of one of the CPUs, in order to be processed by this CPU in a round robin order. Click schedulable elements are *PollDevice*, *ToDevice* and all the *PullToPush* elements. *PollDevice* and *ToDevice* elements represent the interfaces towards network devices. When a *PollDevice* is scheduled by a CPU, it examines the network device for new incoming packets, then pushes them to the downstream element: the whole *push path*, that is, the path starting from the *PollDevice*, and ending with a *Queue* element, is demanded to the CPU which schedules the *PollDevice*. In the same way, the path starting with a *Queue* element and ending with a *ToDevice*, called *pull path*, is demanded to the CPU which schedules the *ToDevice* element, and therefore starts the execution of a *pull* processing. The *PullToPush* elements are generally used in order to break a push path, and to initiate a new push processing.

2.2 CPU Scheduling Management Techniques

SMP Click supports static and adaptive CPU scheduling. When the static CPU scheduling is used, all the schedulable elements are statically distributed between the worklists; in this way each of them is always processed by the same CPU. If adaptive CPU scheduling is used, on the other hand, Click maintains the average processing cost for each schedulable element, in terms of consumed CPU cycles, and the worklist of each CPU is periodically updated according to these costs, in order to keep the load among the different CPUs balanced. The worklist updating interval is a parameter which can be set in the configuration file. More in deep, two different mechanisms can be used in the adaptive scheduling case, according to the algorithm chosen to sort the tasks in the task array before filling the worklists, that is, the increasing order approach or the decreasing order approach. When the updating timer expires, and therefore the load in the worklists has to be re-balanced, the task array is sorted from the element which has the lowest cost, to the element which has the highest cost, in the increasing case, and in the opposite order in the decreasing case. Then, a global scheduler runs through the task array again, by assigning each element to the CPU worklist with the lowest load, in order to achieve a good load balancing. In the following, we will refer to the two adaptive approaches as adaptive increase and adaptive decrease, respectively. According to what we have said so far, the adaptive scheduling algorithms take only into account the load balancing among the CPUs, in order to avoid CPUs idle times. No attention is paid to other important aspects, like cache misses and cache coherency maintenance mechanisms [6], [7], which negatively affect the performance of multiprocessor systems. When packets, buffers or other data structures are handled by more than a single CPU, multiple CPU private caches can have a copy of a given memory location. As soon as one CPU modifies shared data in its private cache, the copies in the caches of the other CPUs must be invalidated. The mechanisms which have to be used in order to ensure the consistence of the caches content (cache coherency protocols), introduce an overhead which degrades router performance. Instead, if a given data structure is always managed by the same CPU, this structure is present in its private cache only. This enforces cache affinity, by reducing the processing time, so increasing router performance.

By taking into account these considerations, a few simple rules can be followed for a good use of parallelism. For example, every mutable data structure, such as a queue, should be managed by as few CPUs as possible. In addition, any single packet, or packets belonging to the same flow, should be processed by a single CPU during its forwarding path. As a consequence of this last suggestion, if we consider that it is likely that a packet coming from a network interface leaves the router from another interface, the *PollDevice* and the *ToDevice* of the same interface should not be scheduled by the same CPU.

3   Performance Analysis

This section presents the results of our performance analysis. The investigation was conducted to strictly respect the RFC1242-2544 specifications proposed by the IETF Benchmarking Methodology Working Group (BMWG). For instance, throughput is specified as the maximum forwarding rate at which none of the offered frames are dropped, and results are obtained as the average of 20 successive 120-second runs. For each case study, the experimental setup will be first described, and then the results will be shown and discussed.

3.1   Case Study I: Hardware and Software Dependencies and Limitations

The testing configuration depicted in Fig. 1 consists of five PC-based systems running the Linux kernel 2.4.18 and Click release 1.3. One PC implements a RFC1812-compliant router (with a basic FIFO output queuing scheme), two are packet generators, and two are packet receivers. During every test, each of the two sources injects, by means of the *FastUDPSource* element, a constant-rate balanced flow of 64-byte UDP packets into the input ports of the router, while receivers collect the traffic coming from the router. The edge router is equipped with a dual Intel Xeon 2.80 GHz CPU (with the HyperThreading technology disabled), an Intel E7501 chipset motherboard with a 533 MHz FSB and four distinct PCI-X slots. The router has four Intel 82544EI Gigabit Ethernet controllers directly connected by copper point-to-point links to the packet generators and receivers. Two of them are plugged to the 64-bit/133 MHz PCI-X bus, while the two others are attached onto the slower 64-bit/100 MHz PCI-X slots. The choice of this NIC allows to use the polling-based driver extensions developed by the MIT for these adapters. Like most of recent network cards, this NIC disposes of an on-board FIFO buffer to store datagrams received from the wire or waiting to be transmitted. In addition, it contains dedicated registers maintaining statistics about its internal state. The MLFFR (Maximum Loss-Free Forwarding Rate) versus input rate for a Click-based router with SMP disabled (the 1-CPU case) is 737,000 64-byte packets per second (Fig. 2). More thorough analyses were done, increasing the input rate further with the aim to investigate system bottlenecks: the forwarding rate no longer increases, while the input interfaces report "FIFO Errors" events through their internal registers (Fig. 3).

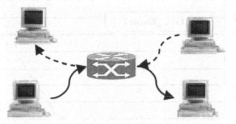

Fig. 1. The Case study I test-bed layout

This denotes that the NIC-to-memory DMA transfer of packets is failing, that is the system is not capable of releasing DMA descriptors to the NIC fast enough to handle all the incoming packets, causing the FIFO overflow. This evidence shows how the main performance problem is the CPU computing limitation. To overcome this limit, SMP Click allows the usage of multiple CPUs (with the adaptive and static scheduling techniques described so far), with automatic or user-defined parallel execution of packet processing tasks to maximize performance, exposing the advantage of multiprocessor architectures. In order to illustrate how the hardware/software choice affects the performance, in Table 1 we further considered the MLFFR values obtained with different configurations. C1 consists of a 1.6 GHz Pentium III platform with 32bit/33MHz bus, while C2-C3-C4 is the dual-Xeon platform described at the beginning of this section, with Click SMP support disabled, and enabled with adaptive increase scheduling and static scheduling, respectively. Comparison between C1 and C2 shows how the technological improvement of FSB

(CPU-memory) and PCI-X buses influences the forwarding rate: C2 CPU is less than twice as fast as C1 CPU, but the MLFFR is more than twice. C3 performs somewhat better than C2, but it seems that in these conditions (i.e. two distinct flows traversing the router) adaptive scheduling algorithm cannot take so much advantage by the presence of an additional CPU . Figure 4 shows how different is the router behavior for C3 in comparison with C2. In fact, despite maximum throughput reaches absolute higher values for C3 (around 1,000 Kpps), packet losses arise in both the NIC and the output queue: thus, the additional computing power is not efficiently utilized.

384   Giorgio Calarco et al.

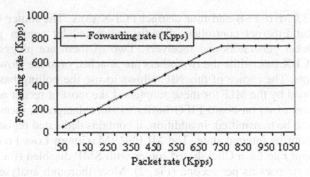

Fig. 2. Forwarding rate vs. input rate for a 1-CPU Click-based RFC-1812 router using 64-byte packets (20 120-second runs)

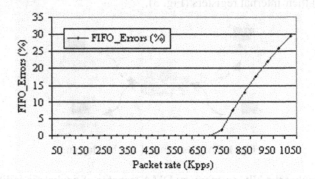

Fig. 3. FIFO errors vs. input rate for a 1-CPU Click-based router using 64-byte packets

Table 1. MLFFR for different router configurations using 64-byte packets

Platform	MLFFR (packet/s)
C1	350,000
C2	737,000
C3	750,000
C4	1,250,000

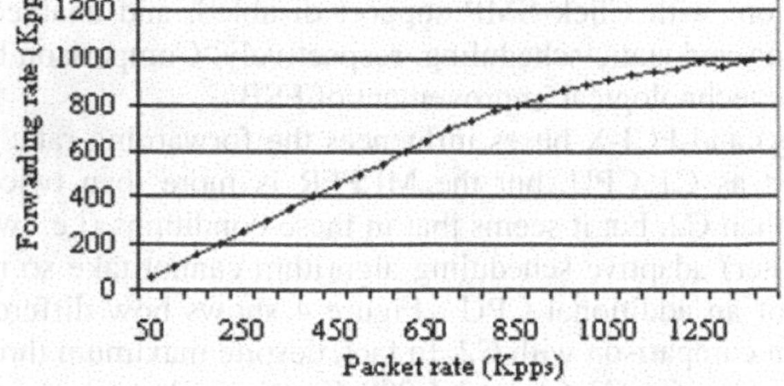

Fig. 4. Forwarding rate vs. input rate for a 2-CPU Click router with adaptive increase scheduling, using 64-byte packets (20 120-second runs)

Since the adaptive load balancing algorithm does not take into account the cost of cache misses, but simply tries to reach a balanced load among the available processors, if two schedulable elements handling the same packet are assigned onto different CPUs, the cost of processing increases, leading to unsatisfactory performance. On the contrary, the static scheduling technique offers the network manager the possibility to improve router behavior starting from the knowledge of traffic statistics. For instance, in our experimental layout, which fundamentally emulates a border router, most of the traffic is exchanged between the inner and outer networks (i.e. different network interfaces). In these circumstances, the *PollDevice* and *ToDevice* elements referring to the same interface rarely handle the same packets and thus can be allocated onto different CPUs, while elements mostly involved in packet forwarding are assigned onto the same CPUs, preventing the expensive cache misses. With these guidelines in mind, Fig. 5 depicts how traffic flows are assigned onto the available processors.

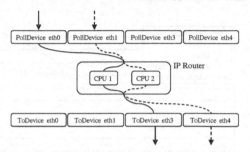

Fig. 5. Assignment of traffic flows to the CPUs, evidencing the parallel processing of forwarding activities

This makes evident the parallel processing of the incoming packets, while numerical performance results are reported in Table 1 (row C4), showing how this technique can take full advantage of the available power computing. A further point to mention is how differences between the two PCI-X buses affect the performance. Figure 6 highlights how the main throughput limitation (using the static scheduling configuration) is due to the card connected onto the 100MHz/PCI-X bus, where "FIFO errors" events appear first, persuading us that a hardware platform equipped with four identical 64-bit/133MHz PCI-X slots should exploit all the available computing power, and could probably reach a maximum loss-free forwarding rate of 1,350 Kpps.

3.2 Case Study II: CPU Scheduling Mechanisms Comparison

The second case study simply consists of two PC-based systems, running Click release 1.3, in a Linux 2.4.20 SMP kernel. The first PC acts as the traffic generator, while the other one performs basic IP routing functionalities, by implementing a RFC1812-compliant router. The hardware is the same for both the PCs, a SuperMicro X5DP8-G2 motherboard, with chipset Intel E7501, dual Intel Xeon CPU at 2.40 GHz (with the HyperThreading technology disabled), FSB at 533 MHz, and 512 MB DDR RAM. The network interfaces involved in the trials are the onboard Intel 82546EB dual port Gigabit network controllers, attached to the 64-bit/133MHz PCI-X bus, used with the polling-based driver extension. The traffic generator sends two flows of

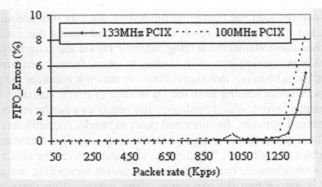

Fig. 6. FIFO errors events for the two distinct network interfaces vs. input rate for a SMP Click router using the "static scheduling" technique

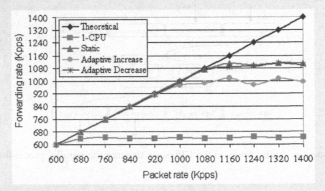

Fig. 7. Forwarding rate vs. input rate comparison for Click-based router, for 1-CPU and 2-CPU using different CPU scheduling techniques

packets, which are, once again, 64-byte UDP packets. A flow is sent from each of its two network interfaces, towards the other one. Thus, the router receives each flow of packets from a network interface, and forwards them to the other. The analysis refers to router performance comparison in terms of forwarding rate, that is, the number of packets per second that the router is able to forward for a given offered load, and the Maximum Loss-Free Forwarding Rate (MLFFR). The compared cases are the single CPU (i.e, SMP disabled) and the dual CPU ones, using all the different CPU scheduling techniques described in Section 2. From the results shown in Fig. 7 and Fig. 8 respectively, the performance gain, obtained when both the first and the second CPU are enabled, can be noticed. In the single CPU case and with the considered hardware, the MLFFR is 644 Kpps, and the forwarding rate keeps at a value of about 650 Kpps while the input rate increases, due to FIFO buffers overflow at the input of the router on-board NICs, as described in the previous case study.

All the dual CPU cases, on the other hand, achieve better forwarding performance than the single CPU case, but with differences due to the different CPU scheduling techniques adopted. The static and adaptive decrease techniques are comparable in terms of forwarding rate, but a little MLFFR gain, of about 80 Kpps, is obtained with the static technique, due to a better management of parallel processing power. The

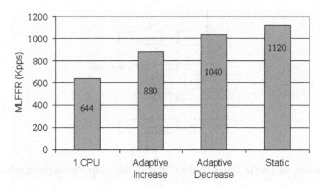

Fig. 8. Maximum Loss-Free Forwarding Rate comparison for Click-based router, for 1-CPU and 2-CPU using different CPU scheduling techniques

static approach outperforms more clearly the adaptive increase mechanism, with 1,120 Kpps and 880 Kpps MLFFR, respectively.

By analyzing the assignment of the schedulable elements to the CPUs worklists, these different behaviors can be explained. Fig. 9 shows the task assignment in the static scheduling experiments: *PollDevice* and *ToDevice* elements of different network devices are here scheduled on the same CPU. This choice ensures that each flow of packets, coming from a network interface and outgoing from the opposite one, is always processed by the same CPU, reducing cache misses by enforcing cache affinity, and therefore reducing the packet processing time.

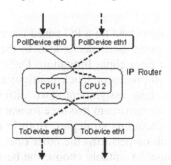

Fig. 9. Assignment of traffic flows to the CPUs in the static scheduling experiments

We have also kept trace of the way the schedulable elements were scheduled by the two CPUs in the adaptive scheduling experiments, by means of the *ThreadMonitor* Click element. This element prints the updated average cost value of each schedulable element, and the thread the same element is assigned to, with a sampling frequency chosen by the user. We want to stress once again that the adaptive scheduling approaches aim is to balance the CPU computing load, without any care for additional costs, like cache misses and coherency costs. Therefore, sometimes the *PollDevice* and the *ToDevice* processing the same flow of packets can be scheduled onto different CPUs. This is what happens in the adaptive increase experiments, where the worklists are filled as in Fig. 10, for roughly all the trials and for the whole duration of each trial, except for a negligible percentage of cases.

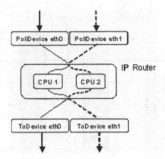

Fig. 10. Assignment of traffic flows to the CPUs in the adaptive increase scheduling experiments

In addition, we have noticed how the use of a different initial sorting algorithm, that is, the increase and the decrease one (see Section 2), affects the tasks-to-CPU assignment. As regards the adaptive decrease experiments, in fact, the same trace technique has shown that the task assignment is roughly as in the static ones. Nevertheless, the little performance degradation in the MLFFR with respect to the static scheduling approach is due to the fact that, in a certain percentage of the trials (remember that each data point is the average of 20 runs) we have carried out for each input rate value, the schedulable elements were scheduled as in Fig. 10, causing the average values to decrease.

4   Conclusions

In this paper measurements performed on a PC-based router have been presented and discussed. Investigation of system bottlenecks lead to the conclusion that the CPU represents the major cause of throughput limitation. Different scheduling techniques have been applied to the multiprocessor Click configuration showing the effect of dynamic task balancing and static assignment. Two different test bed layouts have been considered to evaluate the maximum loss-free forwarding rate in the presence of different flow patterns and scheduling strategies. Both in the first and in the second case study, the static approach outperforms the other ones, because of the possibility offered to the network manager to suitably choose the best task assignment to CPUs in relation to traffic flows characteristics, preventing possible expensive cache misses.

Acknowledgment

Authors wish to thank Prof. Eddie Kohler for his precious suggestions and opinions about data analysis.

References

1. The Click Modular Router Project: http://www.pdos.lcs.mit.edu/click/
2. Tilman Wolf: Design and Performance of Scalable High-Performance Programmable Routers, Dissertation Thesis, Washington University Sever Institute of Technology, Department of Computer Science, August 2002

3. Eddie Kohler: The Click modular router, PhD Thesis, MIT, 2000, available at http://www.pdos.lcs.mit.edu/papers/click:kohler-phd/thesis.pdf
4. E. Kohler, R. Morris, B. Chen, J. Jannotti, M.F. Kaashoek: The Click modular router. ACM Trans. Computer Systems 18, August 2000
5. B. Chen, R. Morris: Flexible Control of Parallelism in a Multiprocessor PC Router. Proceedings of the 2001 USENIX Annual Technical Conference (USENIX '01), June 2001, Boston, Massachusetts
6. J. Archibald, J.L. Baer: Cache Coherence Protocols: Evaluation Using a Multiprocessor Simulation Model. ACM Trans. Computer Systems Vol. 4, No. 4, November 1986
7. M. S. Papamarcos, J. H. Patel: A Low-Overhead Coherence Solution for Multiprocessors with Private Cache Memories. The 11th Intl. Symposium on Computer Architecture, pp. 348-354, June 1984

Implementation of Implicit QoS Control in a Modular Software Router Context[*]

Giorgio Calarco and Carla Raffaelli

D.E.I.S., University of Bologna, Viale Risorgimento 2, 40136 Bologna, Italy
Phone: +39 051 2093776, Fax: +39 051 2093053
{gcalarco,craffaelli}@deis.unibo.it

Abstract. This paper describes a flow-aware QoS model for implicit service differentiation in the Internet to maintain end user best effort interface while differentiating services within the core network. The edge router architecture and functionalities are described and algorithms are proposed for QoS provisioning without the need of explicit on line signaling. The feasibility of the approach is proved with reference to two different service classes (i.e. best effort and real time) and their implementation in a modular software environment on a PC-based test bed. The sensitivity of the system to the main design parameters and the effectiveness of the QoS algorithms in terms of bandwidth usage efficiency and congestion limitation are obtained by practical measurements.

1 Introduction

An intensive research activity has been developed in the last few years on models for service differentiation [1-3] and, more recently, on the related techniques for resource management [5], [6], [8]. Among the latter, the scheme based on the Bandwidth Broker concept has been considered as suitable to cope with the Differentiated Services model proposed for QoS support [4] [9]. Although this approach is fairly centralized it can be made scalable through hierarchical organization [1]of functions as proposed in [6]. This proposal splits the resource management problem into intra-domain and inter-domain functions with different administrative scopes: intra-domain resource management is within the competence of the bandwidth broker of the domain and is typically controlled by a single organization, while inter-domain resource management involves interactions between bandwidth brokers of different organizations and, on the basis of the proposal, it is achieved by bilateral agreements between adjacent domains. A critical discussion is ongoing on suitability of the proposed models to effectively support the convergence of different communication services such as voice, video and data [16]. New concepts for service differentiation are under consideration to better take into account the statistical nature of traffic as evidenced by practical measurements [7].

A still open issue is related to control signaling that impacts on user behavior and protocols to provide quality of service guarantees to legacy application programs. The strategy for graceful evolution towards fully operating QoS networks is to maintain

---

[*] This work was partially funded by the Italian Ministry of Education, Universities and Research through the EURO project 2002098329.

M. Ajmone Marsan et al. (Eds.): QoS-IP 2005, LNCS 3375, pp. 390–399, 2005.

the same user network interface to legacy application program while enhancing network intelligence to manage information flows with different requirements without the introduction of explicit signaling. This approach is referred as implicit service differentiation. The definition of the flow concept seems to easy this task [7] allowing the implementation of efficient algorithms to recognize information produced by different applications and to perform the needed QoS related functions within the network. The role of the bandwidth broker is here played within the core network for call and bandwidth management purposes. In any case users willing to obtain such enhanced service should declare this intention to the network and subscribe a contract in terms of Service Level Agreement (SLA) with the service provider [17].

The implementation of such a concept requires the addition of new functionalities in the edge router that interfaces peripheral QoS unaware network with the QoS enabled core network. This can be done by defining a software architecture to be implemented in a PC software router context.

In this paper an implicit QoS model is proposed, suitable to manage, as an example, real time services, but that can be applied also to a larger variety of service classes. It takes advantage of a call admission procedure to limit the bandwidth requirements of a service class coupled with dynamic bandwidth management performed by the edge router within each class to optimize the bandwidth usage. No explicit signaling is required to legacy application program while the concept of flow is introduced in the edge router to implement implicit QoS concept. In order to achieve QoS targets the implicit admission control is coupled with dynamic bandwidth management algorithms, previously evaluated as stand alone procedures [19]. The evaluation of the effects of the joint application of the admission control and bandwidth management procedures within the proposed QoS model is one of the main target of the paper. The proposed model could cope with the differentiated service model implemented in the core, being its role played at the internetworking between edge and core networks.

The paper is organized as follows. In section 2 the general model for implicit QoS management is described. In section 3 the dynamic bandwidth management algorithm is introduced. In section 4 system implementation and measurements are presented and discussed. In section 5 the conclusions of the work are drawn.

2 The System Model

The basic operations to access QoS enabled networking in the core network are performed by the edge router (ER) at the interface between the legacy networks and the QoS domain. The end user that expects to obtain service differentiation registers a SLA at the pertinent ingress edge router before the start of any communication sessions, with off line procedures that are out of the scope of this work. After then the edge router performs two main functions related to communication sessions:

– Implicit call admission control for real time flows; call admission control acts only for real time flows that belong to a SLA while best effort traffic is considered as an aggregate; implicit admission control was considered in [16];
– Dynamic bandwidth management within the service class with the aim of efficient link bandwidth usage and time constraints. The access link bandwidth is assumed

to be shared among different service classes. Bandwidth allocation management is performed in relation to the aggregate traffic of each class that accesses the QoS domain.

A flow is here identified by packets closely spaced in time [1]. A communication session is a sequence of flows alternating with silence periods. If a silence longer than a time out is detected, the previously related flow is considered finished.

The model can take advantage of the Bandwidth Broker concept as regard bandwidth updates required by the dynamic procedure [6]. Its action is related to core operations and in this case, signaling within the QoS domain could be implemented through protocols like COPS [11].

2.1 Implicit Call Admission

Real time flows are assumed here to adopt the RTP protocol. In any case the model can suit other approaches such as native UDP flow by applying the statistical analysis for real time flow recognition available in the classification process [13]. Under this hypothesis, a new RTP flow is recognized by the ingress edge router through a flow oriented classification process on the basis of a predefined SLA and of the SSRC field of the RTP header. The SLA contains, among its parameters, the specification of quality of service requirements through a DSCP value and the IP source address, thus allowing its correspondence with incoming RTP flows [13]. After a new flow is recognized, it is accepted if the following condition holds:

$$B_o < T_a$$

where B_o is the estimation of the bandwidth currently used by the service class and T_a is a design parameter called admission threshold.

An accepted flow is maintained active until one of the following conditions is verified:

- a time out expires indicating that the flow has been idle for a long time (some seconds) and it has possibly no more information to transmit;
- $B_o > T_d$ that indicates an aggregated bandwidth usage approaching the maximum allowed for the service class. T_d is called dropping threshold. When this condition is true a random choice is performed to drop one of the admitted flows. Other choices could be studied for system optimization.

The value B_o is typically the results of a measurement procedure [18].

2.2 Dynamic Bandwidth Management

A service class is initially equipped with the amount of bandwidth allocated by the network manager. ERs accessing network links are responsible of monitoring bandwidth usage by the aggregate traffic of a class and of asking the related BBs for the necessary increase/decrease of allocated bandwidth; an increase/decrease request is assumed here always followed by a positive answer within a time T_{bb}. The point is to decide when to generate the requests for the broker.

A well-known approach is based on a threshold-based system that behaves as explained in [6][11]. This system has a main drawback consisting in the large number of parameters to set up for system design. Two components characterize this system: the bandwidth increment/decrement mechanism and the calculation of the bandwidth update needed. The choice of the correct parameters' configuration of the system described above is crucial to achieve bandwidth utilization efficiency, scalability and system stability. In fact, under particular conditions of traffic patterns, bandwidth oscillations could arise as a consequence of approximate parameter configuration [12]. Moreover the parameters of the threshold-based model are strictly related to each other and strongly dependent on the traffic behavior thus making the parameters set up a very critical point [12].

3 Dynamic Bandwidth Update and Congestion Control

In order to overcome the main limitations of the threshold-based system our proposal adopts a logarithmic function in order to reduce the number of design parameters. It uses the output of a measurement-based process to know the value B_m of the bandwidth currently used by the aggregate traffic of the service class. The set up of the measurement-based process influences the behavior of the system, but its study is out of the scope of the paper. A suitably dimensioned time-window measurement system is here assumed as explained in [12].

The choice of the logarithmic function has been suggested by the need to carefully increment the bandwidth, when necessary, in order to follow the bandwidth variation as better as possible and, in case of bandwidth decrease, to promptly reduce bandwidth waste. The chosen function intrinsically meets these characteristics when used to calculate the bandwidth update ΔB as

$$\Delta B = K * \ln\left(\frac{B_m}{B_d * S}\right) , \tag{1}$$

being B_m the measured bandwidth, B_d the allocated bandwidth, and having indicated with $\ln(x)$ the natural logarithm of x. The parameter K is the constant of the feedback system and S is a margin to avoid sudden congestion. So the new bandwidth value after the update is given by

$$B'_d = B_d + \Delta B . \tag{2}$$

In order to avoid too frequent requests to the bandwidth broker, the update procedure is applied only if the following condition holds:

$$\Delta B > \Delta B_{min} . \tag{3}$$

So ΔB_{min} must be suitably chosen to assure the scalability of the approach. This system requires three parameters to be defined: K, S and ΔB_{min} whose influence on bandwidth efficiency, scalability and delay was evaluated in [19]. Due to delays in bandwidth estimate and update, congestion can temporarily arise, that causes packets to be queued waiting for transmission resources. The effect of the parameter S is to reduce the occurrence of these events, but a trade off must be reached between system performance and efficiency. So the mechanism can be fruitfully coupled with a congestion resolution mechanism based on the monitoring of the queue occupancy.

The main aim of the algorithm is to limit the time needed for a packet to pass through the system by setting the maximum acceptable time T_s to empty the queue.

So a queue threshold S_q in packets is considered given by $S_q=T_s*B_a/L_p$, being B_a the allocated bandwidth and L_p the packet length in bit. When queue occupancy $S(t)$ at a generic instant t overcomes S_q the system enter a congestion state during which bandwidth is updated to empty the queue within T_s. The new value B_a' is calculated as $B_a'=S(t) L_p/T_s$ that assures that the queue will be emptied within the time constraint. The system stays in the congestion time for at least T_i after then if queue occupancy is less than S_q the regular algorithm is applied.

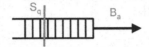

Fig. 1. The congestion control queue

The packet transfer delay is thus bounded by $T_s+2\ T_{bb}$. For example for a target maximum packet transfer delay of 30 ms in the router with $T_{bb}= 10$ ms, T_s must be limited to 10 ms. In all other cases the regular algorithm prevents congestion through S. The frequency of activation of the congestion management algorithm is related to the occurrence of the congestion state that can be reduced through the adoption of a larger guard bandwidth.

4 System Implementation and Measurements

In previous works [13],[15],[19], the Click modular router [14] was adopted for implementing flow-based classification of real time services. In fact it has been demonstrated that this environment is particularly flexible and extensible to enhance router behavior, making much easier to implement new functionalities compared to the well-known Traffic Control framework under Linux [10]. A new set of modules is added here to implement the described QoS model. Figure 2 outlines the main components of the Click configuration:

- *MeterShaper* element implements a time-window measurement algorithm [18] (metering the amount of traffic passing through the shaper) that calculates the value B_o and updates the allocated bandwidth using the logarithmic algorithm previously described. Moreover it implements the implicit admission control and the interaction with the bandwidth broker of the service domain;
- *SLA Manager* is dedicated to select a flow as belonging to a set of existing Service Level Agreements. A time out is set to detect the end of a flow.
- *RoundRobin Scheduler* performs multiplexing of distinct traffic flows pertaining to different SLAs;
- *Priority Scheduler* implements different policies for the real time (high-prioritized) and best effort (low-prioritized) packets.

Other related functions are performed by *ExportBwRatedSplitter,* that limits the bandwidth of flows to the value specified in the corresponding SLAs, and *BwAverageCounter* that collects information about average bandwidth usage that will be processed by the *MeterShaper* element. Measurements have been performed by soliciting the system through steps of bandwidth variation of an RTP flow as shown in Fig. 3. In any case, the effectiveness of this approach with other traffic types such as the superposition of Pareto sources [19] has been verified by simulation, leading to

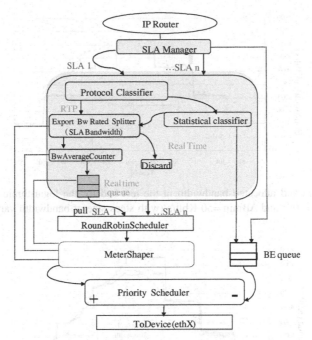

Fig. 2. QoS router internal structure

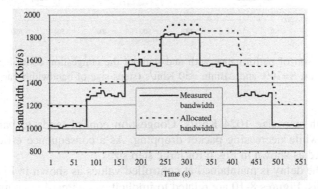

Fig. 3. Measured and allocated bandwidth in the test bed for the logarithmic algorithm with K=600Kbit/s, S=10% and ΔBmin =50 Kbit/s (with steps size of bandwidth variations of 256Kbit/s)

the same conditions. No best effort flow is contemporary injected, the response time of the bandwidth broker is here neglected and the result of its answer is assumed always positive. A good tracking of the measured bandwidth is obtained by the logarithmic bandwidth allocation when the step size is less or equal than 256 Kbit/s as shown in Fig. 3.

When the step size increases the system is no longer so prompt to follow sudden variations as reported in Fig. 4 for step size equal to 1024 Kbit/s. As a consequence the logarithmic algorithm typically tends to loose packets because the bandwidth is not sufficient. Figure 5 shows the improvement introduced by the congestion control

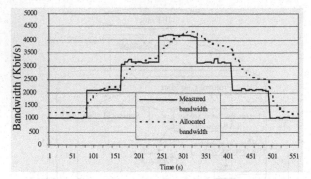

Fig. 4. Measured and allocated bandwidth in the test bed for the logarithmic algorithm with K=600Kbit/s, S=10% and ΔBmin =50 Kbit/s (with steps size of bandwidth variations of 1024 Kbit/s)

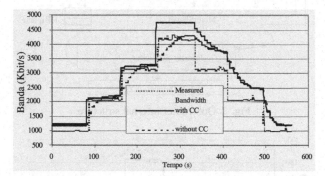

Fig. 5. Comparison of performance with and without the congestion control (CC) algorithm with K=600Kbit/s, S=10% and ΔBmin =50 Kbit/s (steps size of bandwidth variations of 1024 Kbit/s)

algorithm with step size 1024 Kbit/s. Congestion control tends to bandwidth over provisioning while decreasing packet dropping. As a consequence efficiency reduction is expected as shown in Fig. 6 for large step sizes.

However the delay is maintained at controlled values as shown in Fig. 7, obtained with T_s=100 µs. Figures 8-10 are related to implicit call admission control evaluation for 16 RTP flows. They have been obtained by injecting step traffic generated by 3 PCs with step size cyclically assuming the values 0, 8 16 packets/s with step duration uniformly distributed between 0 and 20 seconds and packet size of 256 bytes. Admission and dropping thresholds are given in terms of percentage of the bandwidth assigned to the real time class that is set to 300 Kbit/s. Figure 8 shows the timing behavior of admitted and cancelled calls when the admission and dropping threshold are set to 90% and 95 %. The logarithmic algorithm is applied for dynamic bandwidth management with K=600Kbit/s, S=10% and ΔBmin =50 Kbit/s.

Then, varying the admission threshold, the refused and dropped flows are plotted with dropping threshold set at 95 % (Fig. 9 and 10). It can be seen that the percentage of refused flows decreases as the admission threshold increases while the percentage of dropped flows increases with the admission threshold. This is due to the greater number of flows that are admitted and possibly cancelled after then. An optimum value of the admission threshold should be found to minimize the number of unsuccessful calls.

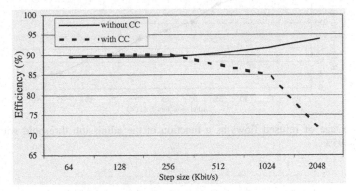

Fig. 6. Efficiency in the bandwidth usage with and without the congestion control (CC) varying the step size, with K=600Kbit/s, S=10% and ΔB_{min} =50 Kbit/s, T_s=100 μs

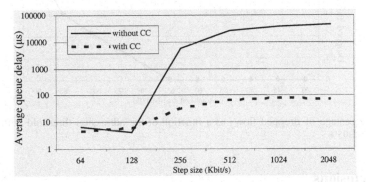

Fig. 7. Average packet latency (μs) with and without the congestion control (CC) algorithm and varying the step size, with K=600Kbit/s, S=10% and ΔB_{min} =50 Kbit/s, T_s=100 μs

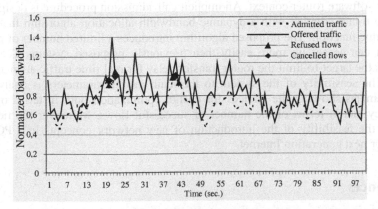

Fig. 8. Timing behaviour of bandwidth normalized to the available bandwidth Bd=300Kbit/s with admission threshold at 90% and dropping threshold at 95%

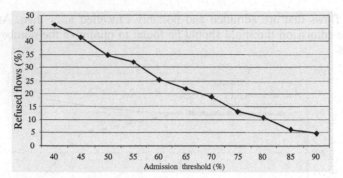

Fig. 9. Percentage of refused flows as a function of the admission threshold with dropping threshold T_d=95%

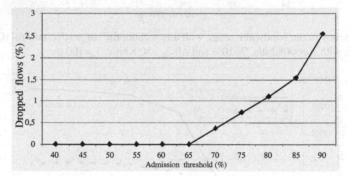

Fig. 10. Percentage of dropped flows as a function of the admission threshold with dropping threshold T_d=95%

5 Conclusions

In this paper a model for implicit QoS support is introduced and implemented in a modular software router context. An implicit call admission procedure is designed for the edge router coupled with a dynamic bandwidth allocation algorithm that uses a logarithmic function. The proposed algorithm introduces a limited number of parameters that can be set up more easily then previously proposed systems The model achieves the target to limit the router transfer delay for real time traffic at an assigned value. The sensitivity of the system to the main model parameters is analyzed by measurements showing the choices for design optimization both in terms of usage efficiency and in terms of the number of successful flows. Moreover the field trial showed the feasibility of the introduction of new network concepts in a PC-based router for next generation Internet.

References

1. W. Roberts: Internet Traffic, QoS and Pricing, Proceedings of IEEE, Vol. 92, No 9, September 2004.

2. R. Braden, D. Clark, S. Sheneker: Integrated Services in the Internet Architecture: an Overview, IETF RFC 1633, June 1994
3. S. Blake et. Al.: An Architecture for Differentiated Services, RFC2475, December 1998.
4. Nichols, V. Jacobson, L. Zhang: A two-bit Differentiated Services Architecture for the Internet, IETF RFC 2638, June 1999
5. E. W. Knightly, N.B. Shroff: Admission Control for Statistical QoS: Theory and practice, IEEE Network, Vol. 13, No. 2, March/April 1999
6. A.Terzis, L. Wang, J. Ogava, L. Zhang: A Two-tier Resource Management Model for the Internet, IEEE Globecom 1999
7. A. Kortebi, S. Oueslati, J. W. Roberts: Cross-protect: implicit service differentiation and admission control, Workshop on High Performance Switching and Routing, Phoenix, Arizona, 2004.
8. E. W. Fulp, D. S. Reeves: On line Dynamic Bandwidth Allocation, IEEE International Conference on Network Protocols, 1997.
9. C.P.W. Kulatunga, P. Malone, M.O.Foghlu: Adaptive Measurement Based QoS Management in DiffServ Networks, First International Workshop on Inter-domain Performance and Simulation (IPS 2003, February 20-21, Salzburg (A).
10. G. Calarco, R. Maccaferri, C. Raffaelli, G. Pau, "Design and Implementation of a Test Bed for QoS Trias", QoSIP 2003, Lecture Notes in Computer Science Vol. 2601, February 2003, Milan (I).
11. R. Mameli, S. Salsano: Use of COPS for Intserv operations over Diffserv: Architectural issues, Protocol Design and Test-bed Implementation, ICC 2001, Helsinky.
12. G. Calarco, C. Raffaelli: Algorithms for inter-domain dynamic bandwidth allocation, First International Workshop on Inter-domain Performance and Simulation (IPS 2003), February 20-21 2003, Salzburg (A).
13. G.Calarco, C.Raffaelli: An Open Modular Router with QoS Capabilities. HSNMC 2003, Lecture Notes in Computer Science, Vol. 2720 pp.146-155, July 2003, Estoril (P).
14. E.Kohler, R.Morris, B.Chen, J.Jannotti, M.F.Kaashoek: The Click modular router. ACM Trans. Computer Systems 18, August 2000.
15. G.Calarco, C.Raffaelli: Implementation of Dynamic Bandwidth Allocation within Open Modular Router, ICN 2004, 3rd IEEE International Conference on Networking, March 1-4 2004, Guadalupe (F)
16. R. Mortier, I. Pratt, C. Clark, S. Crosby : Implicit Admission Control, IEEE Journal on Selected Areas in Communications, vol. 18, pp. 2629-2639, December 2000.
17. Dinesh C. Verma, "Service Level Agreements on IP Networks", Proceedings of IEEE, Vol. 92, No 9, September 2004.
18. S. Jamin, P.B. Danzig, S.J. Shenker, L. Zang: A Measurement-based Admission Control Algorithm for Integrated Services Packet Networks. IEEE/ACM Transactions on Networking, Vol. 5 No 1, February 1997.
19. G.Calarco, C.Raffaelli: Design and Implementation of a New Adaptive Algorithm for Dynamic Bandwidth Allocation, HSNMC 2004, Lecture Notes in Computer Science, Vol. 3079 pp.203-212, July 2004, Toulouse (F).

Should IP Networks Go Multiservices?

Gerard Hébuterne and Tijani Chahed

Institut National des Télécommunications,
Telecommunications Networks and Services Dept.,
9 rue C. Fourier – 91011 EVRY CEDEX – France
{gerard.hebuterne,tijani.chahed}@int-evry.fr

Abstract. Contrary to common trends, we show in this work that multiservices networks do not maximize profit, for a given set of resources available. Networks offering one service only do.

Keywords: multiservices networks, revenue

1 Introduction

Recent and current developments in the networking area argue in favor of designing QoS-capable, multiservices networks able at handling different types of media traffic. The multiservices aspect is clearly the most innovative feature as it comes in contrast with classical networking offers where one network would support one service.

The impacts of mixing several services in a common network are manyfold. They are mainly related with the provision of various levels of QoS, which is done through specific resource allocation mechanisms. Diffserv is the commonly adopted approach through which services are differentiated. This differentiation is based upon real-time mechanisms operating at the packet level.

The main motivation to mixing services is classically in "statistical multiplexing gain". Roughly speaking, the M/M/2 queue works better than 2 M/M/1 queues in parallel. When considering systems with priority levels, this scaling effect does not work the same way. We claim in this paper that such crude understanding of the multiservices concept is not the best option, especially in terms of revenue.

We operate on a crude model of the network, made of with a single multiplexing stage. The model is obviously simplistic, but is captures the essence of the argument. In a real network offering QoS guarantees (whether the network is connection oriented, or connectionless), some kind of acceptance control is performed. Prior to accepting any new flow, the network examines if the flow may be admitted, and possibly rejects it or admits it with a lower QoS level. This amounts to limit the load of each multiplexing stage in the network. In the model used here, the effect of the limitation is taken into account on a single stage: in the real network the acceptance decision would be the concatenation of all decisions of all the nodes involved.

There are many devices able at providing QoS differentiation which have been proposed and experimented. If one puts aside those of the mechanisms oriented "congestion control", which have a clear specific purpose, the basic principle in use is the priority in the access to the bandwidth resource. So we assume that the multiplexing stage differentiates services according to the Head of the Line (HoL) priority.

M. Ajmone Marsan et al. (Eds.): QoS-IP 2005, LNCS 3375, pp. 400–404, 2005.

Note that UMTS offers another specific configuration where to apply the model presented here. Voice and data flows are competing in user terminals of a given cell for accessing the available frequencies. If an optimization process, based upon the operator's revenue is run, the conclusions drawn here apply directly.

2 Model

Consider, without loss of generality, two flows 1 and 2 arriving at a given network element. Extension to more than two flows would be possible, using the same principle. Let the arrival processes follow Poisson distributions with rates λ_1 and λ_2, respectively, and let the service durations have means m_1 and m_2 and second moments S_1 and S_2, respectively. Let $\rho_i = \lambda_i m_i$, $i = 1, 2$, and let $X_i = S_i/2m_i$, $i = 1, 2$. Clearly, $\rho = \rho_1 + \rho_2$.

Let flow 1 have a higher priority than flow 2. Flow 1 may for instance model a flow with real-time constraints, such as voice, or circuit emulation. Flow 2 is of lesser priority and may thus model non-real-time data traffic. By reference to an M/G/1 queue with Head of Line priority discipline [1], the waiting times W_i per flow are given by

$$W_1 = \frac{W_0}{1 - \rho_1},\qquad(1)$$

and

$$W_2 = \frac{W_0}{(1 - \rho_1)(1 - \rho)}\qquad(2)$$

where

$$W_0 = \lambda_1 \frac{S_1^2}{2} + \lambda_2 \frac{S_2^2}{2}\qquad(3)$$
$$= \rho_1 X_1 + \rho_2 X_2.$$

Note that the X_i are constants mainly related to the protocols run by the service: they do not depend on any traffic or other network state parameter ($X_i = m_i$ for the case of exponentially distributed services).

Let D_1 denote the upper bound on the waiting time of flow 1. Eqn. (1) enables us to limit the total flow ρ in terms of the higher priority flow 1 as follows.

$$W_1 \le D_1,\qquad(4)$$

or

$$\frac{\rho_1 X_1 + (\rho - \rho_1)X_2}{1 - \rho_1} \le D_1.\qquad(5)$$

Thus, the total flow ρ is limited by

$$\rho \le \frac{D_1}{X_2} - \rho_1 \frac{D_1 + X_1 - X_2}{X_2}.\qquad(6)$$

To this condition, one must add the classical limitation, which ensure the system is not congested, and guarantees that the average waiting time of flow 2 is finite:

$$\rho = \rho_1 + \rho_2 < 1.\qquad(7)$$

One could impose a limit on the delay of flow 2, which would translate in an additional constraint. For the sake of simplicity one assumes no limitation (or rather the limit is several order of magnitude larger and without actual influence).

3   Analysis

Now, let us turn to revenues. Let G_1 and G_2 denote the revenues from transporting one unit of flows 1 and 2, respectively. The revenue is assumed to be proportional to the volume of data processed by the network. Packets arrive with rates λ_i and carry a volume $m_i C$, C denotes the network bitrate, assumed to be a constant of the problem, so that the volume processed by unit of time is $\lambda_i m_i C = \rho_i C$. In short, the volume carried by time unit is proportional to the loads. The total revenue G is thus proportional with

$$G = \rho_1 G_1 + \rho_2 G_2 \tag{8}$$

Finally, the problem we are dealing with can be formulated according to a basic optimization one:

$$\text{Maximise } G = \rho_1 G_1 + \rho_2 G_2$$
$$\text{under the constraints:} \quad \begin{aligned} \rho = \rho_1 + \rho_2 < 1 \\ \rho_1(X_1 + D_1) + \rho_2 X_2 \le D_1 \end{aligned}$$

Figure 1 below gives a schematic view of this 2-dimension Simplex problem. The full lines correspond to the constraints, while the dotted one is the quantity G to be maximized.

The graphical approach is straightforward: the dotted line is slipped until it reaches the boundaries. Depending on the parameters, the maximum is on A, B, C, or on any point between A and B or between B and C for specific values of the slopes. Note the

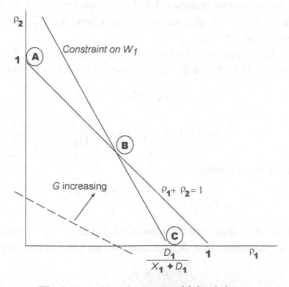

Fig. 1. Variation of revenue with load shares.

line representing the constraint on D_1 cut the vertical axis for $\rho_2 = D_1/X_2$, which is much likely larger than 1 (the X_i are of the same order of magnitude than the service times, and having $D_1 \approx m_1$ or m_2 is unrealistic).

Going on with the geometric formulation of the problem, the analysis depends on the relative slopes of these lines. The objective function has a slope $-G_1/G_2$, the constraint on D_1 a slope $-(X_1 + D_1)/X_2$, and the constraint on total load a slope -1.

- Under the condition

$$-\frac{G_1}{G_2} > -1, \qquad \text{i.e.} \quad G_2 > G_1 \qquad (9)$$

the maximum is reached at point A, i.e. the revenue is maximum for $\rho_1 = 0$. This configuration is unlikely with certainty, as guaranteed services should be charged more than the others.
- If $G_1 = G_2$, the objective function is parallel to the constraint on ρ: the maximum is obtained for all points between A and B. Here too, this configuration is certainly unrealistic.
- Under the condition

$$-1 > -\frac{G_1}{G_2} > -\frac{X_1 + D_1}{X_2}, \qquad \text{i.e.} \quad G_2 < G_1 < G_2\frac{X_1 + D_1}{X_2}, \qquad (10)$$

then the maximum is obtained in B.
- At last, when

$$G_1 > G_2\frac{X_1 + D_1}{X_2}, \qquad (11)$$

the optimum is for $\rho_2 = 0$ and $\rho_1 = D_1/(X_1 + D_1)$.

4 Implications

If we retain only the configurations which seem of realistic use, the analysis above stresses the major role of the ratio $(X_1 + D_1)/X_2$.

As long as $G_1/G_2 > (X_1+D_1)/X_2$, the optimum is in offering only the guaranteed service. Otherwise, the best choice (as long as $G_1 > G_2$) is in the hybrid point B. As soon as $G_1 = G_2$, the most efficient is to restrict the offer to best effort services, conforming to the intuition.

The major implication of this analysis on the network operator is the following. Consider a typical ISP offering a best-effort Internet service. Keeping the same level of resources (capacity installed), it might continue offering this one service, it might want to upgrade its system following the present technological advances and offer a QoS guaranteed service or else it might want to offer a bi-level service : best effort plus guaranteed. The driving force behind those three choices is clearly to maximize the revenues.

In any case, the mean service times X_1 and X_2 are rather fixed (depending on hardware/software capabilities). QoS, in terms of the delay bound D_1 for instance, is required and thus set by the user. The 'only' real choice for the operator is thus based on the costs and revenues, that is what service to offer : best-effort, guaranteed or both.

The operator calculates in this case the value of the ratio$(X_1 + D_1)/X_2$. According to its value and given the current level of the prices (e.g. the offers of competitors), his interest is to offer solely the guaranteed service, preferably in a connection-oriented manner, to transport real-time traffic only (case of equation 3), or to restrict to a mix of guaranteed and a best-effort service, in a packet mode, for the transport of non-real time data traffic. In most cases, a multiservices offer docs not maximize profit.

5   Numerical Application

For the sake of simplicity, let all packets be of the same length, 10 kbits on average. Let the delay constraint be fixed to 10 ms (this is the variable, queueing-related part of the delay in each node).

The following table shows the conditions, in terms of the relative costs G_1 and G_2, under which a QoS guaranteed service is worth offering, rather than a best-effort one.

Throughput	10 Mbps	100 Mbps	10 Gbps
QoS if	$G_1 > 11 G_2$	$G_1 > 10^2 G_2$	$G_1 > 10^4 G_2$

6   Conclusion

Networks that offer one service only maximize the revenues. Needless to say, they are also easier to design, deploy, run and maintain. Now, the results displayed here have to be taken with care. There are implicit assumptions which are worth considering. First, the model assumes a demand of "infinite size" on both kinds of services (QoS guaranteed and best-effort). Second, the optimization problem considers the demands and traffic properties on the one side, and the price structure on the other side. In reality, there should exist a strong relation between these parameters, as the level of demand is clearly related with the offering and with the tariffs in use.

Also, there are obviously other motivations to multiservices offers – and first of all a multiservices approach offers a much greater flexibility, both for operators and users. However, the kind of results obtained here could advocate for reconsidering the traditional view, according to which resource mutualization and statistical multiplexing leads to increasing the efficiency.

Acknowledgments

The authors are grateful to Yves Dallery, Salah Aguir (Ecole Centrale de Paris), Fabrice Chauvet (Bouygues Telecom) and Roland Malhamé (Polytechnique Montréal), for fruitful discussions on the approach used here, which lead to improvements in the paper.

References

1. L. Kleinrock, Queueing Theory, Volume I : Theory, John Wiley and Sons, Inc., 1975.

Architecture and Protocols for the Seamless and Integrated Next Generation IP Networks

Gino Carrozzo[1], Nicola Ciulli[1], Stefano Giordano[2],
Giodi Giorgi[1], Marco Listanti[3], Ugo Monaco[3],
Fabio Mustacchio[2], Gregorio Procissi[2], and Fabio Ricciato[3]

[1] Divisione Informatica e Telecomunicazioni of Consorzio Pisa Ricerche, Italy
{g.carrozzo,n.ciulli,g.giorgi}@cpr.it
[2] Dipartimento di Ingegneria dell'Informazione of University of Pisa, Italy
{s.giordano,fabio.mustacchio,g.procissi}@iet.unipi.it
[3] INFO-COM Dept. of University "La Sapienza" of Rome, Italy
{listanti,monaco,ricciato}@infocom.uniroma1.it

Abstract. The paper presents a novel end-to-end seamless framework to support end-to-end Quality of Service and Traffic Engineering. The network model is based on the MPLS/DiffServ paradigm and addresses the definition of a network architecture according to both users and network providers requirements. A first solution relies on the centralized MPLS/DiffServ based Multi-protocol Access Inter-Domain (MAID) architecture. This architecture allows a seamless QoS-IP service setup through proper Users-Network Interfaces and inter-domain communication through Network-to-Network Interfaces. A fully distributed solution is also presented to address critical scalability issues and to improve network resilience. The overall architecture has been validated by means of functional tests carried out on operational testbeds based on Linux PC platforms.

1 Introduction

The Quality of Service for IP packet flows (QoS-IP) has a long history of standards and tools, both at the Data Plane level (e.g. traffic conditioning) and at the Control Plane level (e.g. signaling and policy protocols). Quality of Service requirements strongly depend on the side in which the interaction users–network is observed. From the user perspective, the basic QoS requirements are the dynamism (e.g. the service should last as long as the user needs), the tailoring (e.g. the network resources allocated for the service should fulfill exactly the end-user requirements), as well as a seamless integration (e.g. the mechanisms involved in QoS support should be transparent to end-user applications). Though some tools for QoS are available in commercial IP routers, their compliancy to these requirements is still far from being a market reality. Indeed, the main obstacles for such a deployment reside in the different backbone networks technologies (e.g. DiffServ, MPLS, IPoATM, etc.), which make hard to guarantee end-to-end QoS, above all when the service has to be deployed across different administrative

M. Ajmone Marsan et al. (Eds.): QoS-IP 2005, LNCS 3375, pp. 419–432, 2005.

domains and in the number of protocols used in the access networks (e.g. RSVP, H.323, SIP, MPEG-4,..), which implies a per-service/per-protocol User-Network-Interface (UNI). From the Service Provider perspective, other requirements drive the evolution of the services offered by the backbone, such as:

- network scalability, which implies a distributed Control Plane, best fitted if based on MPLS;
- traffic engineering both at the flow and at the resource level, best fitted if based on a DiffServ data plane;
- service survivability in case of faults or dynamic network topology changes, easily guaranteed by MPLS recovery strategies;
- interoperation of adjacent domains with the same or different technologies, which implies a Network-to-Network-Interface (NNI);
- interoperation of equipments from different vendors.

The overall objective of the research activity within the TANGO project [1] is to define a novel network architecture able to meet the above summarized users/Service Providers requirements. The crucial point of this architecture will be the definition of proper interfaces between users and network (UNIs) and between network and network (NNIs), respectively.

The paper is organized as follows. Section II presents the paradigm of nested-networks and addresses the concepts of UNIs and NNIs that will be elaborated upon in the next sections. Section III is devoted to the seamless QoS-IP service setup, and includes the definition of the *Multi-protocol Access Inter Domain* (MAID) architecture and of the corresponding UNI (single MAID domain) and NNI (multiple MAID domains). Section IV discuss the key aspects of a MPLS/Diffserv backbone architecture in this scenario. An alternative solution to address critical scalability issues and network resilience is presented. Section V presents the results of functional tests carried out on the experimental test-bed developed in the framework of TANGO project. Finally, Section VI concludes the paper with final remarks.

2   The Nested-Networks Paradigm

The currently operational networks feature a mature IP Data Plane, in which QoS-IP network services are statically configured (and, consequently, under- or over-provisioned) through the Management Plane. These networks are provided with a flat Network Interface (NI) hierarchy, thus in most cases the different NI functions, such as policing and traffic conditioning, are summed up in a single point (the accessing router) even if the network service traverses multiple operators/providers; an example of this single-point Service Level Agreement (SLA) is the Acceptable User Policy (AUP) agreement at the NRENs User-Network Interface (UNI). For these networks, the IETF DiffServ architecture [2] has been largely recognized as a main technology component for QoS-IP networks, due to its native scalability and flexibility.

The DiffServ specifies only mechanisms for packet forwarding (Per Hop Behavior - PHB), flow aggregation rules and traffic conditioning, with a strict Data

Plane scope [3–5]. Internet Service Providers (ISP) configure with internal policies the desired intra-domain services from PHBs, in order to fulfill the Service Level Agreements (SLA) drawn up with their accessing users.

However, although the DiffServ architecture solves, at the Data Plane level, the scalability problems related to QoS provisioning in a single domain, it is not a complete end-to-end solution for the enforcement of a globally dynamic and multi-level chain of SLAs. Different combined solutions (e.g. IntServ/DiffServ, MPLS/DiffServ) have been proposed to overcome the lack of Control Plane procedures and several research projects have been carried out to address this problem (e.g. IST AQUILA [6], IST TEQUILA [7], IST MESCAL [8], IST MOICANE [9], etc.). In general, in those projects where the focus was on intra-domain Control Plane, the inter-domain was out of scope, and vice versa; and this is a major lack for the assessment of an end-to-end seamless framework.

In this scenario, manufacturers can provide network operators with sets of tools (e.g. Bandwidth Broker-like) for managing the QoS parameters of their own network elements. These tools rely on the a common management paradigm, based on standard or, more frequently, on proprietary Management Information Bases (MIBs). However, these tools cannot prove to be effective in multi-region signaling-integrated scenarios, due to the lack of generalized interfaces at the different boundaries of the network (e.g. UNI, NNI), capable of integrating heterogeneous protocols from the access network (e.g. MPEG-4 DMIF, RTP, RSVP or SIP) towards the intra-domain (e.g. RSVP-TE, COPS, SIBS) and, in case, towards the inter-domain. In the following section an architecture aimed at solving the critical seamless interoperabilty issue is presented, with focus on control plane mechanisms and interfaces among different network segments.

3   The Seamless QoS-IP Service Setup

3.1   The MAID Architecture

In current operative IP networks there is a lack of seamless procedures for QoS-IP service setup. The Multi-protocol Access Inter-Domain (MAID) architecture is aimed at providing Network Operators with the robust and user-friendly mechanisms to support QoS in MPLS/DiffServ networks, hiding the underlying complexity of managing all the involved parameters. Concerning the Data Plane this architecture provides the mapping and the forwarding of the IP flows from the access network into the proper DiffServ Label Switched Paths (LSP). On the other hand, the MAID Control Plane is responsible for Admission Control (AC) and policy decisions (taken on a per-flow or per-PHB basis) and for LSPs management. The key elements of the MAID network are the accessing border router (MA-BR), which triggers the setup of QoS-IP services upon receiving QoS requests from the access networks, and the Bandwidth Broker (BB), which manages network resources and policies, as well as inter-domain communications (Fig. 1). The main functionality of the MA-BR is to provide the inter-working between the access network and the backbone. Therefore, the MA-BR manages

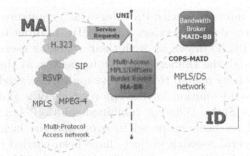

Fig. 1. Multiple Access Inter Domain (MAID) network model.

all the service requests from the access network, defined in terms of protocol-specific QoS semantics, and it conveys them in a generalized and unified service request (i.e. the UNI request) with a unique QoS syntax. If some resources have been provisioned by BB for MA-BR, the processing of the UNI requests for QoS-IP network services (i.e. the AC and policy decisions) can be handled locally to the MA-BR; otherwise, these requests are directly processed by the BB, according to the outsourcing operation model. In any case, BB has pre-emption rights on each MA-BR decision, since it provides a centralized AC and policy that is supposed to be optimal with respect to the local AC provided by MA-BR.

The signaling protocol for the communication between an access network and the MA-BR is application-dependent (e.g. RSVP for IntServ networks, H.323 or SIP for VoIP, DMIF signaling protocol for MPEG-4 services, etc.). Instead, the communications between the MA-BR and the BB are based on a extended version of the Common Open Policy Service protocol (COPS), detailed in the next section. It is possible that BB configures directly core and border routers according to its criteria (e.g. via SNMP protocol). In order to make the entire system scalable, it is desirable to tune an optimum mix of static and dynamic resource allocation (e.g. via MPLS signaling protocols, ref. Fig. 2) to share architectural complexity between BB and other NEs.

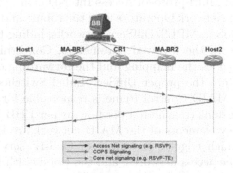

Fig. 2. QoS-IP service setup in a single MAID domain.

3.2   The COPS-MAID Extensions

Relying on the client-server model, the COPS architecture [10] is based on two fundamental elements: a policy server, called Policy Decision Point (PDP), also addressed as COPS server, and one or more policy clients, called Policy Enforcement Points (PEP), addressed as COPS clients. At least one policy server must exist in each administrative domain, in order to implement a complete COPS communication with one or more PEPs. A single PEP is able to support multiple client-types, while, if a client-type is not supported by the PDP, the PDP itself can redirect the PEP to an alternative PDP via COPS. Different applications using different protocols may be viewed as different client types. The trend to define a new client type for each access network protocol results in a hard limit to the system scalability, because of the duplication of the states installed both in the PDP and in the PEP. A possible solution for this issue might be the cluster of PDPs, each supporting one or few client-types; but, all of these COPS servers have either to exchange management information to perform a coherent resource allocation and should refer to a higher level "omniscient" BB.

The novel and original solution we propose through the MAID architecture is to define a unified and extended COPS semantic, which integrate all the QoS information carried out by the different access protocols. This semantic is based on the contents of the UNI service request and it is characterized by a new unique COPS Client Type (i.e. the COPS-MAID one). The proposed extension to the standard COPS specification can be found in [11]. This solution transfers the system complexity on the border routers, in which appropriate Inter Working Units (IWUs) are used to map protocol specific messages into generalized client messages. Moreover, a unique COPS client-type can transmit all the information to a unique PDP, which can be located inside the BB itself.

3.3   The Inter-domain Problem

The research community is dealing with a number of open issues regarding inter-domain communications (e.g. the optimal TE routing, the NNI signaling, etc.). In this context, the MAID architecture arises as an effective and open solution, because of the centralized action of the BB and of the modularity of the MA-BR. Two possible strategies for the inter-domain connection setup are possible and are sketched here to prove the architectural flexibility: Inter-BB communication via COPS-MAID interface (ref. Fig. 3(a)), which has network granularity; Inter BR communications via strict NNI (ref. Fig. 3(b)), which has node granularity. The solution for the inter-domain communication among MAID domains is an inter-BB NNI, based on the COPS-MAID. As shown in Fig. 3(a), the request is processed by MA-BR similarly to the mono-domain case and, thus, propagated to the domain BB. If the destination is out of the BB scope, the COPS-MAID request is propagated to the adjacent BBs, waiting for a response. If a route exist towards the desired destination, it is announced by the downstream BB with a positive COPS-MAID response. Upon receiving this response, the BB configures the internal route from the/an ingress point towards the announced

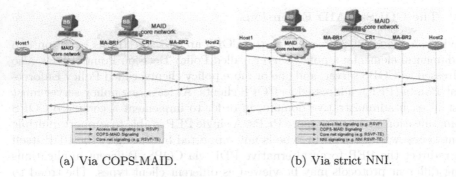

(a) Via COPS-MAID.   (b) Via strict NNI.

Fig. 3. Inter-domain QoS-IP service setup.

egress interface. Thus the inter-domain QoS-IP service setup is provided from the downstream towards the upstream domains. The inter-BB NNI solution proves to be more scalable than the *classical* BGP-based solution, since the space of the solution (e.g. the number of queries for a route) is limited to the adjacent BBs and not to all the possible BRs towards a different domain. A possible solution of the inter-domain communication among heterogeneous domains (such as the alternative architecture proposed in Section IV) relies on Inter BR communications. Such a solution guarantees the interoperability among architectures with centralized and distributed control planes and is still in the process of definition. A proposed approach to address this issue can be found in [12] [13].

4   The MPLS/DiffServ Backbone

In this section the key aspects of the backbone network architecture developed in the the TANGO project framework are discussed. The MPLS/Diffserv backbone here considered is able to support both pre-provisioned and on-demand LSPs establishment and to provide end-to-end LSP protection by single and double fault. Three different backbone operation models has been experimented, namely: i) centralized; ii) partially distributed; iii) fully distributed.

According to the centralized model, the control logic, needed to execute the Admission Control (AC) algorithm, the LSP route selection algorithm and the path protection mechanisms resides in a centralized device, i.e. the Bandwith Broker (BB). In this scenario the centralized device directly interacts with the backbone routers for LSP setup/tear-down and for network link status information retrieving. These interactions could be based on SNMP protocol.

In the partially distributed scheme, the control logic is still centralized, but the signaling processes for the setup and tear down of the LSPs are handled by the Edge Router (ER). In this solution, when a service request is presented at an ER, it informs the BB, via COPS protocol; the BB runs the AC algorithm and searches an available path to support the LSP request. The result of this phase are backward communicated to the ER that begins the LSP signaling phase, e.g. via RSVP-TE protocol.

In the fully distributed scheme, the control logic is completely distributed among the ERs. An ER receiving a request computes the route of the new LSP and locally performs the AC algorithm. The route selection is based on the information stored in a local Network State Database containing network topology and link available bandwidth; such information are disseminated by OSPF-TE flooding. After the route selection, the ER starts the LSP setup by triggering the RSVP-TE signalling. In this phase, AC must be performed by each node along the path because the link load information available at ER may not be synchronized with the current network state.

The following presentation is focused on the AC functions and, coherently with the scope of the paper, mainly refers to the centralized and partially distributed solutions, the specific mechanisms developed for the fully distributed solution for fault protection are given in a companion paper [14].

4.1 LSP Classes of Service

In the MPLS/Diffserv backbone defined within the TANGO project each LSP is associated to a single DiffServ PHB Scheduling Class (PSC) (corresponding to the class type defined in [15]) and to a specific protection class.

EF, AF1x and AF2x and standard Best Effort (BE) DiffServ PSC have been implemented in the test-bed, whereas, five protection classes characterized by different level of resilience and backup bandwidth sharing have been defined. Three levels of resilience have been defined: Unprotected (UP), Single-Fault Protected (SFP), Double-Fault Protected (DFP).

For UP demands only a service circuit is established, and no service continuity is guaranteed after the occurrence of a fault. For SFP demands a service circuit plus a backup one are allocated. When a failure occurs on the service circuit traffic can be readily switched on backup path. In case of DFP demands, a service circuit plus two backup paths are allocated: a primary backup and a secondary one. In this case when a failure occurs the traffic is switched on primary backup path. If a double fault occurs, it is then switched on the secondary one. Note that for DFP the preference order between the two backup paths is fixed a priori. Both SFP and DFP schemes can be implemented as Dedicated or Shared Protection [16]. In our model both Dedicated and Shared alternatives are supported for SFP and DFP, resulting in four different protection classes in addition to the basic unprotected one, as summarized in Table 1.

In the following we will index by i the DiffServ PSC, including BE. In particular, $i=1,2,3,4$ will refer to EF, AF1x, AF2x and BE respectively. For each

Table 1. Protection classes defined within TANGO project.

H	Acronym	Protection Class	Backup Bandwidth
0	UP	Unprotected	No backup bandwidth
1	Sh-SFP	Shared Single Fault Protection	Shared
2	De-SFP	Dedicated Single Fault Protection	Dedicated
3	Sh-DFP	Shared Double Fault Protection	Shared
4	De-DFP	Dedicated Double Fault Protection	Dedicated

generic link, we will denote by v_i the current assigned bandwidth to class i, i.e. the sum of the bandwidth values associated to all current LSPs of class i. Typically, since no bandwidth reservation is associated to BE LSPs, the assigned bandwidth is always zero for BE, nevertheless a counter has been associated to it for notation compactness. The route selection algorithm does not distinguish between the bandwidth assigned to working and backup LSPs and simply prefers the links with the largest residual bandwidth with no regards of working and backup components. Additionally we define u_i as the class i minimum guaranteed bandwidth. This is a configurable parameter that allows to enforce a minimum guaranteed cushion to class i on the specific link. Also if no class i LSPs are established, such a bandwidth cushion can not be taken by other classes and it is preserved for future class i requests. This is useful to apply bandwidth isolation between classes, which is an important requirement as stated in [17]. Whether or not apply such a minimum bandwidth cushion is a provider business policy matter. The default value for u_i is zero, except for BE class. In fact, it is likely that any provider might want to let a percentage of link capacity available to the BE traffic. From v_i and u_i, the generic node responsible for the link will extract the value of the current reserved bandwidth as

$$r_i = max(v_i, u_i) \qquad (1)$$

For each LSP, the r_i counters are used to enforce local AC. The AC function must ensure that the reserved bandwidth components meet a set of constraints. These can be defined on the single values of r_i (e.g. *EF class can not exceed 50% of the link capacity*), on some combinations (e.g. *AF1x and AF2x classes jointly can not exceed 70% of the link capacity*), or on their complete sum (e.g. *the sum of reserved bandwidth for all classes can not exceed the link capacity*). Each of such constraints can be dictated by business related policies or QoS related considerations. The constraints set can be written in a formal way as follows:

$$r'L \leq c \qquad (2)$$

wherein r is the column vector r={r1, r2,..} collecting the reserved bandwidth value for each class. The matrix L is composed of binary elements, each row represents a single constraint, and c is the column vector of associated limits. Similarly to r, we will denote by v and u the vectors collecting the v_i and u_i components. With the above positions, the AC algorithm can be described in a simple way. When a LSP of class j and bandwidth b is requested, the tentative value $v_j^* = v_j + b$ (*update rule*) is computed as well as new value of vector r*. Then, it is checked whether constraint(2) holds for the tentative vector r*. In the affirmative case the request can be accepted and the new value of v_j recorded. Conversely, the request is refused and the counters are not updated. If backup bandwidth sharing is NOT applied, the simple update rule given above is applied to both working and backup LSPs. On the other hand, in case of bandwidth sharing, the update rule for v_j^* for the backup LSP setup must be revised according to the algorithm detailed in [18].

It is evident that, in centralized and partially distributed solutions, the centralized logic that performs the AC algorithm exactly knows the actual reserved bandwidth on each link. Vice versa, in the fully distributed solution, due to the concurrent operation of the ERs, a mechanism to continuously inform each ERs on the status of each link is needed. In such a distributed environment when a LSP is installed/removed from a link, the local node should advertise through OSPF-TE flooding the new unreserved bandwidth value for the specific link. In this way, the ERs can update their local Network State Database and perform a route selection process coherent with the current network state.

More formally, if r is the currently reserved bandwidth and c the maximum reservable bandwidth (not necessarily the link capacity), the value of g=c-r is advertised, where g represents the link residual reservable bandwidth. In order to avoid a large amount of flooding, the new value of g is not advertised upon each LSP setup/tear-down. The Opaque LSA generation process is performed according to local update policies embedding watermark-based algorithms and/or hold-down timers [19]. As a consequence, it follows the variations of g less accurately, but flooding overhead decreases. In our model we extend this approach to a multi-class environment. The *residual bandwidth* becomes a vector $g = \{g1, g2, ...\}$, whose component g_i represents the additional bandwidth that can be assigned to future class i requests when no new requests from other classes are performed. The computation of each component g_i involves a very simple manipulation of r, L and c. A trivial optimization problem with a single variable and linear constraints must be solved. To build the vector g, the computation is repeated independently for each class except BE. This vector can be advertised in the sub-TLV Unreserved-bandwidth of OSPF-TE Opaque LSA in conformance with the semantic defined in [20]. Given the independence between the g_i components, the same flooding reduction policy used for g in the single class environment can be straightforwardly applied to each component separately. In particular, we use the same mechanism described in [21] based on adaptive watermarks. We notice that, due to composed constraints, a LSP setup/tear-down impacts the residual bandwidth of all classes and that the sum of the components g_i can exceed the link capacity, since each element has been computed in absence of new requests from other classes. Such a semantic, coherent with the information needed by the route selection algorithm, greatly simplifies g computation and flooding reduction algorithm application. The proposed model comprises as special cases the models being currently discussed in IETF, [17] [22] [23], and is compliant with the requirements given in [22]. In particular, isolation is provided by u_i setting and inter-class sharing can be obtained by an appropriate constraint design(2). Finally, let us consider the possibility to apply TE to BE traffic. Even if no bandwidth reservation is associated to BE LSPs, it is possible to envisage a model where explicit routing capability is applied to these ones. In order to perform route selection for BE LSPs, a BE *residual bandwidth* (g_4) is defined as the *link measured residual capacity*. This value can be derived, for example, from the local MIB containing the bytes sent in the last measurement interval. In order to disseminate such an information, the value is inserted in Unreserved-bandwidth sub-TLV of OSPF-TE Opaque LSA.

5 The TANGO Platforms Assessment

In this section we describe the experimental activity aimed at assessing a functional validation of the MAID inter-domain mechanisms as well as to test data plane performance. The tests have been carried out on a distributed test-bed made up of two inter-connected domains located, respectively, in the laboratories at the Department of Information Engineering of the University of Pisa and at the META Centre of the Consorzio Pisa Ricerche. The two domains are permanently interconnected through a Gigabit Ethernet optical fiber link. At the network layer, each domain is configured as an independent autonomous system with proper strategies and policies for QoS provisioning and Traffic Engineering. The routers in each domain are prototypal routers based on IA32(PC) Linux OS platforms, equipped with the kernel modules for MPLS and Traffic Control (TC), and with the MAID-specific modules developed in the TANGO project. An overview of the test-bed topology is shown in Fig. 4. Each domain has its own BB, which manages the dynamic configuration of the network resources under its scope, as well as the inter-domain communication by means of COPS-MAID protocol. Concerning TC the scheduler used to realize the different DiffServ PSC is the Hierarchical Token Bucket (HTB) available in the Linux kernel 2.4.20. HTB is a kind of CBQ (Class Based Queuing) algorithm [24], approximating service discipline based on the class concept. The access networks/clients have been configured in order to play the role of source/destination of different kind of QoS-unaware IP traffic. Two traffic typologies have been injected in the test-bed: artificial traffic, generated by a specialized application (BRUTEv1.0 [25]), and real-time traffic, generated by the delivery of multimedia contents (based on Helix DNA platform [26]). Different tests are carried out for assessing the performance of the MAID test-bed with respect to the different source applications and traffic profiles injected into the network. These tests highlight also MAID data plane critical elements, responsible for an unexpected limitation in the overall performance. In all the tests traffic is sent after a configuration phase takes place. This phase is similar to the static resource provisioning provided by the Network Operator for those QoS-unaware access networks that can not use

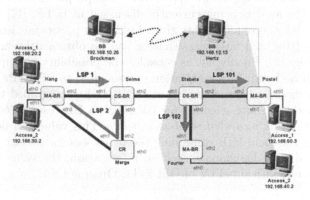

Fig. 4. MAID testbed topology.

the dynamic MAID-UNI features. Configuration consists of the DiffServ LSPs setup and traffic flows mapping into LSPs by means of a WEB interface. For each LSP, a QoS class and a reserved bandwidth are signaled. More details about these tests can be found in [27].

Artificial Traffic. The aim of this test is to identify the critical element on the MAID data plane. Two traffic flows have been generated from the same source client: the first to 192.168.50.3:6970 through an EF LSP with a reservation of 1.5Mbps, the latter to 192.168.40.2:7970 through an AF1x LSP with a reservation of 512kbps. Experiments shows that some packets are dropped even if the reserved rate equals exactly the nominal mean rate. The policer located on the ingress MA-BR is the software element responsible for this packet dropping. This element requires an accurate configuration/tuning of its parameters, in order to achieve the desired performance, above all when operating in quasi-saturation conditions. Fig. 5 shows the results obtained when two flows are generated with a Constant Bit Rate of 1.5Mbps and 512kbps, respectively.

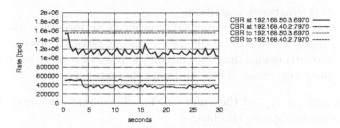

Fig. 5. BRUTE: CBR traffic flows.

Real-Time Traffic. Two types of tests are performed in this testing scenario. The first one is characterized by:

- a fixed amount of bandwidth reserved for each LSP;
- a single encoded version of the multimedia content, streamed at the encoding bit-rate of 768kbps;
- a variable connection type configured on the destination client.

The video streaming is flowed to 192.168.50.3:6970 through an EF LSP with a reservation of 1Mbps and to 192.168.40.2:7970 through an AF1x LSP with a reservation of 512kbps. Clients connection have been configured with a LAN connection speed (e.g. 10Mbps) or with a DSL one (e.g. 768 kbps). In the first case (Fig. 6(a)), after a few seconds in which some packets are dropped on both connections, due to the server attempt to fill the buffer at the full connection speed (e.g. 10Mbps), the client attached to the EF LSP perceives a good video and audio quality. Instead, the client attached to the AF1x LSP experiences a jerky reproduction because of the packet drops induced by a reserved bandwidth (e.g. 512kbps) lower than the encoding rate (e.g. 768kbps). In fact, when both clients have been configured with a DSL connection (Fig. 6(b)), no packet drops

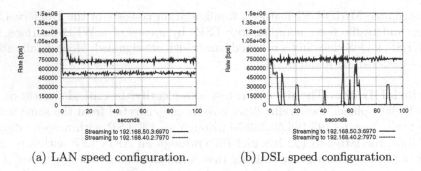

(a) LAN speed configuration.   (b) DSL speed configuration.

Fig. 6. RTSP streaming.

for the traffic flow on the EF LSP is experimented, resulting in a optimal perceived quality throughout the whole streaming. Instead, the client that receives the traffic flow on the AF1x LSP cannot succeed in filling up its buffer at an acceptable rate and it triggers automatically a PAUSE/PLAY mechanism, waiting for possibly better network conditions.

The latter type of tests is performed to evaluate performance when the same multimedia content is available at different encoding rates. In this case, the server chooses the best fitting encoding bit-rate on the basis of connection information, collected in the setup phase. These tests are characterized by:

- two different versions of the same multimedia content streamed at encoding bit-rates of 768kbps or 512kbps;
- a DSL (e.g. 768kbps) connection type configured on the clients.

In Fig. 7, the streaming is flowed to 192.168.50.3:6970 through an EF LSP with a reservation of 1Mbps and to 192.168.40.2:7970 through an AF1x LSP with a reservation of 512kbps. In this case, the clients negotiate the proper rate with the server (e.g. 768kbps for the traffic through the EF LSP and 512kbps for the other). In this situation no packet loss is experienced and the differences on the perceived playing quality are due to the different encoding rates.

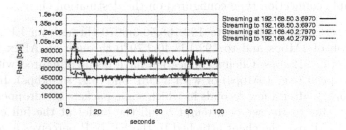

Fig. 7. RTSP streaming with different multimedia content encoding bit-rates.

6   Conclusions

In this paper a novel end-to-end seamless framework to support end-to-end Quality of Service and Traffic Engineering. The objective is to provide a complete end-to-end solution for the enforcement of a globally dynamic and multi-level chain of SLAs. Such a solution, based on the MPLS/DiffServ paradigm, allows a seamless QoS-IP service setup, according to both users and network providers requirements, by means of MAID Control Plane mechanisms. An analysis of key issues of a MPLS/DiffServ backbone have been also presented, leading to the formulation of an alternative core network architecture. The problem of intercommunication between these hetherogenous domains is still under discussion and possible solutions have been detailed. Finally, a functional validation and performance analysis of proposed architecture have been carried out on operational testbeds based on Linux PC platforms.

Acknowledgments

This work was funded by the TANGO project of the FIRB programme of the Italian Ministry for Education, University and Research.

References

1. TANGO project homepage. http://tango.isti.cnr.it.
2. S. Blake et al. An Architecture for Differentiated Services. *IETF RFC 2475*, Dec. 1998.
3. K. Poduri V. Jacobson, K. Nichols. An expedited forwarding PHB. *IETF RFC 2598*, June 1999.
4. J. Heinanen et al. Assured Forwarding PHB Group. *IETF RFC 2597*, June 1999.
5. Y. Bernet et al. A Framework for Integrated Services Operation over DiffServ Networks. *IETF RFC 2998*, Nov. 2000.
6. IST-AQUILA Project. Homepage: http://www-st.inf.tu-dresden.de/aquila/.
7. IST-TEQUILA Project. Homepage: http://www.ist-tequila.org/.
8. IST-MESCAL Project. Homepage: http://www.mescal.org/.
9. IST-MOICANE Project. Homepage: http://www.moicane.com.
10. D. Durham et al. The COPS (Common Open Policy Service) Protocol. *IETF RFC 2748*, Jan. 2000.
11. G. Sergio G. Carrozzo, N. Ciulli. COPS-MAID - COPS Usage for Multi-Access Inter-Domain MPLS-DiffServ Networks. *Internet Draft, work in progress*, Nov. 2003.
12. A. D'Achille, M. Listanti, U. Monaco, F. Ricciato, V. Sharma. Diverse Inter-Region Path Setup/Establishment.
    draft-dachille-diverse-inter-region-path-setup-00.txt, July 2004.
13. F. Ricciato, U. Monaco, A. D'Achille. A novel scheme for end-to-end protection in a multi-area network. *Proc. of 2nd International Workshop on Inter-domain Performance and Simulation, IPS04, Budapest, Hungary.*, March 2004.
14. R. Albanese, D. Alì, S. Giordano, U. Monaco, F. Mustacchio. G. Procissi. Experimental Comparison of Fault Notification and LSP Recovery Mechanisms in MPLS Operational Testbeds. *submitted to QoS-IP 2005*.

15. W. Lai F. Le Faucheur. Requirements for Support of Differentiated Services-aware MPLS Traffic Engineering. *IETF RFC 3564*, July 2003.
16. D. Awduche et al. Requirements for Traffic Engineering Over MPLS. *IETF RFC 2702*, Sept. 1999.
17. W. Lai F. Le Faucheur. Maximum Allocation Bandwidth Constraints Model for Diff-Serv-aware MPLS Traffic Engineering. *Internet Draft, work in progress*, Mar. 2004.
18. M. Listanti F. Ricciato, S. Salsano. An Architecture for Differentiated Protection against Single and Double Faults in GMPLS. *Photonic Networks Magazine*, 2004.
19. G. Apostolopoulos et al. Quality of Service Based Routing: A Performance Perspective. *SIGCOMM*, 1999.
20. D. Yeung D. Katz, K. Kompella. Traffic Engineering Extensions to OSPF Version 2. *Internet Draft, work in progress*, June 2003.
21. A. Botta et al. Traffic Engineering with OSPF-TE and RSVP-TE: Flooding Reduction Techniques and Evaluation of Processing Cost. *CoRiTeL Report*, 2003.
22. J.Ash. Max Allocation with Reservation Bandwidth Constraint Model for MPLS/Diffserv TE Performace Comparison. *Internet Draft, work in progress*, Jan. 2004.
23. J.Ash. Russian Dolls Bandwidth Constraints Model for Diff-Serv-aware MPLS Traffic Engineering. *Internet Draft, work in progress*, Mar. 2004.
24. V. Jacobson S. Floyd. Link-sharing and resource management models for packet network. *IEEE/ACM Transactions on Networking*, 1995.
25. BRUTE (Brawny and Rough UDP Traffic Engine) homepage. http://netgroup-serv.iet.unipi.it/brute/.
26. Helix DNA Server 9.0. http://www.helixcommunity.org/.
27. G. Carrozzo et al. MPLS/DiffServ interworking: preliminary functional tests for TANGO project. *TANGO project Symposium*, 2004.

Bandwidth Management in IntServ to DiffServ Mapping

António Pereira[1,2] and Edmundo Monteiro[2]

[1] Polytechnic Institute of Leiria
Department of Informatics Engineering, ESTG
Morro do Lena, Alto do Vieiro 2411-901 Leiria, Portugal
apereira@estg.ipleiria.pt
http://www.estg.ipleiria.pt
[2] University of Coimbra
Laboratory of Communications and Telematics
CISUC / DEI, Pólo II, Pinhal de Marrocos 3030-290 Coimbra, Portugal
edmundo@dei.uc.pt
http://lct.dei.uc.pt

Abstract. This work presents a bandwidth management mechanism to be used
with mapping mechanisms between Integrated Services (IntServ) and Differentiated Services (DiffServ) domains. The mapping mechanisms have a dynamic
nature and are associated with the Admission Control functions such that the
state of the network is reflected in the admission decisions of new IntServ flows
into the DiffServ network. The work is focused in the mapping between the
IntServ Controlled-Load (CL) service and the DiffServ Assured Forward (AF)
Per-Hop-Behaviour group. The proposed bandwidth management mechanism
evaluates the bandwidth needs of the AF classes in the DiffServ domain taking
into account the behaviour of the previously mapped CL flows for the same
IntServ destination network. The results obtained by simulation show that the
mechanism improves the use of the available bandwidth for a given AF class.
The mapping mechanisms detects Quality of Service (QoS) degradation occurrences and once detected the control mechanism allows the reestablishment of
AF Class QoS.

1 Introduction

The research effort in the area of the Quality of Service (QoS) provisioning in the
Internet has been carried out by the IETF (Internet Engineering Task Force) according to two main approaches: the Differentiated Services (DiffServ) model [1] and the
Integrated Services (IntServ) model [2, 3]. These two models have been developed by
two different IETF work groups [4, 5].

The IntServ model provides individually QoS guarantees to each flow. For such, it
needs to make resource reservation in network elements intervening in the communication. For resources reservation the Resource Reservation Protocol is used (RSVP)
[6, 7]. The IntServ model supports two distinct services: Guaranteed Service (GS) [8]
for applications with strict needs of throughput, limited delay and null losses; Controlled-Load service (CL) [9] that emulates the behaviour of the best-effort service in
an unloaded network. The need of maintenance of state information on the individual
flows is usually pointed as the origin of the scalability problems of the IntServ model.

M. Ajmone Marsan et al. (Eds.): QoS-IP 2005, LNCS 3375, pp. 433–444, 2005.
© Springer-Verlag Berlin Heidelberg 2005

The DiffServ model embodies the second approach where the flows are aggregated in a few Classes of Service (CoS) according to their specific characteristics. The packets belonging to specific classes are forwarded according to their Per Hop Behaviour (PHB) associated with the DiffServ Code Point (DSCP) [10], which is included in the Type of Service (ToS) field of the IP header. Currently the DiffServ model supports Expedited Forwarding (EF) PHB intended to offer a service of type "virtual leased line" with throughput guarantees and limited delay [11]. DiffServ also supports the Assured Forwarding (AF) PHB group that exhibits a similar behaviour to a low loaded network for traffic that is in accordance with the service contract [12].

In order to combine the superior scalability of the DiffServ model with IntServ superior QoS support capabilities, the ISSLL (Integrated Services over Specific Link Layers) working group of the IETF [13] proposed the interoperation between these two models [14]. The defined approach combines the IntServ model features – the capability to establish and maintain resources reservations through the network – with the scalability provided by the DiffServ model. The IntServ model is applicable at the network edge, where the number of flows is small, while the DiffServ model is applicable in the network core to take advantage of its scalability. The boundary routers between these two domains are responsible for mapping the IntServ flows into the DiffServ classes. These functions include the choice of the most appropriate PHB to support the flow and the use of admission control (AC) and policing functions on the flows at the entrance of the DiffServ region.

In DiffServ networks Admission Control is based on Bandwidth Brokers (BBs) and also on pricing schemes associated with Service Level Agreements (SLAs) at the entrance of the DiffServ domains. This solution does not intrinsically solve the problem of congestion control. Upon overload in a given service class, all flows in that class suffer a potential QoS degradation. To solve this and to integrate the DiffServ and IntServ models in a end-to-end service delivery model with the associated task of reservation, a new admission control function, which can determine whether to admit a service differentiated flow along the nominated network is needed [15]. There are several proposals of admission control mechanisms that can be used to address this problem. One approach of admission control developed at the Laboratory of Communications and Telematics of the University of Coimbra (LCT-UC) [16] uses a metric to evaluate a Congestion Index (CI) at each network element to admit or not a new flow [17, 18]. Other approaches use packet probing [19, 20, 21], aggregation of RSVP messages [22, 23] between an ingress egress routers or Bandwidth Brokers (BBs) [24]. The issue of the choice of the admission control mechanisms was left open by the ISSL IETF group [25].

In previous work we proposed a mapping mechanism between the Controlled-Load service of the IntServ model and the Assured Forwarding PHB group of the DiffServ model [26, 27]. The option was due to the less difficulty of the problem when compared with the mapping between GS and PHB EF and to the wider acceptance of IntServ CL service among network equipment manufacturers. The proposed mapping mechanism included a dynamic Admission Control module that takes into account the state of the DiffServ network. In our approach, the decision of mapping and admission of a new IntServ flow in the DiffServ region is based on the behaviour of previous flows to the same IntServ destination network. This behaviour is used to estimate the bandwidth needs of the DiffServ AF classes.

To complement the mapping mechanisms previously proposed, this paper presents a bandwidth management mechanism to be used with dynamic mapping mechanisms between the IntServ Controlled-Load service (CL) and the DiffServ Assured Forward (AF) Per-Hop-Behavior group. The proposed mechanism is based in the continuous adjustment of the bandwidth used by the DiffServ classes. The simulation results show that the proposed mechanism guarantees a good level of bandwidth utilization in the mapping between IntServ CL service and DiffServ AF classes.

Besides the present section the paper has the following structure. In Section 2 the principles and the architecture proposed for dynamic mapping mechanisms are presented. The bandwidth management algorithm that supports the Admission Control functions in IntServ flows mapping into AF classes is described in Section 3. In Section 4 the simulation scenario is presented and the proposed mechanisms are evaluated. Finally, in Section 5, some conclusions and directions for future work are presented.

2 Dynamic Mapping Mechanisms

In the border between the IntServ and DiffServ regions, the network elements must perform the mapping of the requested IntServ service into a DiffServ class of service. The DiffServ class must be selected in a way to support the type of IntServ service requested for the application. Taking into account the already defined IntServ services (CL and GS), the PHBs currently available in DiffServ (AF and EF) and, considering the characteristics of each service and PHB respectively, the choice of mapping between service CL and PHB AF and between service GS and PHB EF is evident.

The mapping of the CL service into the AF PHBs must be based on the burst time of the CL flow [25]. This way, the flows are grouped in an AF class, which provides the better guarantee that the packet average queue delay does not exceed the burst time of the flow. The mapping can be static or dynamic: static mapping is defined by the administrator of the network; dynamic mapping is driven according to the characteristics of the existing traffic in the network.

In the mapping mechanism proposed in previous work [26, 27], the aim is to complement the traffic control functions of the DiffServ network by using a dynamic Admission Control mechanism that reflect the state of network. In the adopted strategy, the decision of mapping and admitting a new flow at the ingress of the DiffServ region is based on the behaviour of previous flows which going to the same IntServ network. This behaviour is evaluated by of delay and losses suffered by the flows in the DiffServ region. The underlying idea is inspired in the congestion control mechanism used by TCP, applied to the admission control and mapping of IntServ flows into DiffServ classes.

The strategy adopted is based on the monitoring of flows at both the ingress and the egress of DiffServ domains to evaluate if the QoS of the mapped flows was degraded or not. In the case where no degradation occurs new flows can be admitted and mapped. On the other hand, if the QoS characteristics have been degraded, no more flows can be admitted into the DiffServ network ingress and the number of active flows must be reduced. By monitoring the flows at the egress of the DiffServ domain, the QoS characteristics are evaluated on the basis of the packet loss, since the queuing delay is less representative [19] and more difficult to treat with passive measurements due to its wide variability and to the difficulty of clock synchronization.

The proposed strategy for mapping IntServ flows into DiffServ classes is based on two mechanisms located in the network elements at the boundary of the DiffServ region: the Mapper and the Meter. In the edge router at the ingress of DiffServ domain, the Mapper maps CL flows into the AF class that better supports the IntServ service. This mechanism acts on the basis of the information supplied by the Meter mechanism located in edge router at the egress of the DiffServ domain.

The Meter mechanism interacts with the modules of the IntServ model, and with the meter module of the DiffServ model (which is responsible for accounting, for each flow, the packets in agreement with the attributed DSCP). Whenever a RSVP message of reserve removal occurs, the collected information is inserted in a new object called *DIFFSERV_STATUS* and is sent to the ingress edge router of the DiffServ domain such that it can be taken into account for the next flow mapping.

3 Bandwidth Management Mechanism

The Mapper mechanism needs to reflect the state of the network. In order to allow this the edge router Mapper mechanism needs to know the available resources for each DiffServ class. The evaluation of these resources is based on what happened to the previously mapped flows for each particular IntServ network destination. This calculation is made by the bandwidth management mechanism when a *DIFFSERV_STATUS* object arrives at the Mapper. The Meter sends this object in a RSVP message when a reserve removal message is received. Is this RSVP message that invokes bandwidth management mechanism.

When the Mapper receives the *DIFFSERV_STATUS* object, the bandwidth management mechanism extracts the number of packets received by the meter in the egress edge router, collects on the local Meter the number of packets sent to the DiffServ domain and compares them to evaluate if QoS degradation has occurred or not. If the difference is less than the threshold of allowed losses then it is considered that no QoS degradation has occurred. In this case, the allowed throughput (number of admitted flows) can be increased. Otherwise, it is considered that degradation occurred and the allowed number of flows is reduced. The increasing of the number of flows is additive and the reduction is multiplicative which allows the AF class of DiffServ Network to recover quickly from the degradation. The additive increment provides a gradually use of the remaining class resources by all the IntServ Networks. Figure 1 shows the algorithm of the bandwidth management mechanism. This algorithm is invoked upon each flow release event.

The concrete values used to increase and decrease the number of flows are defined in the TCA (Traffic Control Agreement) specification. When the flow QoS has not been degraded the resources (throughput) allowed by the algorithm for a specified AF class will be increased by a given amount (10% in the current analysis). All the mapped flows present in the DiffServ network are validated, i.e., any one of these flows can serve as probing to the DiffServ AF class (they not suffered degradation). If the QoS has been degraded the resources allowed by the algorithm will be decreased by a multiplicative factor (50% in the current analysis). If the active flows

```
inc = 0.1   {Throughput increment}
dec = 0.5   {Throughput decrement}
losses = LOSSES {allowed losses threshold}
throughput {throughput of the active CL mapped flows}
r           {TSpec flow throughput}
degradation {QoS flow degradation}
MaxThroughput {max throughput admitted in DiffServ region}
npkts_sent {number of packets sent to the DiffServ region}
npkts_rcv {number of packets received at DiffServ region}

begin
   extract_from (DIFFSERV_STATUS object, npkts_rcv);
   collect_from_local meter (npkts_sent);
   if npkts_sent - npkts_rcv <= LOSSES then
      degradation = FALSE;
      MaxThroughput += inc * MaxThroughput;
      Remove_flow_from_probing_list;
   else
      degradation = TRUE;
      MaxThroughput = dec * MaxThroughput;
      Clean_probing_list;
   end if
   Remove_flow_from_mapping_list;
   throughput -= r;
   Update_policer;
   Send_upstream_RSVP_reserve_removal;
end
```

Fig. 1. Bandwidth management algorithm

have been degraded, only a new flow admitted later can act as probing to the AF class.

After degradation is verified, the flow is removed from the list of admitted flows and the throughput of the active mapped flows is updated. In this case the total throughput of active flows is reduced by r (throughput of removed flow) and the policer is updated. After this, the removal message is sent upstream to complete the flow release process.

When a new CL flow reserve request arrives (*RSVP_RESV* message) at the edge router Mapper, an admission control process is activated. Then the Mapper chooses the AF class that better guarantees that the CL flow QoS is preserved in the DiffServ network. Once the AF class is determined, the AC takes the decision of mapping or not, based on the resources allowed for this class. The algorithm defined above calculates these resources. The CL flow is admitted if the sum of its throughput with that of the active flows does not exceed the maximum throughput determined by the algorithm for the class. If the flow is admitted, the throughput of active flows is updated to take into account this flow, which is then inserted in the mapping and probing lists. The policer is updated and a RSVP_RESV message is sent upstream to the sender. Otherwise, if the flow is not admitted, a *RSVP_RESVERR* message is sent to release the reserve on the downstream network elements.

4 Results and Evaluation

In this section, the bandwidth management mechanism for the dynamic mapping of CL flows into AF classes, previously described, is evaluated. The implementation of the mapping and bandwidth management mechanisms was done in the Network Simulator version 2 environment (NS2) [28] integrated with the available NS2 Int-Serv and DiffServ modules [29].

The aim of this evaluation was, firstly, to verify if there is an improvement in the use of the available bandwidth in the AF classes, that is, if the bandwidth resources are redistributed in accordance with the state of the network by the algorithm. Secondly, to verify if in the occurrence of QoS degradation, this is detected by the Mapper at the DiffServ network entrance and, once detected, to evaluate if the used algorithm allows the reestablishment of the AF Class QoS.

The simulation scenario illustrated in Figure 2 shows four IntServ networks interconnected through a DiffServ network. At the DiffServ domain entrance the CL flows from IntServ A1 are mapped/admitted by the Edge Router Mapper 1 (ERM1) and the CL flows from IntServ A2 are mapped/admitted by the Edge Router Mapper 2 (ERM2). The DiffServ backbone has a bandwidth of 4 Mbps for the resources defined in the profiles for each IntServ network, for the best-effort (BE) traffic and in order to have a bandwidth remainder to test the bandwidth management mechanism on the resources redistribution by the Mapper.

For the AF classes, a profile of 1 Mbps was defined at the DiffServ domain entrance. In order to separate BE traffic of AF traffic, two queues in the DiffServ domain have been defined. The BE queue is a FIFO, while the AF queue is a RIO (Random Early Detection with in and Out) [30]. The latter queue is configured with the values obtained from [29]. Both queues are served by the WFQ (Weighted Fair Queuing) scheduler [31], which is configured such that the profile defined for the AF class is assured.

In the simulation tests, the bandwidth management mechanism for dynamic mapping of CL flows into the AF PHB that takes into account the state of the DiffServ network was evaluated in the presence best-effort flows of 100 Kbps. The delay, the losses and throughput of CL flows were measured. The throughput of existent mapped flows (*Throughput*) in the class AF as well as the maximum throughput allowed (*MaxThroughput*) in the DiffServ network was recorded. These values were obtained from the dynamic Admission Control mechanism using the bandwidth management mechanism whenever a reserve removal of a CL flow previously mapped had occurred.

In the above scenario, each IntServ network, A1 and A2, generates 15 best-effort flows of 100 Kbps. Reserve requests of CL flows of 100Kbps are generated every 15 seconds by IntServ A1 network and are generated every 10 seconds by IntServ A2. A CL flow is mapped and transmitted if resources are available in the IntServ networks and if the dynamic admission control, based on the bandwidth management mechanism, at the DiffServ domain entrance accepts the request.

After 250 seconds of simulation time, and every 50 seconds thereafter, the existent flow reserves into IntServ A1 and B1 networks are removed in the same order they were created. Also, after 225 seconds of simulation time, and every 25 seconds thereafter, the existent flow reserves into IntServ A2 and B2 networks are removed in the

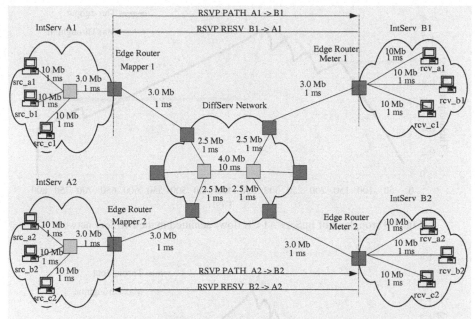

Fig. 2. Simulation scenario

same order they were created. The tests allowed more reserve requests and mappings than reserve releases allowing the evaluation of bandwidth management and dynamic mapping mechanisms. Also, IntServ networks in the scenario generate more traffic and communicate more frequently its state trough the reserve releases and therefore they use a higher slice of AF class throughput in the DiffServ region.

Figures 3 and 4 show the results obtained using the dynamic Admission Control mechanism with the bandwidth management mechanism to map CL flows into the AF classes. The analysis of the figures shows that the flows were admitted until the number of flows of the predefined profile is attained. Afterwards, new flows were admitted only if the reserve of a previous mapped flow was released and if these flows did not suffer any QoS degradation. According to the algorithm, in such case the variable MaxThroughput is incremented by 10%. The figures also show that the resources available for the IntServ A2 network are more frequently updated than the ones for the IntServ A1 network. This happens because there are more reserve releases on the IntServ A2 and B2 networks.

When QoS degradation occurs, the maximum throughput allowed to the CL flows drops by 50% of the current value. This way, the AF class can recover from the degradation. The variable *MaxThroughput* is updated in the edge router Mappers only when the state of the network is verified after the degradation. The state of the network is known when a new mapped flow probes the network. If this new flow does not suffer QoS degradation, the *MaxThroughput* value is incremented 10% and the process of mapping new flows is repeated. Otherwise the *MaxThroughput* value is decremented 50% and will be updated only when a new mapped flow probes the network.

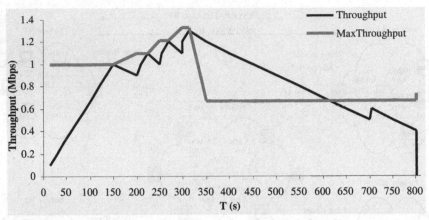

Fig. 3. Throughput of IntServ A1 CL flows admitted in the DiffServ network

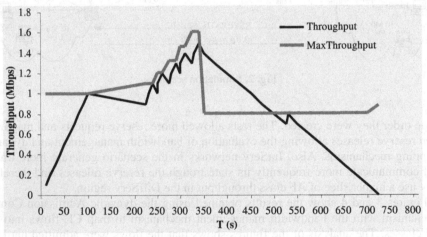

Fig. 4. Throughput of IntServ A2 CL flows admitted in the DiffServ network

The simulation results regarding throughput of the flows generated in the IntServ A1 network, are presented in Figure 5. The delay results are presented in Figure 6 and the loss results are presented in Figure 7. Simulation results for the IntServ A2 flows were also obtained and have a similar behaviour.

From these figures it can be verified that until t = 315s, the algorithm allows an improvement in the use of the available bandwidth by the AF mapped class. In this situation losses do not occur, the throughput of CL flows is the reserved one and the delay only slightly increases with the admission of new flows. BE flows absorb the congestion when the number of admitted CL flows increase, and are degraded in terms of delay throughput and losses. After the admission of an new flow, CL16, at t=315s, QoS degradation occurs. This situation is detected when the next flow, CL4, is released at t=350s. This release brings information about the state of the corresponding AF class and triggers the *degradation* flag of the bandwidth management mechanism.

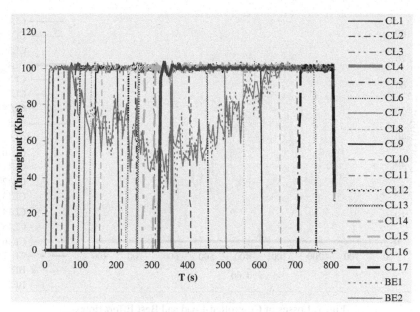

Fig. 5. Throughput of Controlled-Load and Best Effort flows

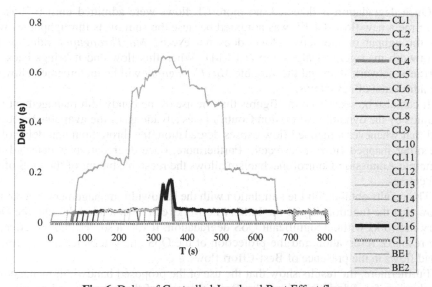

Fig. 6. Delay of Controlled-Load and Best Effort flows

From the analysis of Figure 6 it can be noticed that losses only occur in a small period of time, between t=315 and t=350 seconds. When losses take place the proposed bandwidth control mechanism reacts allowing the rapid QoS reestablishment. From Figure 6 it can also be verified that the delay increases with QoS degradation and returns to previous values when degradation is detected. The breaking in terms of losses and delays at t=325s, is a consequence of a CL flow reserve release by IntServA2 region.

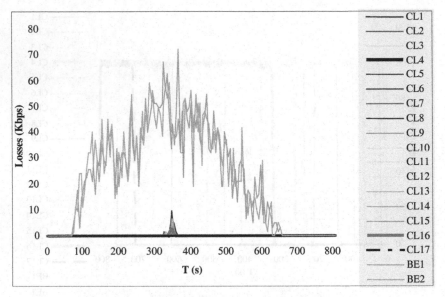

Fig. 7. Losses of Controlled-Load and Best Effort flows

Once degradation is detected, no more CL flows were admitted until t=705s. At this time a new flow, CL17, was admitted because the sum of its throughput (r) with the throughput of the active flows does not exceed *MaxThroughput* value of the bandwidth management algorithm at ERM1. When this flow end it brings back the DiffServ network state and the variable *MaxThroughput* will be incremented allowing the admission of new flows.

It can also be seen from the figures that the use of the bandwidth management mechanics by the dynamic Admission Control takes advantage of the available resources and that whenever a mapped flow causes degradation, the throughput and delay of all the other mapped flows is affected. Furthermore, once degradation is detected, the dynamic Admission Control mechanism allows the reestablishment of the QoS of the AF classes.

The results obtained in the simulation with the bandwidth management mechanism show that the functionality of the IntServ networks can be extended through the DiffServ networks without significant QoS degradation. It was also verified the effect of the resource reservation and the protection of the QoS characteristics of Controlled-Load flows in the presence of Best-Effort flows.

Furthermore, the results show that the use of the proposed bandwidth management mechanism for Admission Control reflects the state of the network and provides an improvement of the available resources of certain AF classes.

5   Conclusions and Future Work

In this work we presented a bandwidth management mechanism to be used in dynamic mapping between the Controlled-Load service (CL) of the IntServ model and the Assured Forward (AF) Per-Hop-Behaviour group of the DiffServ model.

The proposed algorithm takes into account the behaviour of the previous CL flows for the same IntServ destination network. Based on this behaviour, the algorithm calculates the resources to the DiffServ AF classes. If no degradation occurred, the resources (throughput) suffer an additive increment. Otherwise, the resources suffer a multiplicative decrement.

The results obtained by simulation shown that the algorithm improves the use of the available resources for the AF classes. Moreover, the results shown that the mapping mechanisms detect QoS degradation occurrences and once detected the algorithm allows the reestablishment of the QoS characteristics of the AF Classes.

Future work (already ongoing) will address the validation of a dynamic mapping mechanism with this bandwidth management mechanism in more demanding scenarios, with more AF classes and with different types of traffic to be generated in the IntServ network.

Acknowledgments

This work was partially supported by the Portuguese Foundation for Science and Technology (FCT), by European Union FEDER under program POSI (Project QoS-MAP) and by PRODEPIII, Measure 5, Action 5.3.

References

1. D. Black et al., An Architecture for Differentiated Services, RFC 2475, IETF, Dec. 1998.
2. R. Braden et al., Integrated Services in the Internet Architecture: an Overview, RFC 1633, IETF, June, 1994.
3. S. Shenker et al., General Characterization Parameters for Integrated Service Network Elements, RFC 2215, IETF, September 1997.
4. IntServ workgroup charters, http://www.ietf.org/html.charters/IntServ-charter.html.
5. DiffServ workgroup charters, http://www.ietf.org/html.charters/DiffServ-charter.html.
6. J. Wroclawski, The Use of RSVP with IETF Integrated Services, RFC 2210, IETF, September 1997.
7. R. Braden et al., Resource Reservation Protocol (RSVP) – Version 1 Functional Specification, RFC 2205, IETF, September 1997.
8. S. Shenker et al., Specification of Guaranteed Quality of Service, RFC 2212, IETF, Sep. 1997.
9. J. Wroclawski, Specification of the Controlled-load Network Element Service, RFC 2211, IETF, September 1997.
10. K. Nichols et al., Definition of the Differentiated Services Field (DS Field) in the IPv4 and IPv6 Headers, RFC 2474, IETF, December 1998.
11. B. Davie et al., An Expedited Forwarding PHB, RFC 3246, IETF, March 2002.
12. J. Heinanen et al., Assured Forwarding PHB Group, RFC 2597, IETF, June 1999.
13. ISSLL workgroup charters, http://www.ietf.org/html.charters/issll-charter.html.
14. Y. Bernetwork et al., A Framework for Integrated Services Operation over DiffServ Networks, RFC 2998, IETF, November 2000.
15. G. Houston, Next Steps for the IP QoS Architecture, RFC 2990, IETF, November 2000.
16. D. Lourenço et al., "Definição do Mecanismo de Controlo de Admissão para o Modelo de serviços do LCT-UC", in Proc. of CRC2000, FCCN, Viseu, Portugal, Nov. 16-17, 2000.

17. E. Monteiro et al., "A Scheme for the Quantification of Congestion in Communication Services and Systems", in Proc. of SDNE'96, IEEE Computer Society, Macau, June 3-4, 1996.
18. G. Quadros, et al., "An Approach to Support Traffic Classes in IP Networks", in Proceedings of QofIS2000, Berlin, Germany, September 25-26, 2000.
19. L. Breslau et al., "Endpoint Admission Control: Architectural Issues and Performance", in Proceedings of ACM SIGCOM 2000, Stockholm, Sweden, August 2000.
20. V. Eleck et al., "Admission Control Based on End-to-End Measurements", in Proceedings of IEEE INFOCOM 2000, Tel Aviv, Israel, March 2000.
21. G. Bianchi et al., A migration Path to provide End-to-End QoS over Stateless networks by Means of a probing-driven Admission Control, Internet Draft, IETF, July 2001.
22. F. Baker et al., Aggregation of RSVP for IPv4 and IPv6 Reservations, RFC3175, IETF, September 2001.
23. Y. Bernet, Format of the RSVP DCLASS Object, RFC2996, IETF, November 2000.
24. Z. Zhang et al.. "Decoupling QoS Control from Core Routers: A Novel Bandwidth Broker Architecture for Scalable Support of Guaranteed Services", in Proceedings of ACM SIGCOM 2000, Stockholm, Sweden, August 2000.
25. J. Wroclawski et al., Integrated Services Mappings for Differentiated Services Networks, Internet Draft, IETF, February 2001.´
26. A. Pereira et al, "Interligação IntServ DiffServ: Mapeamento do Serviço CL no PHB AF", in Proc. of CRC2002, FCCN, Faro, Portugal, September-2002
27. A. Pereira et al, "Dynamic mapping between the Controlled-Load IntServ service and the Assured Forward DiffServ PHB", in Proc. of the HSNMC2003, pp. 1-10, Portugal, July-2003
28. Network Simulator – NS (version 2), http://www.isi.edu/nsnam/ns/
29. J. F. Rezende, "Assured Service Evaluation", IEEE Global Telecommunications Conference - Globecom'99, Rio de Janeiro, Brazil, December 1999.
30. D. Clark et al., "Explicit Allocation of Best Effort Packet Delivery Service", IEEE/ACM Transactions on Networking, vol. 6, no 4, August de 1998.
31. H. Zhang, "Service Disciplines for Guaranteed Performance Service in Packet-Switching Networks," Proc. IEEE, vol. 83, no 10, October 1995.

Optimizing Routing Decisions
Under Inaccurate Network State Information*

Xavier Masip-Bruin, Sergio Sánchez-López,
Josep Solé-Pareta, and Jordi Domingo-Pascual

Departament d'Arquitectura de Computadors, Universitat Politècnica de Catalunya
Avgda. Víctor Balaguer, s/n, 08800 Barcelona, Spain
{xmasip,sergio,pareta,jordid}@ac.upc.edu

Abstract. Maintaining accurate network state information in the Traffic Engineering Databases of each node along a network is extremely difficult. The BYPASS Based Routing (BBR) mechanism has appeared to reduce the effects produced in the network performance when selecting paths under inaccurate network state information. The BBR mechanism is based on applying the dynamic bypass concept. In this paper the BBR mechanism is modified to extend its applicability, and therefore to increase the benefits of its implementation.

Keywords: QoS routing, inaccurate routing decisions, routing scalability.

1 Introduction

Traditional IP networks are based on the best effort model to transport traffic flows between network clients. Since this model cannot properly support the requirements demanded by the emerging real time applications (such as video on demand, multimedia conferences or virtual reality), some modifications in the network structure, mainly oriented to optimize the network performance providing Quality of Service (QoS) guarantees are required. It is widely known that routing decisions strongly impacts on the QoS network behaviour. Having a powerful routing mechanism which takes into account the QoS parameters when selecting the routes (QoS Routing), targeting to an optimal end-to-end path selection is still a challenge.

Most QoS routing algorithms select paths based on the information contained in the network state databases (named Traffic Engineering Databases, TED, when including QoS parameters) stored in the network nodes. In fact, TED information is not only topology and connectivity but also QoS parameters are included. The management of such QoS parameters is done by the routing protocol, which must include a mechanism to collect, to distribute and to update all the parameters needed by the QoS routing algorithm. Important factors in the global routing behaviour are how and where routing decisions are taken. On the one hand, most QoS routing algorithms utilize the nominal link available bandwidth information to select paths. However, assuming that most clients generating network traffic do not completely use the assigned bandwidth, that is, do not strictly fulfil the Service Level Agreement (SLA), a

---

* This work was partially funded by the MCyT (Spanish Ministry of Science and Technology) under contract FEDER-TIC2002-04344-C02-02, the IST project E-Next under contract FP6-506869 and the CIRIT (Catalan Research Council) under contract 2001-SGR00226.

M. Ajmone Marsan et al. (Eds.): QoS-IP 2005, LNCS 3375, pp. 445–455, 2005.
© Springer-Verlag Berlin Heidelberg 2005

difference exists between the nominal link utilization and the actual link utilization. This gap leads to non-efficient network resource utilization, since the path selection process is performed according to nominal link state information instead of actual link utilization information. This problem is addressed in [1], where authors propose a path selection algorithm named *Available Bandwidth Estimation Algorithm* based on performing an estimation of the future link utilization. This prediction is obtained by sampling the network state with a certain period of time which dynamically changes depending on the traffic characteristics and the conservatism requirements of the network domain.

On the other hand, being aware that most QoS routing algorithms take routing decisions in the source nodes (explicit routing) based on their global network state information, a mechanism to keep this network state information perfectly updated must be included in the routing protocol. This mechanism is mainly based on flooding updating messages to advertise network nodes about network state changes. In a large connection oriented packet switching network scenario, where changes are produced very often, this updating process generates a non-desirable signalling overhead. To reduce such a signalling overhead, a triggering policy is applied. In addition to the simple triggering policy based on sending updating messages at fixed intervals of time (*time based* triggers), usually three triggering policies are applied, namely *Threshold based policy*, *Equal class based policy* and *Exponential class based policy* [2]. The Threshold based policy lies in sending an updating message whenever the actual residual bandwidth differs (is lower or greater) from the last advertised residual bandwidth in a quantity defined by a threshold tv. The other two policies are based on a link bandwidth partitioning, in such a way that the total link capacity is divided into several classes. Being Bw (base class size) a fixed bandwidth value, the *Equal class based policy* establishes its classes according to [$(0, Bw), (Bw, 2Bw), (2Bw, 3Bw),...$], and the *Exponential class based policy* according to [$(0, Bw), (Bw, (f+1)Bw), ((f+1)Bw, (f^2+f+1)Bw),...$], where f is a constant value. Then an updating message is triggered when the link capacity variation implies a change of class.

Most QoS routing algorithms assume as a condition that the network state databases, from which the routing tables are built, represent a current picture of the network state. Therefore, routing algorithms select routes based on this hypothetical accurate network state information. Unfortunately, as a consequence of using triggering policies to reduce the signalling overhead produced by the updating process, maintaining accurate network state information cannot be always guaranteed. Thus, when the information contained in the network state databases is not perfectly updated, i.e. does not represent a current picture of the network state, the routing process might potentially select a path that is unable to support the call requirements. Consequently, a certain incoming connection that is initially allocated to a particular path would be rejected in the path set-up process. This is known as the routing inaccuracy problem.

This paper focuses on addressing the routing inaccuracy problem produced when the path selection process is performed under inaccurate nominal link state information because of the application of a certain triggering policy needed to reduce the signalling overhead produced by the updating process.

Several significant documents exist addressing the routing inaccuracy problem. Most relevant contributions were done by Guerin and Orda [3], Lorenz and Orda [4],

T.Korkmaz and M.Krunz [5], Apostolopoulos et al. [6], Chen and Nahrstedt [7] and X.Masip et al. [8]. Table 1 shows the main characteristics of these contributions. A detailed review of these contributions can be also found in [8].

Table 1. Algorithms comparison

Ref.	Based on	QoS Constr.	Main Goal	Routing	Routing information
[3]	Probabil. appr.	Bw or delay	Reduce RIP* effects	Source	Global
[4]	Probabil. appr.	Delay	Reduce RIP* effects	Source	Global
[5]	Probabil. appr.	Bw/delay	Improve Computational efficiency	Source	Global
SBR [6]	Probabil. appr.	Bw	Reduce RIP* effects	Source	Global
TBP [7]	Multipath sche.	Delay	Reduce RIP* effects	Distributed	Local
BBR [8]	Dynam. bypass	Bw	Reduce RIP* effects	Source	Global

*RIP: Routing Inaccuracy Problem

One of the most recent contributions addressing the routing inaccuracy problem for bandwidth constrained applications is the BYPASS Based Routing (BBR) mechanism [8]. Although it is deeply demonstrated the benefits obtained in the global network performance (in terms of blocking probability) when using any routing algorithm inferred from the BBR mechanism, it is also noticed that the BBR mechanism suffers from two significant weaknesses, that is, inefficient *bypass-paths* utilization and large computational cost, both limiting the potential benefits of its implementation. It is the main goal of this paper to analyze to what extent such weaknesses degrade the global network performance as well as to provide new solutions to reduce such a performance degradation.

The rest of this paper is organized as follows. Section 2 shortly describes the BBR mechanism and its current existing weaknesses. Then, Section 3 provides the BBR with a solution to address them which are evaluated in Section 4. Finally, Section 5 concludes the paper.

2 BYPASS Based Routing Review

The BYPASS Based Routing (BBR) mechanism was proposed to reduce the effects produced by the routing inaccuracy problem in the global network performance for bandwidth constrained applications. All the algorithms inferred from the BBR mechanism are based on the dynamic bypass concept. By implementing such a concept some intermediate nodes along the selected path might reroute the setup message through alternative pre-computed paths, named *bypass-paths*, whenever these intermediate nodes do not have enough resources to cope with the incoming traffic requirements.

2.1 Basic Guidelines in the BBR Understanding

Let $G(N,L,B)$ describe a defined network, where N is the set of nodes, L the set of links and B the bandwidth capacity of the links. Suppose that a set P of source-destination node pairs (s,d) exists, and that all the LSP incoming requests occur between elements of P. Let b_{req} be the bandwidth requested in an element $(s,d) \in P$. Let

$G_r(N_r, L_r, B_r)$ represent the last advertised residual graph, where N_r, L_r and B_r are the remaining nodes, links and residual bandwidths at the time of path setup respectively. The main steps in the BBR performance are:

Obstruct-Sensitive Links. A new parameter is introduced in the path selection process to represent the routing inaccuracy. This parameter is translated into a new class of link. In this way, an *Obstruct-Sensitive Link (OSL)* is a link that potentially would be unable to support the traffic requirements. Specifically, a particular link is defined as *OSL* for an incoming connection request depending on the triggering policy in use. So, let L^{os} be the set of *OSL's*. Let $l^{os} \in L^{os}$ be a link of the residual graph G_r. As stated before a link l_i is defined as *OSL* as follows:

- *Threshold triggering policy*: Let b_r^i be the last advertised bandwidth for a link l_i. This link l_i is defined as *OSL*, that is l^{os}_i if

$$l_i = l^{os}_i \mid l^{os}_i \in L^{os} \iff b_{req} \in (b_r^i(1-tv), b_r^i(1+tv)] \tag{1}$$

- *Exponential class triggering policy*: Let B_{l_j} and B_{u_j} be the minimum and the maximum bandwidth values allocated to class j. A link l_i is defined as *OSL*, that is l^{os}_i if

$$l_i = l^{os}_i \mid l^{os}_i \in L^{os} \iff b_{req} \in (B_{l\_j}, B_{u\_j}] \tag{2}$$

Working Path Selection. The first step implemented by any BBR routing algorithm is to mark those links to be defined as *OSL*. Concerning the path selection process, there are four routing algorithms inferred from the BBR mechanism so far, which are classified in two main groups, depending on the parameters used to select the paths, that is, the number of hops [8] and the last advertised residual bandwidth [9]. On the first group, the *SOSP* (Shortest-Obstruct-Sensitive Path) selects the shortest path among the paths with the minimum number of *OSLs*, and the *OSSP* (Obstruct-Sensitive-Shortest Path) selects the path with the minimum number of *OSLs* among the shortest paths. Then, on the second group the *WSOSP* (Widest-Shortest-Obstruct-Sensitive Path) selects the widest path among those selected by the SOSP (when there is more than one) and the *BOSP (Balanced-Obstruct–Sensitive Path)* selects that path that minimizes the F_p value among those paths that minimize the number of *OSLs*, where F_p, represents the relation between the maximum residual bandwidth and the number of hops along a path p, according to

$$F_p = n \left[\max \left(\frac{1}{b_r^i} \right) \right] \qquad i = 1..n \tag{3}$$

where n is the number of hops and b_r^i is the available residual bandwidth on the link i in the path p. In this way, by using F_p as the cost of each link, the network load and the network occupancy is balanced in the path selection process. Simulation results obtained in [9] show that the *BOSP* algorithm is the best BBR algorithm in terms of blocking probability.

BYPASS Paths Computation. Once the route is selected, the BBR mechanism computes an alternative path that bypasses each *OSL*. These new paths are named *bypass-paths* and are used when an *OSL* cannot cope with the traffic requirements. If more than one *bypass-path* exists, the route that minimizes the number of *OSLs* is chosen (other options, such as either to minimize the number of hops or to maximize the residual available bandwidth, are left for further studies).

BYPASS Paths Usage. Being aware that the routing information updating process is not instantaneous, links that cannot cope with the traffic requirements are only known at the time of path setup. In this way, when a node detects a link i with $b^i_r < b_{req}$ it sends the setup message along the *bypass-path* which bypasses this link. Moreover, it is important to note that the *bypass-path* nodes are included in the setup signalling message as well (i.e., *bypass-paths* are also explicitly routed). In order to minimize the setup message size, *bypass-paths* are removed from the setup message when the links that they are intended to bypass have been traversed. An analysis of the complexity of the BBR mechanism can be found in [8].

2.2 BBR Current Weaknesses

There are two main weaknesses in the current BBR implementation that significantly lead the BBR to a non-optimized performance. The former appears when analyzing the *bypass-paths* utilization. In fact, there are some network scenarios where the BBR mechanism cannot be applied because a *bypass-path* to bypass the link defined as *OSL* cannot be found. This is mainly motivated by the currently hard and close *bypass-path* definition. According to such a definition, the edge nodes of the *bypass-path* to be computed must be the same that the edge nodes of the bypassing *OSL* link. Therefore, whenever a disjoint route between the edges nodes of the link defined as *OSL* cannot be found, the BBR mechanism cannot be applied. In this case, the setup message will be blocked when the selected route lacks enough bandwidth (instead of rerouting the setup message along the *bypass-path* that should have been computed), driving to an inefficient bypass-path utilization. The latter appears when analyzing the cost associated with the BBR mechanism. It is worth noting that if a *bypass-path* must be computed for each link defined as *OSL*, the routing mechanism involves a large computational cost. A straightforward solution to this problem is to reduce the number of computed *bypass-path*. Unfortunately this means either there are links defined as *OSL* for which a *bypass-path* is not computed or there are links that despite fulfilling the requirements to be defined as *OSL* they are not. Consequently a more elaborated solution to improve the BBR performance must be sought.

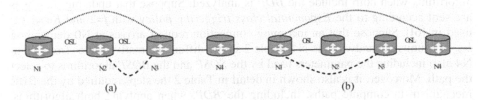

Fig. 1. BDP process

3 Optimizing the BYPASS Based Routing Mechanism

It has been shown in the last Section that the BBR mechanism suffers from two weaknesses, namely the inefficiency in the *bypass-path* utilization and the large computational cost. In this Section we show the impact of such weaknesses on the global network performance as well as we face them by extending the BBR mechanism with two new solutions. Thus, while the *BYPASS Discovery Process* (*BDP*) is proposed to address the inefficient *bypass-path* utilization, the *Network Load Dependent Policy* (*NLDP*) is proposed to reduce the computational cost required by the basic BBR mechanism.

3.1 BYPASS Discovery Process

The *BYPASS Discovery Process* (*BDP*) appears as a solution to extend the BBR applicability therefore increasing its benefits in the global network performance. In short, the BDP addresses the problem of inefficient *bypass-path* utilization by modifying the *bypass-paths* computation process targeting to increase the number of computed *bypass-paths*, so increasing the chances to apply the BBR mechanism.

Let i_j and e_j be the edge nodes of a link $l^{os}_j \in L^{os}$. Let e_{j+k} be the k node adjacent to e_j downstream along the working path. Then, the *BDP* computes *bypass-paths* in accordance with the following rules:

- Look for an alternative and disjoint route between the (i_j, e_j) pair (as done in the normal BBR mechanism).
- If there is not a feasible disjoint route between the (i_j, e_j) pair, then look for a route between the (i_j, e_{j+k}) pair, for $1 \leq k \leq d$, being e_{j+d} the destination node.

There are three main concerns to be deeply analyzed when applying the *BDP*. The first one states that feasible bypass routes cannot include any node belonging to the working path but the egress (destination) node. The two other concerns are analyzed by the examples drawn in Fig. 1. According to Fig. 1 (a) when two adjacent nodes are defined as *OSL* the *BDP* will compute a *bypass-path* to bypass both links and then another *bypass-path* to bypass the second link. The scenario is even more realistic in Fig. 1 (b) where there are two non-adjacent links defined as *OSL*. In such a scenario, despite the fact that the dash line could depict a tentative *bypass-path* to bypass link N1-N2 an optimal *bypass-path* would be that drawn by the dot line from N1 to N4 in order to reduce the number of used *bypass-paths*. Then, a *bypass-path* must also be computed to bypass link N3-N4 (not drawn in Fig. 2 (b)).

Fig.2 is used to make the *BDP* understanding easier. The performance of the *SOSP* (*Shortest-Obstruct-Sensitive Path*) and the *BOSP* (*Balanced-Obstruct-Sensitive Path*) algorithms when both include the *BDP* is analyzed. Suppose that updating messages are sent according to the *Exponential class triggering policy* with $f=2$ and $Bw=1$ (as used in [6]). Suppose that an incoming connection request arrives at N0 demanding b_{req} of 4 units of bandwidth to N4. Table 2 shows different tentative routes from N0 to N4 also including the parameters used by the *SOSP* and the *BOSP* algorithms to select the path. Moreover, it is also shown in detail in Table 2 the steps required by the BBR mechanism to compute paths, including the *BDP*, when applying both algorithms.

Table 2. BBR process

Id	Route (N)	H	OSL	b_r^{min}	F_p		BBR	SOSP	BOSP
a	0-1-2-3-4	4	1	4	1		1th step	Mark OSLs	Mark OSLs
b	0-1-5-6-7-4	5	2	7	0.71		2^{on} step	a,c	a,c
c	0-1-5-2-3-4	5	1	6	0.83		3th step	a	c
d	0-8-9-4	3	3	4	0.75		4th step	1-2 (1,5,2)	5-2 (5,6,7,4)

According to the *SOSP* behaviour (Fig.2 (a)) a *bypass-path* exists (N1-N5-N2), represented by a dash line, to bypass the edges nodes of the link defined as *OSL*, that is N1-N2. However, when applying the *BOSP* algorithm, Fig.2 (b), a *bypass-path* to bypass the edge nodes of the link defined as *OSL*, that is N5-N2, cannot be found, so limiting the potential BBR benefits.

It is shown in the example that the BBR applicability depends on the routing algorithm in use. In fact, while using the *SOSP* does not involve any problem in the BBR applicability, the scenario is quite different when using the *BOSP*. In this case, if the *BDP* mechanism is not implemented, the required *bypass-path* cannot be computed. The *BDP* allows the BBR mechanism to compute a *bypass-path* (to bypass the *OSL*) from N5 to N4 (destination node), represented by a dash line, therefore extending and improving the BBR applicability.

It is worth noting that a collateral and negative effect because of including the *BDP* in the BBR mechanism is its impact on the computational cost, since as already defined, the more the number of computed bypass-paths the more the computational cost. Therefore, the solution also proposed in this paper to reduce the computational cost is even more relevant.

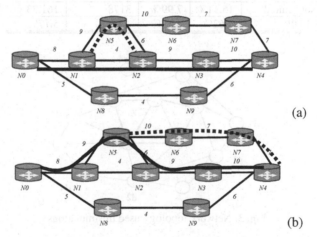

Fig. 2. Illustrative example

3.2 Network Load Dependent Policy

In [9] we carried out a simulation study concluding that the *BOSP* was the BBR routing algorithm showing the best performance in terms of blocking probability. Nevertheless, this algorithm exhibits large computational cost. In order to obtain both low blocking probability and low computational cost, we introduced a new policy named the *Network Load Dependent Policy* (*NLDP*) based on varying the number of *bypass-path* to be computed per route (*n_bp*) value according to the network load. The main *NLDP* concept is based on the following: when the network is not highly loaded no many links will be defined as *OSL*, so a low number of *bypass-paths* per route might be computed; instead, when the network is heavily loaded, a high number of *bypass-paths* per route will be needed to cover *OSL* definition. In order to limit the number of *bypass-paths* to be computed we decide ranging the *n_bp* value from 0 to 5.

4 Simulation Results

In this section we show by simulation the benefits of using both mechanisms, i.e. the *BDP* and the *NLDP*, when applying the BBR mechanism. In order to make the visibility easier we check both mechanisms separately.

The *NLDP* is evaluated over the NSFnet topology, by simulating 2500 incoming traffic connection following a Poisson distribution. Three different routing algorithms are evaluated to analyze the cost reduction together with the obtained bandwidth blocking reduction. Therefore, the *Widest Shortest Path* (*WSP*) [10] is selected as an example of QoS routing algorithm, the *Shortest-Safest Path* (*SSP*) [6] is selected as an example of QoS routing algorithm addressing the routing inaccuracy problem and finally the *BOSP* which has been defined as the BBR algorithm presenting larger benefits on bandwidth blocking reduction.

Table 3. Cost Analysis

n_bp	BW_{WSP}	BW_{SSP}	Cost (LSP)	Cost (time)
3	8.24%	4.62%	282	14.30%
Not limited	15.11%	7.99%	3178	161.23%
NLDP	7.62%	4.13%	102	5.17%

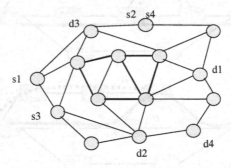

Fig. 3. Network topology used in simulations

Table 3 evaluates the impact on the BBR cost reduction because of applying the network load dependent policy. Three different situations are compared n_bp = 3, n_bp is not limited (a ∞ number of *bypass-paths* might be computed per route) and n_bp fulfils the *NLDP*. According to the *NLDP*, n_bp is in the range from 0 (network highly loaded) to 5 (network not loaded) *bypass-paths* per route. Results shown in Table 3 are obtained assuming tv = 80% (threshold value).

The values BW_{WSP} and BW_{SSP} stand for the difference in the bandwidth blocking obtained when comparing the *BOSP* to the *WSP* and the *SSP* respectively. The cost is represented in terms of both the total number of computed *bypass-paths* and the impact of computing these *bypass-paths* on the computational time. It is easy to note that the second scenario, where n_bp is not limited, is really not affordable because of the extreme cost. Analyzing the first and the last situation, while similar results in bandwidth blocking are obtained, the cost is significantly reduced when distributing n_bp proportionally to the network load.

Based on these results we can conclude that by using the *NLDP* the cost can be significantly reduced without substantially reducing the bandwidth blocking reduction.

Once we have shown the benefits obtained by including the *NLDP* in the BBR mechanism we show the benefits in the global network performance introduced when implementing the *BDP*.

In order to clearly identify the improvement obtained by the BBR mechanism when the *BDP* is also applied, the *BOSP* algorithm is evaluated in both situations, i.e., when the *BDP* is not implemented (named *BOSP*) and when the *BDP* is implemented (named *B/BDP*). Moreover, in order to evaluate the impact on the blocking probability because of the number of *bypass-paths* that can be computed per route, different simulations are carried out as a function of n_bp. The notation used in the figures to denote the n_bp values is *BOSP(n_bp)* and *B/BDP(n_bp)*. The simulations are performed over the network topology shown in Fig. 3, using the ns2 simulator extended with BBR and *BDP* features. We use two link capacities, 622 Mb/s represented by a light line and 2.4 Gb/s represented by a dark line. Every simulation requests 1700 connection demands which arrive following a Poisson distribution where the requested bandwidth is uniformly distributed between 1 Mb/s and 5 Mb/s. The holding time is randomly distributed with a mean of 120 seconds. The *Threshold* and the *Exponential class* (with $f = 2$) *triggering policies* are implemented in the simulations. The results presented in this paper have been obtained after repeating 300 seconds of simulation 10 times.

The parameters used to measure the algorithms behaviour are the routing inaccuracy and the bandwidth blocking ratio. The routing inaccuracy represents the total number of paths incorrectly selected among the total number of requested paths.

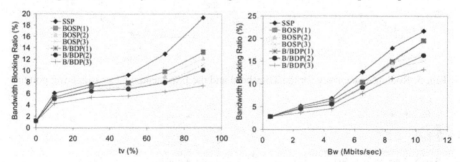

Fig. 4. Bandwidth blocking ratio for the *Threshold and the Exponential Class triggering policies*

The bandwidth blocking ratio for the *Threshold and Exponential class triggering policies* are depicted in Fig. 4. Focusing, for instance, in the *Threshold triggering policy* we can see that in any case a lower blocking is obtained when the BBR mechanism is applied compared to the *SSP* algorithm. Several conclusions can be obtained when deeply analyzing Fig. 4. Firstly, the lowest blocking is obtained when the *BDP* is applied, i.e., by the *B/BDP* algorithm. Secondly, although applying the *BDP* when n_bp = 1 hardly impacts on the blocking probability, we can conclude that the effects produced because of including the *BDP* in the BBR mechanism are completely dependent on the n_bp value. In fact, whereas a similar blocking is obtained for the

BOSP(1) and the *B/BDP(1)* algorithms, (13.27 % and 13.3 % respectively) a significant reduction of 2 % is obtained when $n\_bp = 2$. Increasing the n_bp values leads to
obtain an even more significant blocking reduction. Hence, the larger the number of
computed *bypass-paths* per route the lower the blocking, that is, the blocking depends
on the computational cost introduced in the network. Lastly, the larger the threshold
value, i.e., the larger the inaccuracy, the larger the improvement obtained in the
blocking reduction. In the worst conditions (the threshold *tv* of the triggering policy is
90 %), the bandwidth blocking ratio obtained due to applying the *BDP* process substantially improves that obtained by only applying the *BOSP*. For instance, when
$n\_bp = 3$, the obtained reduction in the blocking probability is 3.75 %.

Fig. 5 depicts the number of routes incorrectly selected for the *Threshold and the
Exponential class triggering policies*. A similar behaviour is obtained for both triggering policies. As expected, the number of routes incorrectly selected decreases with the
number of computed *bypass-paths*. Focusing again on the *Threshold triggering policy*, in the worst conditions (*tv* = 90 %), a reduction of 1.3 % is obtained when applying the *B/BDP(3)* algorithm compared to the *BOSP(1)* algorithm.

Finally, Fig. 6 shows the cost of applying the BBR mechanism with and without
BDP capabilities. Moreover, it is also drawn the variation in the cost produced as a
function of the n_bp value.

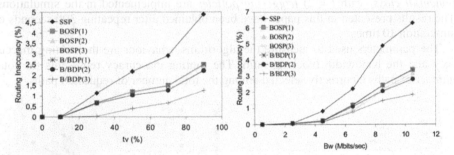

Fig. 5. Routing inaccuracy for the Threshold and the Exponential Class triggering policies

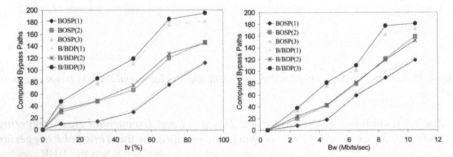

Fig. 6. Computed bypass-paths for the Threshold and the Exponential Class triggering policies

Definitely, we can conclude that extending the BBR with the *BDP* substantially
improves the global network performance. It should be also noticed that a trade-off
exists between the number of computed *bypass-paths* and the reduction obtained in
the blocking probability.

5 Conclusions

This paper describes an optimization in the BBR mechanism to increase even more the benefits of applying this routing mechanism when selecting paths under inaccurate routing information. The optimization lies in on including two new mechanisms, the *BYPASS Discovery Process (BDP)* and the *Network Load Dependent Policy (NLDP)* to address two weaknesses of the BBR mechanism, that is, inefficient *bypass-path* utilization and a large computational cost. It is really important from the point of view of the BBR implementation to address such weaknesses since otherwise the potential benefits of applying the BBR mechanism are substantially reduced.

The *NLDP* appears to reduce the cost associated with the BBR mechanism. The main concept of this policy is based on dynamically limiting the number of *bypass-paths* to be computed according to the network load.

The *BDP* addresses the scenario where a *bypass-path* between the edge nodes of a link defined as *OSL* cannot be found. In this case, the *BDP* looks for a *bypass-path* between different edge nodes, so bypassing a set of links (instead of a single link) along the working path which must include the link defined as *OSL*. As a consequence the BBR can be correctly applied.

Both mechanisms are evaluated by simulation. Obtained results show a significant reduction in the blocking probability when including both mechanisms while keeping a reduced computational cost.

References

1. T.Anjali, C. Scoglio, J. de Oliveira, L.C. Chen, I.F. Akyldiz, J.A. Smith, G. Uhl, A. Sciuto, "A New path Selection Algorithm for MPLS networks Based on Available Bandwidth Estimation", Proc. QofIS 2002, pp.205-214, Zurich, Switzerland, October 2002
2. G.Apostolopoulos, R.Guerin, S.Kamat, S.Tripathi, "Quality of Service based routing: a performance perspective". Proc. SIGCOMM 1998
3. R.Guerin, A.Orda, D.Williams, "QoS routing mechanisms and OSPF extensions", Proc. 2nd Global Internet Miniconference (joint with Globecom'97), Phoenix, AZ, November 1997
4. D.H.Lorenz, A.Orda, "QoS routing in networks with uncertain parameters", IEEE/ACM Transactions on Networking, Vol.6, n°.6, pp.768-778, December 1998
5. T.Korkmaz, M. Krunz, "Bandwdith-Delay Constrained Path Selection Under Inaccurate State Information", IEEE/ACM Transactions on Networking, vol.11, n°:3, pp.384-398, June 2003
6. G.Apostolopoulos, R.Guerin, S.Kamat, S.K.Tripathi, "Improving QoS routing performance under inaccurate link state information", Proc. ITC'16, June 1999
7. S.Chen, K.Nahrstedt, "Distributed QoS routing with imprecise state information", Proc.7th IEEE International Conference of Computer, Communications and Nettworks, 1998
8. X.Masip, S.Sánchez, J.Solé, J.Domingo, "A QoS routing mechanism for reducing the routing inaccuracy effects", Proc. QoS-IP 2003, Milan, Italy, February 2003
9. X.Masip-Bruin, S.Sànchez-López, J.Solé-Pareta, J.Domingo-Pascual,, "QoS Routing Algorithms Under Inaccurate Routing Information for Bandwidth Constrained Applications", Proc. IEEE International Communications Conference, ICC 2003, Anchorage, Alaska, May 2003.
10. R.Guerin, A.Orda, D.Williams, "QoS routing mechanisms and OSPF extensions", Proc. 2nd Global Internet Miniconference (joint with Globecom'97), Phoenix, AZ, November 1997

On the Performance
of Dynamic Online QoS Routing Schemes*

Antonio Capone, Luigi Fratta, and Fabio Martignon

DEI, Politecnico di Milano, Piazza L. da Vinci 32, 20133 Milan, Italy
{capone,fratta,martignon}@elet.polimi.it

Abstract. Several dynamic QoS routing techniques have been recently proposed for new IP networks based on label forwarding. However, no extensive performance evaluation and comparison is available in the literature. In this paper, after a short review of the major dynamic QoS routing schemes, we analyze and compare their performance referring to several networks scenarios. In order to set an absolute evaluation of the performance quality we have obtained the ideal performance of any routing scheme using a novel and flexible mathematical programming model that assumes the knowledge of arrival times and duration of the connections offered to the network. This model is based on an extension of the maximum multi-commodity flow problem. Being an integer linear programming model, its complexity is quite high and its evaluation is constrained to networks of limited size. To overcome the computational complexity we have defined an approximate model, based on the multi-class Erlang formula and the minimum multi-commodity cut problem, that provides an upper bound to the routing scheme performance. The performance presented in the paper has been obtained by simulation. From the comparison of the schemes considered it turns out that the Virtual Flow Deviation routing algorithm performs best and it almost reaches, in several scenarios, the ideal performance showing that no much gain is left for alternate new schemes.

1 Introduction

The current evolution of Internet architecture is towards service differentiation and Quality of Services (QoS) support [1]. In order to offer guaranteed end-to-end performance (as bounded delay, jitter or loss rate), it is necessary to introduce some sort of resource reservation mechanism and traffic control. With classical IP routing, however, when the resources are not available on the shortest path, the connection request is rejected even if sufficient resources exist on alternative paths.

With new label based forwarding mechanisms, such as MPLS (Multi Protocol Label Switching) [2] and GMPLS (Generalized MPLS) [3, 4], per flow path selection is possible and QoS parameters can be taken into account by routing algorithms. The goal of QoS routing schemes is to select a path for each traffic flow

---

* This work has been partially supported under the grant of MURST Tango project.

M. Ajmone Marsan et al. (Eds.): QoS-IP 2005, LNCS 3375, pp. 456–469, 2005.

(micro-flows or aggregated-flows according to routing granularity) that satisfies quality constraints based on the actual available resources in the network.

The QoS requirement of a connection can be given as a set of constraints on link and paths. For instance, bandwidth constraints require that each link on the path has sufficient bandwidth to accommodate the connection.

The QoS routing algorithms proposed in the literature [5–11] can be classified into static or dynamic, and online (on demand) or offline (precomputed) [6]. Static algorithms use only network information that does not change in time, while dynamic algorithms use the current state of the network, such as available link capacity and blocking probability. In online routing algorithms, path requests are considered one by one, and usually previously routed connections cannot be re-routed. Offline routing does not allow new path route computation and it is usually adopted for permanent connections.

From the user point of view QoS routing algorithms must satisfy the QoS requirements, while from the provider point of view they have also to maximize the resource utilization. For online routing schemes the maximum resource utilization is achieved by minimizing connection rejection probability of future requests.

This paper is focused on the performance evaluation of dynamic online QoS routing algorithms [12].

First, we review some of the most popular algorithms proposed in the literature, such as the Min-Hop Algorithm (MHA) [13], the Widest Shortest Path Algorithm (WSP) [14], the Minimum Interference Routing Algorithm (MIRA) [15], the Profile-Based Routing algorithm (PBR) [11] and the Virtual Flow Deviation (VFD) algorithm [17]. We describe in some detail MIRA and VFD algorithms. These algorithms take explicitly into account the topological layout of the ingress and egress points of the network. The VFD algorithm, recently proposed in [17], considers also the traffic statistics. More precisely, VFD exploits the knowledge of the layout of the ingress/egress nodes of the network, and uses the statistics information about the traffic offered to the network in order to forecast future connections arrivals.

Then, to provide a measure of the quality of the performance, we present some theoretical bounds to the performance achievable by any online QoS routing algorithm by means of two novel and flexible mathematical models.

The first one, Ideal Routing (IR), is an Integer Linear Programming model and is based on an extension of the maximum multi-commodity flow problem [18]. It provides an optimal routing configuration capable of accommodating the traffic offered to the network. The model minimizes the number of rejected connections assuming that the connection arrival times and their durations are known. Accepted connections are provided a single path which is maintained for the whole connection lifetime (no re-routing is allowed). The IR model describes an ideal routing scheme that achieves the minimum connection rejection probability. However, due to the complexity of its formulation, the solution of this model requires long computing time and large memory, even with state of the art optimization tools [19]. Therefore, its applicability is limited to small size network scenarios.

The second model, based on the multi-class Erlang formula and on the minimum multi-commodity cut problem [20–22] (Min-Cut model), is an approximate one and provides a looser lower bound to the connection rejection probability. It can be applied to larger and more complex network topologies since its memory occupation and computing time are considerably lower than in the first model.

The numerical results on the performance of the algorithms considered have been obtained by simulating a set of relevant network scenarios. The comparison of these results with the bounds obtained with the IR and Min-Cut models shows that the VFD algorithm performs quite close to the ideal algorithm.

The paper is structured as follows: in Section 2 we address the QoS routing problem and we review some existing routing algorithms. In Section 3 we review the Virtual Flow Deviation algorithm, pointing out its innovating features. In Section 4 we illustrate the IR model, discussing the problem of setting the objective function parameters, and the Min-Cut model. In Section 5 we analyze and discuss the performance of online algorithms under a variety of network scenarios, comparing their performance to the theoretical bounds calculated using the mathematical models. Section 6 concludes the paper.

2   Dynamic Online QoS Routing Schemes

In this section we review some of the most relevant dynamic QoS algorithms proposed in the literature. In the following we assume that all the quality parameters requested by incoming connections can be controlled by defining an equivalent flow bandwidth as discussed in [23, 24]. This assumption allows us to focus only on bandwidth constraints.

Let a network be represented by a graph $G(N, A)$, where the nodes N represent routers and arcs A represent communication links, as shown in Figure 1.

The traffic enters the network at ingress nodes S_i and exits at egress nodes T_i. Each connection requires a path from S_i to T_i. The capacity C_{ij} and the actual flow F_{ij} are associated to each link (i, j). The residual bandwidth of link (i, j) is defined as $R_{ij} = C_{ij} - F_{ij}$.

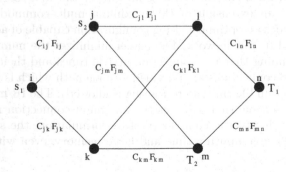

Fig. 1. QoS Network State.

A new connection can be routed only over links with R_{ij} greater or equal to the requested bandwidth. Referring to a new connection k with requested bandwidth d_k, a link is defined as *feasible* if $R_{ij} \geq d_k$. The feasible network for connection k is the sub-graph of G obtained by removing all un-feasible links. A connection can be accepted if at least one path between S_i and T_i exists in the feasible network. The minimum R_{ij} over a path defines the maximum residual bandwidth of that path.

The Virtual Flow Deviation (VFD) is a new routing algorithm, recently presented by the authors [17], that aims to overcome the limitations of the routing algorithms just reviewed by exploiting all the information available when a route selection must be taken.

To better describe the current state of the network and to forecast its future state, VFD exploits the topological information on the location of ingress/egress pairs, used by MIRA, as well as the traffic statistics obtained by measuring the load offered to the network at each source node. This information plays a key role in choosing the best route of a new request in order to prevent network congestion.

To account for the future traffic offered to the network, VFD routes not only the real call, but also some *virtual* calls which represent an estimate (based on measured traffic statistics) of the connection requests that are likely to interfere with the current real call. The number of these virtual calls, as well as the origin, destination, and the bandwidth requested should reflect as closely as possible the real future conditions of the network. These parameters can be estimated based on the past traffic statistics of the various ingress/egress pairs, as detailed in [17].

The accuracy of the measured traffic statistics is an important factor for the performance of the algorithm. However, even if the traffic statistics are not very accurate, the use of this information has shown to be effective and to provide better performance than simply using the topological information about the position of source and destination nodes as performed by MIRA.

All the information on the network topology and the estimated offered load is used to select a path which uses at best the network resources and minimizes the number of rejected calls. Such a path selection is performed in VFD by the Flow Deviation method [25, 26], which allows to determine the optimal routing for all connections entering the network.

3   Mathematical Models

In this Section we introduce two novel mathematical models that provide bounds to the performance achievable by any dynamic online routing algorithm.

The first model, Ideal Routing (IR), assumes the exact knowledge of future traffic. The routing decisions are taken to optimize the operation of the network loaded with the actual present and future traffic. No practical routing scheme can perform better. The model is based on an extension of the maximum multi-commodity problem and its solution obtained by ILP.

The second model, Min-Cut, releases all the constraints due to the network topology and optimizes the call acceptance assuming that the min-cut capacity of the network can be fully exploited. Its solution, based on minimum multicommodity cut problem, is easier to obtain than IR. However, it provides a looser lower bound to the rejection probability that can anyway be used as a performance benchmark for large networks scenarios.

3.1   Ideal Routing Model

The basic assumption of the IR model is the knowledge of future traffic offered to the network. Let $K = \{1, \ldots, N_c\}$ be the set of connections, each one represented by the triplet (S_k, T_k, d_k) that specify source node, destination node and requested bandwidth. Connection k is further characterized by its arrival time, t_k and its duration τ_k. Given N_c connections (Fig. 2 shows an example for $N_c = 4$), the time interval from the arrival of the first connection and the last ending time of a connection is subdivided in a set I of $2N_c - 1$ time intervals.

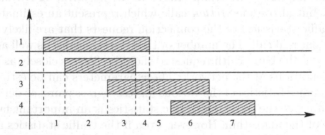

Fig. 2. Arrival time and duration of the connections offered to the network.

In each time interval, t, the number of active connections $M(t)$ remains constant. This number changes by one from interval to interval: it increases if a new connection arrives, and decreases if a connection ends. Let us denote with $B(k)$ the time interval beginning at the arrival time of connection k, and with I_k the set of time intervals in which connection k is active.

Given the function $M(t)$, the optimum routing must minimize the call rejection probability. This optimization problem can be formulated as Integer Linear Programming (ILP) if the following notations and definitions are adopted.

Let $G = (N, A)$ be the direct graph representing the network. Let $n = |N|$ and $m = |A|$ be the number of nodes and arcs, respectively. The capacity C_{ij} is associated to each arc (i, j).

For each connection k, $k \in K$, create two new nodes SS_k and TT_k and two new directed arcs, (SS_k, S_k) and (T_k, TT_k), of infinite capacity. Let SN and TN be the sets of the added nodes containing all SS_k and TT_k, respectively. Similarly, let A_{SN} and A_{TN} be the sets of arcs containing all (SS_k, S_k) and (T_k, TT_k), respectively.

Finally let $G' = (N', A')$ with $N' = N \cup SN \cup TN$ and $A' = A \cup A_{SN} \cup A_{TN}$.

Based on the above definitions and notation, we establish the ILP formulation of the IR model. To this purpose, let us define the following decision variables:

$$x_{ijt}^k = \begin{cases} 1 \text{ if connection } k \text{ is routed on arc } (i,j) \text{ in time slot } t \\ 0 \text{ otherwise} \end{cases}$$

for $(i,j) \in A'$, $k \in K$ and $t \in I$. We force $x_{ijt}^k = 0$, $\forall t \notin I_k$.

Since the goal is to minimize the connection rejection probability, we can equivalently maximize the number of connections accepted by the network. The problem can thus be formulated as follows:

$$Maximize \sum_{k \in K} b_k \cdot x_{SS_k S_k B(k)}^k \tag{1}$$

$$s.t. \sum_{k \in K} d_k \cdot x_{ijt}^k \leq C_{ij} \quad \forall (i,j) \in A, t \in I \tag{2}$$

$$\sum_{(j,l) \in A'} x_{jlt}^k - \sum_{(i,j) \in A'} x_{ijt}^k = \begin{cases} 1 \text{ if } j \in SN \\ 0 \text{ if } j \in N \quad \forall k \in K, j \in N', t \in I \\ -1 \text{ if } j \in TN \end{cases} \tag{3}$$

$$x_{ijt}^k = x_{ijB(k)}^k \quad \forall k \in K, (i,j) \in A', t \in I_k \tag{4}$$

$$x_{ijt}^k \in \{0,1\} \quad \forall k \in K, (i,j) \in A', t \in I_k \tag{5}$$

The objective function (1) is the weighted sum of the connections accepted in the network, where b_k represents the benefit associated with connection k. Different settings of b_k are possible, and they reflect different behaviors of the model as discussed later.

Constraints (2) ensure that, at each time slot, the total flow due to all the connections that use arc (i,j) does not exceed the arc capacity, C_{ij}, for all $(i,j) \in A$.

Constraints (3) represent the flow balance equations expressed for each node belonging to the extended graph G', in each time slot $t \in T$. Note that these constraints define a path for each connection between its source and destination nodes.

Constraints (4) impose that the accepted connections cannot be aborted or rerouted for their entire lifetime.

Finally, requiring that the decision variables in (5) are binary implies that each connection is routed on a single path.

The online QoS routing algorithms we are considering in this paper do not reject a new connection with (S_k, T_k, d_k) if at least one path with a residual available bandwidth greater than or equal to the requested bandwidth d_k exists.

To account for this feature, the objective function (1) must be properly set. To this purpose it is sufficient to set:

$$b_k = 2^{N_c - k} \tag{6}$$

having numbered the N_c connections from 1 to N_c according to their arrival times. With such a setting of b_k the benefit to accept connection k is always greater than the benefit of accepting, instead, all the connections from $k + 1$ to N_c, since:

$$2^{N_c-k} > \sum_{i=k+1}^{N_c} 2^{N_c-i} \tag{7}$$

This choice of the weights b_k allows the mathematical formulation to model very closely the behavior of real online routing algorithms. To verify the accuracy of the model we have considered a simple scenario where a single link connects a source-destination pair. We have obtained the performance in the case of channel capacity equal to 20 bandwidth units and assuming the bandwidth b_k to be uniformly distributed between 1 and 3 units and the lifetime τ_k to be exponentially distributed with mean 15 s.

In this simple case all the routing algorithms provide the same performance since only one path exists between source and destination. The rejection probability shown in Fig. 3(a) has been computed using the multi-class Erlang Formula. The bound provided by the IR model completely overlaps the online routing performance. Note that different choices of b_k provide different IR Model performance. For instance, selecting $b_k = 1$ for all k we obtain the performance shown in Fig. 3(b). The large reduction in rejection probability is expected since the optimization of the objective function will result in rejecting connections with high bandwidth requirements and long lifetime in favor of smaller and shorter ones. The difference between the bound and the real performance increases as the network load increases.

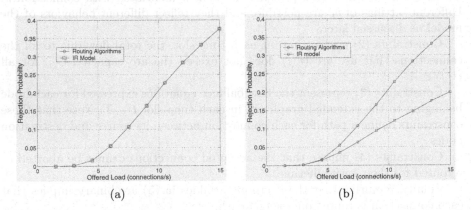

Fig. 3. Connection rejection probability versus the average total load offered to a single link with (a) $b_k = 2^{N_c-k}$ (b) $b_k = 1$.

3.2   Min-Cut Model

In this Section we propose a second mathematical model that allows to determine a lower bound to the connection rejection probability. The solution of this model has computing times and memory occupation considerably lower than the previous one. However, in some scenarios, the bound obtained can be quite lower than the value provided by IR.

Let us consider a directed graph $G = (N, A)$ defined by a set of nodes, N, and a set of arcs, A each one characterized by a capacity C_{ij}. A set of source/destination pairs $K = \{1, ..., N_s\}$, indicated by S_i and T_i, respectively, $i \in K$, is also assigned. Each source S_i generates a flow αf_i, towards destination T_i, that can be split over multiple paths. The problem is to find the maximum α, indicated by α^*, such that for all $i \in K$ the flow quantities $\alpha^* f_i$ can be routed to their destinations.

The solution to this problem, obtained via linear programming techniques, provides the maximum multi-commodity flow $F_{max} = \sum_{i \in K} \alpha^* f_i$. Note that F_{max} represents a lower bound to the capacity of the minimum multi-commodity cut of the network, as discussed in [21, 22].

Once F_{max} has been obtained, the connection rejection probability for the given network scenario is obtained by using the multi-class Erlang formula with F_{max} servers [20] that is briefly reviewed in the following.

Let us consider N different traffic classes offered to a network system with C servers. The connections belonging to the class i request d_i bandwidth units. The connections arrival process is a Poisson process with average λ_i, while the connections duration is distributed according to a generic distribution $f_{\Theta_i}(\theta_i)$. Let $\Lambda = \sum_{i=1}^{N} \lambda_i$ be the total load offered to the network.

An appropriate state description of this system is $n = (n_1, ..., n_N)$, where $n_i, i = 1, ..., N$ is the number of connections belonging to the class i that occupy the servers. The set of all the possible states Ω is expressed as $\Omega = \{n | X \leq C\}$, where X, the total occupation of all the servers, is given by $X = \sum_{i=1}^{N} n_i d_i$.

If we indicate with $A_i = \lambda_i E[\Theta_i]$ the traffic offered to the network by each class, the steady state probability of each state is simply given by the multi-class Erlang formula:

$$\pi(n) = \frac{1}{G} \prod_{i=1}^{N} \frac{A_i^{n_i}}{n_i!} \tag{8}$$

where G is the normalization constant that ensures that the $\pi(n)$ sum to 1 and it has therefore the following expression:

$$G = \sum_{n \in \Omega} \pi(n) = \sum_{n \in \Omega} (\prod_{i=1}^{N} \frac{A_i^{n_i}}{n_i!}) \tag{9}$$

Using the steady state probability calculated with equation (8) we can derive the loss probability of the generic class i, Π_i, as follows:

$$\Pi_i = \sum_{n \in B_i} \pi(n) \tag{10}$$

where B_i is the set of the blocking states for the class i, defined as $B_i = \{n | C - d_i < X \leq C\}$. The overall connection rejection probability, p_{rej}, is then given by:

$$p_{rej} = \sum_{i=1}^{N} \frac{A_i \Pi_i}{\sum_{i=1}^{N} A_i} \tag{11}$$

If we substitute C with the maximum multi-commodity flow value F_{max} in all the above expressions, we can compute the connection rejection probability using equation (11).

In network topologies with high link capacities, F_{max} can assume high values, and the enumeration of all the allowed states becomes computationally infeasible, since the cardinality of Ω is of the order of F_{max}^N [27]. In these network scenarios, equations (8)-(11) are computationally too complex so we propose to apply the algorithm described in [27, 28] that computes recursively the blocking probability based on the peculiar properties of the normalization constant G. For network topologies with very high link capacities we implemented the inversion algorithm proposed in [29] to compute the blocking probabilities for each class.

4   Numerical Results

In this Section we compare the performance, measured by the percentage of rejected calls versus the average total load offered to the network, of the Virtual Flow Deviation algorithm, the Min-Hop Algorithm and MIRA with the bounds provided by the mathematical models presented in the previous Section referring to different network scenarios in order to cover a wide range of possible environments.

The first scenario we consider is illustrated in Figure 4. In this network the links are unidirectional with capacity equal to 120 bandwidth units. In the following capacities and flows are all given in bandwidth units. The network traffic, offered through the source nodes S_1, S_2 and S_3, is unbalanced since sources S_2 and S_3 generate a traffic four times larger than S_1. Each connection requires a bandwidth uniformly distributed between 1 and 3. The lifetime of the connections is assumed to be exponentially distributed with average equal to 15s.

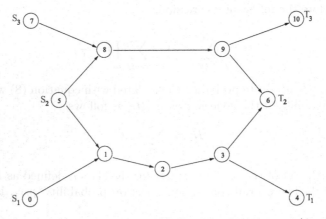

Fig. 4. Network topology with unbalanced offered load: the source/destination pairs S_2-T_2 and S_3-T_3 offer to the network a traffic load which is four times higher than that offered by the pair S_1-T_1.

In this simple topology connections S_1-T_1 and S_3-T_3 have one path only, while connections S_2-T_2 have two different paths.

The rejection probability versus the offered load for MIRA, MHA, VFD, IR and Min-Cut models are shown in Fig. 5. The poor performance of MIRA is due to its lack of considering any information about the load distribution in the network. In this particular topology, due to critical links (1,2), (2,3) and (8,9), S_2-T_2 connections are routed on the path (5-8-9-6) that contains the minimum number of critical links. MHA, that selects for connections S_2-T_2 the path with the minimum number of hops, routes the traffic as MIRA and their performances overlap. Better performance is achieved by VFD. Since its behavior depends on the number of virtual connection N_v' used in the routing phase, we have considered three cases: $N_v' = 0$, $N_v' = 0.5 \cdot N_v$ and $N_v = \lfloor (N_{max} - N_A) \rfloor$. In the first case, even if no information on network traffic statistics is taken into account, the VFD algorithm achieves much better performance than previous schemes due to the better traffic balance provided by the Flow Deviation algorithm. Only when the offered load reaches very high values the improvement reduces. The third case corresponds to the VFD version described in Section 3.2 that takes most advantage from traffic information. The best performance has been measured and the gain over existing algorithms is provided even at high loads. An intermediate value of N_v' (case 2) provides, as expected, intermediate performance. As far as the performance of the two mathematical models, we observe that the approximate Min-Cut model curve overlaps that of the IR model. Note that VFD performs very close to the theoretical bounds in this scenario.

To investigate the impact of connection lifetime distribution, we have considered a Pareto distribution with the same average as the previous exponential distribution and several shape parameters ($\alpha = 1.9, 1.95, 2.1, 3$). The performance observed in all cases are within 1% of those shown in Fig. 5.

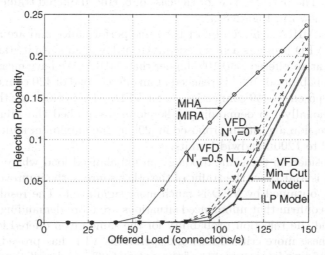

Fig. 5. Connection rejection probability versus the average total load offered to the network of Figure 4.

To test the sensitivity of the performance to the network capacity, we have considered, for the network in Fig. 4, different parameters. The results, shown in Fig. 6, are very similar to those of Fig. 5. It is worthwhile to observe that in all the different scenarios considered the approximate model provides results very close to IR. This validates the approximate model that can be easily evaluated even in more complex networks.

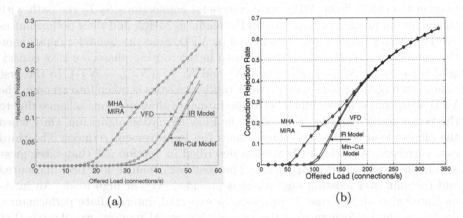

(a)   (b)

Fig. 6. Connection rejection probability versus the average total load offered to the network of Fig. 4 (a) with link capacity equal to 24 and bandwidth requests always equal to 1 (b) with link capacity equal to 60 and bandwidth requests always equal to 1.

A more realistic scenario that was first proposed in [15] is shown in Fig. 7. The links marked by heavy solid lines have a capacity of 480 while the others have a capacity equal to 120, in order to replicate the ratio between OC-48 and OC-12 links. The performance for the case of balanced offered traffic, considered in [15], are shown in Figure 8(a).

VFD and MIRA achieve almost the same performance and are much better than MHA. VFD presents a slight advantage at low load since it starts rejecting connections at an offered load 10% higher than MIRA. We have measured that a rejection probability of 10^{-4} is reached at an offered load of 420 connections/s by MIRA as opposed to 450 connections/s for VFD. Also in this case the IR model is computationally too demanding. Therefore, we applied the Min-Cut model with the inversion algorithm proposed in [29], as the maximum multicommodity flow is equal to 1200 bandwidth units.

If we consider on the same topology an unbalanced load where for instance traffic S_1-T_1 is four times the traffic of the other sources, the improvement in the performance obtained by VFD is much more significant. The results shown in Figure 8(b) confirm that unbalanced situations are more demanding on network resources and the rejection probability for the same given offered load is much higher. In these more critical network operations VFD has proved to be more effective providing improvements of the order of 20% and well approaching the lower bound provided by the Min-Cut model.

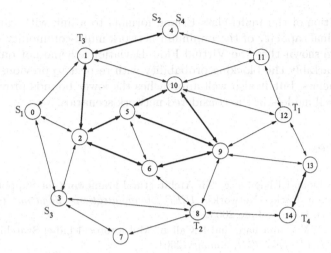

Fig. 7. Network Topology with a large number of nodes, links, and source/destination pairs.

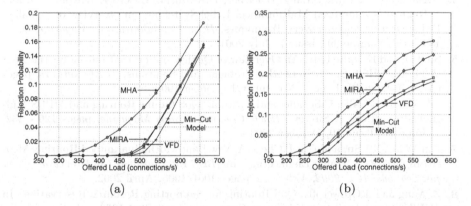

Fig. 8. Connection rejection probability versus the average total load offered to the network of Fig. 7 (a) where all sources produce the same amount of traffic; (b) where the traffic between S_1-T_1 is four times higher than the traffic produced by the other pairs.

5   Conclusions

We have discussed and analyzed the performance of online QoS routing algorithms for bandwidth guaranteed connections in MPLS and label switched networks.

To provide a theoretical bound on the performance achievable by dynamic online QoS routing algorithms we have proposed two novel mathematical models. The first is an Integer Linear Programming model that extends the well known maximum multi-commodity flow problem to include connections arrival-times and durations, while the second, which has a much lower complexity, is based on the application of the multi-class Erlang formula to a link with capacity equal to the residual capacity of the minimum network multi-commodity cut.

We have shown that the Virtual Flow Deviation scheme not only allows to reduce remarkably the blocking probability with respect to previously proposed routing schemes, but it also well approaches the lower bounds provided by the mathematical models in the considered network scenarios.

References

1. Hui-Lan Lu and I.Faynberg. An Architectural Framework for Support of Quality of Service in Packet Networks. *IEEE Communications Magazine*, pages 98-105, Volume 41, Issue 6, June 2003.
2. E.Rosen, A.Viswanathan, and R.Callon. Multiprotocol Label Switching Architecture. In *IETF RFC 3031*, January 2001.
3. Ed. L. Berger. Generalized Multi-Protocol Label Switching (GMPLS) Signaling Functional Description. In *IETF RFC 3471*, January 2003.
4. A.Banerjee, L.Drake, L.Lang, B.Turner, D.Awduche, L.Berger, K.Kompella, and Y.Rekhter. Generalized Multiprotocol Label Switching: an Overview of Signaling Enhancements and Recovery Techniques. *IEEE Communications Magazine*, pages 144 - 151, Volume 39 , Issue 7, July 2001.
5. S.Chen and K.Nahrstedt. An Overview of Quality-of-Service Routing for the Next Generation High-Speed Networks: Problems and Solutions. *IEEE Network*, pages 64-79, Volume 12, Issue 6, November/December 1998.
6. J.L.Marzo, E.Calle, C.Scoglio, and T.Anjali. QoS Online Routing and MPLS MultilevelProtection: A Survey. *IEEE Communications Magazine*, pages 126 - 132, Volume 41 , Issue 10, October 2003.
7. Bin Wang, Xu Su, and C.L.P.Chen. A New Bandwidth Guaranteed Routing Algorithm for MPLS Traffic Engineering. In *IEEE International Conference on Communications, ICC 2002, Volume 2*, pages 1001–1005, April 2002.
8. Z.Wang and J.Crowcroft. QoS Routing for Supporting Resource Reservation. In *IEEE Journal on Selected Areas in Communications*, Sept.1996.
9. A.Orda. Routing with End to End QoS Guarantees in Broadband Networks. In *IEEE INFOCOM'98*, Mar.1998.
10. B.Awerbuch et al. Throughput Competitive On Line Routing. In *34th Annual Symp. Foundations of Computer Science*, Palo Alto, CA, Nov.1993.
11. S.Suri, M.Waldvogel, D.Bauer, and P.R.Warkhede. Profile-Based Routing and Traffing Engineering. *Computer Communications*, pages 351 - 365, Volume 26 , Issue 4, March 2003.
12. A.Capone, L.Fratta, and F.Martignon. Dynamic Online QoS Routing Schemes: Performance and Bounds. *submitted to Computer Communications*.
13. D.O.Awduche, L.Berger, D.Gan, T.Li, V.Srinivasan, and G.Swallow. RSVP-TE: Extensions to RSVP for LSP Tunnels. In *IETF RFC 3209*, December 2001.
14. R.Guerin, D.Williams, and A.Orda. QoS Routing Mechanisms and OSPF Extensions. In *Proceedings of Globecom*, 1997.
15. Murali S. Kodialam and T. V. Lakshman. Minimum Interference Routing with Applications to MPLS Traffic Engineering. In *Proceedings of INFOCOM*, pages 884–893, 2000.

16. Koushik Kar, Murali Kodialam, and T.V. Lakshman. Minimum interference routing of bandwidth guaranteed tunnels with MPLS traffic engineering applications. *IEEE Journal on Selected Areas in Communications*, December 2000, Volume 18, Issue 12.

17. A.Capone, L.Fratta, and F.Martignon. Dynamic Routing of Bandwidth Guaranteed Connections in MPLS Networks. *International Journal on Wireless and Optical Communications*, pages: 75-86, Volume 1, Issue 1, July 2003.

18. R.K.Ahuja, T.L.Magnanti, and J.B.Orlin. *Network Flows*. Prentice-Hall, 1993.

19. *ILOG CPLEX*. Available at http://www.ilog.com/products/cplex/.

20. J.F.P. Labourdette and G.W. Hart. Blocking probabilities in multitraffic loss systems: insensitivity, asymptotic behavior, and approximations. *IEEE Trans. on Communications*, 40:1355–1366, 1992.

21. Tom Leighton and Satish Rao. Multicommodity Max-Flow Min-Cut Theorems and their Use in Designing Approximation Algorithms. *Journal of the ACM*, Volume 46, Issue 6, November 1999.

22. Y.Aumann and Y.Rabani. An O(log k) Approximate Min-Cut Max-Flow Theorem and Approximation Algorithm. In *SIAM Journal on Computing*, pages 291–301, Volume 27, Issue 1, February 1998.

23. R.Guerin, H.Ahmadi, and M.Naghshineh. Equivalent capacity and its application to bandwidth allocation in high speed networks. In *IEEE Journal on Selected Areas in Communications*, pages 968–981, September 1991.

24. J.A.Schormans, J.Pitts, K.Williams, and L.Cuthbert. Equivalent capacity for on/off sources in ATM. In *Electronic Letters*, pages 1740–1741, Volume 30, Issue 21, 13 October 1994.

25. L.Fratta, M.Gerla, and L.Kleinrock. The Flow Deviation Method: An Approach to Store-and-forward Network Design. In *Networks 3*, pages 97–133, 1973.

26. D.Bertsekas and R.Gallager. *Data Networks*. Prentice-Hall, 1987.

27. J.S.Kaufman. Blocking in a Shared Resource Environment. *IEEE Transactions on Communications*, pages 1474 - 1481, Volume 29, Issue 10, 1981.

28. A.A.Nilsson, M.Perry, A.Gersht, and V.B.Iversen. On Multi-rate Erlang-B Computations. *in 16th International Teletraffic Congress; ITC'16*, pages 1051 - 1060, Elsevier Science, 1999.

29. G.L.Choudhury, K.K.Leung, and W.Whitt. An Inversion Algorithm to Compute Blocking Probabilities in Loss Networks with State-Dependent Rates. *IEEE/ACM Transactions on Networking*, pages 585 - 601, Volume 3, Issue 5, 1995.

A Dynamic QoS-Aware Transmission
for Emergency Traffic in IP Networks

Kazi Khaled Al-Zahid and Mitsuji Matsumoto

Global Information and Telecommunication Studies, Waseda University,
1011 Okuboyama Nishitomida, Honjo, Saitama 367-0035, Japan

Abstract. The Internet Emergency Preparedness (IEPREP) working group of the Internet Engineering Task Force (IETF) is investigating to provide solutions for emergency systems based on IP networks. With the emergence of new applications in heterogeneous network environment, the need of adaptation to serve in emergencies with acceptable Quality of Service (QoS) is urgently required. With the help of IETF's integrated and differentiated services extending over best effort Internet, it is possible to reach certain Service Level Agreement (SLA) but not sufficient to serve all applications. In this paper we proposed a dynamic differentiated service approach for emergency traffic so that during emergency situation the routers in the source-destination shortest path tune themselves using Software based Router Agent (SRA) to serve the emergency traffic with optimum effort. With QoS assurance infrastructure, applications specify what they want and network tries to honor their request. Here we presented how SRA and XML based Quality of Service sPecification ($XQSPec$) played important role to tune IP routers in access and transit network to get better QoS for time sensitive emergency traffic over the best effort Internet.

1 Introduction

An important factor during emergencies is to contact the service professional rapidly to minimize the loss of life and property. Communication systems can help in three different roles: emergency calling, emergency communications, and emergency alerting [1]. Public systems like PSTN and mobile telephones are already used for these purposes. But Next Generation Internet (NGI) demands system which uses only IP. We hope future networks will be a heterogeneous access medium where IP will be the common denominator [2]. Internet based communications offer new challenges as since there are no such acceptable systems running that exploits end-to-end IP devices. Internet does not use signaling to transfer message as fixed telecommunication does. But to ensure the call setup we need signaling mechanism which can establish the communication within an agreeable time limit. Different research works are under consideration to solve these problems. SIP, H.323 is one of such signaling protocol used in internet. Ensuring the call setup is not necessarily solving the overall problem.

Different applications can be used to reach emergency center and among them IP telephone is the most usable and user friendly. Internet telephony has

M. Ajmone Marsan et al. (Eds.): QoS-IP 2005, LNCS 3375, pp. 470–480, 2005.

a significant growth in last few years due to its low price calls compared to PSTN and mobile telephones. Different VoIP application services like net2phone is also very popular for their low cost and easily accessible from any multimedia PC connected with the Internet. PSTN based service is already equipped but emergency service using IP still lacking some infrastructural components to provide ultimate user satisfaction especially in emergency cases. The voice traffic in PSTN network suffers propagation delay but becomes stable as soon as the circuit establishes. On the other hand Internet is still the best effort service and provides no end to end guarantee unless it is carefully considered. With the advancement of technology the call setup delay is reduced in recent years but no preferential treatment is taken in contrast to service guarantees. The overall problem of service assurances for Internet emergency communications consists of addressing the following technical areas "unpublished" [3].

1. Authentication, authorization, and privacy for users of emergency service capabilities.
2. Application level signaling and service control for voice, conferencing, instant messaging, video, web browsing, etc.
3. Network level packet delivery and performance.

To address this problem, here we propose an on demand *XQSPec* for DiffServ network architecture which works as a dynamic probe to establish the QoS path between the source-destination peer based on the application preference. The type of service (ToS) byte in IPv4 or traffic-class byte in IPv6 or DS-Code Point (DSCP) provides packet marking mechanisms which we adopted here to differentiate the emergency traffic from traditional Best Effort (BE) traffic.

This paper is organized as follows. In Section 2, existing QoS mechanism is given. In Section 3 and 4, the detail system design is described. Finally, some considerable issues, conclusion and future directions are given in Section 5 and 6.

2   Preliminaries

2.1   Quality of Service

IETF proposed three models to facilitated true end to end QoS viz. best effort, integrated and differentiated services. The primary goal of QoS is to provide priority including dedicated bandwidth, controlled jitter and latency (required by some real-time and interactive traffic), and improved loss characteristics [4]. For the best effort service, the network delivers data without any kind of assurance for reliability. Integrated Service (IntServ) is a multiple service model that can accommodate multiple service requirements. In this type of model, the application requests a specific kind of service from the network before it sends data. Intserv explicitly relies on Resource Reservation Protocol (RSVP) to signal the network about its traffic profile and reserve the required QoS for each flow in the network. Transmission takes place only after the confirmation of reservation from the network is confirmed. Differentiated Service approach works in a way to set up elements of the network to serve multiple classes of service. It operates in the network layer of the Open Systems Interconnection (OSI) model and defines a packet marking technique to distinguish between packets belonging to different classes of service based on QoS [5].

2.2   QoS vs. Signaling

Implementing QoS for internet's packet handling is not as simple as we sketch it. However the separation of signaling at the application layer from the QoS signaling at the IP layer does not reduce the interaction between the two signaling layers. Because messages of Session Initiation Protocol (SIP) [6] which is widely used for signaling in Internet can take different paths than the media packet travel and as a consequence SIP servers have no access to the media path. Previously, call signaling on a circuit-switched network was sufficient to provide preferential call setup and a clear sounding call. This is because signaling reserves resources for a voice channel along the path between caller and callee [7]. It is assumed that the network infrastructure works correctly and that emergency communications are to be given preferential treatment to remaining resources. However, IP networks do not function in the same way and hence new analysis is required. In general the access to Internet can be classified in three top categories. Source access network, core transit network and destination access network. The access devices in these areas are called Edge, Core and Border router respectively. Different research stated that there are no significant changes in call setup time even if we use QoS signaling with SIP during call initiation phase. Actually session establishment is independent of QoS with generic call flow of SIP methods. Therefore, prioritized SIP signaling does not solve the call quality issues rather it requires special mechanism to handle the packets in the media stream. IP QoS assurance or over provisioning network resources can give better call quality. QoS is available in IP network but no integration is done yet over multiple ISP for guaranteed flow [7] . So a new business rule is necessary to integrate the Service Provider (SP) to create QoS responsive system for emergency traffic.

Various reservation oriented approaches can be adopted to provide preferential treatment for emergency traffic such as resource conservation approach, preemption approach, delay based policy, etc. [3]. But in the most cases we faced some disadvantage of adopting the policy to deploy it in practical situation. On the other hand, changes in packet handling will help provide QoS support in IP network. So rather using per flow signaling, it is desirable to use packet level signaling, so that the router can handle it dynamically.

2.3   QoS Monitoring

Existing QoS estimation techniques either make use of administrative control abilities or assume no cooperation from the core network and attempt to perform estimations from the network edges by sending probes to other end. SNMP (Simple Network Management Protocol) and RMON (Remote Monitoring) are most widely used administrative methods for observing traffic characteristics in large domain. Edge-based estimation is the inference of network QoS by sending probes across the network and observing the treatment they receive in terms of variability of delay and loss in delivery to the destination. A large variety of such tools exists to estimate performance in terms of delay, jitter, packet loss, and bandwidth. They generally either use Internet Control Messaging capabilities or packet dispersion techniques [8].

2.4   Why XML with QoS?

The Extensible Markup Language (XML) is a textual markup language subset of SGML that is defined in 1996 by W3C (World Wide Web Consortium) [9]. The XML-based information channel isolates the platform from the various database structures. It enables the platform independent interchange to any database via a unified language. It is human friendly and easily deployable irrespective to device capability and also platform independent. It is also a universal hub for information exchange. The XML specification has the following properties:

1. It can be used to store any kind of structured information, and to enclose or encapsulate information in order to provide different abstraction level.
2. XML makes it easy to generate new data automatically and dynamically (i.e. transformation rules) and to ensure that the data structure is unambiguous.
3. XML is self describing and offers a modular data structure.

In this paper, XML is used to manage QoS profile which enables network to provide guaranteed service in on demand basis.

3   Design Overview

Our proposal has following breakdowns. Packet marking for emergency dispatch in both application and network layer, QoS based route selection between source-destination pair, and adjust the DiffServ architecture that allows the Emergency Traffic (ET) with guaranteed QoS.

3.1   Traffic Marking

There are different ways to marking IP packets for respective classes. Among them one technique is to set TOS bit in the IP header for each traffic flow and another one is to specify traffic trunks with the specified class of service and associate the traffic trunk at the ingress router. To ensure SLA for any class of service, SP needs to define Per Hop Behavior (PHB) policy for different classes. In IP DiffServ architecture with different queuing scheme such as WFQ, PQ, FIFO, RED/WRED etc. IP packet is allowed to process differently by PHB rules. But it requires the agreement between the SP along the passage where the packet travels to reach the destination. Therefore a predefined agreement is necessary to treat the various classes in different way based on their priority. But

```
<?xml version="1.0" encoding="UTF-8"?>
<!--XQSpec for End-to-End QoS-->
<Session xmlns:xsi=http://www.w3.org/2001/XMLSchemainstance>
<Application Information>
     <flow type="IP telephony" source_addr="133.9.109.218"
     Source_port="2005" destination_addr="200.10.109.200"
     Destination_port="8000"
     </flow>
</Application Information>
<Network Parameter
     resource_id="XYZ"  Protocol="UDP"
     Reliability="full"     Priority="High"
     DSByte="11111100"  Bandwidth="r Mbps"
     Latency="None" Jitter="None"  Delay="None"
     Packet Loss="2%"  QueuingMethod="WFQ"
     Scheduling="non-Preemptive"
</Network Parameter>
<Cost value="X"  Condition="Y"</Cost>
</Session>
```

Fig. 1. XML based QoS Specification.

such a static agreement between different ISP does not show satisfactorily consistent with all sorts of traffic. The SP also not interested to do it permanently. Therefore we classified the traffic in two major classes- ET and non-emergency traffic (nET). IP's TOS byte (or DSCP) is used for the packet classification. The classification is performed by application that process emergency requests to a specific destination. Next a scheme is proposed to generate XML based probe packet which dynamically configure routers those indeed take part in the transmission of ET between source-destination pair during media/data packet transmission.

3.2   XML Based QoS Specification (XQSPec)

Generating QoS specification from user end emergency application is one of the important schemes in our proposal. According to this proposal ET will create its own *XQSpec* based on ET type, class of service(COS), TOS or DS Code Point (DSCP) information, queuing methods, priority over other traffic, dropping or shaping policy, required resources etc. The application attributes and relevant network parameters that we used to form XML based QoS specification can be grouped as shown in Table 1.

The *SRA* adjusts them on time when they receive *XQSpec* to handle ET. The processing of the *XQSpec* is distributed and depends on the authoritative *SRA*.

Table 1. Classification of data and network parameter.

Parameter Classification	Description
Media control parameter	sampling rate, sampling bit, frame rate, resolution, etc.
Traffic control parameter	bandwidth, delay, lost, rate, bit error rate, jitter, etc.

Whenever router receives *XQSpec, SRA* handovers it to the XSLT processor and retrieves the detail description of the session and takes the decision instantly for either acceptance or rejection. So the role of SP is vital for better QoS agreement which is solved here by creating a trade off between the network QoS and cost. But in future the government and standard body should regulate a minimum rule to serve ET using Internet. A typical *XQSpec* in XML serialization could be an example shown in Fig. 1.

3.3   User Interface (UI) with QoS Support

Emergency application generates traffic and sends them through sockets to transport layer (TCP or UDP) and then on to the network layer (IP). In this IP layer, the kernel looks up the route to the host in either from the routing cache or it's Forwarding Information Base (FIB). For our case, we know the destination IP during the call setup phase through the signaling method. Here UI generates different service classes depending on user's application preference. To behave properly, all packets are given tag by the application to differentiate them from BE traffic before they put in the physical medium. UI also generates QoS sensible *XQSpec* probe with required parameters to satisfy the transmission of ET for ensuring specified QoS and SLA between the communication peer.

3.4   Dynamic QoS Routing Algorithm

Initialization: Let, assume the initial QoS aware path is **R**[partial]={Source, intermediate path, Destination}, while the intermediate path is empty. To find a complete path, source requests with the *XQSpec* to the next hop router that falls in the so called shortest path to destination and do the same in each node at intermediary routers with the help of traditional routing table until it reaches the destination.

When Destination Reached: Whenever *XQSPec* probe reaches the destination, it completes the path and sends back the complete QoS route information **R**[complete] using the reverse unicast route, it discovered from source to destination.

Visiting Intermediary Routers: As in (Fig. 2) i, j,k,l,m,n are intermediary routers between the source-destination pair and let the current visiting node is i. Using the existing shortest path algorithm, $s \rightarrow i \longrightarrow l \longrightarrow d$ is the shortest path from source to destination. In our proposed system, *SRA* is tightly coupled

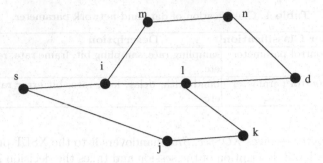

Fig. 2. How XQSpec discovers path.

with existing router entity and works based on the existing routing information. As soon as router i (in Fig. 2) gets the *XQSpec* from last hop router (any router exclusive destination) s (predecessor of i in the shortest path) two things may happen:

1. If i agrees with the requested *XQSpec,* it sends *ACCEPT* reply to the last requested router and insert the node identity in route vector R. Or
2. If i is incapable to accept the offer then replies *DENY* message back to the requester router.

The *SRA* of requester router (here s) waits time, $T_w = RTT_{s \longrightarrow i}$ (round trip time from node s to i) + internal processing time (required time to check *XQSpec* using XSLT processor) for either *ACCEPT* or *DENY* or $XQSpec_n$ (the network/logical resource parameters the current visiting node can support other than the requested one) reply. If the *ACCEPT/DENY* or negotiable $XQSpec_n$ reply does not reach within time T_w, then the requester router (here s) performs one of the following below.

If Visiting Router Agree with the Offer: If the current visiting router i agrees to accept the offer then it inserts its address in $\mathbf{R}[s, i, ..., d]$ and forwards the information $(\mathbf{R}[s, i, ..., d], XQSpec)$ to the immediate next hop router l along the shortest path to destination.

If the Visiting Router Rejects: If i *REJECT* the offer from s, two things may happen,

1. If *SRA* finds totally unable to serve the requested *XQSpec* then it simply sends back *DENY* reply. Then the last hop will try for alternate path to destination in next available shortest path. But if there is no available shortest path from current visiting node (s) after the *DENY* reply from i, then it trace back until it finds any alternate path to destination from corresponding visiting router. Every time when it backs, it purges the duplicate node information from the route **R.**
2. But if SRA_v (visited) finds some option which is near to satisfy the offered *XQSpec,* then it sends the $XQSpec_{ni}$ (which is a new *XQSPec* at node i, that

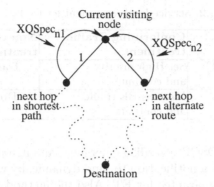

Fig. 3. New XQSpec with alternate path.

can be supported by the currently visited node) back to last immediate hop. The SRA of the requester router holds this information until it finds any feasible alternate path. When it faces same experience from other visiting router in alternate path, it compares $XQSpec_{n1}$ and $XQSpec_{n2}$ to find the QoS path as shown in Fig. 3 and selects the best one. SRA ensures one more thing that ET will never blocked even if the $XQSpec$ is not strictly served by all routers in the source to destination route.

Although we are thinking about the rejection from any node due to resource unavailability but the chance to happen this is very small in NGI while most of the links will be Gigabit enabled. But the concept of all-IP network merges almost all access technology in a single point. But bandwidth is still a scarce resource for wireless network. In our future work we will provide detail algorithm for wireless medium with simulated result using the same concept.

To find the QoS route using $XQSpec$ properly it requires a set of standards which is necessary to implement SRA in all router hardware before it is deployed. The minimal functionality requirements of SRA are to check the $XQSpec$ and take decision immediately and intelligently based on the request/reply from the neighbor hops. Extensive simulation can provide optimal configuration to choose best $XQSpec$ by emergency user application. Another advantage of the approach is that SRA can tune only those sub interface where ET actually flows and others IP interface can be remain untouched. When a live session terminates, SRA can readjust the router state for usual operation according to SP's need. Moreover SRA can send advance warning message to other same priority or low priority streams to inform them to choose either different path or slow down their transmission to avoid congestion and loss of transmission quality.

4 Router's Functionality

The overall system architecture is scalable. It requests for necessary changes in runtime with related parameters and data format to only those nodes who take part in ET transmission. It is possible that all routers in transit network may

Table 2. Service level agreement for traffic classes.

Traffic class	Conforming Traffic treatment	Non-conforming Traffic treatment
Emergency Traffic (ET)	Set High priority and continue	Transmit
Non Emergency traffic (nET)	Mark as BE traffic and continue	Drop

not aware of TOS or DSCP regarding emergency and non-emergency traffic. The *XQSpec* in this case is a notification to bind dynamically with the desired policy so it can ensure the guarantee for ET. Also the method described here is non preemptive priority model with dynamic queuing mechanism which prevents the unprecedented drop of nET. The policy that is maintained for ET and nET is given in Table 2. Edge or core router can change their configuration based on *XQSpec* while others in neighbor network remain untouched. Routers those receive *XQSpec,* check each packets pass through them and can remark if it is impersonated to take extra advantage. Token bucket is used with high priority queue so that no ET is dropped even in high traffic load.

4.1   Edge Router

Although packet tagging is initially done by the application layer for emergency traffic before it placed in the access link, they are checked in the edge router again and retagged if found incorrect. Non emergency packets are always placed in normal, medium or low priority queue based on how they classified for transmission. The ET is never discarded as it passes through a token bucket interface while a leaky bucket interface is used for nET in edge router. So in case of overflow they will be discarded at least during the high volume of ET transmit. But most of the cases, *SRA* informs the inbound nET to slow down their transmission until it completes the transmission of the ET. For separate queues exhaustive scheduling mechanism is chosen for ET but FIFO scheduling is followed for individual queue. The exhaustive in this case means emergency queue is served and finished before any other queues in the system. There is a possibility of context switching overhead due to irregular packet arrival and one of the bottlenecks of performance of this method. But in most cases ET comes in succession and dispatched as soon as their arrival due to exhaustive scheduling method. Non preemptive scheduling prevents dropping of nET whose services started before the arrival of ET and consequentially reduces the packet loss ratio of nET that may occur in preemptive scheduling. The other rules and policies like shaping and dropping, bandwidth allocation, latency, jitter, queue type etc. can be decided in run time by *SRA* if corresponding router entity agrees with the *XQSpec.*

4.2   Core Router

The job of core router is bit simpler. Because a dedicated path based on *XQSpec* is already set and traffic is also controlled. It further checks the packet tag using its PHB (Per Hop Behavior) rule and place packets in appropriate queue. It also retags packets if it finds impersonated. Other functionality of *SRA* is almost same.

4.3 Border Router

The processing of the best effort nET and ET in border router is similar to edge and core router. The *SRA* in border router informs all other routers in QoS route **R**, when the destination node terminates the session.

5 Few Factors

Our proposal requires less overhead than those methods that keep detail network information (like delay, bandwidth, cost, etc.) for the whole network. *SRA* keeps information about the outgoing links and its related parameters only. The processing is distributed and scalable as all the *SRA* works independently and does not depend on others to make any decision. The main short coming of the proposed system is the implementation of *SRA* in routers hardware. But this problem will no longer exist as hardwares are becoming cheaper and also the processing power is increasing day by day.

Therefore, the performance of the system depends on the two key factors. A good *XQSpec* selection and good design of *SRA*. Common standard for a class of service is necessary for ET as a set of standards. The developer of router will follow the rules to give special attention to ET by *SRA* at run time. So a common regulation is needed which will promote the development of *SRA* and mass implementation so in NGI we can use IP as communication medium for SOS (Save our Soul). Internet in future will be so pervasive, reliable and transparent that it will be a seamless part of our life like electricity. The proposal opens a new dimension of business horizon including dynamic charging facilities which will interest the SP to make necessary arrangement for QoS based transmission. Information highway in near future will be treated as an expressway similar to highway of developed countries where user can choose path according to their choice and demands.

6 Conclusion and Future Work

We have described a dynamic QoS mechanism for high priority traffic especially for emergencies in IP network. We have identified the necessity for such a system. With the presentation of the framework we have shown how to achieve end-to-end QoS guarantee for time critical information based on user's selection. The article also emphasizes the importance of software based dynamic routing which can support QoS transmission. In NGI SOS through public Internet is obvious. We have described one positive indication of how IP can serve as lifeline. At the end the system should be realized by mass people so that they can take forward step to secure their life.

In further work we will simulate all queuing mechanisms to verify which methods work best for ET along with the traditional traffic. We have also a plan to develop a robust dynamic QoS based algorithm which can maximize the throughput as well as minimize the loss of non emergency traffic in integrated wired and wireless environment.

References

1. Henning Schulzrinne, Knarig Arabshian: Providing Emergency Services in Internet Telephony . Technical report (2002)
2. Maniatis, S.I.; Nikolouzou, E.V.I.: End-to-end QoS specification issues in the converged all-IP wired and wireless environment. Volume 42. (2004) 80–86
3. Cory Beard: Mechanism for providing internet emergency services. (Technical report)
4. Quality of service networking. (Technical report)
5. De, B.S.; Joshi, P.S.V.C.D.: End-to-end voice over ip testing and the effect of qos on signaling. In: Proceedings of the 35th Southeastern Symposium on System Theory, 2003. (2003) 142 – 147
6. M. Handley, H. Schulzrinne, E.S., Rosenberg, J.: SIP: Session Initiation Protocol. RFC 2543, IETF (1999)
7. J. Polk : Internet Emergency Preparedness (IEPREP) Telephony Topology Terminology. Technical report (2003)
8. Matta, J.M.; Takeshita, A.: End-to-end voice over ip quality of service estimation through router queuing delay monitoring. In: Global Telecommunications Conference, 2002. GLOBECOM '02. IEEE. Volume 3. (2002) 2458 – 2462
9. Bray, T.: Extensible markup language (XML) 1.0 (third edition). (2004)

Unicast and Multicast QoS Routing with Multiple Constraints

Dan Wang[1], Funda Ergun[1], and Zhan Xu[2]

[1] School of Computer Science, Simon Fraser University,
Burnaby BC V5A 1S6, Canada
{danw,funda}@cs.sfu.ca
[2] EECS Dept, Case Western Reserve University,
Cleveland OH 44106, USA
zxx10@po.cwru.edu

Abstract. We explore techniques for efficient Quality of Service Routing in the presence of multiple constraints. We first present a polynomial time approximation algorithm for the unicast case. We then explore the use of optimization techniques in devising heuristics for QoS routing both in the unicast and multicast settings using algorithmic techniques as well as techniques from optimization theory. We present test results showing that our techniques perform very well in the unicast case. For multicast with multiple constraints, we present the first results that we know of where one can quickly obtain near-optimal, feasible trees.

1 Introduction

As diverse applications such as online conferences, video broadcast, online auctions appear in the Internet, the demands by the application owners regarding the delivery of data have become more extensive and varied. Currently the delivery of such data primarily focuses on minimizing the path length, or obeying a given policy. However, the needs of the applications involve other issues such as latency, packet loss, jitter avoidance, etc. Such requirements of the applications from the network can be formally expressed in terms of *Quality of Service (QoS) constraints*. The satisfaction of these constraints comes at a cost of using valuable network resources such as buffer space, bandwidth, etc. The focus of QoS routing is to select the routes by taking into account the requirements of the applications while being efficient in terms of link costs.

Most QoS constraints are additive, such as delay, packet loss etc, i.e., they accumulate along the path. Given such constraints, in this paper, we explore QoS routing in unicast and multicast settings. Unicast QoS routing involves finding a min-cost path from a source to a destination node satisfying a set of constraints generally given as upper bounds that the path must respect. These problems are NP-complete when the number of constraints is one or higher[5]. In multicast QoS routing, given a source node and a set of multicast nodes, we seek to find a min-cost tree such that the constraints are satisfied along the path from the source to each multicast node. Even without constraints, this problem is NP-complete [10].

M. Ajmone Marsan et al. (Eds.): QoS-IP 2005, LNCS 3375, pp. 481–494, 2005.

The prohibitive hardness of these problems, and the requirements of a high-traffic network makes it necessary to develop techniques that efficiently generate near-optimal solutions. In this paper we investigate provably good, as well as practically feasible schemes. First, we present an ϵ-approximation algorithm for unicast routing with K constraints that runs in polynomial time for small K. We then develop heuristics for unicast and multicast QoS routing using algorithmic and optimization techniques with the goal of finding good solutions efficiently regardless of the number of constraints. We give out two heuristics for the unicast case, one of which is flexible with respect to optimality or feasibility. This is desirable since finding a feasible path is NP-complete if we have more than one constraint. Our algorithm explores the trade-off between cost and feasibility. As shown through our simulation results, our algorithms are very fast and obtain over 92% success rate for the unicast case. We also show with multicast multiconstraint routing, which is far more difficult than unicast routing, how to obtain a feasible solution with a high success rate.

1.1 Previous Work

Since this field is quite mature, we give a sample of the related work. Samples of abundant recent work can be found in [11, 2, 6, 3, 9], and their references. Recently several related work have appeared that use optimization techniques. In [11] a simple application of the technique is used for multiconstraint unicast routing and in [9] the technique is discussed for one constraint unicast problems. Our approximation algorithm builds on top of several work; for a full description of the algorithm see [14] and [4].

The Steiner Tree Problem is the simplest form of multicast routing and is well-studied [12, 8]. The heuristic KMB that we use as a building block is given in [12], and performs always within a factor 2 and usually within 10% of the optimal [15]. There are several results on single constraint QoS multicast routing, examples are in [13, 15], and their references.

2 The Multivariate QoS Routing Problem

We model our network as an undirected graph $G = (V, E)$, where V is the set of nodes, and E is the set of links; we assume $|V| = n$, and $|E| = m$ throughout. Each link e is associated with a cost $c(e)$ and K different QoS parameters, denoted $w_1(e) \ldots w_K(e)$ representing end-to-end restrictions on the routes such as delay, packet loss, etc. Nodes s and t refer to the source and destination; for multicast the set $V_t = \{t_1, t_2, \ldots, t_L\}$ refers to the set of multicast destinations. We denote by $P(s, t)$ the set of $s \to t$ paths in G and by $T(s, V_t)$ the set of multicast trees rooted at s with destinations V_t in G.

A constraint is an upper bound on the value of a QoS parameter for each destination[1]: W_{ij} refers to constraint i for destination t_j. When there is a single

---

[1] We use the terms "constraint" and "upper bound" interchangeably.

destination (unicast), we simplify constraint i as W_i. A path or a tree that satisfies all of the K constraints is said to be *feasible*; the least cost feasible path/tree is called *optimal*.

Unicast QoS Routing. Given source and destination nodes s, t, we aim to find the minimum $s \to t$ path p such that the sum of parameter w_i along p is upper bounded by W_i (for all $i = 1, \ldots, k$). The above can be formulated as follows.

$$\min_{p \in P(s,t)} \sum_{e \in p} c(e)$$

$$s.t. \sum_{e \in P} w_i(e) \le W_i \ \ where \ i \in 1 \cdots K$$

Multicast QoS Routing. Given a source node s and a set of destinations $V_t = \{t_1, \cdots t_L\}$, we aim to find the min-cost multicast tree T rooted at s such that the sum of parameter w_i along path $s \to t_j$ is upper bounded by W_{ij} for each i, j. Formally, multicast QoS routing problem is as follows.

$$\min_{t \in T(s,V_t)} \sum_{e \in t} c(e)$$

$$s.t. \sum_{e \in P_t(s,t_j)} w_i(e) \le W_{ij} \ \ for \ i \in 1 \cdots K, j \in 1 \cdots L$$

3   An ϵ-Approximation Algorithm for Unicast

In this section we give a polynomial time approximation algorithm that solves unicast QoS routing with K constraints. When $K = 1$, the problem is called Restricted Shortest Path (RSP), for which there are ϵ-approximation algorithms (see [14], [4] for the latest results).

Our algorithm has the following specifications. If a feasible path exists and the cost of the optimal feasible path is OPT, the algorithm is guaranteed to return a path p of cost $\le (1 + \epsilon)OPT$ such that (i) the total value for parameter w_1 on p is at most W_1, and (ii) for $i = 2 \ldots K$, the total value for QoS parameter w_i across p is at most $(1 + \epsilon)W_i$. Note that a slight violation of all constraints except w_1 is allowed.

We construct an approximation algorithm for $K = 2$; the technique generalizes easily to any K. We express the problem as a dynamic program. For node v, let $V(cc, d, v)$ denote the minimum value of parameter w_2 along an s-v path with total cost $\le cc$ and total value of parameter w_1 at most d. Then, $V(cc, d, v) = min_{e=(v',v) \in E}\{V(cc - c(e), d - w_1(e), v') + w_2(e')\}$.[2] The base cases are that $\forall d, cc, V(cc, d, s) = 0$ (s-s path is trivial), $\forall cc, v \ne s$, $V(cc, 0, v) = \infty$ (parameters are positive), and $\forall d, v \ne s, V(0, d, v) = \infty$ (costs are positive). To solve the problem, we need to compute the smallest cost cc

---

[2] We assume all parameters are integers.

where $V(cc, W_1, t) \leq W_2$; the actual path is implicit in the formulation and can be obtained from the steps leading to this particular value.

Our approximation algorithm generalizes [7] by first approximately testing whether a feasible solution for a fixed budget V exists and using this test to search for the optimal solution. We proceed below by scaling down costs and w_2, and running the dynamic program on the scaled parameters.

procedure TestMult(V, ϵ)
Step 1 $\forall d, \forall cc, V(cc, d, s) = 0; \forall d, \forall v \neq s, V(0, d, v) = \infty;$
$\quad\quad \forall cc, \forall v \neq s, V(cc, 0, v) = \infty;$
Step 2 $\forall e \in E, c'(e) = \lfloor \frac{c(e)n}{\epsilon V} \rfloor; w_2'(e) = \lfloor \frac{w_2(e)n}{\epsilon W_2} \rfloor.$
Step 3 **for** $cc = 1 \ldots n/\epsilon$
$\quad\quad V(cc, d, v) = min_{e=(v',v) \in E}\{V(cc - c'(e), d - w_1(e), v') + w_2'(e')\}.$
Step 4 **if** $V(cc, W_1, t) \leq W_2$ **return** "SMALLER"
$\quad\quad$ **else return** "GREATER".

We now show that TestMult is efficient and effective.

Lemma 1. *If TestMult (V, ϵ) returns "SMALLER" there is a path of cost $\leq V(1 + \epsilon)$, total value of $w_1 \leq W_1$, and total $w_2 \leq W_2(1 + \epsilon)$. If it returns "GREATER", then there is no path of cost at most V where parameter w_1 is at most W_1 and w_2 at most W_2. The running time of TestMult is $O(mn^2/\epsilon^2)$.*

Proof. If the procedure returns "SMALLER", then it must have found a path whose scaled total cost and scaled w_2 are $\leq n/\epsilon$ and sum of w_1 is $\leq W_1$. That W_1 is satisfied follows from the dynamic program. The scaled cost and w_2 values can be restored in the end by multiplying the values on the output path by $\epsilon V/n$ and $\epsilon W_2/n$ respectively. This way, on a path implicitly found feasible by TestMult, the cost is underestimated by at most $\epsilon V/n$ per link and the sum of the w_2 values are underestimated by at most $\epsilon W_2/n$. Since a path can have at most n links, the actual cost can be at most $(n/\epsilon) \cdot (\epsilon V/n) + n(\epsilon V/n) = V(1+\epsilon)$. A similar argument can be made for w_2. Now consider cases where TestMult returns "GREATER". Since the inaccuracy caused by the scaling only results in underestimation, clearly there cannot be a feasible solution with cost $\leq V$. The running time follows from the size of the table that one needs to keep for the dynamic program.

Equipped with a test, we can now search for the optimal cost with a multiplicative binary search. For this, we establish upper and lower bounds UB and LB for the cost, and search between them for the smallest cost that will make the test return a positive ("SMALLER") answer.

procedure ApproxMult(ϵ)
Step 1 Determine UB, LB
Step 2 **while** $UB/LB > 2$
$\quad\quad V = \sqrt{UB \times LB}.$
$\quad\quad$ **if** TestMult$(V, \epsilon)=$ "GREATER" $LB = V$ **else** $UB = V(1 + \epsilon)$
Step 3 Run a modified TestMult(LB, ϵ) with the loop in Step going up to $\frac{2n}{\epsilon}$,
$\quad\quad$ **return** path with smallest cc where $V(cc, \frac{n}{\epsilon}, t) \leq W_1$

Next we show that this is an ϵ-approximation algorithm[3].

Theorem 1. *ApproxMult is an ϵ-approximation algorithm for QoS routing with two parameters which returns a path P that satisfies the following. The sum of w_1 along P is at most W_1, the sum of w_2 along P is at most $W_2(1 + \epsilon)$, and the total cost is at most $OPT(1 + \epsilon)$, where OPT is the cost of the cheapest path. ApproxMult runs in time in time $O(mn^2 \log \log(UB/LB)/\epsilon^2)$.*

A trivial initial value for LB is 1; one for UB is nC where C is the highest link cost in the network G. If one is willing to do some extra work, one can establish tighter lower and upper bounds. In fact, a generalization of the upper-bounding technique in [14] helps reduce the overall running time. The technique proceeds as follows. Initially we run Restricted Shortest Path on the network, trying to find a path that minimizes the sum of parameter w_2 such that the parameter w_1 adds up to W_1. (Recall that Restricted Shortest Path is the one constraint QoS routing problem and can be approximated in time $O(mn/\epsilon)$[4].) If the total value of w_2 on the path returned exceeds W_2, we conclude that there is no feasible solution and stop.

Otherwise we sort all links based on their costs. After that, we start removing links from the graph in order of their cost, starting from the one with the highest cost. After each link is removed, we run RSP as above, minimizing total w_1 and bounding the sum of w_2. At some point, it will become impossible to find a path whose total w_1 is under W_1. This means that the last link removed, say e, is a *bottleneck* link. We now make the following observation regarding the upper and the lower bounds with respect to the bottleneck link.

Lemma 2. *Let e be the bottleneck link. (1) There exists an s-t path P such that the sum of w_1 across P is at most W_1, the sum of w_2 is at most $(1 + \epsilon)W_2$ and the total cost of P is at most $nc(e)$. (2) There exists no path of cost less than $c(e)$ that satisfies both constraints W_1, W_2 exactly.*

Proof. (1) By definition, RSP must have returned a path of total w_1 value at most W_1 and total w_2 at most $(1 + \epsilon)W_2$. In addition, since the cost of the most expensive link in the graph where P was found was $c(e)$ the overall cost of the path was at most $nc(e)$. (2) RSP could not find a solution after link e was removed, and could find one before. Thus, e must be part of the solution, and thus, the total cost of the solution must be at least $c(e)$.

One can easily see that, by setting $LB = c(e)$ and $UB = nc(e)$ as above, we guarantee that a feasible solution (where, as described in the specification of the approximation algorithm, W_1 is satisfied and W_2 is not exceeded by more than an ϵ fraction) is known to exist for all costs above UB and an exactly feasible path does not exist with a total cost under LB. This initial step involves running at most m RSP executions, which are dominated by the other terms in the running time, giving the following running time.

Corollary 1. *ApproxMult has running time $O(mn^2/\epsilon^2) \log \log n$.*

---

[3] The proof of the following theorem can be found in the full version of this paper[16].

One can easily generalize this result to K parameters, by scaling down the costs and $K - 1$ the parameters $w_2, \ldots, w_K$, obtaining a running time of $O(mn^K/\epsilon^K) \log \log n$.

4   Optimization Techniques for QoS Routing

Since the intrinsic difficulty of QoS routing is due to the constraints, we consider heuristics whose performance are independent of constraints in this section.

4.1   Optimization Techniques for Unicast QoS Routing

QoS routing problems are instances of integer programming, which are in general NP-complete to solve, with efficient solutions only under certain restrictions. Our goal thus is to transform unicast QoS routing into one or multiple integer programming problem/s which can be solved in polynomial time, while ensuring that the answer that we obtain is good enough for the original problem.

For efficiency, we will make sure that our integer programs are free of any "difficult" constraints, by using new cost functions which incorporate "penalties" for violating the (now omitted) constraints. The choice of the set of constraints to be relaxed into the objective function is made by the assumption that after relaxing them, the problem becomes easy. First there is a hidden constraint in the integer programming described in the above section, i.e. the solution for the problem must be a path. Therefore, instead of using e, our unicast problem can also be formalized as follows:

$$P : \qquad \min c(p) \quad s.t. \ w_i(p) \leq W_i, \ i \in 1 \ldots K$$

where $c(p)$ is the total cost, and $w_i(p)$ is the sum of constraint i along path p.

To solve this problem, we give out two heuristics.

Algorithm LRA. We relax all the constraints and construct a new problem as follows:

$$P' : \qquad max_\lambda \ L(\lambda) = min \ c(p) + \sum_{i=1}^{K} \lambda_i(w_i(p) - W_i)$$

Intuitively, λ_i determines how much we penalize the violation of ith constraint. A simple observation is that $L(\lambda)$ is a lower bound for P for $\lambda \geq 0$ since $\sum_{i=1}^{K} \lambda_i(w_i(p) - W_i) \leq 0$. Clearly, finding a suitable λ to maximize the above expression will bring us closer to the optimal solution.

First, to ensure that the links we choose indeed form a path, we run Dijkstra's shortest path algorithm as the basic building block of our algorithm. Second, for quick convergence on an iterative search for the best λ, we need a good starting value. For this, we would like a quick upper bound on the cost, which, unfortunately, is usually as computationally difficult to solve as the original problem. As a solution, we use a two-phase approach.

In Phase 1 we inject a feasible path q into the input such that $c(q) = \infty$. (q is an upper bound for the solution.) The algorithm then efficiently improves q to use it as the initial state in Phase 2.

In Phase 2, the artificial path is removed, and the parameters are reset. Using the output of Phase 1 as the initial state, it is now possible to process the parameters in a fine-grained manner, to converge to a good solution quickly.

procedure LRA
Step 1 Find the min-cost path p_c using **Dijkstra**. if p_c is feasible, **return** p_c.
Step 2 **for** each parameter w_i find the minimum weight path p_{w_i} w.r.t. w_i using
    Dijkstra. if total weight for $p_{w_i} > W_i$, **return** "no solution".
Step 3 Find a feasible path as a starting point.
    if failed **return** "no solution".
Step 4 Set the combined cost $c_\lambda(e) = c(e) + \sum_{i=1}^{K} \lambda_i^{k+1} w_i(e)$. Find the min cost
    path by **Dijkstra** using c_λ.
    Adjust λ^k and **repeat** Step 4.

Step 3 corresponds to Phase 1. It starts off by establishing (for > 1 constraint) an artificial feasible path. This result is then refined by running a shorter instance of Phase 2 (Only step 4 is performed), resulting in a nested repetition of Phase 2 with different parameters. In Step 4 we adjust λ iteratively as $\lambda^{k+1} = \lambda^k + \theta^k(w(p) - W)$ where the step size is $\theta^k = \frac{L(\lambda^{k+1}) - L(\lambda^k)}{\|w_i(p^*) - W_i\|^2}$.[4]

Algorithm SRA. In the above algorithm we relax all the constraints into the object function. Even though the solutions are often quite satisfactory, the problem tends to become oversimplified. For example, given a reasonable amount of running time, LRA can not guarantee to return a solution. For applications which may require a high chance of finding a solution, a trade-off between low cost and feasibility is preferred. Motivated by the above, we develop a variant of LSA which relaxes all the constraints into one single constraint. Our reasoning is that for one-constraint QoS routing, finding a feasible path is easy, whereas for two or more constraints, this task is NP-complete. Therefore, reducing the number of constraints to one (and not to any higher value) makes the feasibility aspect of the problem easier to handle. In addition, research on one-constraint QoS routing is mature enough that we are given a choice of several approximation algorithms and heuristics. Our approach involves using a solution for one-constraint QoS routing as a building block in our algorithm to solve the multi-constraint case.

Again consider our problem formulation P as shown in the previous subsection. We can derive a new problem $P'(\lambda) : \min c(p)$ s.t. $\sum_{i=1}^{K} \lambda_i(w_i(p) - W_i) \leq 0$ where $\lambda = (\lambda_1, \lambda_2, \cdots, \lambda_K)$ is a vector of norm 1 with all $\lambda_i > 0$. Intuitively λ represents the weight of different constraints. In contrast to LRA, we start with an initial (normalized) weight vector λ which gives equal weight to each constraint since we do not know how each parameter behaves. Instead of using Dijkstra, we use QOSONE(λ) as a building block which returns an optimal path p for problem $P'(\lambda)$, i.e. the algorithm that solves the one-constraint, min-cost problem. Notice that in practice QOSONE can not return a real optimal solution. In our simulation, we test different QOSONE, focusing on feasibility or optimality or both.

---

[4] The derivation of this parameter is in the full version of this paper[16].

procedure SRA

Step 1. $\lambda \leftarrow (\frac{1}{\sqrt{K}}, \frac{1}{\sqrt{K}}, \cdots, \frac{1}{\sqrt{K}})$

Step 2. Combine the constraints by λ, find a min cost path by **QOSONE**.

    if p == NULL **return** "No Solution"

    if p satisfies all the constraints **then return** p

Step 3. Adjust λ and **repeat** Step 2.

4.2   Optimization Techniques for Multicast QoS Routing

Algorithm LRATree. For multicast QoS routing, we simplify the integer program (which is similar to that for unicast except with K constraints/multicast node) and build a cost function which encompasses penalties for QoS constraint violations. In addition, we can think that each destination is an additional constraint to the integer problem and we also associate the penalties to these "constraints". Our algorithm requires that we solve a minimum-Steiner tree problem. Since the problem is NP-complete, instead we use a heuristic KMB [12]. Our multicast algorithm also uses the two phases approach as described in the unicast case. Here we point out a few notable differences[5]. To obtain an initial feasible tree, we use our unicast algorithm to find a feasible path to each destination and then combine these paths to obtain a tree. If this results in a cycle, we insert an artificial feasible tree into the graph, and then improve this tree to obtain a feasible tree that exists in the original graph. When updating λ, we take into account how many paths share a given link. This makes sure that, if a particularly "bad" link was being shared by many paths in the tree obtained in the previous iteration, the new tree will likely not include this link, or some of the branches which do not satisfy the constraint upper bounds will avoid using this link.

5   Observations and Analysis

Efficiency of the Heuristics. As seen in Figure 3, a constant number of iterations yields high accuracy, since the improvement reduces exponentially at each iteration. As a result, we show below that our algorithms scale well in terms of both the number of constraints and the network size.

Theorem 2. *Let K be the number of constraints. The running time of LRA is $O(K+Dijkstra)$. The running time of LRATree is $O(LK + KMB)$, where L is the number of multicast nodes. The running time of SRA is $O(QOSONE)$.*

Quality of the Heuristics. We now argue about the quality of our algorithms. First notice that for unicast and multicast routing with a single constraint, LRA and LRATree are guaranteed to return a feasible path or tree if one exists.

    We next try to obtain an intuition about why our technique is expected to work well. For the next lemma, assume that all costs and weights (we will call them delays) are uniformly chosen from the same range (if the ranges are different, they can be normalized). It shows that "better" paths will be favored.

---

[5] Pseudocode for LRATree can be found in the full version of the paper [16].

Lemma 3. *For single constraint unicast routing, let the minimum cost path p_0 have cost C_0 and delay d_0. Let p_1 and p_2 be two paths, with costs $C_1 < C_2$ and total delay $d_1 < d_0$ and $d_2 < d_0$. Assume that $d_0 > W$, thus, the minimum cost path is not feasible. Let λ be the step size. Then, for $\lambda = 1$, if one of p_1 and p_2 is returned at the next step (if both are feasible), the probability that p_1 will be returned rather than p_2 is at least 75%. In addition, there exists a $\lambda > 0$ which will cause p_1 to be picked over p_2 with probability 1.*

Proof. After adjusting λ, the new costs for p_1 and p_2 will be $C_1' = C_1 + \lambda d_1$ and $C_2' = C_2 + \lambda d_2$ respectively. Since the delays are uniformly chosen, with probability 0.5, $d_2 > d_1$, which trivially satisfies $C_1' < C_2'$. If $d_2 \le d_1$ (with probability 0.5), however, we must have $C_1 + \lambda d_1 < C_2' = C_2 + \lambda d_2$, i.e., we need $\frac{C_2 - C_1}{d_1 - d_2} > 1$. Since the delays and the costs were uniformly picked from the same range, by symmetry the likelihood of this is at least 0.5. Thus, the probability that $C_1' < C_2'$ is at most $0.5 + 0.25 = 0.75$. Note that $C_1' < C_2'$ can always be satisfied if $\lambda < \frac{C_2 - C_1}{d_1 - d_2}$.

This simplified analysis seems to support small λ. However, note that $C_0 < C_1 < C_2$, and $d_0 > W$. Thus, if $\lambda < \frac{C_1 - C_0}{d_0 - d_1}$, then under the new cost function, p_0 will have cost C_0', less than C_1', thus, p_0 will seem better than both p_2 and p_1, even though it is not feasible. Also note that the likelihood of this scenario depends on $C_0 - C_1$, thus the need to adjust λ depending on how close to the solution we are. This analysis generalizes to any number of paths. For example, if there are 3 paths with $C_0 < C_1 < C_2 < C_3$ one can show that the probability p_1, p_2, and p_3 will be preferred to the other paths after the adjustment is $\frac{11}{18}, \frac{5}{18}, \frac{2}{18}$ respectively. This tells us that the likelihood that one of the least costly paths will be chosen is high. Our simulations reaffirm this property.

Theorem 3. *If QOSONE returns an exact solution to P', every time a feasible solution is found by SRA, it must also be optimal.*

Proof. Assume a feasible path p is found and p is not optimal. Let opt denote the optimal path. Then $c(opt) \le c(p)$, p and opt satisfy all the upper bounds. By design, our algorithm will next find a positive λ. Since opt is feasible for P'', i.e. $\forall i, w_i(opt) - W_i \le 0$, summing up, we have $\sum_{i=1}^{K} \lambda_i(w_i(opt) - W_i) \le 0$. Thus opt is also a feasible path for the one-constraint QoS routing, $P'(\lambda)$. Since p is the optimal path for $P'(\lambda)$, then $c(p) \le c(opt)$. Thus $p = opt$.

Corollary 2. *If the output of SRA is a feasible path, the corresponding path from QOSONE is a non-optimal feasible path.*

The above theorem and corollary clearly show that our algorithm can control the performance of QoS routing by the choice of QOSONE. If QOSONE is more likely to find an optimal path, SRA is more likely to find an optimal solution. This gives us a chance of improving multi-constraint QoS routing by focusing on improving single-constraint QoS routing. On the other hand, if QOSONE is focusing on finding a feasible path, the probability of finding a feasible path in a certain amount of time for multi- constraint problem is also improved.

Multicast. For multi-constraint multicast routing, the number of the destinations has a significant effect on the accuracy. This is because early in the algorithm we need to find feasible paths for each destination. As the number of destinations increases, the likelihood that LRA will succeed on all of these destinations drops exponentially, thus increasing the probability that LRATree will fail. Instead we notice from the above analysis that SRA can obtain high feasibility by choosing QOSONE as an algorithm focusing on feasibility. Therefore in the first phase of our algorithm where the goal is to find a feasible tree as a starting point for further minimization, we use SRA in our experiments.

6 Simulation Results

Simulation Environment. To test our technique, extensive simulations have been carried out. We used two different methods to generate the network topology. First we used *ANSNET* [3], shown in Figure 1. The cost of each link was set uniformly in the range [1, 1000]. The delays (constraints) of each link were set uniformly in the range [0, 100]. The delay upper bounds were set to a random number uniformly in range [100, 200], [100, 300], ... [100, 500] for different simulations. Next we used the widely adopted Waxman Model [17] to generate random networks. The network consisted of 40 to 90 nodes where the two parameters α and β were set to 0.3 and 0.2 respectively, and the grid was 30 × 30. The cost on each link was generated by the model itself as the Euclidean distance. The delays and delay upper bounds on each link were chosen the same as *ANSNET* model. Each data point obtained in our figures represents the average of 150 runs on different random networks.

For SRA we chose different algorithms for single constraint problem as a building block for our algorithm. The sample building blocks represent the tradeoff between accuracy and efficiency, and are as follows. 1) SRA-Dijkstra: we find the shortest path based on the single combined constraint. 2) SRA-LARAC: a relaxation heuristic algorithm for one constraint unicast QoS routing [9]. 3) SRA-Exact: we use dynamic programming to obtain an exact solution to one constraint unicast QoS routing. Here, Dijkstra and Exact are two extremes in that Dijkstra has no claim to optimality, but is efficient, whereas Exact always returns the optimal solution for QOSONE, but runs in super-polynomial time. We use Exact just to show how our algorithm perform. In practice, people can use approximation algorithm, e.g. [4] to achieve similar result instead or use LARAC which is an in-between algorithm.

We compare our algorithm with the optimal solution both for unicast and multicast routing. However, obtaining an optimal solution for multivariate multicast routing in a large network has proven to be an unconquerable task, as mentioned in [15]. For a 10-node graph, the computation for every single point in our figures took many hours. To overcome this problem, for large multicast problems we insert a random feasible tree into the network and test our algorithm based on the modified network.

Interpreting the Results. The outcome of the algorithm will fall within one of the following cases. (S) Optimal answer found, or no feasible solution exists. (F1) Feasible but not optimal answer found. (F2) Feasible answer exists, but not found. Note that by design the algorithm never returns a non-feasible path. We evaluate our algorithms with respect to the following criteria: (i)*Full success* = S/(S+F1+F2), the ratio of fully correct answers. (ii) *Partial success* = (S+F1)/(S+F1+F2), the ratio of fully correct and feasible answers. (iii) *Excess*, the percentage excess (over optimal) cost of a returned solution.

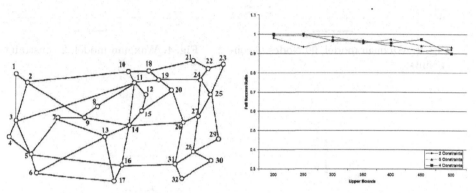

Fig. 1. ANSNET Network.   **Fig. 2.** ANSNET model.

6.1   Unicast Routing Simulations

LRA. In our simulations, we observed that partial or full success rates were very much independent of the number of parameters and above (usually, well above) 92% and 90% respectively. Increasing the range for the constraints led to a small (around 2-5%) drop in the full success rate, which we attribute to the increase in the number of feasible paths in the network. This can be seen in Figure 2. Figure 3 shows that increasing the number of the iterations yielded diminishing returns due to the exponentially decreasing step size with no gain beyond 16 iterations. We also explored the effects of the correlation (Figure 4) between the QoS parameters with two (as well as four, grouped into two) random, positively correlated (with a small random difference between the values), and negatively correlated (two parameters adding up to a constant plus a small random number). Even though our partial success rates was high throughout, correlation affected the total success rate (over 97%, 96%, 90% for positive, random, negative correlation). Finally, in Figure 5, we observed that the percentage value of the excess as a measure of the effectiveness of our partial successes compared to the full successes, and found average excess rates are extremely low.

SRA. We then investigated the effects of using different one-constraint QoS routing algorithms. In Figure 6, we observe the *full success ratio*. This indicates how well the algorithm performs in achieving optimality. As expected, SRA-Exact outperformed all other QOSONEs in terms of full success rates, followed

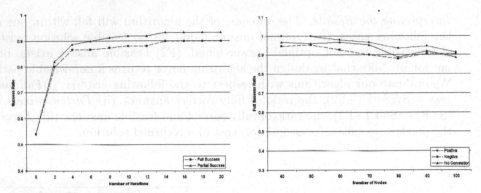

Fig. 3. Waxman model, 90 Nodes, 3 constraints.

Fig. 4. Waxman model, 2 constraints.

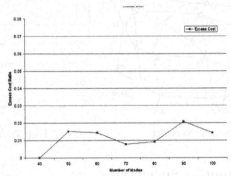

Fig. 5. Waxman model, 3 constraints.

by SRA-LARAC. LRA (Lagrange) was a very close third, and SRA-Dijkstra performed the worst. With the partial success rate (Figure 7), SRA-Dijkstra was the best with a near-perfect performance in finding a feasible path. SRA-LARAC was the second and LRA the third. As we expected, SRA-Exact performed the worst. We can conclude that SRA maintains the properties of the one-constraint algorithm; therefore one can choose different one constraint unicast algorithm based on their requirements.

6.2   Multicast Routing Simulations

Our experiments for multivariate multicast routing were similar to those with unicast routing. Due to the prohibitive running time of finding the optimal solution we tested our algorithms on small graphs. The number of iterations was set to be 8 for the main program, and 8 for finding an upper bound.

In general, while the running times remained fast, the accuracy dropped when we wanted to satisfy all of the constraints for all of the nodes. Part of this was due to cases where the KMB algorithm failed. In addition, as argued before, with more destinations a fully feasible multicast tree is difficult to find. We thus used SRA in some experiments. Test results are given in Figures 8-11.

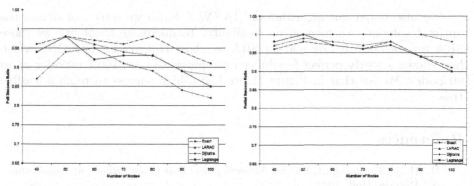

Fig. 6. Upper bounds [100, 400], 2 constraints, full success ratio.

Fig. 7. Upper bounds [100, 400], 2 constraints, partial success ratio.

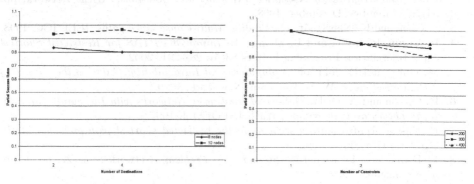

Fig. 8. Waxman, 2 constraints.

Fig. 9. Waxman, 10 nodes 6 destinations.

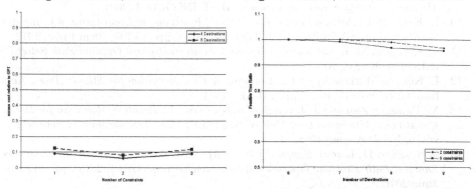

Fig. 10. Waxman model, 10 nodes.

Fig. 11. ANSNET, SRA with Dijkstra.

Even though our success rates were lower than those for unicast, our results remained desirable: one must consider the enormous difficulty of solving this problem exactly, our extremely low running times, as well as the absence of known efficient algorithms for this problem. The figures show that network size, the number of multicast nodes, and the number of constrains all lead to a fairly small drop in accuracy. The excess cost is upper bounded by 17%, higher for increased number of multicast nodes.

We also tested our algorithm on *ANSNET*. Since we could not obtain the optimal solution, we focused on feasibility. To maximize feasibility we chose SRA for each destination and for SRA we chose Dijkstra for QOSONE which had shown a nearly perfect feasibility ratio previously, i.e. 100% ratio for up to 90 nodes. We see that in Figure 11, we had a high success to reach a feasible tree.

References

1. R. Ahuja, T. Magnanti, J. Orlin *Network flows: theory, algorithms, and applications* NJ Prentice Hall, 1993.
2. S. Chen and K. Nahrstedt, *An Overview of Quality of Service Routing for Next-Generation High-Speed Networks: Problems and Solutions*, IEEE Network, pp. 64-79, November/December, 1998.
3. S. Chen and K. Nahrstedt, *On Finding Multi-Constrained paths*, IEEE ICC, 1998.
4. F. Ergun, R. Sinha and L. Zhang *An Improved FPTAS for Restricted Shortest Path* Information Processing Letters 83(5): 287-291, 2002.
5. M. Garey and D. Johnson, *Computers and Intractability: Guide to the Theory of NP-Completeness*, W. H. Rreeman, New York, 1979.
6. R. Guerin and A. Orda, *QoS-based Routing in Networks with Inaccurate Information: Theory and Algorithms*, IEEE INFOCOM, 1997.
7. R. Hassin *Approximation scehems forthe restricted shortest path problems* Math. Op. Res. 17(1):36-42,1992.
8. F. Hwang and D. Richards, *Steiner Tree Problem*, Networks, vol. 22. no. 1, pp. 55-89, January 1992.
9. A.Juttner, B. Szviatovszki, I. Mecs and Z. Rajko, *Lagrange Relaxation Based Method for the QoS Routing Problem*, IEEE INFOCOM, 2001.
10. R. Karp, *Reducibility among Combinatorial Problems*, in Complexity of Computer Computatioins (R. Miller and J. Thatcher, eds.), pp. 85-103, Plem Press, 1972.
11. T. Korkmaz, M. Krunz and S. Tragoudas *Multiconstrained Optimal Path Selection* IEEE INFOCOM, 2001.
12. L. Kou, G. Markowsky and L. Berman, *A Fast Algorithm for Steiner Trees*, Acta Infomatica, vol. 15, no. 2, pp. 141-145, 1981.
13. M. Parsa, Q. Zhu and J. J. Garcia-Luna-Aceves, *An Iterative Algorithm for Delay-Constrained Minimum-Cost Multicasting* IEEE/ACM Transactions on Networking, vol. 6, No. 4, August 1998.
14. D. Raz and D. Lorenz *Simple Efficient Approximation Scheme for the Restricted Shortest Path Problem* Operational Research Letters (ORL), 28(5), pp. 213-219, June 2001.
15. H. F. Salama, D. S. Reeves, Y. Viniotis, *Evaluation of Multicast Routing Algorithms for Real-Time Communication on High-Speed Networks*, IEEE Journal on Selected Areas in Communication, vol 15, no. 3, pp 332-345, April 1997.
16. http://vorlon.cwru.edu/~dxw49
17. B. M. Waxman, *Routing of multipoint connections*, IEEE Journal on Selected Areas in Cummunications, 6(9): 1617-1622, 1988

Q-MEHROM:
Mobility Support and Resource Reservations for Mobile Hosts in IP Access Networks

Liesbeth Peters*, Ingrid Moerman, Bart Dhoedt, and Piet Demeester

Department of Information Technology (INTEC), Ghent University – IMEC,
Sint-Pietersnieuwstraat 41, B-9000 Gent, Belgium
Tel.: +32 9 264 99 70, Fax: +32 9 264 99 60
{Liesbeth.Peters,Ingrid.Moerman,Bart.Dhoedt,Piet.Demeester}
@intec.UGent.be

Abstract. The increasing use of wireless networks and the popularity of multimedia applications, leads to the need of Quality of Service support in a mobile IP-based environment. This paper presents Q-MEHROM, which is the close coupling between the micromobility protocol MEHROM and a resource reservation mechanism. In case of handoff, Q-MEHROM updates the routing information and allocates the resources for receiving mobile hosts simultaneously. Invalid routing information and reservations along the old path are explicitly deleted. Resource reservations along the part of the old path that overlaps with the new path are reused. Q-MEHROM uses information calculated by QOSPF. Simulation results show that the control load is limited, mainly QOSPF traffic and influenced by the handoff rate in the network. Q-MEHROM uses mesh links and extra uplinks, present to increase the link failure robustness, to reduce handoff packet loss and improve the performance for highly asymmetric network loads.

1 Introduction

Today, wireless networks evolve towards IP-based infrastructures to allow a seamless integration between wired and wireless technologies. Most routing protocols that support IP mobility, assume that the network consists of an IP-based core network and several IP domains (also referred to as access networks), each connected to the core network via a domain gateway. This is illustrated in Fig. 1.

In contrast to wired networks, the user's point of attachment to the network changes frequently due to mobility. Since an IP address indicates the location of the user in the network as well as the end point of its connections, user mobility leads to several challenges. Mobile IP (IPv4 [1],IPv6 [2]), which is standardized by the IETF, is the best known routing protocol that supports host mobility.

While Mobile IP is used to support macromobility, i.e. the movements from one IP domain to another, much research is done to support the local movements

---

* Liesbeth Peters is a Research Assistant of the Fund for Scientific Research – Flanders (F.W.O.-V., Belgium).

M. Ajmone Marsan et al. (Eds.): QoS-IP 2005, LNCS 3375, pp. 495–508, 2005.
© Springer-Verlag Berlin Heidelberg 2005

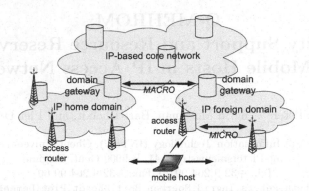

Fig. 1. General structure of an all IP-based network.

within one IP domain, called micromobility. Examples of micromobility protocols are per-host forwarding schemes like Cellular IP [3] and Hawaii [4], and tunnel-based schemes like MIPv4 Regional Registration [5] and Hierarchical MIPv6 [6], proposed within the IETF, and the BCMP protocol [7], developed within the IST BRAIN Project. These protocols try to solve the weaknesses of Mobile IP by aiming to reduce the handoff latency, the handoff packet loss and the load of control messages in the core network. Low Latency Handoff protocols, like Low Latency MIPv4 [8] and Fast MIPv6 [9], were developed by the IETF to reduce the amount of configuration time in a new subnet by using link layer triggers in order to anticipate on the movement of the mobile host.

Most research in the area of micromobility assumes that the links of the access network form a tree topology or hierarchical structure. However, for reasons of robustness against link failures and load balancing, a much more meshed topology is required. Although the above mentioned micromobility protocols work correctly in the presence of mesh links, the topology has an important influence on their performance. E.g. Cellular IP and MIPv4-RR do not use the mesh links, while Hawaii takes advantage of the mesh links to reduce the hand-off latency and packet loss. However, the use of Hawaii in a meshed topology results in suboptimal routes and high end-to-end delays after several handoffs. In our previous work, MEHROM (Micromobility support with Efficient Hand-off and Route Optimization Mechanisms) was developed, resulting in a good performance, irrespective of the topology, for frequent handoffs within an IP domain. This handoff scheme consists of two phases. During the first phase, a fast handoff is made in order to limit the handoff latency and packet loss. If necessary, the second phase is executed, which guarantees the set up of an op-timal new path between the domain gateway and the current access router. For a detailed description of MEHROM and a comparison with Cellular IP, Hawaii and MIPv4-RR, we refer to [10] and [11].

In a mobile IP-based environment, users want to receive real-time applica-tions with the same QoS (Quality of Service) as in a fixed environment. Sev-eral extensions to RSVP under macro- and micromobility are proposed in [12] and [13]. However, the rerouting of the RSVP branch path at the cross-over node under micromobility again assumes an access network with tree topology. Moreover, for mobile receivers, the cross-over node triggers a PATH message after a route update message was received, introducing some delay. Current work within the IETF NSIS (Next Steps in Signaling) working group includes the analysis of some of the existing QoS signaling protocols for an IP network [14] and the listing of Mobile IP specific requirements of a QoS solution [15].

In this paper, we present Q-MEHROM, which is the close coupling between MEHROM and a resource reservation mechanism. Hereby, the updating of routing information and the allocation of resources for mobile receivers is performed at the same time, irrespective of the topology. Information gathered by QOSPF (Open Shortest Path First protocol with QoS extensions [16], [17]) is used. In what follows, we assume that the requested resource is a certain amount of bandwidth.

The rest of this paper is structured as follows. Section 2 presents the framework used. Section 3 describes which information, obtained by the QOSPF protocol, is used by Q-MEHROM. In section 4, the operation of Q-MEHROM is explained. Simulation results are presented in section 5. The final section 6 contains our concluding remarks.

2 Micromobility and Resource Reservations

In Fig. 2, the framework, used for a close interaction between micromobility routing and resource reservations, is presented. A micromobility protocol updates the routing tables to support data traffic towards mobile hosts. In this paper, we consider the resource reservations for data flows towards mobile hosts.

The central building block is responsible for the propagation of mobility and requested resources information through the access network. Frequent handoffs within one IP domain are supported by MEHROM [11]. This is a per-host scheme and every router along the path between the domain gateway and the current access router has an entry in its routing table indicating the next hop to the mobile host. At the time of handoff, the necessary signaling to update the routing

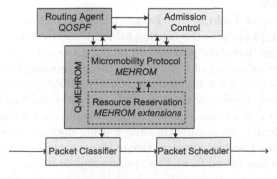

Fig. 2. Framework for micromobility routing and QoS support in an IP access network. The grey building blocks are the focus of this paper.

tables is kept locally as much as possible. New entries are added and obsolete entries are explicitly deleted, resulting in a single installed path for each MH (mobile host). This micromobility scheme is very suitable to be closely coupled with a resource reservation mechanism for traffic towards the mobile hosts. When a mobile host performs a handoff, we want to propagate the necessary QoS information as fast as possible to limit the degradation of the delivered QoS. Hereby, we want to restrict the signaling to the part of the new path that does not overlap with the old path and explicitly release reserved resources along the unused part of the old path. By defining the resource reservation mechanism as an extension of MEHROM, resources can be re-allocated at the same time that the routing tables are updated after handoff. The combination of MEHROM and its resource reservation extensions is referred to as Q-MEHROM.

In order to obtain information about the topology of the access network and the state of the links, we use QOSPF, as described in [17]. This is presented by the upper left building block of the framework.

The interaction between both grey building blocks is as follows. QOSPF computes paths in the access network that satisfy given QoS requirements and provides this information to Q-MEHROM. Every time Q-MEHROM reserves or releases resources on a link, QOSPF is informed so that updated link state advertisements can be spread throughout the access network.

The upper right block represents the admission control policy. Although a variety of mechanisms is possible, we have chosen for a simple admission control priority mechanism: when a new request for resources is made by a mobile host, the access router decides whether the access network has sufficient resources to deliver the required QoS. If not, the request is rejected. At the time of handoff, priority is given to handoff requests above new requests. When the required resources for an existing connection can not be delivered via the new access router at the time of handoff, the handoff request is not rejected, but the delivered service is reduced to best-effort. For simplicity, only at the time of a next handoff, the availability of resources is checked and reservations can be made again. The admission control mechanism gives QOSPF information about the resources that were reserved by the mobile host before handoff (these resources can be reused). QOSPF in its turn provides information about paths with sufficient resources in the access network.

Finally, the lower building blocks are the packet classifier and scheduler. When Q-MEHROM updates the routing tables, these routes are installed in the packet classifier so that data packets are mapped towards the correct outgoing link. As scheduler, a variant of CBQ (Class Based Queueing) is used [18]. When resources are reserved for a mobile host, Q-MEHROM inserts a class for that mobile host. When these resources are released after an handoff, the class is removed again. One class that is never removed is the class for all control traffic. To reduce complexity, we assume that the routing decision is only based on the mobile host's care-of address, irrespective of the amount of flows. For simplicity, we consider one class for all flows to a specific mobile host.

3   Information Calculated by QOSPF

QOSPF advertises link metrics, like link available bandwidth and link delay, across the access network by Link State Advertisements (LSAs). As a result, the domain gateway, all routers and all access routers have an updated link-state database that reflects the access network state and topology. To provide useful information to Q-MEHROM, QOSPF calculates, in each node of the access network, several QoS routing tables from the database, see also Table 1.

Delay Table. The Delay table is used during the first phase of the Q-MEHROM handoff algorithm, when the new access router sends a route update message towards the old access router as fast as possible. The Delay table of a router has this router as source and an entry for every access router (AR) as destination. Therefore, the router calculates the path with smallest delay to a specific AR, using a single objective path optimization. The next hop to that AR is then put in the Delay table. As the delay characteristics are not expected to change frequently, the Delay table needs no frequent recalculation.

Bandwidth Table. A Bandwidth table is used when resources are reserved for a new traffic flow. Hereby, the available bandwidth on the links of the access network is taken into account. As we consider traffic towards the mobile hosts, a router calculates a Bandwidth table with the domain gateway as source and itself as destination. The Bandwidth table gives the last node on the path before reaching the router. A double objective path optimization is used, taking also the hop count into account. For a certain value of the hop count, the path with at most this amount of hops and with maximum bandwidth (BW) is calculated. When the amount of available bandwidth in the access network changes, the Bandwidth tables should be recalculated.

Reuse-Bandwidth Table. A Reuse-Bandwidth table is used for handoff reservations. When a mobile host performs handoff, a new path must be set up. It is

Table 1. Delay, Bandwidth and Reuse-Bandwidth table for router R2 of Fig. 3.1.

Delay table		Bandwidth table (for new traffic flow)		
destination	next hop	max. hops	last hop	max. bandwidth
oAR	R1	1	–	–
nAR	nAR	2	R3	2 Mbit/s
		3	R3	2 Mbit/s

Reuse-Bandwidth table (if MH performs handoff)		
max. hops	last hop	max. bandwidth
1	–	–
2	R3	5 Mbit/s
3	R3	5 Mbit/s

possible that the new path partly overlaps with the old path. Resources along this common part should be reused and not be allocated twice. Therefore, QOSPF must consider the resources, reserved for the mobile host before handoff, also as available resources and calculate another bandwidth table, called the Reuse-Bandwidth table. This table can only be calculated at the time of handoff and needs as input the old path and the amount of reserved bandwidth for a specific mobile host. This table has the same structure as the Bandwidth table.

4   Q-MEHROM Mechanism

The basic operation of Q-MEHROM is explained using the example in Fig. 3. There are three successive situations:

Bandwidth Reservations. When a MH powers up in the access network, only basic MEHROM is used to set up a (best-effort) path for the MH in the access network. This path is a path with minimum hop count between the domain gateway (GW) and the current access router.

When a traffic flow is set up towards a MH, this MH requests an amount of bandwidth (information obtained from the application layer) for that flow at its current access router. After a positive admission control, the access router starts the resource reservation mechanism, using its Bandwidth table. If the MH is the receiver of multiple flows, the requested bandwidth is the sum of the individually amounts of bandwidth requested by the applications.

Handoff Phase 1 – Fast Handoff. After the receipt of a Mobile IP registration request, the new access router (nAR) (we describe the case of a positive admission control) adds a new entry for the MH in its routing cache and sends a route update message towards the old access router (oAR) in a hop-by-hop way. Like basic MEHROM, Q-MEHROM is capable to detect if the new path will be optimal, using an optimal_path parameter in the route update messages.

When R2 processes the route update message, it checks the optimal_path parameter in this message. This parameter has value 1, i.e. the node is part of the optimal path between the GW and the nAR. Therefore, the necessary resources are reserved on the link between R2 and the nAR. As a next step, it looks up the next hop on the path with smallest delay towards the oAR in the Delay table (i.e. R1). Furthermore, it retrieves in the Reuse-Bandwidth table the last node before reaching R2 on the optimal path between the GW and R2 (i.e. R3). As both nodes are different, the optimal_path parameter is set to 0, indicating the path will be suboptimal. Once this parameter contains the value 0, it remains 0 and only the Delay table must be consulted. The route update message is forwarded using the next hop from the Delay table.

R1 has an old entry for the MH. Therefore R1 is considered as the cross-over node (CN), and the route update message is not forwarded further. The CN sends a suboptimal handoff acknowledgment (ACK) back to the nAR. The CN also deletes the reservation on the link from R1 to oAR and sends a route delete message to the oAR, to delete the invalid entry in its routing cache.

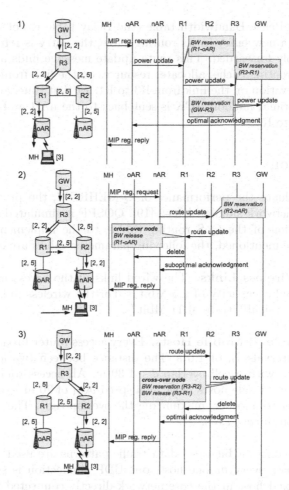

Fig. 3. Example of bandwidth reservation for a mobile host. 1) the MH requests resources for its traffic flows, 2) the MH performs handoff which consists of phase 1 and 3) phase 2. The arrows indicate entries for the mobile host and point to the next hop in the path. A dashed arrow indicates an entry without resource reservations on the next link. [x,y] = [link delay (ms), available bandwidth (Mbit/s)]; [z] = [requested bandwidth (Mbit/s)].

The nAR is informed through a suboptimal handoff ACK that a suboptimal path is found during phase 1. As data packets can reach the MH via this suboptimal path, the nAR sends first a Mobile IP registration reply to the MH. Next, the nAR starts the route optimization phase.

Handoff Phase 2 – Route Optimization. The nAR sends a new route update message towards the GW, also in a hop-by-hop way, to establish an optimal path. During phase 2, the optimal_path parameter is set to value 2, indicating that the route update message must be forwarded towards the node retrieved in the Reuse-Bandwidth table (the Delay table is not consulted). As R2 has already a new entry in its routing table, the entry is refreshed and the message is simply forwarded. This route update message finds a new CN, R3. R3 updates its routing cache, allocates resources on the link from R3 to R2 and deletes the reservation on the link from R3 to R1. At that time, an optimal path is set up: an optimal handoff ACK is sent back to the nAR and a route delete message is sent to R1.

5   Evaluation

In order to evaluate the performance of Q-MEHROM, the protocol is implemented in the network simulator ns-2 [19]. QOSPF is simulated by extensions, for the calculation of the QoS routing tables, to the implementation of [20]. Unless otherwise mentioned, the following parameter values are chosen:

Wired and Wireless Links. The wired links of the access network have a delay of 2 ms and a capacity of 2.5 Mbit/s. For the wireless link, IEEE 802.11 is used with a physical bitrate of 11 Mbit/s.

Access Routers and Mobile Hosts. Every access router broadcasts beacons at fixed time intervals T_b of 1.0 s. The distance between two adjacent access routers is 200 m, with a cell overlap d_o of 30 m. All access routers are placed on a straight line. Mobile hosts move at a speed v_{MH} of 20 m/s and travel from one access router to another, maximizing the overlap time. They reside thus in an overlap region during 1.5 s.

Traffic. CBR (constant bit rate) data traffic patterns are used, with a bitrate of 0.5 Mbit/s. For every mobile host, one UDP connection is set up between the sender (a fixed host in the core network directly connected to the domain gateway) and the receiver (the mobile terminal).

Access Network Topology. Tree, mesh and random topologies are investigated. The topologies that are used for the simulations, are given in Fig. 4.

5.1   Load of Control Traffic in the Access Network

For the different topologies of Fig. 4, the average amount of control traffic on a wired link of the access network is investigated. The leftmost figure of Fig. 5 gives the results for an increasing number of mobile hosts, while the rightmost figure shows the influence of the speed of a single mobile host. The mobile hosts move randomly from one access router to another during 300 s, requesting 0.5 Mbit/s. Randomly means that a mobile node moves towards a randomly chosen access router. When arriving, the mobile host stays there for maximum 1 second before a new access router is chosen as destination. The Q-MEHROM control load consists of Route Update, Route Delete, Acknowledgment and Power Up messages.

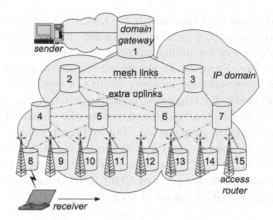

Fig. 4. Tree, mesh and random topology of the access network, used during the simulations. The mesh topology consists of the tree structure (full lines) with the indicated additional mesh links (dashed lines), while the random topology is formed by adding extra uplinks (dotted lines) to the mesh topology.

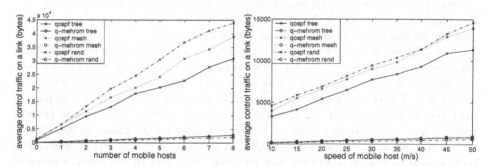

Fig. 5. Average amount of control traffic on a wired link of the access network as a function of the number of mobile hosts (left figure) and the speed of a single mobile host (right figure).

The QOSPF control traffic consists of two parts. The first part is formed by the QOSPF messages at the start of the simulation, advertising the initial link state information through the network. The second part consists of QOSPF traffic caused by resource reservations and resource releases during the simulation as they change the available bandwidth of some links.

The load of control messages specific to Q-MEHROM is very low compared to the load of QOSPF traffic. The QOSPF traffic in the absence of mobile hosts contains only the first part, with messages to spread the initial link state information. This part depends purely on the topology and the number of links. The second part of the QOSPF load, with messages triggered by resource reservations and releases, is also determined by the number of handoffs and their location in the topology.

The required amount of bandwidth for QOSPF control traffic increases for higher handoff rates, i.e. the number of handoffs in the network per time unit. This handoff rate grows for higher number of mobile hosts, illustrated in the leftmost figure of Fig. 5 and for higher speeds as shown in the rightmost figure. Note that this is the worst case, as every change in available bandwidth triggers QOSPF LSAs. Periodically triggering of LSAs or triggering only after a significant change in available bandwidth can reduce the control load, at the cost of the accuracy of link state information. The required amount of bandwidth for control traffic remains low compared to the link capacities ($\pm 0.05\%$ for 8 mobile hosts with a speed of 20 m/s). In what follows, we reserved 1% of the given link capacity for the control traffic class of CBQ. This is sufficient to support more than 100 mobile hosts with a speed of 20 m/s.

5.2   Handoff Performance

To investigate the handoff performance, the following scenario is considered: during 1 simulation, a single mobile host moves from the leftmost to the rightmost access router of Fig. 4, performing 7 handoffs. The results in Fig. 6 are average values of a set of 200 independent simulations, i.e. there is no correlation between the sending of beacons by the access routers, the movements of the mobile host and the arrival of data packets in the access routers.

The leftmost figure of Fig. 6 shows the loss of data packets as a function of the speed of the mobile host. The data flow has a packet size of 500 bytes and a rate of 1 Mbit/s. Results are given for beacons sent at time intervals $T_b = 1.0$ s, 0.75 s and 0.50 s. The packet loss is independent of the velocity, as long as the mobile host resides long enough in the overlap region to receive at least one beacon from a new access router while still connected to its previous access router. Otherwise, packet loss increases rapidly. In the considered situation, where the mobile host moves on a straight line through the center of the overlap regions, this requirement is fulfilled when

$$v_{\mathrm{MH}} \leq d_o/T_b \ . \tag{1}$$

In this case, Q-MEHROM takes advantage of mesh links and extra uplinks to reduce the packet loss compared to the tree topology.

The rightmost figure of Fig. 6 gives the results for the data packet loss as a function of the data rate. The mobile host moves at a speed of 20 m/s and the access routers send beacons at time intervals of 1.0 s. The used packet sizes are 500 bytes, 1000 bytes and 1500 bytes. For a given packet size, the packet loss increases for higher data rates and is higher for the tree topology compared to the other topologies. The use of IEEE 802.11 and collisions on the wireless link limit the achievable throughput. For a wireless bitrate of 11 Mbit/s, the use of the RTS/CTS mechanism and a data packet size of x bytes, the TMT (Theoretical Maximum Throughput) is given by [21]:

$$\mathrm{TMT}(x) = \frac{8x}{ax + b} \cdot 10^6 \text{ bps} \quad \text{with } a = 0.72727 \text{ and } b = 1566.73 \ . \tag{2}$$

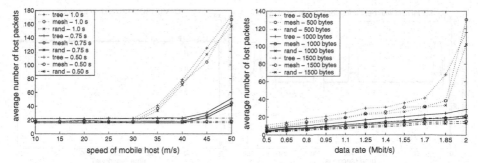

Fig. 6. Average packet loss as a function of the mobile host's speed for several beacon time intervals (left figure) and as a function of the data rate for several packet sizes (right figure).

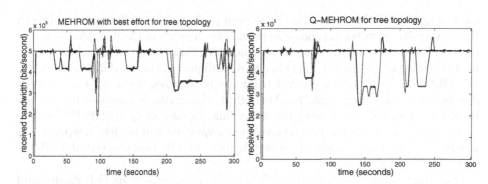

Fig. 7. Received bandwidth by the mobile hosts in highly loaded networks for the tree topology.

This explains the rapid increase of packet loss for a packet size of 500 bytes and data rates higher then 1.85 Mbit/s.

5.3 Highly Loaded Access Networks

In order to investigate how well Q-MEHROM can handle the mismatch between traffic load and network capacity, we consider the following scenario: at the start of the simulation, the network is symmetrically loaded by a single mobile host in each of the eight cells of Fig. 4. During one simulation, those mobile hosts move randomly from one access router to another during 300 s. Each mobile host is the receiver of a CBR data flow with a packet size of 1500 bytes and a bitrate of 0.5 Mbit/s. These values are chosen so that the wireless channel is not a bottleneck for the received bandwidth (TMT = 4.515 Mbit/s), even if all mobile hosts move into the same cell. Figure 7 gives the average bandwidth received by each mobile host during the last 3.0 seconds as a function of time for the tree topology, Fig. 8 for the random topology.

For the tree topology, only one possible path between the domain gateway and a specific access router exists. So both in case of best-effort (left figure) as in

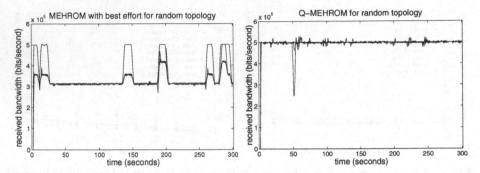

Fig. 8. Received bandwidth by the mobile hosts in highly loaded networks for the random topology.

case of Q-MEHROM (right figure) the data packets for mobile hosts residing in the same cell are routed via the same path. The capacity of the links closest to the domain gateway limits the amount of bandwidth that can be allocated for mobile hosts residing in the underlying cells. As a small part of the total link bandwidth, 2.5 Mbit/s, is reserved for control traffic, only four reservations of 0.5 Mbit/s can be made on a single link. For best-effort, the available bandwidth is shared between all mobile hosts using the same link. In case of Q-MEHROM, mobile hosts that made a reservation, receive their requested bandwidth irrespective of newly arriving mobile hosts. As a result, more mobile hosts receive 0.5 Mbit/s compared to best-effort.

For the random topology, the difference between best-effort (left figure) and the use of Q-MEHROM (right figure) is much more significant. Due to the presence of extra uplinks, more than one path with minimum hop count may be found between the domain gateway and a specific access router. In case of best-effort, one of these possible paths is chosen without taken into account the available bandwidth. Only for mobile hosts in cell 15, ns-2 uses the path via the right link closest to the domain gateway. Q-MEHROM chooses a path with enough resources, if available. Due to mesh links and extra uplinks, the capacity of the links closest to the domain gateway forms no longer a bottleneck. Q-MEHROM uses the presence of these extra links, to support asymmetric network loads and to spread the data load over the network.

6   Conclusions

In this paper we presented Q-MEHROM, which is the close coupling between the micromobility protocol MEHROM and a resource reservation mechanism. In case of handoff, the updating of routing information and resource reservations for traffic flows towards the mobile host are performed at the same time. Hereby, invalid routing information and reservations along the old path are explicitly deleted. Resource reservations along the part of the old path that overlaps with the new path are reused. Q-MEHROM uses information gathered and calculated by QOSPF.

Simulation results showed that the amount of control overhead caused by Q-MEHROM and QOSPF is mainly QOSPF traffic. The control load increases for higher handoff rates in the network, which is influenced by the number of mobile hosts and their velocity. For the studied topologies and scenarios, this load remains very low compared to the link capacities. The first phase of Q-MEHROM takes advantage of extra links, present to increase the robustness against link failures, to reduce the handoff packet loss compared to a pure tree topology. This handoff packet loss is independent of the speed of the mobile hosts and increases with the data rate. For high network loads, simulations indicated that by the use of Q-MEHROM, mobile hosts can make reservations and protect their received bandwidth against newly arriving mobile hosts. Furthermore, Q-MEHROM takes the available bandwidth into account by choosing a path and is able to use mesh links and extra uplinks to improve the balancing of data load in the access network.

The study of resource reservations for traffic flows sent by mobile hosts and the study of different classes per mobile host are subject of future work. Also the influence of inaccurate link state information on the Q-MEHROM performance is an important issue that will be investigated.

Acknowledgments

Liesbeth Peters is a Research Assistant of the Fund for Scientific Research – Flanders (F.W.O.-V., Belgium). Part of this research is funded by the Belgian Science Policy Office (BelSPO, Belgium) through the IAP (phase V) Contract No. IAPV/11, and by the Institute for the promotion of Innovation by Science and Technology in Flanders (IWT, Flanders) through the GBOU Contract 20152 "End-to-End QoS in an IP Based Mobile Network".

References

1. Perkins, C. (ed.): IP mobility support for IPv4. IETF RFC 3344, Augustus 2002
2. Johnson, D., Perkins, C., Arkko, J.: Mobility support in IPv6. IETF RFC 3775, June 2004
3. Valkó, A.: Cellular IP: a new approach to internet host mobility. ACM Computer Communication Review, January 1999
4. Ramjee, R., La Porta, T., Salgarelli, L., Thuel, S., Varadhan, K.: IP-based access network infrastructure for next-generation wireless data networks. IEEE Personal Communications, August 2000, pp. 34-41
5. Gustafsson, E., Jonsson, A., Perkins, C.: Mobile IPv4 Regional Registration. draft-ietf-mobileip-reg-tunnel-07.txt, October 2002 (work in progress)
6. Soliman, H., Catelluccia, C., El Malki, K., Bellier, L.: Hierarchical Mobile IPv6 mobility management (HMIPv6). draft-ietf-mipshop-hmipv6-02.txt, June 2004 (work in progress)
7. Boukis, C., Georganopoulos, N., Aghvami, H.: A hardware implementation of BCMP mobility protocol for IPv6 networks. GLOBECOM 2003 – IEEE Global Telecommunications Conference, Vol. 22, no. 1, December 2003, pp. 3083-3087

8. El Maki, K. (ed.): Low latency handoffs in Mobile IPv4. draft-ietf-mobileip-lowlatency-handoffs-v4-09.txt, June 2004 (work in progress)
9. Koodli, R. (ed.): Fast handovers for Mobile IPv6. draft-ietf-mipshop-fast-mipv6-02.txt, July 2004 (work in progress)
10. Peters, L., Moerman, I., Dhoedt, B., Demeester, P.: Micro-Mobility support for random access network topologies. IEEE Wireless Communication and Networking Conference (WCNC 2004), 21-25 March, Georgia, USA, ISBN 0-7803-8344-3
11. Peters, L., Moerman, I., Dhoedt, B., Demeester, P.: MEHROM: Micromobility support with efficient handoff and route optimization mechanisms. 16th ITC Specialist Seminar on Performance Evaluation of Wireless and Mobile Systems (ITCSS16 2004), 31 August-2 September, Antwerp, Belgium, pp. 269-278
12. Moon, B., Aghvami, A.H.: Quality-of-Service mechanisms in all-IP wireless access networks. IEEE Journal on Selected Areas in Communications, Vol. 22, No. 5, June 2004, pp. 873-887
13. Moon, B., Aghvami, A.H.: RSVP extensions for real-time services in wireless mobile networks. IEEE Communications Magazine, December 2001, pp. 52-59
14. Manner, J., Fu, X., Pan, P.: Analysis of existing quality of service signaling protocols. draft-ietf-nsis-signalling-analysis-04.txt, May 2004 (work in progress)
15. Chaskar, H. (ed.): Requirements of a quality of service (QoS) solution for Mobile IP. IETF RFC 3583, September 2003
16. Moy, J.: OSPF Version 2. IETF RFC 2328, April 1998
17. Apostolopoulos, G., Williams, D., Kamat, S., Guerin, R., Orda, A., Przygienda, T.: QoS routing mechanisms and OSPF extensions. IETF RFC 2676, August 1999
18. Floyd, S., Van Jacobson: Link-sharing and resource management models for packet networks. IEEE/ACM Transactions on Networking, August 1995, Vol. 3, No. 4
19. NS-2 Home Page, www.isi.edu/nsnam/ns
20. QoSR in ns-2, www.netlab.hut.fi/tutkimus/ironet/ns2/ns2.html
21. Jun, J., Peddabachagari, P., Sichitiu, M.: Theoretical maximum throughput of IEEE 802.11 and its applications. IEEE International Symposium on Network Computing and Applications (NCA-2003), Cambridge, USA, 2003

802.11 MAC Protocol with Selective Error Detection for Speech Transmission*

Antonio Servetti[1] and Juan Carlos De Martin[2]

[1] Dipartimento di Automatica e Informatica,
Politecnico di Torino,
10129 Torino – Italy
antonio.servetti@polito.it
[2] IEIIT–CNR,
Politecnico di Torino,
10129 Torino – Italy
demartin@polito.it

Abstract. The IEEE 802.11 standard currently does not offer support to exploit the unequal perceptual importance of multimedia bitstreams. All packets affected by channel errors, in fact, are simply discarded, irrespective of the position and percentage of corrupted bits. The objective of this paper is to investigate the effect of bit error tolerance in WLAN speech communications. More specifically, we introduce QoS support for sensitive multimedia transmissions by differentiating the scope of the standard MAC error detection step in order to discard multimedia packets only if errors are detected in the most perceptually sensitive bit class. Speech transmission using the GSM-AMR speech coding standard is simulated using a set of experimental bit-error traces collected in various channel conditions. Perceived speech quality, measured with the ITU-T P.862 (PESQ) algorithm, is consistently improved with respect to standard link layer technique. In other words, the results show that the negative effect of errors in the less perceptually important bits is clearly counterbalanced by the lower number of speech packets discarded because of retransmission limits. In fact, the number of received packets is consistently doubled throughout all the simulation conditions with quality gains that reach 0.4 points of the MOS scale in noisy scenarios.

1 Introduction

In recent years the IEEE 802.11 wireless standard has been adopted in an increasing number of places: many shopping malls, airports, train stations, universities, have now wireless infrastructures to provide people with tetherless access to the Internet. This emerging scenario is creating the basis of a new set of services based on the communication opportunities offered by ubiquitous network accessibility. In particular a great deal of interest is focusing on interactive voice communication applications over Wi-Fi links.

---

* This work was supported in part by MIUR, project FIRB–PRIMO, http://primo.ismb.it.

M. Ajmone Marsan et al. (Eds.): QoS-IP 2005, LNCS 3375, pp. 509–519, 2005.

However several challenges need to be addressed to provide successful speech services over a network originally intended for generic data traffic and characterized by potentially high error rates. While for data transfers, in fact, throughput is the main parameter for measuring the network performance, multimedia services depend on strict quality of service (QoS) requirements in terms of packet loss and delay. Moreover, in current data-oriented WLAN's, packet losses are also due to the need of data integrity during transfers, i.e., every hop discards all packets affected by channel errors, irrespective of the percentage of corrupted data. This approach does not exploit what modern multimedia compression algorithms offer, namely, a certain degree of error resilience, so that the decoder can still benefit from corrupted packets. As a consequence a new error tolerant extension to the MAC layer is advisable for multimedia transmission in wireless environments.

In this paper we analyze the advantages of modifying the link layer of IEEE 802.11 networks allowing partially corrupted speech packets to be forwarded (and not discarded) without requiring additional retransmissions. We find that this new functionality enables differentiated treatment of voice streams and can be tuned to meet speech QoS requirements of low delay and losses.

IEEE 802.11 Medium Access Control (MAC) layer [1] provides a checksum to prevent forwarding of erroneous frames: if a bit or more are corrupted the packet is discarded and the sender will retransmit the data until a maximum retransmission limit is reached. Since speech data bits are known to have different perceptual importance [2], they can be packed in sensitivity order and a checksum applied only to the most important subset. Partial checksum will prevent useful frames to be dropped and will reduce the number of retransmissions, thus reducing the network load and delay when the error probability is high.

Previous work addressed the problem of multimedia transmission over lossy networks suggesting to apply selective checksums to the UDP transport protocol [3]: if the packet is received with errors, it is delivered to the application only if the checksummed bits are correct, otherwise the packet is dropped. First proposed by Larzon in [4], for progressive coding schemes, PCM audio, and MPEG video, UDP-Lite has then been studied by Singh [5] and Reine [6]. UDP-Lite has been shown capable of providing less end-to-end delay, reduced jitter, higher throughput, less packet losses, and better video quality than plain UDP.

According to the current IEEE 802.11 MAC standard UDP-Lite cannot be effectively employed on wireless networks because erroneous frames are dropped by the link layer before reaching the UDP layer. Also the recent 802.11e extension to the MAC standard [7], specifically proposed for multimedia applications, does not allow the adoption of partial checksum techniques. However the idea of an error tolerant 802.11 network has already been discussed in the literature. The performance of the UDP-Lite protocol is, in fact, evaluated for a WLAN scenario in [8] where the checksum coverage is limited to protocol headers, and the 802.11 MAC level error checking feature is completely disabled (thus disabling the retransmission mechanism too). [9] and [10] take a step forward with the idea of reflecting the UDP-Lite policy of sensitive and insensitive data in the

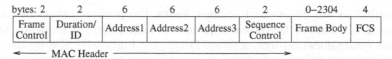

bytes: 2	2	6	6	6	2	0–2304	4
Frame Control	Duration/ ID	Address1	Address2	Address3	Sequence Control	Frame Body	FCS

◄——— MAC Header ————————————————————►

Fig. 1. Frame format of an 802.11 data frame MPDU. Each data-type frame consists of a MAC Header, a variable length information Frame Body, and a Frame Check Sequence (FCS). The MAC overhead due to the MAC header and the FCS is 28 bytes in total.

MAC protocol. That permits to detect, discard, and retransmit heavily "perceptually" corrupted packets not only at the receiver-end, but also at every wireless hop. As stated in these papers, the link-layer implementation of selective error detection can be more effective especially in case of high end-to-end delay scenarios, where end-to-end retransmission would not be applicable. Hop-by-hop retransmission is, in fact, a prompter solution that enables the delivery of an "acceptable" packet with lower delay. The results obtained by means of network simulations in infrastructure [9] and in ad-hoc [10] scenarios show that error checking only the most sensitive part of the multimedia payload is very effective when speech transmission is considered: delay, packet loss and network load are significantly reduced. Substantial gains are also reported for video transmission, notably in [11], where the H.264 coder is considered.

In comparison with previous works, the original contribution of this paper is in the simulation and evaluation method. Instead of using a model to generate a realistic 802.11 bit-error behavior, we use experimental 802.11b error traces collected under various network conditions. Moreover our performance analysis is not limited to consider the overall packet loss rate, but the perceptual quality of GSM AMR-WB [12] coded speech transmission is measured using the ITU-T P.862 standard, i.e., the Perceptual Evaluation of Speech Quality (PESQ) [13]. Network simulations show that the speech distortion introduced by decoding partially corrupted packets is clearly lower than the distortion that would have been caused by their discarding.

The paper is organized as follows. In Section 2, we introduce the wireless Voice over IP scenario and we describe the selective bit error checking MAC protocol for wireless multimedia. Performance evaluation and quality results are presented in Section 3 followed by the conclusions in Section 4.

2 Selective Error Detection in 802.11 WLAN's

The wireless case is a quite challenging environment for packet communications. Hop-by-hop transmissions are affected by high error probability because of interference with others signal sources as well as fading. To overcome this problem the current 802.11 medium access control implementation forces a receiver station to discard every erroneous packet and a simple retransmission control policy is used to reduce end-to-end packet loss.

In this scenario Voice over IP applications can tolerate few losses ($< 3\%$); above that threshold, the communication quality becomes unacceptable. If many retransmissions are employed to reduce losses, the positive effects of receiving more packets is undone by the consequences of higher end-to-end delays that reduce the communication interactivity. Delays are influenced also by the shared nature of the wireless link, in fact, stations contend for a transmission opportunity for each packet transmitted, a severe problem in congested scenarios.

2.1 IEEE 802.11 MAC Protocol

The IEEE 802.11b physical layer describes a Direct Sequence Spread Spectrum (DSSS) system with an 11 Mbps bit-rate [1] operating in the industrial, scientific, and medical (ISM) band at 2.4 GHz. The fundamental transmission medium defined to support asynchronous data transfer on a best effort basis is called Distributed Coordination Function (DCF). It operates in a contention mode to provide fair access to the channel for all stations. If two or more stations initiate their transmission at the same time, a collision occurs. A collision avoidance mechanism based on a random backoff procedure is meant to reduce this problem.

Because of collisions and channel noise, each station must be informed about the success of its transmission. Therefore each transmitted MAC protocol data unit (MPDU) requires an acknowledgment (ACK). Acknowledgment are sent by the receiver upon successful and error free packet reception. To verify the data-unit integrity the frame format, shown in Fig. 1, provides a 4-byte Frame Check Sequence (FCS) field. The destination station compares the packet FCS with a new one computed over all the received MAC bits. Only if all the bits are correct the two FCS's match each other and the packet is positively acknowledged by sending an ACK frame back to the source station. If this ACK frame is not received right after the transmission the sending station may contend again for the channel to transmit the unacknowledged packet until a maximum retry limit is reached. For each retransmission the random backoff time increases because drawn from a double-sized time window (up to a maximum defined value), so that the probability of repeated collisions is reduced.

2.2 Selective Error Detection

The adoption of selective error detection in the IEEE 802.11 MAC layer, provided that packet discarding can be avoided as long as errors do not affect important bits of the multimedia payload, can produce several positive effects on the network.

To illustrate the behavior of the proposed technique let us consider an n-bit long frame with m bits covered by the checksum. For simplicity's sake, only in the current section, we assume that the bit error probability is uniform and equal to p, and each bit is independent of the others. The packet loss rate (PLR) of a packet with m checksummed bit is expressed by the equation $PLR_m = [1 - (1 - p)^m]$, independently by its size n. For a full checksum scheme, m is equal to n. If selective error detection is implemented then $m < n$ and a partial checksum, that do not covers all data bits, is employed.

If a maximum of $N - 1$ retransmissions are allowed then the packet loss rate, that is the probability of N unsuccessful transmissions, is $PLR_m(N) = (PLR_m)^N$. Finally, consider the expected number of transmissions T for each packet when maximum N attempts are allowed:

$$E\{T\} = \sum_{i=1}^{N-1} i(1 - PLR_m)PLR_m^{i-1} + N \cdot PLR_m^{N-1}$$

$$= \frac{1 - PLR_m^N}{1 - PLR_m} = \sum_{n=0}^{N-1} PLR_m^n. \tag{1}$$

As expected, the reduction of checksummed bits effectively decreases the packet loss and the total number of transmissions. Consequently end-to-end delay and network load are reduced with positive effects on the quality of service that can be achieved by all the transmissions in the network.

Among the effects of selective error detection we should also introduce the concept of *corrupted* packet that refers to a packet where errors occur only outside the checksum coverage. The packet corruption rate (PCR) is then defined as [11]

$$PCR_m(N) = (1 - p)^m \cdot [1 - (1 - p)^{(n-m)}] \sum_{i=0}^{N-1} [1 - (1 - p)^m]^i. \tag{2}$$

For the case of $N = 1$ the equation simply becomes: $PLR_n(1) = PLR_m(1) + PCR_m(1)$. With no retransmissions, in fact, the sum of corrupted and lost packets for selective detection shall be the same as the number of lost packets if the whole data is checksummed. The complex effect of retransmissions will be further investigated in Section 3. Retransmissions play a fundamental role in favoring full or partial checksum techniques. The latter solution guarantees a lower PLR often accepting lightly corrupted packets, while the former presents more packet losses, but it ensures the integrity of the received ones.

Performance evaluation of the proposed technique should not only be limited to considering the packet loss rate, but should also include the effect of decoding partially corrupted packets. Hence, in the following, we drop the simplistic assumption of uniform bit error probability, and we derive bit error rate and location from actual transmission measurements to prove the effectiveness of the proposed technique in a real scenario.

3   Performance Evaluation for Speech Transmission

Selective error detection performance is evaluated for speech transmission over a wireless channel using the wideband GSM Adaptive Multi Rate (AMR) coder [12]. Different channel conditions and maximum number of retransmissions are considered. Objective speech quality measures are then employed to compare the perceived speech quality of the received streams.

Figure 2 shows the error detection coverage in a 802.11 data MPDU with speech payload. At the application level voice is encoded by the 23.85 kb/s GSM

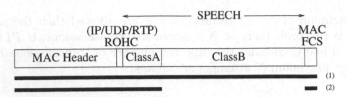

Fig. 2. Standard (1) and modified (2) MAC checksum coverage.

AMR-WB coder in frames of 477 bits. Speech encoder output bits are ordered according to their subjective importance and divided in two classes: Class A and Class B (as defined in the standard [14]). Class A (first 72 bits) contains the bits most sensitive to errors and any error in these bits typically results in a corrupted speech frame which should not be decoded without applying appropriate error concealment. Class B contains bits where increasing error rates gradually reduce the speech quality, but decoding of an erroneous speech frame is usually possible without annoying artifacts.

Real-time Transport Protocol (RTP) is used according to RFC 3267 specification for AMR encoded speech signals: ten control bits are present at the beginning of the speech payload and additional bits are added to the end as padding to make the payload byte aligned for a total of 61 bytes. For the RTP, UDP, and IP headers, Robust Header Compression is assumed that allows the 40-byte header to be compressed in 2 bytes. Then each data-type MPDU comprises a 24-byte header plus a 4 byte checksum. To support partial checksum in the modified MAC proposal an additional two-byte fixed-length field should be introduced to specify the number of covered bit, starting from the beginning of the MAC data unit.

3.1   Wireless Bit-Error Trace Collection

Characterizing the error behavior of the 802.11 channel is a fundamental issue that strongly influences wireless network simulations. While it is well known that wireless links have typically higher error rates than their wired counterparts, the detailed characteristics of wireless errors are not easy to reproduce in a computer simulation [15]. Lacking of a widely acknowledged model for simulating the wireless 802.11b bit-error behavior, we decided to use an experimental, trace-based approach.

The basic idea, for bit-error trace collection, has been to transmit a well-known packet stream over an 11 Mbps 802.11b wireless network using specially formatted UDP packets that include in the payload information for error detection such as a redundant sequence number and a repeated signature. For our modeling of packet errors, we decided to send sequences of 100-byte packets every 20 ms.

All the bit-error traces were collected at the receiver by modifying the wireless device drivers. More specifically, the receiver was a Linux box using a Prism2 wireless 802.11b PCMCIA-Card and Wlan-Ng (ver. 0.2.1-pre20) device drivers. When in monitoring mode the modified device drivers passed all packets to the

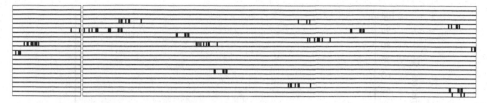

Fig. 3. Measured 802.11 error pattern mapped on transmitted speech data. Each line corresponds to a 23.85 kb/s GSM AMR-WB frame (the block on the left represents Class A bits). Erroneous bits are marked as black points.

upper network layers, thus the traces collected at the client by the network sniffer (ethereal ver. 0.10.0a, with libpcap 0.7.2) included successful (i.e., packets with no errors) and unsuccessful (i.e., packets failing the 802.11 standard MAC layer checksum) transmissions.

Due to our interest in analyzing bit errors inside corrupted packets, our traces provided bit-level information about all transmissions by bit-wise comparing sent and received packets. These bit-level traces were analyzed to study the wireless channel error characteristics and then used in the simulations to generate bit errors. Packet losses are simulated by using non overlapping windows on the bit error traces: any window with one or more errors (in the checksummed bits) is classified as a lost packet.

The bursty nature of the wireless channel is visible in Figure 3, where part of a trace is illustrated and bit errors are represented by black boxes. Error bursts clearly appear as bit sequences with errors at their ends but that allow few corrected bits within them. Bit-level errors are, in fact, a consequence of errors in decoding symbols of the Complementary Code Keying (CCK) modulation scheme used for the 11 Mbps data rate, where each symbol represents 8 bits of information. Thus, assuming a sequence of at least nine correct bits as a burst delimiter, the burst length distribution for an entire trace is as in Figure 4. The plot confirms what is already noticeable in Figure 3, the vast majority of bit error bursts last from 13 to 16 bits.

3.2 Results

MPDU transmission is simulated between two wireless 802.11 hops. Packet loss and corruption are modeled using the collected bit error traces and retransmissions are scheduled based on the error detection technique under examination. Corrupted speech frames are decoded as is without further processing by the AMR-WB decoder, lost frames are concealed as defined in the standard [16]. Perceived quality results are expressed my means of the objective quality measure given by the PESQ-MOS. Speech samples have been taken from the NTT Multilingual Speech Database. We chose 24 sentence pairs spoken by 2 English speakers (male and female). 9600 packets are transmitted in the simulations for a total of 192 seconds of speech.

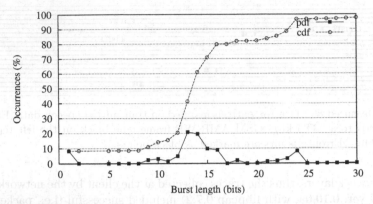

Fig. 4. Cumulative distribution and probability density function for burst error lengths in a trace. Mean burst length is 14.81 bits. An error burst begins and ends with an error and does not contain more than eight consecutive correct bits.

In the first simulation scenario two error detection strategies with different data coverage (see Figure 2) are tested without any link-level retransmission: the standard MAC checksum technique drops packets wherever a bit error occurs, while the modified MAC with partial checksum forwards packets only if errors occur in the perceptually least important part of the speech payload. Figure 5 presents the simulation results in terms of lost and corrupted packets *(top)* and objective speech quality *(bottom)* for different error traces. The percentage of lost frames is not always proportional to the bit-error rate because of the non-uniform nature of wireless errors, i.e., for the same BER, a lower number of packets is lost if bit errors have a higher burstiness. However, with selective error detection the percentage of lost frames is consistently lower. The PESQ score of the corresponding decoded speech also confirms that, for the scenarios under consideration, it is better to receive and decode partially corrupted packets than to lose them altogether. Improvements that range from 0.2 to 0.5 on the PESQ-MOS scale are clearly noticeable, proving modified MAC checksum particularly efficient at high error rates. This quality gain is motivated by the presence of a great number of bits only lightly sensitive to errors in the speech frame.

In the second simulation scenario, see Figure 6, a maximum of three retransmissions are allowed: the two techniques under analysis present different loss behavior *(top)* and quality *(middle)* , but also different number of retransmissions *(bottom)* given a particular BER. The modified MAC with partial checksum on speech data reduces the network load because in case of slightly corrupted packets it does not require a retransmission but forwards them to the next hop. Accepting a damaged packet proves to be a better solution than relaying on another uncertain transmission (especially at high bit error rates), provided that the most sensitive speech bits are correct and useful. The benefits are twofold: the perceived speech quality at the receiver is significantly increased, 0.3 on the MOS scale with 8% of packet losses, and the channel utilization due to retransmissions is roughly reduced by one third, with also positive effects on the end-to-end delays.

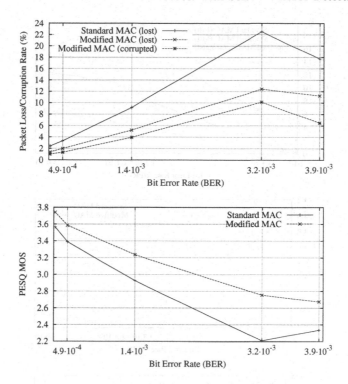

Fig. 5. Performance in terms of packet loss rate *(top)* and PESQ score *(bottom)* of IEEE 802.11 GSM AMR-WB speech transmission at 23.85 kb/s, with standard and modified MAC checksum, for different bit error rates. Link-level retransmissions are disabled.

4   Conclusions

A selective error detection technique to enhance the IEEE 802.11 link-layer effectiveness in supporting QoS sensitive speech communications has been presented. Since speech decoders can often deal with corrupted frames better than with lost ones, the standard 802.11 MAC-level error detection on the whole packet has been limited to the most perceptually sensitive bits of a speech frame. Retransmissions are then not required when bit errors occur outside the checksum coverage, on the assumption that they will result in only minor distortion. Full and partial checksum strategies have been simulated using 802.11b bit-error traces measured in different scenarios. Speech transmission shows a clear quality improvement if error detection is applied only to the most sensitive part of the payload. Furthermore, the proposed error detection technique also reduces end-to-end delays and network load, since successful packet delivery requires, on average, less retransmissions.

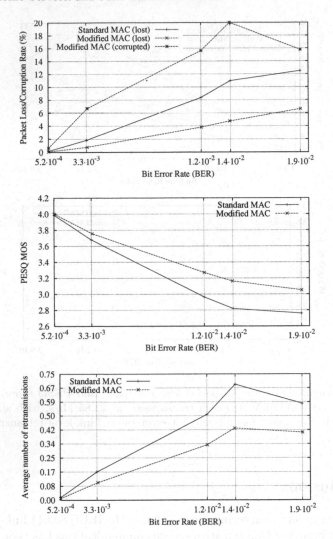

Fig. 6. Performance in terms of packet loss rate *(top)*, PESQ score *(middle)* and average number of retransmissions *(bottom)* of IEEE 802.11 GSM AMR-WB speech transmission at 23.85 kb/s, with standard and modified MAC checksum, for different bit error rates. The maximum number of link-level retransmissions is set to three.

References

1. ISO/IEC: Wireless LAN medium access control (MAC) and physical layer (PHY) specifications. ANSI/IEEE Std 802.11 (1999)
2. Swaminathan, K., A.R. Hammons Jr., Austin, M.: Selective error protection of ITU-T G.729 codec for digital cellular channels. In: Proc. IEEE Int. Conference on Acoustics, Speech, and Signal Processing, Atlanta, Georgia, USA (1996) 577–580

3. Larzon, L.A., Degermark, M., Pink, S., Jonsson, L.E., Fairhurst, G.: The UDP-lite protocol. draft-ietf-tsvwg-udp-lite-02.txt (2003)
4. Larzon, L.A., Degermark, M., Pink, S.: UDP-lite for real-time multimedia applications. In: Proc. QoS mini-conference of IEEE Int. Conference on Communications (ICC), Vancouver, Canada (1999)
5. Singh, A., Konrad, A., Joseph, A.: Performance evaluation of UDP-lite for cellular video. In: Proc. Int. Workshop on Network and Operating System Support for Digital Audio and Video (NOSSDAV), New York, USA (2001) 117–124
6. Reine, R., Fairhurst, G.: MPEG-4 and UDP-lite for multimedia transmission. In: Proc. PostGraduate Networking Conference (PGNet), Liverpool, UK (2003)
7. ISO/IEC: Draft supplement to standard for telecommunications and information exchange between systems – LAN/MAN specific requirements – part 11: Wireless medium access control (MAC) enhancements for quality of service (QoS). IEEE 802.11e/D5.0 (2003)
8. Khayam, S., Karande, S., Radha, H., Loguinov, D.: Performance analysis and modeling of errors and losses over 802.11b LANs for high-bit-rate real-time multimedia. Signal Processing: Image Communication 18 (2003) 575–595
9. Servetti, A., J.C. De Martin: Link-level unequal error detection for speech transmission over 802.11 networks. In: Proc. Special Workshop in Maui – Lectures by Masters in Speech Processing, Maui, Hawaii, USA (2004)
10. Dong, H., Chakares, D., Gersho, A., Belding-Royer, E., Gibson, J.: Selective bit-error checking at the MAC layer for voice over mobile ad hoc networks with IEEE 802.11. In: Proc. IEEE Wireless Communications and Networking Conference (WCNC), Atlanta, GA, USA (2004) 1240–1245
11. Masala, E., Bottero, M., J.C. De Martin: Link-level partial checksum for real-time video transmission over 802.11 wireless networks. In: Proc. 14th Int. Packet Video Workshop, Irvine, CA, USA (2004)
12. ETSI: AMR speech codec, wideband; general description. ETSI TS 126 171 version 5.0.0 (2002)
13. ITU-T, Recommendation P.862: Perceptual evaluation of speech quality (PESQ), an objective method for end-to-end speech quality assessment of narrowband telephone networks and speech codecs. (2001)
14. ETSI: AMR speech codec, wideband; frame structure. ETSI TS 126 201 version 5.0.0 (2002)
15. Khayam, S., Radha, H.: Markov-based modeling of wireless local area networks. In: Proc. 6th ACM Int. Workshop on Modeling, Analysis, and Simulation of Wireless and Mobile Systems, San Diego, CA, USA (2003) 100–107
16. ETSI: AMR speech codec, wideband; error concealment of lost frames. ETSI TS 126 191 version 5.1.0 (2002)

Performance Evaluation of a Feedback Based Dynamic Scheduler for 802.11e MAC⋆

Gennaro Boggia, Pietro Camarda, Luigi Alfredo Grieco,
Saverio Mascolo, and Marcello Nacci

DEE – Politecnico di Bari, Via E. Orabona, 4 – 70125 Bari Italy
{g.boggia,camarda,a.grieco,mascolo,m.nacci}@poliba.it

Abstract. IEEE 802.11 Wireless Local Area Networks (WLANs) are a fundamental tool for enabling ubiquitous wireless networking. Their use has been essentially focused on best effort data transfer because the basic access methods defined in the 802.11 standard cannot provide delay guarantees to real-time flows. To overcome this limitation, the 802.11e working group has recently proposed the Hybrid Coordination Function (HCF), which is an enhanced access method that allows service differentiation within WLAN's. While the 802.11e proposal gives a flexible framework to address the issue of service differentiation in WLAN's, it does not specifies effective algorithms to really achieve it. This paper compares the performance of standard access methods proposed by the 802.11e working group with a novel Feedback Based Dynamic Scheduler (FBDS), which is compliant to 802.11e specifications. Simulation results, obtained using the *ns*-2 simulator, have shown that FBDS guarantees bounded delays to real-time flows for a very broad set of network loads and packet loss probabilities, whereas, analogous algorithms proposed by the 802.11e working group fail in presence of high network load.

1 Introduction

The continuous growth of the Wireless LAN (WLAN) market is essentially due to the leading standard IEEE 802.11 [1], which is characterized by easy installation, flexibility and robustness against failures. At present, IEEE 802.11 has been essentially focused on best effort data transfer because its radio access methods cannot provide delay guarantees to real-time multimedia flows [2–5]. In particular, the 802.11 Medium Access Control (MAC) employs a mandatory contention-based channel access scheme called Distributed Coordination Function (DCF), which is based on Carrier Sense Multiple Access with Collision Avoidance (CSMA/CA) and an optional centrally controlled channel access scheme called Point Coordination Function (PCF) [1, 6]. With PCF, the time is divided into repeated periods, called *SuperFrames* (SFs), which consist of a Contention Period (CP) and a Contention Free Period (CFP). During the CP, the channel is accessed using the DCF whereas, during the CFP, is accessed using the PCF.

⋆ This work was funded by the TANGO project of the FIRB programme of the Italian Ministry for Education, University and Research.

M. Ajmone Marsan et al. (Eds.): QoS-IP 2005, LNCS 3375, pp. 520–532, 2005.

Recently, in order to support also delay-sensitive multimedia applications, such as real-time voice and video, the 802.11e working group has enhanced the 802.11 MAC with improved functionalities. Four Access Categories (ACs), with different priorities, have been introduced and an enhanced access function, which is responsible for service differentiation among different ACs and is referred to as Hybrid Coordination Function (HCF), has been proposed [7]. The HCF is made of a contention-based channel access, known as the Enhanced Distributed Co-ordination Access (EDCA), and of a HCF Controlled Channel Access (HCCA). The use of the HCF requires a centralized controller, which is called the Hybrid Coordinator (HC) and is generally located at the access point. Similarly to the PCF, the time is partitioned in a infinite series of superframes. Stations operating under 802.11e specifications are usually known as enhanced stations or QoS Stations (QSTAs).

The EDCA method operates as the basic DCF access method but using different contention parameters per access category. In this way, a service differentiation among ACs is statistically pursued [8].

EDCA parameters have to be properly set to provide prioritization of ACs. Tuning them in order to meet specific QoS needs is a current research topic. A method to set these parameters to guarantee throughput of each TC has been described in [9]. Regarding the goal of providing delay guarantees, several papers have pointed out that the EDCA behavior is very sensitive to the value of the contention parameters [10, 11] and that the Interframe-space-based priority scheme used by the EDCA mechanism can provide only a relative differentiation among service classes, but not absolute guarantees on throughput/delay performance [8]. Finally, in the paper [12] it has been shown that that EDCA can starve lower priority flows. To overcome these limitations, adaptive algorithms that dynamically tune EDCA parameters have been recently proposed in [13, 14], however, the effectiveness of these heuristic schemes have been proved only using simulations and no theoretical bounds on their performance in a general scenario have been derived.

The HCCA method combines some of the EDCA characteristics with some of the PCF basic features as it is shown in Fig. 1. Each superframe starts with a beacon frame after which, for legacy purpose, there could be a contention free period for PCF access. The remaining part of the superframe forms the CP, during which the QSTAs contend to access the radio channel using the EDCA mechanism. After the medium remains idle for at least a PIFS interval during the CP, the HC can start a Contention Access Phase (CAP)[1]. During the CAP, only QSTAs polled and granted with a special frame, known as *QoS CF-Poll frame*, are allowed to transmit during their TXOPs. Thus, the HC implements a prioritized medium access control. CAP length cannot exceed the value of the system variable *dot11CAPLimit*, which is advertised by the HC in the Beacon frame when each superframe starts [7].

---

[1] HCCA can be also enabled during the CFP with a procedure similar to the one described in this Section.

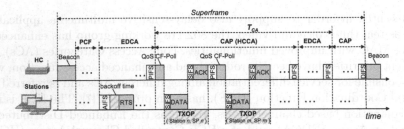

Fig. 1. Scheme of a superframe using the HCF controlled access method.

The IEEE 802.11e specifications allow QSTAs to feed back queue lengths of each AC to the HC. This information is carried on a 8 bits long subfield contained in the *QoS Control Field* of each frame header, during data transmission both in the CAPs and in the CPs; queue lengths are reported in units of 256 octets. This information can be used to design novel HCCA-based dynamic bandwidth allocation algorithms using feedback control [15, 16], as will be shown in this paper. In fact, the 802.11e draft does not specify how to schedule TXOPs in order to provide the required QoS; it only suggests a simple scheduler that uses static values declared in TSPECs for assigning fixed TXOPs (more details about this scheduler can be found in [7] and [17]). In particular, the simple scheduler designed in the draft [7] states that the $TXOP_i$ assigned to the ith queue should be computed as follows:

$$TXOP_i = max\left(\frac{N_i \cdot L_i}{C_i} + O, \frac{M}{C_i} + O\right) \qquad (1)$$

where L_i is the nominal MAC Service Data Unit (MSDU) size associated with the ith queue; C_i is the physical data rate at which the data of the ith queue are transmitted over the WLAN; O is the overhead due to ACK packets, SIFS and PIFS time intervals; M is the maximum MSDU size; and $N_i = \left\lceil \frac{T_{SI} \cdot \rho_i}{L_i} \right\rceil$, where ρ_i is the Mean Data Rate associated with the ith queue and T_{SI} is the Service Interval. This scheduler does not exploit any feedback information from mobile stations in order to dynamically assign TXOPs. Thus, it is not well suited for bursty media flows [17]. An improved bandwidth allocation algorithm has been proposed in [18], which schedules transmission opportunities by taking into account both the average and the maximum source rates declared in the TSPECs. However also this scheme does not consider the dynamic behavior of the multimedia flows. An adaptive version of the simple scheduler, which is based on the Delay-Earliest Due-Date algorithm, has been proposed in [17]. However, this algorithm does not exploit the explicit queue length to assign TXOPs, but implements a trial and error procedure to discover the optimal TXOP to be assigned to each station.

Recently, a Feedback Based Dynamic Scheduler (FBDS) has been proposed in [19] in order to address the first-hop bandwidth allocation issue in WLAN's. It is based on a closed-loop control scheme based on feeding back queue levels [20]. Preliminary investigations reported in [19, 21] have shown that FBDS is able to provided bounded delays to real-time multimedia flows over an ideal loss-free WLAN channel. The present paper will evaluate the performance of FBDS also in the presence of a noisy WLAN channel affected by bursty frame losses. The rest of the paper is organized as follows: Section 2 proposes and analyzes the dynamic bandwidth allocation algorithm, and illustrates the proposed CAC scheme; Section 3 shows simulation results; finally, the last Section draws the conclusions.

2   FBDS: Theoretical Background

FBDS distributes the WLAN bandwidth among all the multimedia flows by taking into account the queue levels fed back by the QSTAs [19]. Bandwidth allocation is pursued by exploiting the HCCA functionalities, which allows the HC to assign TXOPs to the ACs by taking into account the specific time constraints of each AC.

We will refer to a WLAN system made of an Access Point (AP) and a set of quality of service enabled mobile stations (QSTAs). Each QSTA has N queues, with $N \leq 4$, one for any AC in the 802.11e proposal. Let T_{CA} be the time interval between two successive CAPs. In every time interval T_{CA}, assumed constant, the AP must allocate the bandwidth that will drain each queue during the next CAP. We assume that at the beginning of each CAP, the AP is aware of all queue levels q_i, $i = 1, \ldots, M$ at the beginning of the previous CAP, where M is the total number of traffic queues in the WLAN system. The latter is a worst case assumption, in fact, queue levels are fed back using frame headers, as a consequence, if the ith queue length has been fed at the beginning of the previous CAP, then the feedback signal might be delayed up to T_{CA} seconds.

The dynamics of the ith queue can be described by the following discrete time linear model:

$$q_i(k+1) = q_i(k) + d_i(k) \cdot T_{CA} + u_i(k) \cdot T_{CA}, \qquad i = 1, \ldots, M, \qquad (2)$$

where $q_i(k) \geq 0$ is the ith queue level at the beginning of the kth CAP; $u_i(k) \leq 0$ is the average depletion rate of the ith queue (i.e., the bandwidth assigned to drain the ith queue); $d_i(k) = d_i^s(k) + d_i^{RTX}(k) - d_i^{CP}(k)$ where $d_i^s(k) \geq 0$ is the average input rate at the ith queue during the kth CAP, $d_i^{RTX}(k) \geq 0$ is the amount of data retransmitted by the ith queue during the kth CAP divided by T_{CA}, and $d_i^{CP}(k) \geq 0$ is the amount of data transmitted by the ith queue during the kth CP divided by T_{CA}.

The signal $d_i(k)$ is unpredictable since it depends on the behavior of the source that feeds the ith queue and on the number of packets transmitted during the contention periods. Thus, from a control theoretic perspective, $d_i(k)$ can be modelled as a disturbance. Without loss of generality, the following piece-wise constant model for the disturbance $d_i(k)$ can be assumed [15]:

$$d_i(k) = \sum_{j=0}^{+\infty} d_{0j} \cdot 1(k - t_j) \qquad (3)$$

where $1(k)$ is the unitary step function, $d_{0j} \in \mathbb{R}$, and t_j is a time lag.

Due to the assumption (3), the linearity of the system (2), and the superposition principle that holds for linear systems, the feedback control law can be designed by considering only a step disturbance: $d_i(k) = d_0 \cdot 1(k)$.

2.1   The Control Law

The goal of FBDS is to drive the queuing delay τ_i experienced by each frame going through the ith queue to a desired target value τ_i^T that represents the QoS requirement of the AC associated to the queue.

In particular, we consider the following control law:

$$u_i(k+1) = -k_i \cdot q_i(k) \tag{4}$$

which gives the way to compute the $\mathcal{Z}$-transform of $q_i(k)$ and $u_i(k)$ as follows:

$$Q_i(z) = \frac{z \cdot T_{CA}}{z^2 - z + k_i \cdot T_{CA}} D_i(z); \qquad U_i(z) = \frac{-k_i \cdot T_{CA}}{z^2 - z + k_i \cdot T_{CA}} D_i(z) \tag{5}$$

whit $D_i(z) = \mathcal{Z}[d_i(k)]$. From eq. (5) the system poles are $z_p = \frac{1 \pm \sqrt{1 - 4k_i \cdot T_{CA}}}{2}$, which give an asymptotically stable system if and only if $|z_p| < 1$, that is:

$$0 < k_i < \frac{1}{T_{CA}}. \tag{6}$$

In the sequel, we will always assume that k_i satisfies this asymptotic stability condition stated by (6).

To investigate the ability of the control system to provide a queuing delays approaching the target value τ_i^T, we apply the final value theorem to Eq. (5). By considering that the $\mathcal{Z}$-transform of the step function $d_i(k) = d_0 \cdot 1(k)$ is $D_i(z) = d_0 \cdot \frac{z}{z-1}$, the following results turn out:

$$u_i(+\infty) = \lim_{k \to +\infty} u_i(k) = \lim_{z \to 1}(z-1)U_i(z) = -d_0; \qquad q_i(+\infty) = \frac{d_0}{k_i},$$

which implies that the the steady state queueing delay is:

$$\tau_i(+\infty) = \left| \frac{q_i(+\infty)}{u_i(+\infty)} \right| = \frac{1}{k_i}. \tag{7}$$

As a consequence, the following inequality has to be satisfied in order to achieve a steady-state delay smaller than τ_i^T:

$$k_i \geq \frac{1}{\tau_i^T}. \tag{8}$$

By considering inequalities (6) and (8) we obtain that the T_{CA} parameter has to fulfill the following constraint:

$$T_{CA} < \min_{i=1..M} \tau_i^T. \tag{9}$$

2.2 TXOP Assignment

We have seen in Sec. 1 that every time interval T_{CA} the HC allocates TXOPs to mobile stations in order to meet the QoS constraints. This sub-section shows how to transform the bandwidth $|u_i(k)|$, defined above, into a $TXOP_i$ assignment. In particular, if the ith queue is drained at data rate C_i, the following relation holds:

$$TXOP_i(k) = \frac{|u_i(k) \cdot T_{CA}|}{C_i} + O \qquad (10)$$

where $TXOP_i(k)$ is the TXOP assigned to the ith queue during the kth service interval and O is the time overhead due to ACK packets, SIFS and PIFS time intervals (see Fig. 1). The extra quota of TXOP due to the overhead O depends on the number of MSDUs corresponding to the amount of data $|u_i(k) \cdot T_{CA}|$ to be transmitted. O could be estimated by assuming that all MSDUs have the same nominal size specified into the TSPEC. Moreover, when $|u_i(k) \cdot T_{CA}|$ does not correspond to a multiple of MSDUs, the TXOP assignment will be rounded in excess in order guarantee a queuing delay always equal or smaller than the target delay τ_i^T.

2.3 Channel Saturation

The above bandwidth allocation algorithm is based on the implicit assumption that the sum of the TXOPs assigned to each queue is smaller than the maximum CAP duration, which is the *dot11CAPLimit*; this value can be violated when the network is saturated. When a network overload happens, it is necessary to reallocate the TXOPs to avoid exceeding the CAP limit. Thus, when $\sum_{i=1}^{M} TXOP_i(k_0) > dot11CAPLimit$, each computed $TXOP_i(k_0)$ is decreased by an amount $\Delta TXOP_i(k_0)$ to satisfy the following capacity constraints:

$$\sum_{i=1}^{M} [TXOP_i(k_0) - \Delta TXOP_i(k_0)] = dot11CAPLimit. \qquad (11)$$

In particular, the generic amount $\Delta TXOP_i(k_0)$ is evaluated as a fraction of the total amount $\sum_{i=1}^{M} TXOP_i(k_0) - dot11CAPLimit$, as follows:

$$\Delta TXOP_i(k_0) = \frac{TXOP_i(k_0)C_i}{\sum_{i=1}^{M} TXOP_i(k_0)C_i} \left(\sum_{i=1}^{M} TXOP_i(k_0) - dot11CAPLimit \right). \qquad (12)$$

Notice that Eq. (12) provides a $\Delta TXOP_i(k_0)$, which is proportional to $TXOP_i(k_0)C_i$, in this way connections transmitting at low rates are not penalized too much.

3   Performance Evaluation

In order to test FBDS in realistic scenarios, computer simulations involving voice, video and FTP data transfers have been run using the *ns*-2 simulator [22]. We have considered the reference scenario shown in Fig. 2 where a 802.11a WLAN network is shared by a mix of 3α voice flows encoded with the G.729 standard [23], α video flows encoded with the MPEG-4 standard [24], α videos encoded with the H.263 standard [25], and α FTP best effort flows. For the video flows, we have used traffic traces available from the video trace library [26]. For voice flows, we have modelled G.729 sources using Markov ON/OFF sources [27], where the ON period is exponentially distributed with mean value 350 ms, and the OFF period is exponentially distributed with mean value 350 ms [28]. During the ON period, the source sends 20Bytes sized packets every 20ms (i.e., the source data rate is 8 kbps) however, by considering the overheads of the RTP/UDP/IP protocol stacks, the rate becomes 24 kbps during the ON periods. During the OFF period the rate is zero because we assume a Voice Activity Detector (VAD).

We have used the HCCA access method during the CFP, and the EDCA access method during the CP. EDCA parameters have been set as in [7].

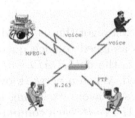

Fig. 2. Scenario with multimedia flows.

In the *ns*-2 implementation, the T_{CA} is expressed in Time Unit (TU), which in the 802.11 standard [1] is equal to $1024\mu s$. We assume a T_{CA} of 29 TU. The proportional gain k_i is set equal to $1/\tau_i^T$. Main characteristics of the considered multimedia flows are summarized in Table 1.

To take into account also the effect of random bursty losses, we have considered a Gilbert two state discrete Markov chain to model the loss process affecting the WLAN channel [29]. In particular, we assume a frame loss probability equal to 0, when the channel is in the Good state, and ranging from 0 to 0.1 when

Table 1. Main features of the considered multimedia flows.

Type of flow	Nominal (Maximum) MSDU Size	Mean (Maximum) Data Rate	Target Delay
MPEG-4 HQ	1536(2304) byte	770 (3300) kbps	40 ms
H.263 VBR	1536(2304) byte	450 (3400) kbps	40 ms
G.729 VAD	60(60) byte	8.4 (24) kbps	30 ms

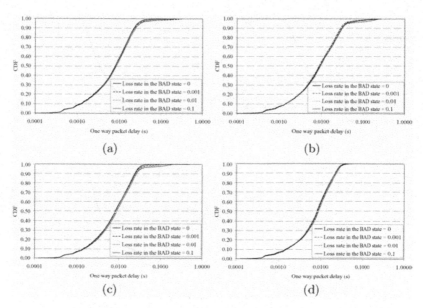

Fig. 3. CDF's of the one way packet delay experienced by the MPEG4 flows when $\alpha = 5$, obtained using the (a) DCF; (b) EDCA; (c) Simple scheduler; (d) FBDS.

the channel is in the Bad state. The permanence times in the Good and Bad states are geometrically distributed with mean values equal to 0.1 s and 0.01 s, respectively. The transition probabilities between the two states have been set equal to 0.1. When a MAC frame gets lost due to interference or channel unreliability, it is retransmitted up to a maximum number of times equal to 7.

We have evaluated the performance of FBDS, Simple scheduler, EDCA, and DCF for a very broad set of values for the load parameter α. Herein we report significant comparisons obtained for $\alpha = 5$ ($\alpha = 10$), which represents the case of a low(high) loaded WLAN. Fig. 3 shows the CDF of the one-way-packet delay experienced by the MPEG flows for all the considered medium access schemes when $\alpha = 5$. In particular, it shows that all the considered schemes are able to ensure a bounded delay, which is less than 30ms for the 90% of the packets. This result is due to the low load offered to the WLAN, which is able to drain all the transmittng queues also with the basic DCF access function. For the same reason, the impact of the loss rate on the one-way packet delay is negligible.

Similar results have been obtained for the H.263 flows (see Fig. 4). Results for the G.729 flows are reported in Fig. 5, which shows that the one-way packet delay is less than 6 ms for the 90% of the packets. In this case, we observe delays smaller than those obtained for the video flows because G.729 packets are smaller than those generated by the video streams.

Results are very different when $\alpha = 10$ due to the higher load. In this case, only the FBDS scheduler ensures bounded delays for all kinds of flow. In fact, Fig. 6 shows that, when DCF or ECDA or the Simple scheduler are used, the MPEG flows experience delays larger than hundreds of milliseconds for more

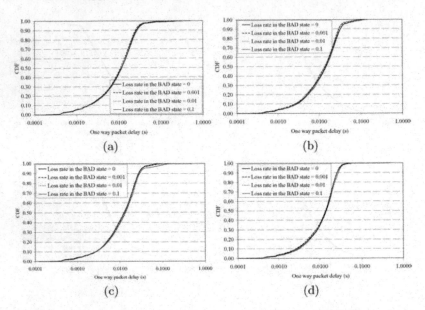

Fig. 4. CDF's of the one way packet delay experienced by the H.263 flows when $\alpha = 5$, obtained using the (a) DCF; (b) EDCA; (c) Simple scheduler; (d) FBDS.

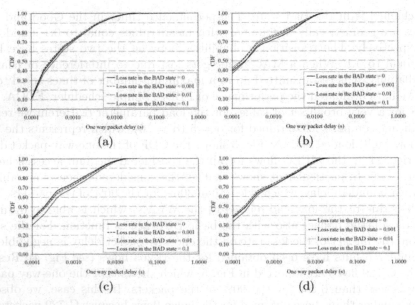

Fig. 5. CDF's of the one way packet delay experienced by the G.729 flows when $\alpha = 5$, obtained using the (a) DCF; (b) EDCA access function; (c) Simple scheduler; (d) FBDS.

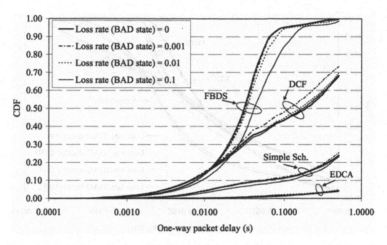

Fig. 6. CDF's of the one way packet delay experienced by the MPEG4 flows when $\alpha = 10$, obtained using the DCF, EDCA, Simple scheduler, and FBDS.

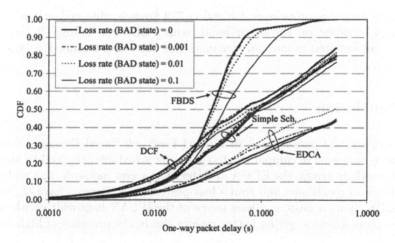

Fig. 7. CDF's of the one way packet delay experienced by the H.263 flows when $\alpha = 10$, obtained using the DCF, EDCA, Simple scheduler, and FBDS.

than a half of the received packets. On the other hand, FBDS provides bounded delays. It is worth to note that delays provided by FBDS increase with the loss rate because a fraction of the channel capacity is used for retransmissions at the MAC layer. Similar results have been obtained for the H.263 flows as shown in Fig. 7. Delays experienced by the G.729 flows exhibit a different behavior with respect to the video flows (see Fig. 8). The reason is that when the EDCA access method is used, a higher priority is given to the voice flows [7]. This effect was not evident in the previous simulations with $\alpha = 5$ because the network load was small, so that also flows with a lower priority got enough bandwidth to drain their respective transmission queues. It is worth to point out that with EDCA

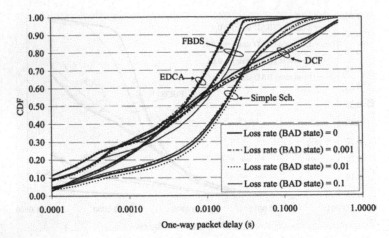

Fig. 8. CDF's of the one way packet delay experienced by the G.729 flows when $\alpha = 10$, obtained using the DCF, EDCA, Simple scheduler, and FBDS.

the performance improvement for the G.729 flows is obtained at the expense of the video flows; on the other hand, with FBDS the WLAN bandwidth is properly distributed among all streams, thus obtaining bounded delays for all kind of flows.

4   Conclusion

In this paper, the performance of a novel Feedback Based Dynamic Scheduler have been evaluated in comparison to the simple scheduler proposed by the IEEE 802.11e working group, the EDCA and the DCF access methods, in a wide range of traffic load conditions and frame loss probabilities. Results have shown that the FBDS allows a more cautious usage of the WLAN bandwidth with respect to the others schemes, giving better results also in the presence of high network load.

References

1. IEEE 802.11: Information Technology – Telecommunications and Information Exchange between Systems – Local and Metropolitan Area Networks – Specific Requirements – Part 11: Wireless LAN Medium Access Control (MAC) and Physical Layer (PHY) Specifications. ANSI/IEEE Std. 802.11, ISO/IEC 8802-11. First edn. (1999)
2. Mangold, S., Choi, S., May, P., Klein, O., Hiertz, G., Stibor, L.: IEEE 802.11e Wireless LAN for Quality of Service. In: European Wireless Conference 2002, Florence, Italy (2002)
3. Bianchi, G.: Performance Analysis of the IEEE 802.11 Distributed Coordination Function. IEEE Journal of Selected Areas in Communications **18** (2000) 535–547

4. Koepsel, A., Ebert, J.P., Wolisz, A.: A Performance Comparison of Point and Distributed Coordination Function of an IEEE 802.11 WLAN in the Presence of Real-Time Requirements. In: Proc. of 7th Intl. Workshop on Mobile Multimedia Communications (MoMuC2000), Tokio,Japan (2000)
5. Zhu, H., Chlamtac, I., Prabhakaran, B.: A survey of Quality of Service in IEEE 802.11 networks. IEEE Wireless Communications (2004) 6–14
6. Prasad, N., Prasad, A., eds.: WLAN Systems and Wireless IP. Artech House universal personal communications series. Artech House (2002)
7. IEEE 802.11 WG: Draft Amendment to Standard for Information Technology – Telecommunications and Information Exchange between Systems – LAN/MAN Specific Requirements – Part 11: Wireless Medium Access Control (MAC) and Physical Layer (PHY) Specifications: Medium Access Control (MAC) Quality of Service (QoS) Enhancements. IEEE 802.11e/D8.0. (2004)
8. Bianchi, G., Tinniriello, I.: Analysis of priority mechanisms based on differentiated inter frame spacing in CSMA-CA. In: IEEE 58th Veichular Technology Conference (VTC03), fall, Orlando (2003)
9. Banchs, A., Perez-Costa, X., Qiao, D.: Providing throughput guarantees in IEEE 802.11e Wireless LANs. In: Providing Quality of Service in Heterogeneous Environments, ITC 2003, Berlin, Germany (2003) 1001–1010
10. Gu, D., Zhang, J.: QoS enhancement in IEEE802.11 wireless local area networks. IEEE Commun. Mag. (2003) 120–124
11. Garg, P., Doshi, R., Greene, R., Baker, M., Malek, M., Cheng, X.: Using IEEE 802.11e MAC for QoS over Wireless. In: 22nd IEEE International Performance Computing and Communications Conference (IPCCC 2003), Phoenix, Arizona (2003)
12. Lindgren, A., Almquist, A., Schelén, O.: Quality of service schemes for IEEE 802.11 wireless LANs – an evaluation. Mobile Networks and Applications **8** (2003) 223–235
13. Romdhami, L., Ni, Q., Turletti, T.: Adaptive EDCF: Enhanced service differentiation for IEEE 802.11 wireless Ad-Hoc networks. In: IEEE Wireless Communications and Networking Conference (WCNC), New Orleans, Louisiana, USA (2003)
14. Xiao, Y., Li, H., Choi, S.: Protection and guarantee for voice and video traffic in IEEE 802.11e Wireless LANs. In: IEEE Infocom, Hong Kong (2004)
15. Mascolo, S.: Congestion control in high-speed communication networks using the smith principle. Automatica, Special Issue on Control methods for communication networks **35** (1999) 1921–1935
16. Priscoli, F.D., Pietrabissa, A.: Control-theoretic bandwidth-on-demand protocol for satellite networks. In: Proceedings of the 2002 International Conference on Control Applications, 2002. Volume 1. (2002) 530 –535
17. Grilo, A., Macedo, M., Nunes, M.: A scheduling algorithm for QoS support in IEEE 802.11e networks. IEEE Wireless Communications (2003) 36–43
18. Lo, S.C., Lee, G., Chen, W.T.: An efficient multipolling mechanism for IEEE 802.11 wireless LANs. IEEE Transactions on Computers **52** (2003) 764–778
19. Boggia, G., Camarda, P., Zanni, C.D., Grieco, L.A., Mascolo, S.: A dynamic bandwidth allocation algorithm for IEEE 802.11e WLANs with HCF access method. In Karlsson, G., Smirnov, M.I., eds.: Fourth COST 263 International Workshop on Quality of Future Internet Services (QoFIS 2003). Number 2811 in LNCS. Springer (2003) 142–151
20. Priscoli, F.D., Pietrabissa, A.: Load-adaptive bandwidth-on-demand protocol for satellite networks. In: Proceedings of the 41st IEEE Conference on Decision and Control, 2002. Volume 4. (2002) 4066 –4071

21. Annese, A., Boggia, G., Camarda, P., Grieco, L.A., Mascolo, S.: A HCF-based bandwidth allocation algorithm for 802.11e MAC. In: Proc. of the IEEE Veichular Technology Conference (VTC 2004-Spring), Milan, Italy (2004)
22. Ns-2: Network simulator. available at http://www-mash.cs.berkeley.edu/ns (2003)
23. International Telecommunication Union (ITU): Coding of Speech at 8 kbit/s using Conjugate-Structure Algebraic-Code-Excited Linear Prediction (CS-ACELP). ITU-T Recommendation G.729. (1996)
24. MPEG-4 Video Group: Mpeg-4 overview. Available at http://mpeg.telecomitalialab.com/ (2002)
25. International Telecommunication Union (ITU): Video coding for low bit rate communication. ITU-T Recommendation H.263. (1998)
26. Telecommunication Networks Group: Video trace library. Available at http://www.tkn.tu-berlin.de/research/results.html (2003) Technical University of Berlin, Germany.
27. Chuah, C., Katz, R.H.: Characterizing Packet Audio Streams from Internet Multimedia Applications. In: Proc. of International Communications Conference (ICC 2002), New York, NY (2002)
28. Xu, Y., Guerin, R.: On evaluating loss performance deviation: a simple tool and its pratical implications. In: Quality of Service in multiservice IP networks (QoSIP2003), Milan, Italy (2003) 1–18
29. Barakat, C., Altman, E.: Bandwidth tradeoff between TCP and link-level FEC. Computer Networks 39 (2002) 133–150

QoS Routing in Multi-hop Wireless Networks: A New Model and Algorithm

Antonio Capone[1], Luca Coletti[2], and Marco Zambardi[2]

[1] Politecnico di Milano; piazza L. da Vinci 32, 20133, Milano, Italy
capone@elet.polimi.it
[2] Siemens Mobile Communications S.p.A,
Via Monfalcone 1, 20092 Cinisello Balsamo – Milan, Italy
Luca.Coletti@siemens.com, Marco.Zambardi@icn.siemens.it

Abstract. The definition of new radio network topologies and deployment concepts capable of providing a ubiquitous radio coverage area for beyond 3G systems is one of the main objectives of the WINNER project [1]. Within this context, routing and forwarding algorithms for infrastructure-based multi-hop networks are investigated. We first present a new mathematical model of the routing problem in infrastructure-based multi-hop wireless networks that takes into account QoS requirements considering bandwidth constraints. Since the characteristics of such new networks allow to control the path selection of each flow, we then propose a new routing algorithm based on a heuristic. Finally, to show the performance of the proposed algorithm several simulation results are presented.

1 Introduction

Actual 3G wireless technologies seems not able to provide different kind of wireless services and applications to end-user with ubiquitous coverage and complete transparency. For this reason, research community has been started a huge effort in order to develop a new radio access network able to provide wireless access for a wide range of services and applications across all environments with one single adaptive system for all application scenarios and radio environments [1]. A promising new architecture envisioned is the Fixed Relay Network (FRN), where many advantages are expected in terms of coverage, flexibility, throughput and QoS provisioning. A FRN is constituted by fixed transceivers, called relays, that establish a wireless mesh topology connected through Access Points (AP) to a core network. In such network support of multi-hop is mandatory and efficient routing algorithms are required to exploit at best network resources and to efficiently manage internal connections and load from and to the core network.

The problem of routing in multi-hop wireless networks has been deeply investigated in the research area of Mobile Ad-hoc NETworks (MANET). These networks are characterized by a varying topology and require a completely distributed operation of routing protocols. Routing mechanisms for MANETs usually aim at minimizing the protocol overhead needed to manage frequent topology changes by means of on-demand route formation [2][3][4] or hierarchical schemes [5]. Therefore, route selection is based on non accurate network state information, and the metrics adopted try to optimize the energy consumption rather then the use of radio resources.

M. Ajmone Marsan et al. (Eds.): QoS-IP 2005, LNCS 3375, pp. 533–544, 2005.

In this hostile network environment the challenge of providing Quality of Service (QoS) guarantees to traffic flows has attracted attention by the research community [6][7] and several algorithms have been proposed [8]-[14]. QoS routing algorithms proposed so far for MANETs are tailored to specific MAC (Medium Access Control) layers able to provide information on resource availability and to control resources assigned to traffic flows. The more common approach is to consider a TDMA (Time Division Multiple Access) based ad-hoc network [13]. Each connection specifies its QoS requirement in terms of time slots needed on its route from source to destination. For each connection, the QoS routing protocol selects a feasible path based on the bandwidth availability of all links and then modifies the slot scheduling of all nodes on the paths. This task is not easy and even the calculation of the residual bandwidth along a path is NP-complete [13] since it can be easily shown that the slot assignment problem is equivalent to the graph coloring problem [15].

Even if FRNs are multi-hop wireless networks like MANETs, their characteristics make the routing problem quite different [16]. FRNs have a defined topology and topology changes are mainly due to node failures that are usually not frequent. Therefore, the distribution of network state information is not much costlier than in wired networks, and even a centralized control of route selection can be adopted [17]. Finally, energy consumption is not a problem for network nodes.

In this paper we propose a new model for the QoS routing problem in multi-hop wireless networks with bandwidth constraints and present an algorithm for its solution suitable for FRNs. The model is an extension of the well known multi-commodity flow problems [18] where link capacity constraints are replaced with new ones that takes into account interference constraints among different radio links. It guarantees that the rates of routed flows are compatible with radio channel capacity, but does not require to explicitly solve the scheduling problem. To solve the proposed problem we present a routing algorithm, Wireless Fixed Relay routing (WiFR), based on an ad-hoc heuristic. Then, we show how it is possible to derive an admissible scheduling of packet transmissions on the paths provided by the routing algorithm.

The paper is structured as follows. In Section 2 the new mathematical programming model is presented and discussed. Section 3 is devoted to the new routing algorithm, while Section 4 to the scheduling procedure. Section 5 presents numerical results on the performance of the proposed approach. Finally, Section 6 concludes the paper.

2   Wireless Multi-hop Routing Problem

The model of the wireless multi-hop routing problem here presented is an extension of the well known multi-commodity flow problem where the network is represented by a graph with a capacity associated with each link and connections are represented as flows that compete for the limited link capacities. The objective is to maximize the fraction of offered traffic admitted in the network. A survey on multi-commodity flow problems can be found in [18]. Differently from wired networks, with wireless multi-hop networks we cannot associate a capacity to each link in the graph, since parallel transmissions on different links may be prevented due to interference and the inability of stations to transmit and receive at the same time. We assume that when a transmission occurs on a link between a couple of nodes, the transmissions of nodes directly connected to the transmitting or the receiving nodes are prevented. Such an assumption assure that the hidden terminal problem does not occur both for the considered transmission and that of the acknowledgement packet, like for IEEE 802.11 systems.

Let us consider a graph $G=(V,E)$ whose N vertexes, $V=\{1,2,...,N\}$, represent the wireless stations and the M edges (i,j) connect stations within transmission range.

Consider now a set of K pairs of vertexes (s^k,t^k) for k=1,2 ...K representing the source and destination node associated with K commodities for which a path should be found on the graph G. For each commodities K can be introduced the following variables:

$f^k_{i,j}$ = Units of flow of commodity k routed on link from i to j;

F^k = Total units of flow to be sent from s^k to t^k;

$f_{i,j} = \sum_{k=1}^{K} f^k_{i,j}$ = Total units of flow on arc from i to j.

Then for each node n, represented in G by a vertex n, the following sets are defined:

$$A(n)=\{j\in V|(n,j)\in E\} \tag{1}$$

$$B(n)=\{j\in V|(j,n)\in E\} \tag{2}$$

where $A(n)$ represents the set of nodes that can be reached from node n while $B(n)$ is the set of nodes that can reach node n with their transmission.

As mentioned before, the objective in solving routing problem is to maximize the fraction of total commodities that is routed through the network from source to destination. This can be expressed introducing a parameter ⍺ that represents for all commodities the fraction of F^k admitted in the network. Therefore objective is to find the optimum α, denoted as α^*, such that for each one of the given K commodities it is possible to route in the network exactly $\alpha^* \cdot F^k$ units of flow from the source node s^k to destination t^k. The optimum α^* is that particular value of α that satisfy the following objective function:

$$Max\left\{\sum_{k=1}^{K}\alpha\cdot F^k\right\} \tag{3}$$

Further, conservation equations and non negativity must be satisfied by all flows:

$$\sum_{j\in A(n)} f^k_{n,j} - \sum_{j\in B(n)} f^k_{j,n} = \begin{cases} \alpha\cdot F^k \ if \ s^k = n \\ -\alpha\cdot F^k \ if \ t^k = n \\ 0 \ \ otherwise \end{cases} \tag{4}$$

$$f^k_{i,j} \geq 0 \ \text{for all} \ (i,j)\in E \tag{5}$$

$$\alpha \in [0,1] \tag{6}$$

The equation (4) states for each commodity k that the difference between incoming flow and outgoing flow on a given node n is positive if node n is the source of the commodity, is negative if node n is the destination and is equal to zero if node n is an intermediate node of path k. Equation (5) states that commodities must be non-negative.

To model the capacity and interference constraints, new sets of nodes are introduced: the set S_i^1 is made up by all links that are adjacent to node i:

$$S_i^1 = \{(i,j), j \in V\} \tag{7}$$

and for each node $j \in A(i)$, the set $S_{i,j}^2$ including all links that have one of their end in $A(i)$ and meanwhile do not belong to S_i^1:

$$S_{i,j}^2 = \{(j,k) \mid k \neq i\} \tag{8}$$

Obviously for the given node i, exactly $|A(i)|$ set of this type exist.

Starting from these sets, it is possible to build up the set G_i formed by all relays within two hops from relay i, this set can be expressed as:

$$G_i = \left\{ S_i^1 \cup \bigcup_{j \in A(i)} S_{i,j}^2 \right\} \tag{9}$$

It is possible to characterized each set G_i by a theoretical capacity C_i that represents the maximum aggregate flow that can be routed over the whole set of links belonging to that group. All flows routed on links belonging to set S_i^1 contributes in consuming common resources associated to the whole group G_i and to bound possibilities for new transmissions/receptions of relay i. About the various set $S_{i,j}^2$ build on each neighbor of node i, only the heaviest loaded one should be considered for writing the new QoS constraint, since it is the one that gives the more restrictive condition about G_i's resources consumption and about limitation for further transmissions/receptions of node i. It is hence possible to write the constraint to be inserted in the mathematical model as follow:

$$\sum_{k=1}^{K} \sum_{(i,j) \in S_i^1} f_{i,j}^k + \max_{j \in A(n_i)} \left\{ \sum_{k=1}^{K} \sum_{(j,l) \in S_{i,j}^2} f_{j,l}^k \right\} \leq C_i \forall G_i \tag{10}$$

It is possible to transpose the non linear constraint (10) to a set of linear constraints without any approximation obtaining in this way a linear model for our problem. This is done eliminating the max operator that gives the non linearity and extending the constraint to all set $S_{i,j}^2$ related to relay i comprising in this way also the heaviest loaded one that gives the more restrictive condition. This set can be compacted in the following expression:

$$\sum_{k=1}^{K} \sum_{(i,j) \in S_i^1} f_{i,j}^k + \sum_{k=1}^{K} \sum_{(j,l) \in S_{i,j}^2} f_{j,l}^k \leq C_i \forall S_{i,j}^2 \in G_i, \forall G_i \tag{11}$$

It is now possible to write the mathematical model used to describe problem of QoS routing in a network composed by fixed/movable relays equipped with omni-directional antennas and a unique radio interface:

$$Max\left\{\sum_{k=1}^{K} \alpha \cdot F^k\right\} \tag{12}$$

so that

$$\sum_{j \in A(n)} f_{n,j}^k - \sum_{j \in B(n)} f_{j,n}^k = \begin{cases} \alpha \cdot F^k \; if \; s^k = n \\ -\alpha \cdot F^k \; if \; t^k = n \\ 0 \;\; otherwise \end{cases} \tag{13}$$

$$f_{i,j}^k \geq 0 \text{ for all } (i,j) \in E \tag{14}$$

$$\alpha \in [0,1] \tag{15}$$

$$\sum_{k=1}^{K} \sum_{(i,j) \in S_i^1} f_{i,j}^k + \sum_{k=1}^{K} \sum_{(j,l) \in S_{i,j}^2} f_{j,l}^k \leq C_i \forall S_{i,j}^2 \in G_i, \forall G_i \tag{16}$$

This new approach results in the complete separation between construction of route that satisfy the required QoS and the definition of proper scheduling at MAC level. This means it could be even applied to MAC that does not support strict resource control, like e.g. IEEE 802.11, taking a proper margin on the radio link capacity in order to take into accounts all overheads due to the protocol and the contention on the channel. In this case if strict control on the flows entering the network is enforced, the new approach assures that no persistent congestion occur in the network.

3 WiFR: A New QoS Routing Algorithm for FRN

To solve the problem presented in previous section we propose a new QoS routing algorithm named Wireless Fixed Relay routing (WiFR). We present WiFR assuming that route computation can performed at a central controller, even if using a proper routing protocol to distribute link usage information, the algorithm could be executed in a distributed way. Since path selection is performed flow-by-flow, the algorithm can be used for on-line routing of connection requests.

The following information is considered and maintained by the routing controller:

– Topology matrix T that is a $N \times N$ matrix where N is the number of relays and the generic element t_{ik} is equal to 1 if j and k are connected and 0 otherwise.
– The set neighbors of any relay j, $H_1(j)$, and the set of relays that are two hops away, $H_2(j)$.
– The residual capacity matrix RC that is a $N \times N$ matrix. The generic element $rc_{j,k} \in [0,1]$ of the matrix represents the fraction of residual capacity on radio link (j,k). The bandwidth B provided by lower layers is an input parameter at routing level.
– Weight matrix W that is a $N \times N$ matrix representing a set of weights associated with network links. These weights are dynamically updated by the algorithms and used for route computation. Weights are defined as:

$$w_{j,k} = \frac{1}{rc_{ik}^2} \qquad (17)$$

if (i,k) is an edge of the network graph, and 0 otherwise.

- Status tab $\underline{st}$ that is a list of record that has N entries used to stored information on relay states. For a given relay j the parameters stored in its record are: $ub(j)$, the fraction of bandwidth consumed either for relay j transmissions or for receptions of other signals addressed to j or not, $fr(j)$, the fraction of bandwidth that relay j has still available to receive without conflicts with its own transmissions or with other received signals in the set $H_1(j)$, $ft(j)$, the fraction bandwidth that relay j has still available to transmit without conflicts with transmissions of other relays in sets $H_1(j)$ and $H_2(j)$.

Suppose now that a route request from relay s to relay t with a request bandwidth β (normalized to the provided bandwidth B) is delivered to the central entity. The central entity applies the route searching routine that selects, just among the feasible paths between source node and destination one, the path that has the minor impact on network global saturation level. We use as basis for our route searching routine the mechanism of Dijkstra algorithm, where the metric is based on weights w_{jk}. However, route selection is performed only among feasible paths in a greedy manner. This can be done with the following exploring routine that we developed to determinate if, starting form generic relay j, a potential next hop relay k should be explored by Dijkstra algorithm or not. The control steps executed by our routine are:

- If j is the source node and $ft(j)<\beta$ then connection is refused. In this case source j has not enough transmission resources.
- If j is the source node, k is not the destination node, and $ub(j)+2\beta>1$ connection is refused. In this case source j has not enough bandwidth to transmit to k and then receive k transmission to next node without conflicts.
- If node k is the destination node it can be selected.
- If $ft(k)<\beta$ then k is discarded. In this case k has not enough bandwidth to forward the flow.
- If i is a common neighbor of j and k and $fr(i)<2\beta$ then k is discarded. In this case node i cannot receive without conflicts the transmissions of j and k.
- If destination node t is a neighbor of k and $ub(k)+2\beta<1$ then node k is selected, otherwise it is discarded.
- If destination node t is not a neighbor of node k and $ub(k)+3\beta<1$ then node k is selected, otherwise it is discarded.

If the routine states that node k can support the new flow then it is explored by Dijkstra algorithm using the metric defined by weights w_{jk}. If the routine instead states that relay k is not able to support the new flow no further actions are taken on relay k. In both cases, after having executed proper actions basing on routines response, a new neighbor of the actual selected node j is chosen, if exist, and routine is run again. Once all neighbors of j have been considered, according to Dijkstra algorithm, a new node j^* is selected to be included in the shortest path tree. If this node is the destination t, then the algorithm terminates, otherwise it is restarted considering the node j^*. Algorithm steps are repeated until destination is reached or until no node can be added. If a feasible path p is found, central controller updates the bandwidth usage information:

$$ub(j) = ub(j) + \beta \quad \forall j \mid j \in p, j \neq t \tag{18}$$

$$ub(k) = ub(k) + \beta \quad \forall k \mid k \in H_1(j), j \in p, j \neq t \tag{19}$$

$$fr(j) = 1 - ub(j) \quad \forall j \in V \tag{20}$$

$$ft(j) = 1 - \max\left\{ \max_{k \in H_1(j)} ub(k), ub(j) \right\} \tag{21}$$

Equation (18) represents resources consumption of relay forming the new path for transmitting the new flow. Equation (19) represents resource consumption due to the physical broadcast: all neighbors of transmitting relay receive the signal. Equation (20) updates the residual capacity for receptions of each relay. Equation (21) updates the residual capacity for transmission of each relay. Also residual capacity values are updated:

$$rc_{jk} = \min\{ft(j), fr(k)\} \quad \forall (j,k) \in E \tag{22}$$

and weight matrix W is updated accordingly.

This basic heuristic can be executed online when a new connection request arrives without re-routing already routed flows. In the case the routing algorithm is performed offline on a set of flows, we have developed an optimization routine that tries to reroute the other flows when no feasible path is found for the considered one.

4 Optimized Scheduling

We have presented a new mathematical model which route bandwidth guaranteed flows without defining at the same time a scheduling scheme for each node in the network. This allows to apply the new routing approach even to networks based on random access schemes at the MAC layer, like IEEE 802.11, provided that the radio link bandwidth B considered by the routing scheme is properly reduced to take into account all overheads due to collisions and protocol headers. Moreover, if an optimized scheduling scheme can be adopted for a TDMA based MAC layer, it can be obtained in a second phase based on the transmission requirements of each node as defined by the solution provided by the routing procedure. To derive numerical results we have considered and solved this last problem.

Since the path selection has been already performed by the routing algorithm, here the problem is that of assigning slots to network nodes according to the total bandwidth required by the node for originated or relayed flows.

We model this slot assignment problem as a variation of the known Minimum Order Frequency Assignment Problem (MO-FAP) which has been proposed for frequency planning in cellular networks [19].

Let $m(j)$ be the number of slots required for relay j. The assumption needed to calculate the number of slots requited by each node is that the bandwidth of flows is a multiple of the minimum bandwidth corresponding to a single slot per frame.

The number of slots in the frame is denoted by N_s. Binary variables y_c defines the slots assigned with $y_c=1$ if slot c is assigned to at least one relay and $y_c=1$ otherwise, $c=1, 2, ...,N_s$. The objective function is that of minimizing the number of slots in the frame:

$$\min\left\{ \sum_{c \in N_s} y_c \right\} \qquad (23)$$

Binary decision variables x_{jc} define if slot c is assigned to relay j ($x_{jc}=1$) or not ($x_{jc}=0$). Obviously, we require that:

$$x_{jc} \leq y_c \qquad (24)$$

$$\sum_{c=1}^{N_s} x_{jc} = m(j) \qquad (25)$$

for all $j \in V$ and $1 \leq c \leq N_s$. Constraints (24) state that a slot c can be assigned to a relay j only if the corresponding variable y_c is set to 1, and constrains (25) that the number of colors assigned to each relay must be equal to the number of slot required.

We need now to introduce the fundamental constraints that forbid assignment of the same slots to conflicting relay. To define if relay j and k cannot use the same slots, let us define the conflict graph, i.e. a graph $G_c(V,E_2)$ where an edge between relay j and k exists if and only if they are prevented to use the same slots. The graph G_c for the interference model adopted in previous section can be easily generated. The compatibility constraints are defined as:

$$x_{jc} + x_{kc} \leq 1 \quad \forall (j,k) \in E_2, 1 \leq c \leq N_s \qquad (26)$$

Constraints (26) state that if a link between j and k exists in the conflict graph a slot c cannot be assigned to j and k at the same time ($x_{jc}=1$ and $x_{kc}=1$).

The slot assignment problem is solved through a heuristic algorithm [19]. The algorithm aims at minimizing the overall number of slot (colors) used so to keep the MAC frame short and the end to end delay low. It is based on a greedy procedure that sorts relays according to their connectivity degree in the graph G_c. For the sake of brevity the details of the scheduling algorithm are omitted.

5   Simulation Results

To evaluate the new model and algorithm presented in this paper, the WiFR algorithm, a benchmark QoS routing algorithm based on pure hop-count metric named WSPF (Wireless Shortest Path First) and the optimized TDMA MAC layer as been added into the event-driven network simulator NS-2 [20]. Simulation results presented here has been obtained using two ray channel model, a bandwidth of 2 Mbit/sec, packets of 1 kbyte, exponential On/Off traffic sources with differentiated bandwidth request resembling voice traffic.

Figure 1 shows the throughput obtained in a FRN of 60 relays random deployed following a uniform distribution over a 1000×1000 meters area for various values of relays radio range. For high values of radio range the network become fully connected

and both algorithm gives the same throughput. With networks with low/medium connectivity degree, our routing algorithm not only outperform the WSPF of about 13%,
but also rises network capacity of about 19% with respect to fully connected situation.
As the radio range increases, the throughout firstly is reduced till a minimum because
every single transmission impact on a set G_i greater and greater rising resources consumption. However, increasing radio range diminishes the route length. For this reason after the minimum the throughput rises till settling to the value corresponding to
the fully connected network.

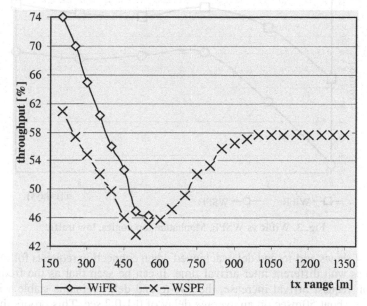

Fig. 1. WiFR vs WSPF, random topologies, 60 relays, high traffic

WiFR performance has been deeply investigated also in Manhattan topologies, see
Figure 2, starting from a basic grid of 4×4 relays used as sources and/or destinations
of the connections and adding a number n, from n=0 (basic grid) to n=5, of additional
relays that has only the task of forwarding packets.

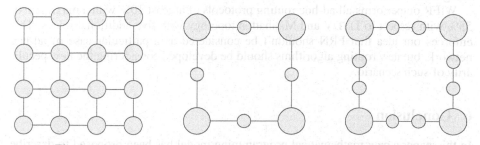

Fig. 2. Manhattan basic grid (n=0) and particular for n=2 and n=5 topologies

Figure 3 shows that WiFR outperforms WSPF in all Manhattan topologies from
6% to 10%; further, impact of additional allows to increase throughput thanks to pos-

sibility of increasing spatial reuse of shared radio resource, with only one additional relay capacity gain is about 16%, with two additional relays gain is about 20%. It can be noticed that with three and four additional relays throughput decreases since route becomes longer and re-transmissions negative effect are not completely balanced by higher spatial reuse. It seems hence that one or two additional relays is the optimal compromise between performance and costs.

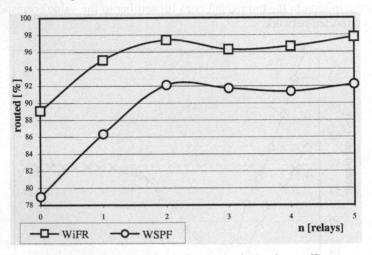

Fig. 3. WiFR vs WSPF, Manhattan topologies, low traffic

Figure 4 shows end-to-end delay achieved when connection requests follow a Poisson process with different inter-arrival time. It can be seen that as the frequency or connection request arrival increase, the end-to-end delay remains stable with slight variation of about 50msec on an average delay of 0,1-0,2 sec. This proofs that WiFR is really able to offer QoS maintaining stable the network condition independently from external offered load.

Finally Figure 5 shows a comparison between WiFR and ad-hoc routing protocols implemented in Ns2, i.e. DSR, DSDV and AODV. Since the ad-hoc routing algorithms do not offer QoS guarantees, simulations has been done with a low traffic level using IEEE 802.11 as MAC layer.

WiFR outperforms all ad-hoc routing protocols of at least 10% with a maximum of 55% with respect to DSDV and Manhattan topology with five additional relays. This enforces our idea that FRN shouldn't be considered as a particular case of ad-hoc network, but new routing algorithms should be developed considering the new peculiarity of such scenario.

6 Conclusion

In this paper a new mathematical programming model has been proposed to describe QoS routing problem for infrastructure-based multi-hop wireless networks. The model is an extension of the well known multi-commodity flow problems where link capacity constraints are replaced with new ones that takes into account interference

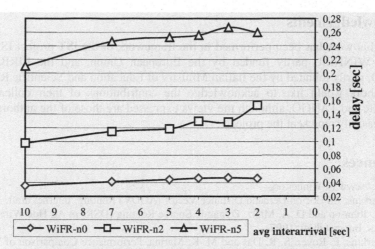

Fig. 4. WiFR, end to end delay with Poisson arrival of connection request

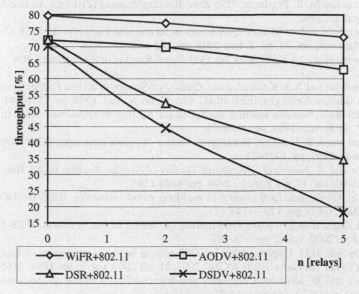

Fig. 5. WiFR vs ad-hoc routing protocols

constraints among different radio links. Feasible solutions of the problem guarantee that the rates of routed flows are compatible with radio channel capacity, but does not require to explicitly solve the scheduling problem. We have also proposed a routing algorithm, Wireless Fixed Relay routing (WiFR), based on an ad-hoc heuristic to find feasible solutions in reasonable time. Numerical results, obtained with simulation, have show the effectiveness of WiFR to provide a network capacity much higher than the routing schemes proposed for ad-hoc networks and even higher than that of WSPF.

Acknowledgements

Part of this work has been performed in the framework of the IST project IST-2003-507581 WINNER, partly funded by the European Union, and the FIRB project TANGO, partly founded by the Italian Ministry of Education and Scientific Research. The authors would like to acknowledge the contributions of their colleagues in WINNER and TANGO, although the views expressed are those of the authors and do not necessarily represent the projects.

References

1. https://www.ist-winner.org
2. C. Perkins, Ad Hoc on Demand Distance Vector (AODV) Routing, Internet draft, 1997.
3. D. B. Johnson and D. A. Maltz, Dynamic Source Routing (DSR) in Ad Hoc Wireless Networks, Internet draft, 1997.
4. C. Perkins, E. Royer, S. R. Das and M. K. Marina, Performance Comparison of Two On-Demand Routing Protocols for Ad Hoc Networks, IEEE Personal Communications, vol. 8, no. 1, Feb. 2001, pp. 16-28.
5. Z. J. Haas and M. R. Pearlman, "The Zone Routing Protocol (ZRP) for Ad Hoc Networks," Internet draft, 1997.
6. S. Chakrabarti and A. Mishra, QoS Issues in ad hoc wireless network, IEEE Communications Magazine, vol. 39, no. 2, Feb 2001, pp. 142-148.
7. E. Crawley et al., "A Framework for QoS-Based Routing in the Internet," RFC 2386, Aug. 1998.
8. J.L. Sobrinho and A.S. Krishnakumar, Quality-of-Service in ad hoc Carrier Sense Multiple Access Wireless Networks, IEEE JSAC, vol. 17, no. 8, Aug. 1999, pp. 1353-1414.
9. A. Iwata et al., Scalable Routing Strategies for Ad Hoc Wireless Networks, IEEE JSAC, vol. 17, no. 8, Aug. 1999, pp. 1369–79.
10. C.R. Lin and J.-S. Liu, QoS Routing in Ad Hoc Wireless Networks, IEEE JSAC, vol. 17, no. 8, Aug. 1999, pp. 1426–38.
11. S. Chen and K. Nahrstedt, Distributed Quality-of-Service Routing in Ad Hoc Networks, IEEE JSAC, vol. 17, no. 8, Aug. 1999, pp. 1488-1505.
12. C.R. Lin, On-demand QoS routing in multihop mobile networks, IEEE INFOCOM 2001. April 2001, vol. 3, pp. 1735-1744.
13. Chenxi Zhu and M.S. Corson, QoS routing for mobile ad hoc networks, IEEE INFOCOM 2002, June 2002, vol. 2, pp. 958-967.
14. Hongxia Sun, H.D. Hughes, Adaptive QoS routing based on prediction of local performance in ad hoc networks, IEEE WCNC 2003, March 2003, vol. 2, pp. 1191- 1195.
15. M.Garey and D. Johnson. Computers and Intractability: a Guide to the Theory of NP-Completeness. W.H. Freeman, 1979.
16. H. Li, M. Lott, W. Zirwas, M. Weckler, E. Schulz, Multihop Communications in Future Mobile Radio Networks, IEEE PIMRC 2002, Sept. 2002.
17. H. Li, M. Lott, W. Zirwas, M. Weckler, E. Schulz, Hierarchical Cellular Multihop Networks, EPMCC 2003, March 2003.
18. A.A. Assad, "Multicommodity network flow – a survey". Networks, vol. 8 pagg. 37-91, John Wiley & Sons, Inc.
19. K. Aardal, S.P.M. van Hoesel, A. Koster, C. Mannino, and A. Sassano, Models and Solution, Techniques for Frequency Assignment Problems, 4OR , (1) 4, 261-317, 2003.
20. http://www.isi.edu/nsnam/ns

Transient QoS Measure for Call Admission Control in WCDMA System with MIMO*

Cheol Yong Jeon and Yeong Min Jang

School of Electrical Engineering,
Kookmin University,
861-1, Jeongneung-dong, Songbuk-gu, Seoul 136-702, Korea
{feon77,yjang}@kookmin.ac.kr

Abstract. This paper presents an efficient capacity evaluation algorithm of the WCDMA with multiple-input multiple-output (MIMO) system for quality of service (QoS) support. To define the capacity of the system, we derive the E_b/N_o gain taking into account MIMO concept and the outage probability as the QoS measure using central limit approximation, Chernoff bound and the refined large deviation approach. Based on the QoS measures, we propose an efficient transient call admission control (CAC) algorithm. Numerical results show that there is a substantial increment in system capacity by adopting MIMO system and the theory of the refined large deviation approach is a good approach for transient QoS support.

1 Introduction

Third-generation mobile wireless systems are often referred to as universal mobile terrestrial telecommunication systems (UMTS's). The UMTS system intends to integrate all forms of mobile communications, including terrestrial, satellite, and indoor communications. Consequently, UMTS must support a number of different air interfaces [1]. One of the main air interfaces for this system is referred to as wideband code division multiple access (WCDMA), which is the topic of this paper. WCDMA is based on CDMA scheme by which multiple users are assigned radio resource using spread spectrum techniques. Although all users are transmitting in the common bandwidth, individual users are separated from each other via the use of orthogonal codes. If total energy from all users, however, is over the given threshold (signal-to-noise ratio (SNR) or energy per bit per bandwidth), the system can not admit any more users [2, 3].

Here we focus on MIMO system to increase the system capacity. Wireless channel using multiple antennas at each ends is commonly referred to as a MIMO channel. Technology built around MIMO channels resolves the fundamental issue of having to deal with two practical realities of wireless communications: a user terminal of limited battery power, and a channel of limited bandwidth. Given fixed values of transmit power and channel bandwidth, this new technology offers

* This work was supported by the KOSEF through the grant No. R08-2003-000-10922-0.

M. Ajmone Marsan et al. (Eds.): QoS-IP 2005, LNCS 3375, pp. 545–558, 2005.
© Springer-Verlag Berlin Heidelberg 2005

a sophisticated approach to exchange increased system complexity for increasing the capacity (i.e., the spectral efficiency of the channel, measured in bits per hertz) up to a value significantly higher than that attainable by any known method based on a single-input single-output (SISO) channel. More specifically the MIMO channel capacity is roughly proportional to the number of transmitter or receiver antennas, whichever is smaller. That is to say, we have a spectacular increase in spectral efficiency, with the channel capacity being roughly doubled by doubling the number of antennas at both ends of the link [1, 4, 5].

Up to now, the capacity analysis was investigated individually MIMO and CDMA. The MIMO spectral efficiency increases linearly with antenna number and the capacity of CDMA system decreases by transmitted user's power [2]. In this paper, we focus on the MIMO technology and on the calculating simultaneous user numbers in the WCDMA systems with MIMO. Then this paper presents how much more increase users in this system by numerical result. We apply the fluid flow queuing model for CAC in WCDMA system with MIMO. Traffic control is necessary to avoid possible congestion at each network node and achieve the QoS requirement by each connection. Due to real time constraints and the dynamic behavior in the system, preventive (e.g. predictive) control is more suitable then reactive control [6]. CAC is form of the preventive traffic control. The CAC is responsible for deciding whether or not a new call can be accepted while maintaining QoS in the network. We choose the transient outage probability as measure of QoS and desire a scheme which optimizes both the transient and steady state performance [7, 8]. So, we provide a complete transient solution of the system starting from a given initial condition. This paper present a general theory to deal with the transient analysis of multiple types of traffic. In order to cope with the computational complexity, a general theory of approximations and bounding approaches should be explored. In this paper, we propose new approximations and bounds for the transient fluid model [9] to deal with requirement of real time computation.

This paper is organized as follows: Section II discusses the spectral efficiency of WCDMA system with MIMO. We discuss CAC algorithm and derive the outage probability and cell loss ratio for CAC in section III. Section IV presents numerical result and finally come to conclusion in section V.

2   Spectral Efficiency of the WCDMA System with MIMO

2.1   MIMO in Gaussian Channel

When we use n_T transmit antennas and n_R receive antennas, $m =\min(n_T, n_R)$ and $n=\max(n_T, n_R)$, the MIMO capacity increases linearly with m. The average SNR per n_R receiver antenna [10, 11] is given by

$$SNR = g\frac{E[|Hx|^2]}{E[|n^2|]} \tag{1}$$

where $1/g$ indicates the average path loss. Assume AWGN, the ergodic capacity in flat fading channel is

$$C = E[log_2 \, det(I_{n_R} + \frac{SNR}{n_T}HH^\dagger)] \tag{2}$$

where $\dagger$ is transpose conjugate and I_{n_R} denote $(n_R \times n_R)$ identity matrix. The ergodic capacity of a MIMO link involves the matrix product $HH^\dagger$. We assume that transmission at each antenna performs with constant power over the burst of vector symbols and that each antenna transmits with the same power. Then we make the statistical assumption that the transmitted symbols are identically and independently distributed (*i.i.d.*). According to [1], we may rewrite Eq. (2) in the equivalent form

$$det(I_{n_R} + \frac{SNR}{n_T}HH^\dagger) = \prod_{i=1}^{n_R}(1 + \frac{SNR}{n_T}\lambda_i) \tag{3}$$

where λ_i is *i-th* eigenvalue of finally, Eq. (3) into Eq. (2)

$$C = E[\sum_{i=1}^{n_R} log_2(1 + \frac{SNR}{n_T}\lambda_i)] = n_R log_2(1 + \frac{SNR}{n_T})[bits/s/Hz]. \tag{4}$$

For simplicity, in the ideal case the eigenvalue $\lambda_i = 1$ (which means an ideal propagation environment) [4, 5], we can achieve maximum data rate

$$C = n_R log_2(1 + \frac{SNR}{n_T})[bits/s/Hz]. \tag{5}$$

Eq. (5) shows spectral efficiency of MIMO system. If the number of antenna increases, the spectral efficiency increases.

2.2 E_b/N_o Gain

Naturally, Shannon's formula about capacity in AWGN channel is

$$C = log_2(1 + SNR)[bits/s/Hz]. \tag{6}$$

Here, SNR is same to received power P per noise N (i.e. $SNR = P/N$). In band-limited channel, we assume that transmit information k bits using bandwidth W Hz and time T sec and define E_b, N_o as energy per bit and one sided noise spectral density, respectively. So it is able to transpose

$$P = \frac{E}{T} = \frac{E_b k}{T}[watt], \tag{7}$$

$$N = N_o W[watt], \tag{8}$$

and use following this

$$C = \frac{k}{WT} \ [bits/s/Hz]. \tag{9}$$

We can achieve

$$SNR = \frac{P}{N} = \frac{E_b k/T}{N_o W} = \frac{E_b}{N_o} \cdot \frac{k}{WT} = \frac{E_b}{N_o} \cdot C. \tag{10}$$

Therefore we obtain this

$$\frac{E_b}{N_o} = \frac{SNR}{C(SNR)} \tag{11}$$

where $C(SNR)$ is the capacity which is function of SNR [12]. Then the required E_b/N_o in the MIMO system, $(E_b/N_o)_{MIMO}$, can be achieved by substituting Eq. (5) to Eq. (11)

$$(\frac{E_b}{N_o})_{MIMO} = \frac{SNR}{n_R} log_2(1 + \frac{SNR}{n_T}). \tag{12}$$

When the SNR level is equal, we get the E_b/N_o gain, M, when MIMO system is considered

$$M = \frac{(\frac{E_b}{N_o})_{MIMO}}{(\frac{E_b}{N_o})_{SISO}} = \frac{log_2(1 + SNR)}{n_R log_2(1 + \frac{SNR}{n_T})}. \tag{13}$$

The antenna number increases, the gain M decreases. At same SNR, therefore, the MIMO user requires small E_b/N_o. It means that total energy decreases in the MIMO system and the system can accept more users in the points of system capacity.

2.3 Calculating the Maximum Number of Connections

While there are many models of CDMA capacity in the current literature, we present a description of WCDMA system capacity using the amount of user interference in the band. The actual capacity of a WCDMA cell depends on many different factor, such as power-control accuracy, and actual interference power introduced by other users in the same cell and in neighboring cells. We assume that reference cell is in the multi-cell environment and sectorized cell. Also we assume that the user in the reference cell is uniformly distributed and perfect power control is adopted. In digital communication, we are primarily interested in a link metric called E_b/N_o, or energy per bit per noise power density. This quantity can be related to the conventional SNR. The SNR is

$$SNR = (\frac{E_b}{N_o})_{SISO} \cdot \frac{R_m}{W} \tag{14}$$

where R_m is data rate[bits/sec]of class-m users.

In the MIMO system E_b/N_o get gain M by Eq. (13) when the SNR is fixed, the maximum number of class-m user in the MIMO system, $N_{max, m}$, is

$$N_{max, m} = 1 + \frac{1}{M} \cdot \frac{W/R_m}{(\frac{E_b}{N_o})_{SISO}}(\frac{1}{1+\eta})\lambda(\frac{1}{a}) \tag{15}$$

where W is chip rate[Hz], η is loading factor, respectively. The inverse of the factor $(1+\eta)$ is known as the frequency reuse factor, λ is sectorization gain which is typically around 2.55 for three-sector configured systems. a is activity factor, speech statistics shows that a user in a conversation typically speaks around 45% of the time. This equation shows that the WCDMA system with MIMO antenna is able to accept more users.

3   Call Admission Control

It is very important that the decision to accept or reject connections is made in real time. To do this, we need a simple and approximated CAC scheme.

3.1   Model Assumptions

For simplicity, we assume that the received power at the Node B which is same base station transceiver system would be the same for each user equipment (UE) and that the number of UEs and active connections at each cell are same. UEs are interconnected via a Node B and a wired IP network with output buffering located in the radio network controller (RNC). The RNC provides bridging functions for connecting to IP packet-switched networks. These platforms can share the back-end infrastructure, which includes home location registers (HLR), 3G service GPRS support nodes (SGSN) and gateway GPRS support modes (GGSN). Suppose that N users are connected to the WCDMA uplink. A virtual path can be modeled as a single server system with uplink capacity. A QoS predictor at the RNC predicts outage probability. We assume that the uplink distance between the UEs in the reference cell and the RNC is same. Fig.1 shows the WCDMA radio access network architecture. We consider that the network consists of a RNC and multiple users associated with RNC. CAC is performed by the RNC when one of the associated users generates a request to open a new connection. At connection set-up the RNC knows what type of connection is requested and thus knows the typical peak rate for this kind of application. A new connection is admitted if the available resource can admit the new calls peak rate. It is calculated every time by RNC's QoS predictor. Although the traffic characteristics of future wireless IP networks are hard to predict with complete accuracy, there are a number of voice and video models reported as On-Off source models, and this On-Off model has been commonly used to model a voice source with speech activity detection [9]. Suppose that $N(= N_1 + \ldots + N_m + \ldots + N_M)$ independent heterogeneous On-Off sources (calls) are connected to a uplink (from user to Node B), where N_m denotes the number of connections of class-m. Therefore it can be modeled as a single server system with uplink capacity of $C = min(N_{max,\, 1} \cdot R_1,\, N_{max,\, 2} \cdot R_2, \cdots,\, N_{max,\, m} \cdot R_m)$, where the maximum number of users in class-m, $N_{max,\, m} \cdot R_m$, which is calculated from Eq. (15). So we can find the link capacity, in the worst case, to support QoS for all traffic classes.

A QoS predictor at the uplink predicts outage probability for a time t ahead. We assume that a series of packets arrive in the form of a continuous stream of

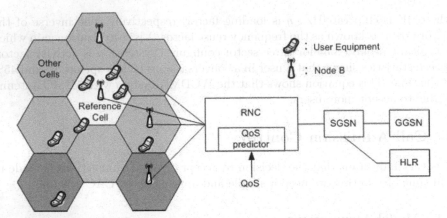

Fig. 1. IP-based WCDMA radio access network architecture.

bits or a fluid. We also assume that 'ON' and 'OFF' periods of sources of class-m are both exponentially distributed with parameters μ_m and, λ_m, respectively. The transitional flow rate from the 'ON' state to the 'OFF' state is μ_m and from 'OFF' to 'ON' is λ_m. In this traffic model, when a source is in the 'ON' state, it generates packets with a constant inter-arrival time, $1/R_m$ seconds/bit. When the source in the 'OFF' state, it does not generate any packets. We assume that N_m, class-m sources of the calls sharing a uplink have the same traffic parameters (μ_m, λ_m, R_m).

3.2   CAC Algorithm

We propose the predictive CAC algorithm. The CAC function is network specific. Each user who requests a connection is subject to CAC. The connection is admitted into the radio access network if the QoS requirements can be met for both the new connection and the existing connections. The CAC algorithm is implemented in the RNC and executed for each radio cell. The algorithm is executed whenever we have a new connection request. The connection acceptance decision must be made within the required allowable connection set-up latency time in accordance with the received connection set-up request. This means that the CAC must operate in real-time, even when various kinds of traffic are multiplexed. A flow chart of the proposed CAC algorithm is depicted in Fig. 2.

3.3   Calculating Exact QoS Measure [13]

We use a statistical bufferless fluid model to predict the $P_{out}(t)$ at a future point in time t. Let $\Lambda_m(t)(= R_m Y_m(t))$ be aggregate arrival rate from $Y_m(t)$ active sources. In a bufferless system, outages occur when $\Lambda(t)(= \sum_{m=1}^{M} \Lambda_m(t))$ exceeds the link capacity of C. The number of active class-m sources forms a birth-death process, with birth and death rates dependent on the state of

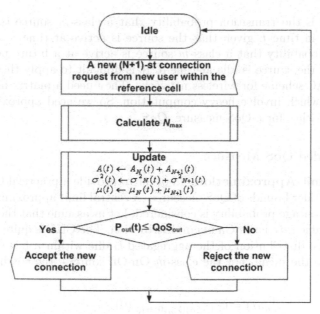

Fig. 2. Flow chart of the CAC algorithm for a new $(N + 1)$-st connection request.

the process $\lambda_{km} = (N_m - k)\lambda_m$ and $\mu_{km} = k\mu_m$. Without loss of generality, we assume that we have two classes traffic. Let $P_{k,j}(t)$ denote the probability that k class-1 and j class-2 sources are active. The conditional transient outage probability is then given by

$$P_{out}(t) = P(\Lambda(t) > C|Y_1(0), Y_2(0))$$

$$= \sum_{k,j \in k,j | (kR_1 + jR_2 - C) \geq 0}^{N_1 R_1 + N_2 R_2} P_{k,j}(t) \tag{16}$$

where $\Lambda(t) = kR_1 + jR_2$ denotes the aggregate arrival rate of the k active class-1 sources and the j active class-2 sources at time t. Taking into consideration the fact that each of $N(= N_1 + \ldots + N_m + \ldots + N_M)$ existing calls belongs to one of the M call classes, given by an arbitrary initial condition $Y(0) = I = [Y_1(0) = i_1, Y_2(0) = i_2, \ldots, Y_M(0) = i_M]$, we obtain the conditional moment generating function of $\Lambda_m(t), s \geq 0$:

$$G_{\Lambda_m(t)|Y_m(0)}(s) = E[e^{sR_m Y_m(i)}|Y_m(0) = i_m]$$
$$= [p_m(t)(e^{sR_m} - 1) + 1]^{N_m - i_m}[q_m(t)(e^{sR_m} - 1) + 1]^{i_m} \tag{17}$$

where $p_m(t)$ and $q_m(t)$ are defined [13] as

$$p_m(t) = \frac{\lambda_m}{\lambda_m + \mu_m}[1 - e^{-(\lambda_m + \mu_m)t}] \tag{18}$$

$$q_m(t) = \frac{\lambda_m}{\lambda_m + \mu_m} + \frac{\mu_m}{\lambda_m + \mu_m}e^{-(\lambda_m + \mu_m)t} \tag{19}$$

where $p_m(t)$ is the transition probability that a class-m source is active at a future point in time t, given that the source is active at time 0. $q_m(t)$ is the transition probability that a class-m source is active at a future point in time t, given that the source is idle at time 0. It is difficult to apply these results to real time CAC scheme for wireless networks. They need a matrix inversion and convolution which involve heavy computation. So we need approximation and bound approaches for a QoS measure, QoS_{out}.

3.4   Bounded QoS Measure

Central Limit Approximation: In order to provide a practical CAC mechanism, we consider bounds of QoS measures. A central limit approximation (CLA) for transient outage probability is considered. Let us assume that the conditional aggregate traffic rate has a Gaussian distribution. The CLA is quite accurate for predicting the distribution of the aggregated traffic within a few standard deviations from the mean. For the class-m On-Off sources, the conditional mean arrival rate is

$$
\begin{aligned}
A_m(t) &= G'_{\Lambda(t,m)|Y_m(0)=i_m}(0) \\
&= R_m[i_m q_m(t) + (N_m - i_m)p_m(t)],
\end{aligned}
\tag{20}
$$

and conditional variance of arrival rate is

$$
\begin{aligned}
\sigma_m^2(t) &= G''_{\Lambda(t,m)|Y_m(0)=i_m}(0) - [G'_{\Lambda(t,m)|Y_m(0)=i_m}(0)]^2 \\
&= R_m^2[(N_m - i_m)p_m(t)(1 - p_m(t)) + i_m q_m(t)(1 - q_m(t))].
\end{aligned}
\tag{21}
$$

By the CLA, $\Lambda(t)$ is approximated by a normal random process with conditional mean and variance,

$$
A(t) = \sum_{m=1}^{M} A_m(t), \qquad \sigma^2(t) = \sum_{m=1}^{M} \sigma_m^2(t).
\tag{22}
$$

Namely,

$$
\Lambda(t) = N(A(t), \sigma^2(t)).
\tag{23}
$$

With established traffic model, $P_{out}(t)$ may now be easily computed from the tail of the normal distribution [14, 15]. Given a specific QoS requirement $P_{out}(t) \leq QoS_{out}$, where QoS_{out} is small number such as 10^{-3}, the QoS requirement $P_{out}(t) \leq QoS_{out}$ is met and only if.

$$
P_{out}(t) = Q(\frac{C - A(t)}{\sqrt{\sigma^2(t)}}) \leq QoS_{out}
\tag{24}
$$

where Q(.) denotes the Q-function defined as

$$
Q(a) = \frac{1}{\sqrt{2\pi}} \int_a^{\infty} e^{\frac{-x^2}{2}} dx.
\tag{25}
$$

Using this admission rule Eq. (24), a new *(N+1)*-st connection is established. We update $A(t) \leftarrow A(t) + A_{N+1}(t)$, $\sigma^2(t) \leftarrow \sigma^2(t) + \sigma^2_{N+1}(t)$, where $A_{N+1}(t)$, $\sigma^2_{N+1}(t)$ are computed from traffic descriptor specified by new user. We then admit the new connection if and only if the connection in Eq. (24) is met.

Chernoff Bound: To predict small $P_{out}(t)$, which is several standard deviations away from the mean, other approximations such as the Chernoff bound (CB) and the large deviations theory can be used. For small tail probabilities, the CLA can underestimate the outage probability. In the subsection, we provide an upper bound for $P_{out}(t)$. To this end, for any random process $A(t)$ and constant C, let

$$\mu_{A(t)|Y(0)}(s) = \ln G_{A(t)|Y(0)}(s) = \ln E[e^{sA(t)}|Y(0) = I]$$

$$= \sum_{m=1}^{M} \mu_{A(t,m)|Y_m(0)=i_m}(s). \qquad (26)$$

Note that $\mu_{A(t)|Y(0)}(s)$ is the logarithm of the conditional moment generating function for $A(t)$. $\mu_{A(t,N+1)|Y_{N+1}(0)=1}(s)$ is the logarithm of the conditional moment generating function for $A(t, N+1)$, associated with a new connection request. The CB gives an upper bound for $P(A(t) > C|Y(0) = I)$. Assume that $E[A(t)]$ exists and $C \geq E[A(t)] \geq -\infty$. Then the supreme in Eq. (27) is obtained within values $s^* \geq 0$, and minimizing the right hand side of Eq. (27) with respect to s yields the Chernoff upper bound

$$P_{out}(t) = P(A(t) > C|Y(0) = I) \leq e^{-s^* C} G_{A(t)|Y(0)}(s^*)$$

$$= e^{-s^* C + \mu_{A(t)|Y(0)}(s^*)}, \quad for \ s^* \geq 0 \ and \ C \geq E[A(t)] \qquad (27)$$

where s^* is the unique solution to

$$\mu'_{A(t)|Y(0)}(s^*) = \frac{\frac{d}{ds^*} G_{A(t)|Y(0)}(s^*)}{G_{A(t)|Y(0)}(s^*)} = \sum_{m=1}^{M} \frac{(N_m - i_m) p_m(t) R_m e^{s^* R_m}}{p_m(t)(e^{s^* R_m} - 1) + 1}$$

$$+ \sum_{m=1}^{M} \frac{i_m q_m(t) R_m e^{s^* R_m}}{q_m(t)(e^{s^* R_m} - 1) + 1} = C. \qquad (28)$$

The CB can be used to predict the number of connections that is needed to satisfy a given the conditional outage probability bound, QoS_{out}, at the downlink, i.e., $P(A(t) > C|Y(0) = I) \leq QoS_{out}$. The CB gives the admission control condition:

$$e^{-s^* C + \mu_{A(t)|Y(0)}(s^*)} \leq QoS_{out}. \qquad (29)$$

Using Eq. (29), the admission control condition scheme operates as follows. The logarithms of conditional moment generating function, $\mu_{A(t,m)|Y(0)}(s)$, are first calculated off line for all connections. Now, suppose a new connection $N + 1$ is requested. We update $\mu_{A(t)|Y(0)}(s) \longleftarrow \mu_{A(t)|Y(0)}(s) + \mu_{A(t,N+1)|Y_{N+1}(0)=1}(s)$ and find the s^* that satisfies Eq. (28). Finally, we admit the connection if and only if Eq. (29) is satisfied.

Refind Large Deviation Approximation: The upper bound given by Eq. (27) is often in the correct order of magnitude. But the bound can be converted to an accurate approximation by applying the theory of large deviation and the CLA [7, 16, 17]. The shifted distribution can be approximated very accurately by a normal distribution around its own mean.

$$P_{out}(t) \approx \frac{1}{s^* \sqrt{2\pi \mu''_{\Lambda(t)|Y(0)}(s^*)}} e^{-s^* C + \mu_{\Lambda(t)|Y(0)}(s^*)} \tag{30}$$

where

$$\mu''_{\Lambda(t)|Y(0)}(s^*) = \sum_{m=1}^{M} \frac{(N_m - i_m)p_m(t)R_m^2 e^{s^* R_m}(2p_m(t)e^{s^* R_m} - p_m(t) + 1)}{(p_m(t)(e^{s^* R_m} - 1) + 1)^2}$$

$$+ \sum_{m=1}^{M} \frac{i_m q_m(t)R_m^2 e^{s^* R_m}(2q_m(t)e^{s^* R_m} - q_m(t) + 1)}{(q_m(t)(e^{s^* R_m} - 1) + 1)^2}. \tag{31}$$

A refined large deviation approximation (RLDA) for $P_{out}(t)$ gives the following admission control condition:

$$P_{out}(t) \approx \frac{1}{s^* \sqrt{2\pi \mu''_{\Lambda(t)|Y(0)}(s^*)}} e^{-s^* C + \mu_{\Lambda(t)|Y(0)}(s^*)} \leq QoS_{out}. \tag{32}$$

4   Numerical Results

For the $N \times N$ MIMO system, we evaluate the performance of the proposed CAC based on the transient outage probability. So we consider the multiple classes of traffic. For this, two different traffic classes are examined. Speech data rate is 9.6Kbps and packet data bit rate is 384Kbps, respectively. Consider the following system factors shown in Table 1. Actually required E_b/N_o in the SISO system for voice communication is 7dB. The simulation results in Fig. 3 are given for class-1 voice service. And the simulation results for each class is given in Table 2. The 4×4 MIMO system in the voice service can accept more calls

Table 1. Simulation parameter.

Parameter	Value
W	$3.84M chips/sec$
R	$9.6Kbps$, $384Kbps$
η	0.5
λ	2.55
a	0.45
E_b/N_o	$7dB$
The number of antenna	1×1, 2×2, 3×3, 4×4

Table 2. Maximum number of connections.

Traffic type	Data rate	QoS_{out}	1×1	2×2	3×3	4×4
Speech	$9.6Kbps$	0.05	748	1,066	1,264	1,402
Data	$384Kbps$	10^{-5}	7	11	14	16

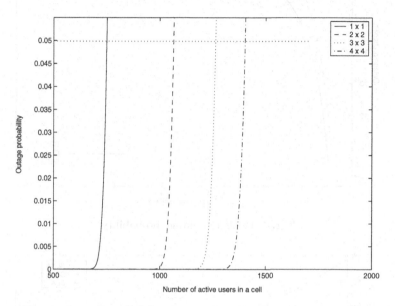

Fig. 3. Outage probability vs. number of active users on speech.

around 87% than SISO system. The 4×4 MIMO system in the data service can accept more calls around 128% than SISO system. This result shows that if we adopt the MIMO concept, there can be a significant increment in capacity for WCDMA system with MIMO technology. Also we consider a voice traffic sources, for example data rate=9.6kbps, $QoS_{out} = 0.05$, $\lambda = 0.5$ and $\mu = 0.833$. Fig. 4 depicts the proposed transient outage probability as a function of the prediction time for various values of the initial conditions, $Y(0)$. We observe that after approximately 4 seconds the predicted outage probability will converge to the steady state value, $P_{out}(\infty)$. We observe differences in the results obtained as a function of the different initial conditions. In Fig. 4, at $t=1$ sec, we observe that the predicted $P_{out}(t)$ at $Y(0) = 200$ is around 3.1×10^{-4}, while the steady state $P_{out}(\infty)$ is around 4.8×10^{-2} when CLA is adopted. The transient approaches are more complex than the steady state approaches. The nonlinear equation, Eq. (28), can be solved by the standard Newton-bisection method. Whenever we increase the initial number of active sources, the value s^* is decreased. The steady state outage probability using the CLA and CB give us the upper bound of RLDA. The RLDA is an optimistic admission policy. We, therefore, conclude that the computation of QoS measures using the RLDA is more accurate and can lead to significant different values of the QoS measures and consequently of

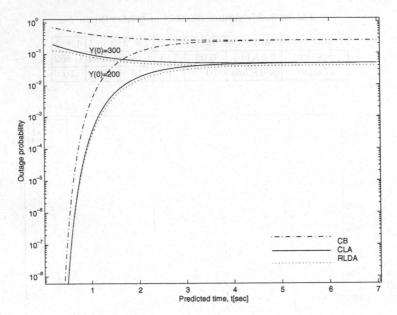

Fig. 4. Predicted outage probability.

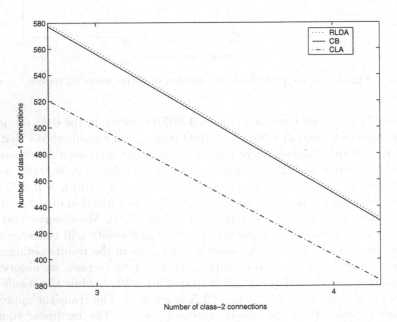

Fig. 5. The maximum number of class-1 connections vs. the number of class-2 connections.

the connection control decisions. We also obtain the performance of the proposed connection control algorithm to obtain the maximum number of connections. We assume the following parameters: $t = 4$, R_1 =9.6kbps, R_2 =384kbps, $QoS_{out} =$

10^{-3}, $\lambda_1 = \lambda_2 = 0.5$ and $\mu_1 = \mu_2 = 0.833$. In Fig. 5 we observe that as the number of class-1 connections decreases, the maximum number of class-2 connections will decrease. Recall that we already assumed to find the worst link capacity to support QoS. The CB is always conservative connection control policy under the light load situation. For the large number of active sources, CLA may be used because we can avoid the solution for Eq. (28) which has been used for the RLDA and CB. The RLDA very closely coincides with *CLA* for transient and steady states but, It can always admit more connections. We, therefore, conclude that the connection control using the RLDA is more accurate and simple for QoS support.

5   Conclusions

In this paper, we have evaluated the capacity for WCDMA with MIMO system and introduces the efficient CAC algorithm for WCDMA systems with QoS support. It is necessary to study outage performance of the system. Therefore, we focused on the reverse link performance of WCDMA systems with MIMO. So we derived the spectral efficiency of MIMO system and showed the effect of the E_b/N_o gain. We evaluated the transient outage probability using the proposed CAC algorithm using CLA, CB and RLDA. Our analytical results show that there can be a substantial increment in system capacity by adopting MIMO and that the optimal number of connections can be utilized for CAC and the RLDA based CAC is more accurate and simple algorithm.

References

1. S. Haykin and M. Moher, *Modern Wireless Communications*, Pearson Prentice Hall, pp. 339-376, 2005.
2. Tero Ojanrera and Ramjee Prasad, *WCDMA: Towards IP Mobility and Mobile Internet*, Artech House, INC., 1998.
3. Samuel C. Yang, *CDMA RF System Engineering*, Artech House INC., pp. 75-83, 1998.
4. Martone and Massimiliano, *Multiantenna Digital Radio Transmission*, Artech House, INC., pp. 113-119, 2002.
5. P. J. Smith and M. Shafi, 'On a Gaussian approximation to the capacity of wireless MIMO systems,' *IEEE ICC*, May 2002.
6. *IEEE Comm. Magazine*, July 1997.
7. J. Roberts, U. Mocci, and J. Virtamo, (Eds.) 'Broadband network teletraffic: performance evaluation and design of broadband multiservice networks,' *Final report of Action COST 242*, 1996.
8. A. Elwalid, D. Mitra, and R. Wentworth, 'A new approach for allocating buffers and bandwidth to heterogeneous, related traffic in an ATM node,' *IEEE JSACs*, August 1995.
9. G. Mao and D. Habibi, 'Heterogeneous On-Off sources in the bufferless fluid flow model,' *IEEE ICON*, pp. 307-312, 2000.
10. A. Lozano, A. M. Tulino and S. Verdu, 'Correlation number: a new design criterion in multi-antenna communication,' *IEEE VTC*, April 2003.

11. A. M. Tulino, S. Verdu and A. Lozano, 'Capacity of antenna arrays with space, polarization and pattern diversity,' *IEEE ITW*, April 2003.
12. S. Verdu, 'Spectral efficiency in the wideband regime,' *IEEE Trans. on Information Theory*, Vol. 48, No. 6, June 2002.
13. Y. M. Jang, 'Estimation and prediction based connection admission control in broadband satellite systems,' *ETRI Journal*, pp. 40-50, 2000.
14. Alberto Leon-Garcia, *Probability and Random Processes for Electrical Engineering*, Addison-Wesley, 1994.
15. X. Luo, I. Thng and S. Jiang, 'A simplified distributed call admission control scheme for mobile cellular networks,' *ICT*, 1999.
16. J. Hui, *Switching and Traffic Theory for Integrated Broadband Networks*. Boston, MA; Kluwer, 1990.
17. Martin Reisslein, 'Measurement-based admission control: a large deviations approach for bufferless multiplexers,' *IEEE Symposium on Computers and Communications*, pp. 462-467, July 2000.

Adaptive Bandwidth Partitioning
Among TCP Elephant Connections
over Multiple Rain-Faded Satellite Channels*

Nedo Celandroni[1], Franco Davoli[2,3], Erina Ferro[1], and Alberto Gotta[1]

[1] ISTI-CNR, Area della Ricerca del C.N.R., Via Moruzzi 1, I-56124 Pisa, Italy
{nedo.celandroni,erina.ferro,alberto.gotta}@isti.cnr.it
[2] Italian National Consortium for Telecommunications (CNIT)
University of Genoa Research Unit, Via Opera Pia 13, 16145 Genova, Italy
franco.davoli@cnit.it
[3] National Laboratory for Multimedia Communications, Via Diocleziano 328, Napoli, Italy

Abstract. The assignment of a common bandwidth resource to TCP connections over a satellite channel is considered in the paper. The connections are grouped according to their source-destination pairs, corresponding to up- and down-link channels traversed, and each group may experience different fading conditions. By exploiting the tradeoff between bandwidth and channel redundancy (as determined by bit and coding rates) in the maximization of TCP goodput, an overall optimization problem is constructed, which can be solved by numerical techniques. Different relations between goodput maximization and fairness of the allocation are investigated. The allocation strategies are tested and compared in a real dynamic fading environment.

1 Introduction

The variability in the characteristics of the satellite channel, due to variable traffic loads and to weather conditions that affect the signal attenuation, is the main problem to be faced, together with the large propagation delay. Adaptive network management and control algorithms are therefore necessary to maintain the Quality of Service (QoS) of data transmitted over AWGN (Additive White Gaussian Noise) links (a reasonable assumption concerning geostationary satellites with fixed earth stations), with high bandwidth-delay product. There is, indeed, a vast literature on performance aspects related to the adaptation of TCP congestion control mechanisms over such channels ([1], [2]), as well as on resource allocation and QoS control in broadband packet networks [3], even in the satellite environment ([1], [2], [4]). In particular, cross-layer optimization approaches are becoming widespread for wireless networks in general [5], even though careful design is necessary [6].

In our study we assume that a number of long-lived TCP connections, (*elephants* [7]), are active on a geostationary satellite network, which consists of N earth stations, among which a master one, in addition to sending data as any other station, exerts the

---

* This work was funded by the *TANGO* project of the FIRB programme of the Italian Ministry for Education, University and Research, and partially by the European Commission, in the framework of the NoE project *SatNEx* (contract No. 507052).

M. Ajmone Marsan et al. (Eds.): QoS-IP 2005, LNCS 3375, pp. 559–573, 2005.
© Springer-Verlag Berlin Heidelberg 2005

control on the access to the satellite bandwidth. The main goal of our study is to investigate the behaviour of TCP connections within a cross-layer resource allocation approach in this type of environment: thus we do not consider guaranteed bandwidth real-time traffic, whose presence would only introduce dynamic variations in the amount of bandwidth available for TCP data. As such, the presence of real-time traffic can be simply emulated by a time-varying bandwidth reduction. We then assume that the system has a certain amount of bandwidth to be utilized at its best for TCP traffic. Moreover, we do not take into account the presence of short-lived connections (*mice*), whose characterization in this environment presents some additional difficulty, caused by their transient behaviour; this study is currently under investigation. Our typical configuration consists of fixed numbers of long-lived TCP Reno connections any station may have with any other in the network. We act on the transmission parameters of the satellite link, which are appropriately tuned-up, in order to achieve the maximum goodput, leaving unaltered the TCP source code.

This paper is the continuation of our study on TCP connections over rain-faded geostationary satellite channels [8-10]. Over wireless links, any gain in the *bit error rate* (BER) (i.e., in the packet loss) is generally obtained at the expenses of the *information bit rate* (IBR), and the end-to-end transfer rate of a TCP connection (*goodput*) increases with the IBR and decreases with the BER. In [8] we have shown that, given an available radio spectrum, antenna size, and transmission power, the selection of a modulation scheme and a FEC (forward error correction) type allows choosing the BER and the IBR of the link that maximize the goodput of a single TCP connection. The FEC techniques do not interfere with the normal behaviour of the TCP stack, as they are applied just before the transmission over the satellite link. In [9] we extended the results obtained in [8] to multiple TCP connections; in [10] we studied the trade-off between bandwidth and channel redundancy for maximizing the TCP throughput of multiple connections in static fading conditions. In this paper we adopt the same philosophy and operate at the physical level as in the mentioned previous works, but we extend the study to dynamically varying fading conditions, we introduce the concept of the bandwidth-redundancy optimization in the context of the Nash bargaining problems, and we compare different bandwidth allocation methods, combined with adaptive bit and coding rate for TCP elephants. Connections take place between different source-destination (SD) pairs over satellite links, which may be generally subject to diverse fading conditions, according to the atmospheric effects at the originating and receiving stations (i.e., on the up- and down-links). Connections on the same SD pair, which experience a specific channel condition, belong to the same "class"; they feed a common buffer at the IP packet level in the traffic station. The bandwidth allocated to serve such buffers is shared by all TCP connections in that class, and, once fixed, it determines the "best" combination of bit and coding rates for the given channel conditions. The goal of the allocation is to satisfy some global optimality criterion, which may involve *goodput*, *fairness* among the connections, or a combination thereof. Therefore, in correspondence of a specific channel situation and of a given traffic load, we face a possible two-criteria optimization problem, whose decision variables are the service rates of the above mentioned IP buffers for each SD pair, and the corresponding transmission parameters. We refer to these allocation strategies as TCP-CLARA (*C*ross *L*ayer *A*pproach for *R*esource *A*llocation). Specifically, we consider five allocation criteria within this general philosophy. In all cases, the indexes chosen for the performance evaluation of the system are the TCP connections' *goodput* and the *fairness* of the allocations. The optimal allocations are derived numerically on the basis of an analytical model, under different fade patterns.

2 Goodput Estimation of Long-Lived TCP Connections

When a number of long-lived TCP sources share the same link, it was empirically observed in [11] (by making use of simulation) that, if all connections have the same latency, they obtain an equal share of the link bandwidth. This is strongly supported by our simulations, as well, obtained by using ns-2 [12], where we assume that the bottleneck is the satellite link. As the latency introduced by a geostationary satellite is quite high (more than half a second), it is reasonable to assume that the additional latency introduced by the network to access the satellite is negligible with respect to the satellite one, and that all connections have the same latency. In order to avoid time consuming simulations, reasonable estimations have been done for the goodput of a TCP Reno agent. A first relation is the one taken from [13] which, being estimated for infinite bottleneck rate, is thus valid far apart the approaching of the bottleneck rate itself. Let μ be the bottleneck (the satellite link) rate expressed in segments/s, n the number of TCP sources, and τ the delay between the beginning of the transmission of a segment and the reception of the relative acknowledgement, when the satellite link queue is empty. Moreover, let us assume the segment losses to be independent with rate q. We have $\tau = c_l + 1/\mu$, where c_l is the channel latency. TCP connections that share the same link also share an IP buffer, inserted ahead of the satellite link, whose capacity is at least equal to the product $\mu\tau$. Let also b be the number of segments acknowledged by each ACK segment received by the sender TCP, and T_o the timeout estimated by the sender TCP. Then, by exploiting the expression of the send rate derived in [13], dividing it by μ/n for normalization and multiplying it by $1-q$ for a better approximation, the relative goodput (normalized to the bottleneck rate) can be expressed as

$$T_g = \frac{1-q}{\frac{\mu}{n}\left[\tau\sqrt{\frac{2bq}{3}} + T_o \min\left(1, 3\sqrt{\frac{3bq}{8}}\right) q(1+32q^2)\right]}. \qquad (1)$$

Relation (1) is rather accurate for high values of q, i.e., far apart the saturation of the bottleneck link. For low values of q, it is found by simulation that, given a fixed value of c_l, for fixed values of the parameter $y = q(\mu/n)^2 \tau^5$, the goodput has a limited variation with respect to individual variations of the parameters q, μ and n. Owing to the high number of simulations needed to verify this observation, we employed a fluid simulator [14], which was validated by means of ns2, for values of $y \leq 1$. Simulation results have been obtained for the goodput estimation, with a 1% confidence interval at 99% level, over a range of values of μ/n between 20 and 300, and n between 1 and 10. For $0 \leq y \leq 1$, goodput values corresponding to the same y never deviate for more than 8% from their mean. We then interpolated such mean values with a 4-th order polynomial approximating function, whose coefficients have been estimated with the least squared errors technique. Assuming a constant c_l, equal to 0.6 s (a value that takes into account half a second of a geostationary satellite double hop, plus some processing time), in the absence of the so-called *Delayed ACKs* option ($b=1$), the polynomial interpolating function results to be

$$T_g = a_o + a_1 y + a_2 y^2 + a_3 y^3 + a_4 y^4 \; ; \qquad y \le 1 \,, \tag{2}$$

where

$$a_o = 0.995; \; a_1 = 0.11 \; [s^{-3}]; \; a_2 = -1.88 \; [s^{-6}]; \; a_3 = 1.98 \; [s^{-9}]; \; a_4 = -0.63 \; [s^{-12}].$$

For $y=1$, $T_g = 0.575$. For $y>1$, we adopt instead relation (1), with $b=1$.

As we assume to operate on an AWGN channel, the segment loss rate q can be computed as [8]:

$$q = 1 - (1 - p_e/l_e)^{l_s} \,, \tag{3}$$

where p_e is the bit error probability (BER), l_s is the segment length in bits, and l_e is the average error burst length (*ebl*). We took p_e data from the Qualcomm Viterbi decoder data sheet [15] (standard NASA 1/2 rate with constraint length 7 and derived punctured codes), while l_e was obtained through numerical simulation in [8]. The complete set of data is plotted in [8] versus E_c / N_0 (channel bit energy to one-sided noise spectral density ratio). In order to make q computations easier, we interpolated such data, and expressed p_e and l_e analytically as functions of the coding rate and the E_c / N_0 ratio. We have:

$$
\begin{aligned}
p_e(1/2) &= 10^{-(1.6 E_c/N_0 + 3)} \; ; & 0 \le E_c/N_0 \le 5 \; dB \\
p_e(3/4) &= 10^{-(1.6 E_c/N_0 - 2.04)} \; ; & 4 \le E_c/N_0 \le 8 \; dB \\
p_e(7/8) &= 10^{-(1.6 E_c/N_0 - 5)} \; ; & 6 \le E_c/N_0 \le 10 \; dB \\
l_e(1/2) &= e^{-0.32 E_c/N_0 + 1.87} \; ; & 0 \le E_c/N_0 \le 5 \; dB \\
l_e(3/4) &= e^{-0.4 E_c/N_0 + 3.45} \; ; & 4 \le E_c/N_0 \le 8 \; dB
\end{aligned} \tag{4}
$$

$$l_e(7/8) = 63.6 - 19(E_c/N_0) + 1.94(E_c/N_0)^2 - 0.067(E_c/N_0)^3 \; ; \quad 6 \le E_c/N_0 \le 10 \; dB$$

and, for the uncoded case [16],

$$p_e(1/1) = \frac{1}{2} erfc\left(10^{(E_c/N_0)/20}\right); \qquad l_e(1/1) = 1 \tag{5}$$

where $erfc(x)$ is the complementary error function $\dfrac{2}{\sqrt{\pi}} \displaystyle\int_x^\infty e^{-t^2} \, dt$.

The TCP goodput relative to the bottleneck rate is a decreasing function of the segment loss rate q, which, in its turn, is a decreasing function of the coding redundancy applied in a given channel condition C/N_0 (carrier power to one-sided noise spectral density ratio; see Section 5 for the relation between C/N_0 and E_c/N_0), and for a given bit rate R. The combination of channel bit and coding rates gives rise to a "redundancy factor" $r \ge 1$, which represents the ratio between the IBR in clear sky and the IBR in the specific working condition. The absolute goodput of each TCP connection, $\hat{T}_g$, is obtained by multiplying the relative value by the bottleneck rate:

$$\hat{T}_g = T_g \frac{\mu}{n} = T_g \frac{1}{n} \cdot \frac{B}{r} \tag{6}$$

where B is the link rate in segments/s in clear sky conditions.

In [8] we showed that, for a given hardware (modulation scheme/rate, FEC type/rate), a set of transmission parameters maximizes the absolute goodput for each channel condition. Given B and C/N_0 (resulting from a link budget calculation), we compute T_g for all allowable bit and coding rates; then, we select the maximum value. The value of T_g is taken from (1) or (2), and q is computed with (3) and (4) or (5).

3 The Bandwidth Allocation Problem

We assume that our satellite network operates in single-hop, so that the tasks of the master station are limited to resource assignment and synchronization. Note that, in this respect, we may consider a private network, operating on a portion of the total available satellite capacity, which has been assigned to a specific organization and can be managed by it (as a special case, this situation could also represent a service provider, managing the whole satellite capacity).

If we have a number of links across the satellite network, representing the physical channels between possible SD pairs, and a total satellite capacity available for the traffic stations, there are different ways of assigning this capacity to the links, according to the target goals. In any case, since the choice of the transmission parameters influences the TCP goodput, optimized assignments must take into account the particular fade situation of each link, as well as its relative load in terms of ongoing TCP connections. The problem we address in the following is thus the assignment of bandwidth, bit and coding rates to the IP buffers which serve each specific link, given the fading conditions and the load of the network.

We make the following assumptions.

1. The end-to-end delay of the TCP connections is the same for each station. This means that the TCP connections are opened in the traffic station itself; they do not come from remote sources, where bottlenecks, possibly present in the terrestrial network, may introduce additional random delays. This corresponds to the case of having the users directly connected to the satellite earth station's router or through a local access network.
2. In each station, there is an IP buffer for each SD pair, and we assume that the TCP connections that share it belong to the same class. Obviously, they experience the same up-link and destinations' down-link conditions. Irrespective of the station they belong to, let $N(N-1)$ be the total number of connection classes, corresponding to all possible SD pairs. Some of them may experience the same fading, but they are anyway distinguished, as they share different buffers.
3. In defining and solving the optimal bandwidth-redundancy assignment problems, we consider the system in static conditions. In other words, given a certain number of ongoing connections, distributed among a subset F of SD pairs, we find the optimal assignment as if the situation (load and fading) would last forever. For some of the assignment strategies we consider, this static situation has already

been numerically investigated in [10]. Clearly, in a dynamic environment (as the one we consider in Section 5), under variable fading conditions and with starting/ending TCP sessions, a possibility is to perform our calculations at each change of parameters. It is worth noting that this simple form of adaptation, though involving some cross-layer interaction, does indeed maintain a flavor of separation principle, as it is common in adaptive control and, as such, represents a "cautious" approach, in the sense of [6], not violating any basic layering concept.

We assume that, if the fading conditions of an active class i ($i = 1,2,..,F$) are such that a minimum goodput $T_{g,thr}^{(i)}$ cannot be reached by its connections, the specific SD pair is considered in outage, and no bandwidth is assigned to it.

Let $B_i \in [0,W]$ (where W is the total bandwidth to be allocated, expressed in segments/s), $r_i \in \{\mathcal{R}^{(1)},...,\mathcal{R}^{(P)}\}$, and $n_c^{(i)}$, $i = 1,...,F$, be the bandwidth, the redundancy factor, chosen in the set of available ones (each value $\mathcal{R}^{(k)}$, $k = 1,...,P$, corresponds to a pair of bit and coding rates), and the number of connections, respectively, of the i-th SD pair. If different bit and coding rates yield the same redundancy factor, the pair that gives rise to the minimum BER is selected.

We start by considering the following two complementary goals to be achieved:
G1) To maximize the global goodput, i.e.,

$$\max_{B_i \in [0,W],\, r_i \in \{\mathcal{R}^{(1)},...,\mathcal{R}^{(P)}\},\, i=1,...F} \sum_{j=1}^{F} n_c^{(j)} \hat{T}_g^{(j)} \tag{7}$$

$$\text{subject to } \sum_{i=1}^{F} B_i = W \cdot \tag{8}$$

- G2) To reach global fairness, i.e., to divide the bandwidth (and to assign the corresponding transmission parameters) in such a way that all TCP connections achieve the same goodput.

Note that, even though the goodput expressions derived in Section 2 are applied in both cases, the two goals are different and would generally yield different results in the respective parameters. Maximizing the global goodput may turn out in an unfair allocation (some SD pairs may receive a relatively poor service), whereas, in general, a fair allocation in the above sense does not achieve globally optimal goodput.

The relative calculations relevant to the single goals may be effected as follows.

- The maximization in (7) is over a sum of separable nonlinear functions (each term in the sum only depending on its specific decision variables, only coupled by the linear constraint (8)). Maximization can be computed efficiently by means of Dynamic Programming [17, 18], if the bandwidth allocations are expressed in discrete steps of a *minimum bandwidth unit* (*mbu*).
- The goodput-equalizing fair allocation can be reached by starting from an allocation proportional to the number of TCP connections per SD pair, computing the average of the corresponding optimal (in the choice of transmission parameters) goodputs, then changing the *mbu* allocations (under constraint (8)) by discrete steps, tending to decrease the absolute deviation of each SD pair's goodput from the average, and repeating the operation with the new allocations. A reasonable convergence, within a given tolerance interval, is obtained in few steps.

In order to achieve a reasonable combination of goodput and fairness, we propose the following two strategies (termed *Tradeoff* and *Range*, respectively).

Tradeoff Strategy. The following steps are performed:

1. To compute the pairs $\left(B_i^*,\ r_i^*\right)$ $i=1,...,F$, by maximizing the global goodput (7), under constraint (8);
2. To compute the pairs $\left(\overline{B}_i,\ \overline{r}_i\right)$ $i=1,...,F$, which correspond to the goodput-equalizing fair choice;
3. To calculate the final allocation as $\tilde{B}_i=\overline{B}_i\,\rho+B_i^*(1-\rho)$, $i=1,...,F$, where $0<\rho\le1$ is a tradeoff parameter, along with the relevant bit and coding rates.

Range Strategy. The following steps are performed:

1. To compute the pairs $\left(\overline{B}_i,\ \overline{r}_i\right)$ $i=1,...,F$, which correspond to the goodput-equalizing fair choice;
2. To choose a "range coefficient" $\beta\ge0$;
3. To compute the global goodput-maximizing allocation, by effecting the constrained maximization in (7), with $\overline{B}_i$ which varies in the range $\left[\max\left(\overline{B}_i\,(1-\beta),0\right)\ \min\left(\overline{B}_i\,(1+\beta),W\right)\right]$, instead of $[0,W]$, $i=1,...,F$.

As terms of comparison, we also consider two other possible strategies, termed **BER Threshold** and **Generalized Proportionally Fair (GPF)**, respectively. The first one only assigns the transmission parameters (bit and coding rates) to each SD pair in order to keep the BER on the corresponding channel below a given threshold. The bandwidth is then assigned proportionally to the number of connections per link, multiplied by the relative redundancy. The second one is derived from the concept of *Generalized Proportional Fairness* [19], which will be discussed in Section 4.

In order to comparatively evaluate the different options, we define the following indexes for comparison:

$$Goodput\quad Factor\quad \varphi_g=\sum_{i=1}^{F}n_c^{(i)}\hat{T}_g^{(i)}(B_i,r_i)\bigg/\sum_{i=1}^{F}n_c^{(i)}\hat{T}_g^{(i)}\left(B_i^*,r_i^*\right),\qquad(9)$$

where (B_i,r_i) is a generic choice and $\left(B_i^*,r_i^*\right)$ is the global goodput-maximizing one.

$$Fairness\quad Factor\quad \varphi_f=1-\sum_{j=1}^{L}\left|\hat{T}_g^{(j)}-\overline{T}_g\right|\bigg/2\overline{T}_g(L-1),\qquad(10)$$

where $L=\sum_{i=1}^{F}n^{(i)}$ is the total number of ongoing TCP connections, and

$\overline{T}_g=\dfrac{1}{L}\sum_{k=1}^{L}\hat{T}_g^{(k)}$ is the average goodput. Note that $\varphi_f=1$ when all goodputs are equal, and $\varphi_f = 0$ when the imbalance among the connections' goodputs is maximized, i.e., the goodput is $\overline{T}_g \cdot L$ for one connection and 0 for the others (yielding a deviation from the average $\left(\overline{T}_g \cdot L - \overline{T}_g\right) + \left|0 - \overline{T}_g\right| \cdot (L-1) = 2\overline{T}_g \cdot L - 2\overline{T}_g$, which is the denominator of (10)).

4 Bandwidth-Redundancy Optimization in the Context of Nash Bargaining Problems

Different concepts of *fairness* have been used in the literature: the max-min fairness [20], (as well as the more general concept of weighted max-min fairness), the proportional fairness [21, 22], the harmonic mean fairness [23], and the notion of fairness intrinsic in Nash Bargain Solutions (NBS) [24, 25]. All these criteria give rise to Pareto optimal solutions. However, as noted by Kelly [21], max-min fairness gives absolute priority to smaller flows, and does not make efficient use of the available bandwidth. In contrast, proportional fairness allows more resources to be used by non-minimal flows, and provides better efficiency. An assignment is proportionally fair if any change in the distribution of the assigned rates would result in the sum of the proportional changes to be non-positive. As we are interested in *goodput* maximization, we may use a concept of fairness that is defined directly in terms of utilities of the users (the goodputs, in our case), rather than in terms of the bandwidths assigned to them, as done in [19]. This is obtained as the solution of a utility maximization problem, where the performance index to be maximized is the product of the individual utilities, yielding a NBS. Proportional fairness agrees with the NBS in case that the object that is fairly shared is the bandwidth (and the minimum required rate is zero) [19, 22, 25].

Our goal is to find the NBS with respect to the bandwidth assignment for F classes of TCP connections, with $n_c^{(j)}$ connections per class, $j = 1,...,F$, with initial agreement point 0 and performance functions $f_j^{(i)}(.)$, $i = 1,...,n_c^{(j)}$, $j = 1,...,F$, given by the connections' optimal goodputs (in the sense of Section 2, i.e., by choosing the goodput-maximizing redundancy for each given value of bandwidth). More formally:

$$f_j^{(i)}\left(B_j^{(i)}\right) = \max_{r_j \in \left\{\mathcal{R}^{(1)},...,\mathcal{R}^{(P)}\right\}} \hat{T}_g^{(j)}\left(r_j, B_j^{(i)}\right) \tag{11}$$

where $B_j^{(i)}$ is the bandwidth assigned to the i-th connection in the j-th class. We assume in the following that the bandwidth assignments can be varied continuously. Functions $f_j^{(i)}(.)$ are given by the envelope of the goodput curves defined in Section 2 for each value of redundancy. By defining $\overline{B} = \left[\overline{B}_1,...\overline{B}_F\right]$; $\overline{B}_j = \left[B_j^{(1)},...B_j^{\left(n_c^{(j)}\right)}\right]$,

and $\overline{F} = \sum_{j=1}^{F} n_c^{(j)}$, the set $B_o = \left\{\overline{B} \in [0,W]^{\overline{F}} : \sum_{j=1}^{F} \sum_{i=1}^{n_c^{(j)}} B_j^{(i)} = W\right\}$ is a convex and compact subset of $R^{\overline{F}}$, wherein the individual goodputs satisfy the initial agreement point performance. If $f_j^{(i)}(.)$ were concave upper-bounded functions of $B_j^{(i)}$, by theorem (2.1) in [11], there would exist a unique NBS that satisfies

$$(P_{\overline{F}}) \qquad \max_{B \in B_o} \prod_{j=1}^{F} \prod_{i=1}^{n_c^{(j)}} f_j^{(i)}\left(B_j^{(i)}\right). \tag{12}$$

Moreover, if the functions $f_j^{(i)}(.)$ are injective on B_o, by theorem 2.2 in [25], problem $(P_{\overline{F}})$ is equivalent to

$$(P_{\overline{F}}') \qquad \max_{B \in B_o} \sum_{j=1}^{F} \sum_{i=1}^{n_c^{(j)}} \ln f_j^{(i)}\left(B_j^{(i)}\right). \tag{13}$$

In our situation, the part of functions $f_j^{(i)}(.)$ that derives from (1) (i.e., for values of $B_j^{(i)}$ such that the corresponding value of y – under given parameter values – is greater than 1) can be easily verified to be concave. Unfortunately, the same cannot be affirmed in general in the cases of bandwidth values yielding $0 \le y \le 1$. Therefore, in general, it is not true that the functions are concave in all the range of interest (whereas they are everywhere injective). This does no longer guarantee the uniqueness of the NBS. Nevertheless, it makes sense to investigate the behavior of a cost function like (13), and to analyze the goodput and fairness factors of the corresponding assignments. By discretizing the bandwidth space and applying dynamic programming to solve the problem, as we have done in the previous section, we would apply anyway a global search method.

The relevant NBS $\overline{B}^*$ is Generalized Proportionally Fair, in the sense of [19], i.e.,

$$\sum_{j=1}^{F} \sum_{i=1}^{n_c^{(j)}} \frac{f_j^{(i)}\left(B_j^{(i)}\right) - f_j^{(i)}\left(B_j^{(i)*}\right)}{f_j^{(i)}\left(B_j^{(i)*}\right)} \le 0, \tag{14}$$

for any other assignment $\overline{B}$. At this point, we can note that:

1. as the TCP connections in each class share the same buffer at the IP packet level, we have no explicit control over their bandwidth assignment, which is determined by the TCP congestion control algorithm, given the rate at which the IP buffer is being served;
2. in steady state, the effect of the congestion control algorithm is that of dividing the bandwidth available for a given class equally among its connections, thus achieving equal goodput for all connections in the same class.

Then, we can do the following, which corresponds, in general, to a suboptimal assignment. We let the bandwidth of the j-th class be

$$B_j = \sum_{i=1}^{n_c^{(j)}} B_j^{(i)*} \quad , \quad j = 1, \dots, F \tag{15}$$

In the following, we will refer to this as *GPF* assignment.

5 Numerical Results

5.1 The Satellite Network Characterization

The fully meshed satellite network considered uses bent-pipe geo-stationary satellite channels and operates in TDMA (Time Division Multiple Access) mode. The master station maintains the system synchronization, other than performing capacity allocation to the traffic stations. The master station performance is the same as the others; thus, the role of master can be assumed by any station in the system. This assures that the master normally operates in pretty good conditions because, when the current master's attenuation exceeds a given threshold, its role is assumed by another station that is in better conditions. To counteract the signal attenuation the system operates bit and coding rates changing. Traffic stations transmit in temporal slots assigned by the master. In order to compute the link budget, we considered a portion of the Ka band (20/30 GHz) transponder of the Eutelsat satellite Hot Bird 6, and took data from the file "*Hot Bird 6 data sheet.fm*", which is downloadable from [26]. We consider exploiting 1/4 of the transponder power. Our carrier is modulated in QPSK (quadrature phase shift keying) at 5, 2.5 or 1.25 Msymbols/s; thus, the resulting uncoded bit rates range from 10 to 2.5 Mbit/s. A 1/2 convolutional encoder/Viterbi decoder is employed, together with the punctured 3/4 and 7/8 codes for a possible total 12 combinations of bit/coding rates. The net value of about 7.5 dB of E_c/N_0 (C/N_0=77.5 dBs^{-1}), with the maximum modulation rate and the 7/8 coding rate, is assumed as the clear sky condition. In clear sky, after the Viterbi decoder, the bit error rate is about 10^{-7}. The *mbu* size, i.e., the minimum bandwidth unit that can be allocated, has been taken equal to 5 kbit/s; this value is referred to clear sky conditions.

The resulting net values of E_c/N_0 at the earth station's receiver input is given by

$$E_c/N_0 = C/N_0 - 10Log_{10}R - m_i \tag{16}$$

where R is the uncoded data bit rate in bit/s and m_i is the modem implementation margin (taken equal to 1 dB). We have assumed $b=1$ (no *Delayed ACKs* option) and T_o =1.5 s when using relation (1). We also considered l_s = 4608 bits (576 bytes), which is the default segment length assumed by sender and receiver TCPs, when no other agreement has been possible.

Actually, not all combinations of bit and coding rates must be probed to find the maximum goodput, because some of them result inefficient (i.e., they yield higher BER with the same redundancy). The possible cases are then limited to the following 7 ones: 10 Mbits/s, with code rates 7/8, 3/4, and 1/2; 5 Mbits/s, with code rates 3/4, and 1/2; 2.5 Mbits/s, with code rates 3/4 and 1/2. The uncoded case results inapplicable with the values of C/N_0 available in our situation, even in clear sky.

5.2 The Real Case Study

TEAM (*TCP Elephant bandwidth Allocation Method*) is the free software (available at the address http://www.isti.cnr.it/ResearchUnits/Labs/wn-lab/software-tools.html) specifically developed to implement the mechanisms proposed in this paper. The simulative analysis reported in the following is the result of the adaptive application of the previously described strategies, when real-life fading attenuation samples (in

Fig. 1) are dynamically applied. The attenuation data are taken from a real-life data set chosen from the results of the propagation experiment, in Ka band, carried out on the Olympus satellite by the CSTS (Centro Studi sulle Telecomunicazioni Spaziali) Institute, on behalf of the Italian Space Agency (ASI). We have preferred to use real fading traces rather than rely upon a model for rain fade generation, as no thoroughly satisfactory models have been devised so far.

The results have been obtained by means of the TEAM software and the ns2 simulator. TEAM calculates both the bandwidth allocations for each class, according to the chosen strategy, and the relative segment loss rates, according to the C/N_0 values of the SD pairs. Ns2 performs a dynamic simulation according both to the input trace files, performed by TEAM, and the attenuation patterns. The situation investigated in Fig. 2 shows a faded class and a clear-sky class out of ten active ones. Five of the classes are in clear sky condition, while the other five experience different patterns of fading. The total number of TCP connections is 30 and the number of connections per class is plotted in Table 1.

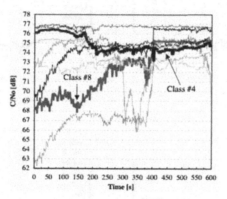

Fig. 1. C/N_0 values vs time represented for the ten classes

Table 1. Number of TCP Connections per Class

Class numb.	1	2	3	4	5	6	7	8	9	10
Connections	2	4	3	2	2	5	3	3	2	4

Three different allocation strategies are considered: *Merge*, *GPF* and *BER Threshold*, with the threshold set to 10^{-6}. Actually, the **Merge** strategy simply is a merge between the *Range* and the *Tradeoff* ones, in the sense that, for each record of the input file, the two strategies are separately run for a certain fairness factor threshold; the allocation values to be passed to ns2 are then chosen according to the strategy that better performs in terms of goodput factor.

Each simulation run gives an observation window of 600 s. In each chart, we trace the behaviors of the goodput and the segment loss rate, respectively, as functions of time. The allocations of the *Merge* strategy present many oscillations, as highlighted in Fig. 2 (a), in the range 400-600 s. This tendency is due to the optimal choice in a range in which the cost function is quite flat and variations of the bandwidth allocation do not produce a sensible change in the cost value in terms of goodput. However, oscillations are evidenced because the goodput values are averaged over a pretty short

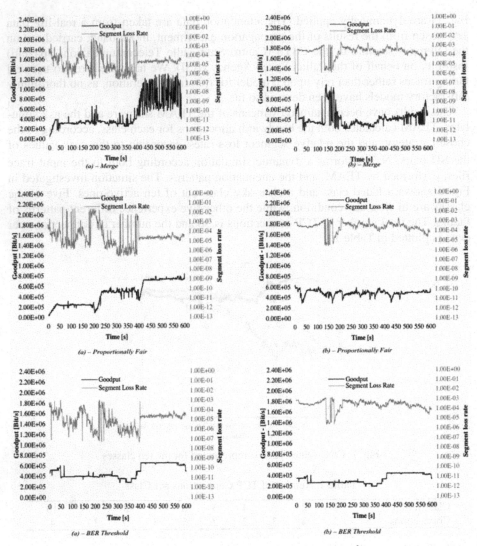

Fig. 2. *Merge, Proportionally Fair* and *BER Threshold* (threshold=10^{-6}) strategies for class #8 in fading (a) and for class #4 in clear sky (b)

interval (1 s). Figure 3 shows the percentage gain in total goodput, normalized with respect to the worst case, namely, the *BER Threshold* strategy with the threshold set to 10^{-7}. In the *Range* and *Tradeoff* strategies, the fairness factor is targeted to 0.85, which is the *GPF* value, averaged over the simulation time. The absolute goodput value per strategy is given by the sum of the goodputs of all classes obtained by averaging the values of the dynamic simulation over 600 s. All the allocation strategies are based on an optimization technique, so we do not reasonably expect great differences in performances. Anyway, the merging between *Range* and *Tradeoff*, in the dynamic case, gives a further gain over both the threshold policies, and over the *GPF* one, with a 10% gain over the latter and a gain of 25%-30% over the former ones.

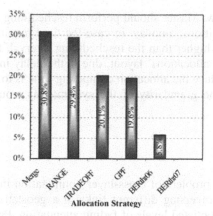

Fig. 3. Total goodput gain averaged over all classes compared with the worst case

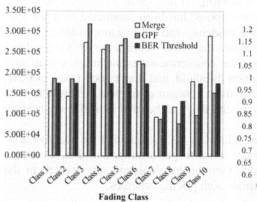

Fig. 4. Goodput per TCP class per connection for the most significant strategies

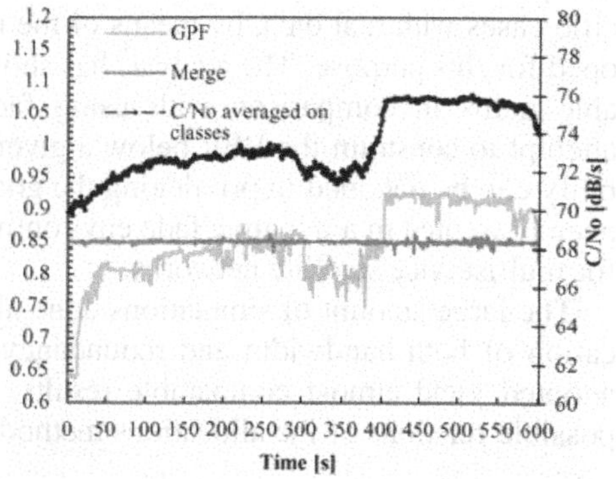

Fig. 5. Fairness factor of *Merge* and *GPF* strategy vs time, with an indicative value of C/N_0, averaged over classes

Figure 4 shows the goodput per connection (averaged over the simulation time) in the three most significant cases considered, for all classes, with the selected fairness factor of 0.85. The *Merge* strategy privileges the faded classes, i.e. classes 6, 7, 8, 9, 10, reducing the allocations and therefore the goodputs of the classes in clear-sky. In the attempt to equalize goodputs all over the classes also in the presence of hard fading condition, *Merge* gives better results than *GPF*. The goodput gain of the *Merge* strategy for the faded classes is furthermore evident if compared to that of the threshold policy. Finally, Figure 5 shows that the *Merge* strategy gives a further gain in term of fairness, compared with the *GPF* one, over a simulation window. The *Merge* strategy does not experience fall effects in fairness, as *GPF* does, and the fairness factor is kept to the target value.

Among all strategies, *Merge* indeed requires the highest computing power; however, in the presented case we did not observe any critical situations from the computing power point of view. Master should perform rescheduling at any change in traffic load or in fading conditions; anyway, in case the time required by the rescheduling computation would be higher than the rescheduling interval, this would yield only a temporary sub-optimal allocations' layout, due to the delay in the rescheduling time. Another solution to reduce the allocation computing time is to increase the size of the *mbu*, thus enlarging the granularity of the system; this would yield a sub-optimal solution as well.

6 Conclusions

We have considered a problem of cross-layer optimization in bandwidth assignment to TCP connections, traversing different links in a geostationary satellite network, characterized by differentiated levels of fading attenuation. Provided that there exists a tradeoff between bandwidth and data redundancy that influences TCP goodput, we have proposed optimization mechanisms that can be used to control the QoS of the TCP connections (in terms of goodput and fairness) that share a satellite bandwidth. The performance analysis of the methods proposed has been conducted on a few specific cases with real data, by means of the modeling and optimization software developed for this purpose. The analysis has shown that these mechanisms obtain considerable gains in comparison with using fade countermeasures techniques that only attempt to constrain the BER below a given threshold, and that a good range of flexibility can be obtained in privileging the goals of goodput or fairness. The results have been presented in a dynamic fade environment; they can be used in traffic engineering for multiservice satellite networks.

The large amount of simulations done allows us to conclude that the adaptive allocation of both bandwidth and redundancy is worth performing, and that all criteria adopted yield almost comparable results. Current developments are devoted to the possible relations of the allocations methods with pricing schemes.

References

1. S. Kota, M. Marchese, "Quality of Service for satellite IP networks: a survey", *Internat. J. Satell. Commun. Network.*, vol. 21, no. 4-5, pp. 303-348, July-Oct. 2003
2. A. Jamalipour, M. Marchese, H.S. Cruickshank, J. Neale, S.N. Verma (Eds.), Special Issue on "Broadband IP Networks via Satellites – Part I", *IEEE J. Select. Areas Commun.*, vol. 22, no. 2, Feb. 2004.
3. H. J. Chao, X. Guo, *Quality of Service Control in High-Speed Networks*, John Wiley & Sons, New York, NY, 2002.
4. F. Alagoz, D. Walters, A. Alrustamani, B. Vojcic, R. Pickholtz, "Adaptive rate control and QoS provisioning in direct broadcast satellite Networks," *Wireless Networks*, vol. 7, no. 3, pp. 269-281, 2001.
5. S. Shakkottai, T. S. Rappaport, P. C. Karlsson, "Cross-layer design for wireless networks", *IEEE Commun. Mag.*, vol. 41, no. 10, pp. 74-80, Oct. 2003.
6. V. Kawadia, P. R. Kumar, "A cautionary perspective on cross layer design", to appear in *IEEE Wireless Commun. Mag.*,
   On-line: http://black.csl.uiuc.edu/~prkumar/ps_files/cross-layer-design.pdf.

7. M. Ajmone Marsan, M. Garetto, P. Giaccone, E. Leonardi, E. Schiattarella, A. Tarello, "Using partial differential equations to model TCP mice and elephants in large IP networks", *Proc. IEEE Infocom 04*, Hong Kong, March 2004.
8. N. Celandroni, F. Potortì, "Maximising single connection TCP goodput by trading bandwidth for BER", *Internat. J. Commun. Syst.*, vol. 16, no. 1, pp. 63-79, Feb. 2003.
9. N. Celandroni, E. Ferro, F. Potortì, "Goodput optimisation of long-lived TCP connections in a faded satellite channel", *Proc. 59th IEEE Vehic. Technol. Conf.*, Milan, Italy, May 2004.
10. N. Celandroni, F. Davoli, E. Ferro, A. Gotta, "A dynamic cross-layer control strategy for resource partitioning in a rain faded satellite channel with long-lived TCP connections", *Proc. 3rd Internat. Conf. on Network Control and Engineering for QoS, Security and Mobility (Net-Con 2004)*, Palma de Mallorca, Spain, pp.83-96, Nov. 2004.
11. T. V. Lakshman, U. Madhow, "The performance of TCP/IP for networks with high bandwidth-delay products and random loss", *IEEE/ACM Trans. Networking*, vol. 5, no. 3, pp. 336-350, June 1997.
12. The Network Simulator – ns2. Documentation and source code from the home page: http://www.isi.edu/nsnam/ns.
13. J. Padhye, V. Firoiu, D. F. Towsley, J. F. Kurose, "Modeling TCP Reno performance: a simple model and its empirical validation", *IEEE/ACM Trans. Networking*, vol. 8, pp. 133-145, 2000.
14. http://wnet.isti.cnr.it/software/tgep.html.
15. "Q1401: K=7 rate 1/2 single-chip Viterbi decoder technical data sheet", *Qualcomm, Inc.*, Sept. 1987.
16. J. G. Proakis, *Digital communications*, McGraw Hill, New York, NY, 1995.
17. K. W. Ross, *Multiservice Loss Models for Broadband Telecommunication Networks*, Springer, London, UK, 1995.
18. N. Celandroni, F. Davoli, E. Ferro, "Static and dynamic resource allocation in a multiservice satellite network with fading", *Internat. J. Satell. Commun. Network.*, vol. 21, no. 4-5, pp. 469-487, July-Oct. 2003.
19. C. Touati, E. Altman, J. Galtier, "On fairness in bandwidth allocation", Research Report no. 4296, INRIA, Sophia Antipolis, France, Sept. 2001.
20. D. Bertsekas, R. Gallager, *Data Networks*. Prentice-Hall, Englewood Cliffs, NJ, 1987.
21. F. P. Kelly, "Charging and rate control for elastic traffic", *Europ. Trans. Telecommun.*, vol. 8, pp. 33-37, 1998.
22. A. Maulloo, F. P. Kelly, D. Tan, "Rain control in communication networks: shadow prices, proportional fairness and stability", *J. Oper. Res. Society*, vol. 49, pp. 237-252, 1998.
23. L. Massoulì, J. W. Roberts, "Bandwidth sharing and admission control for elastic traffic", *Telecommun. Systems*, vol. 15, pp. 185-201, 2000.
24. R. Mazumdar, L. G. Mason, C. Douligeris, "Fairness in network optimal flow control: optimality of product forms", *IEEE Trans. Commun.*, vol. 39, pp. 75-782, 1991.
25. R. R. Mazumdar, H. Yaïche, C. Rosenberg, "A game theoretic framework for bandwidth allocation and pricing in broadband networks", *IEEE/ACM Trans. Networking*, vol. 8, no. 5, pp. 667-677, 2000.
26. http://www.eutelsat.com/satellites/13ehb6.html

On the Performance of TCP over Optical Burst Switched Networks with Different QoS Classes*

Maurizio Casoni and Maria Luisa Merani

Department of Information Engineering, University of Modena and Reggio Emilia,
Via Vignolese, 905, 41100 Modena, Italy
{casoni.maurizio,merani.marialuisa}@unimore.it

Abstract. In this paper we investigate the end-to-end performance of an optical burst switched (OBS) network that adopts a core router architecture with no fiber delay lines and limited set of wavelength converters. In particular, by evaluating both the node and a reference network performance in terms of burst blocking probability we want to determine the end-to-end performance when a transport protocol like TCP is employed. Core node analysis is employed to identify the blocking experienced by incoming bursts, under the assumption of exponentially distributed burst interarrival times and arbitrarily distributed burst durations.

1 Introduction

The huge increase of capacity in optical transmissions allowed by Dense Wavelength Division Multiplexing (DWDM), capable of accommodating hundreds of wavelengths per fiber, and the advances in integrated optics, for passive and active optical components design, have pushed for the study and the development of transfer modes suitable to deal with very high bit rates. The ultimate goal is the development of a fully optical Internet, where signals carried within the network never leave the optical domain [1]. A first important step in this direction is to have optical networks transparent at least for data, with the control part converted and processed in electronics. In this paper the focus is on a particular transfer mode, the Optical Burst Switching (OBS) solution [2–4] where data never leave the optical domain: for each data burst assembled at the network edge, a reservation request is sent as a separate control packet, well in advance, and processed within the electronic domain. The key idea behind OBS is to dynamically set up a wavelength path whenever a *large* data flow is identified and needs to traverse the network: a separate control packet, carrying relevant forwarding information, therefore precedes each burst by a basic offset time. This offset time is set to accommodate the non-zero electronic processing time of control packets inside the network. In addition, offset time allows the core switches to be bufferless, avoiding thus the employment of optical memories, e.g. fibre delay lines, required on the other hand by optical packet switching.

* This work has been partially supported by MIUR within the framework of the national project "INTREPIDO: Traffic Engineering and Protection for IP over DWDM".

M. Ajmone Marsan et al. (Eds.): QoS-IP 2005, LNCS 3375, pp. 574–585, 2005.

In a previous paper [5] the authors have evaluated the performance of OBS networks, when the Just Enough Time (JET) reservation mechanism is employed, in terms of burst loss and delay with several mechanisms for service differentiation. This study was basically carried on by simulation evaluating separately the edge switches with the assembly function from the core network.

The aim of this paper is to present some results on the performance of TCP over an OBS network. We also propose a simple yet significant analytical model for bufferless core switches. Limited range wavelength converters [6] have been investigated in [7] and [8]. Here we investigate the behavior of a core node, whose key feature is to be equipped with a limited set of full range optical wavelength converters. The metric figure we focus onto is the burst blocking probability, where the event of blocking occurs whenever the request of bandwidth at the traversed optical node cannot be fulfilled.

The rest of the paper is organized as follows. In Section II we present the examined architecture and describe the blocking probability analysis while in Section III the core network with the related routing algorithms is shown. Section IV provides the framework for the end-to-end performance evaluation when a transport protocol like TCP is employed. Analytical and simulation results are then collected in Section V while concluding remarks are in Section VI.

2   Core Node Analysis

The core optical router examined is equipped with $M \times M$ optical interfaces capable of supporting N wavelengths each. We suppose that the size of the pool of converters is limited and that the optical node is bufferless, i.e., no fiber delay lines (FDLs) are present in order to resolve contention for an output fiber, output wavelength (Figure 1). It is however reasonable to assume the use of a set of input FDLs [9], whose exclusive task is to re-align the OBS data burst and its control packet, so as to guarantee a minimal offset time at any intermediate node.

Let wc be the number of wavelength converters the node is equipped with. Whenever an incoming burst enters the node on the λ_j wavelength ($j \in 1 \ldots N$),

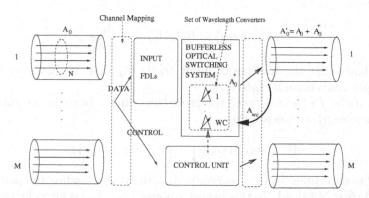

Fig. 1. General core router architecture.

an attempt is made to forward it to the desired output fiber on that same wavelength λ_j. If this is not available, the core router looks for a free output wavelength λ_i, $i \neq j$, on that same output fiber, but has to resort to a converter, in order to successfully forward the burst. We aim at analytically determining the blocking probability a generic burst experiences in traversing such a bufferless optical node, when a limited number of wavelength converters is available.

The assumptions we work under can be summarized as follows:

– equal traffic on each incoming wavelength
– the burst arrival process to the core router is modelled by a Poisson process, whereas the burst duration is arbitrary.

Let $M/G/m/m$ be the queueing system with exponentially distributed interarrival times, service times obeying a generic distribution, m servers and no queueing available. For such system, denote by $B(m, A) = \frac{A^m/m!}{\sum_{i=1}^{m} A^i/i!}$ the blocking probability a generic customer experiences, provided A represents the traffic load, defined as the ratio between the mean service time and the mean interarrival time.

Let A_0 be the (Poisson) traffic on the single *incoming* wavelength. As all routes are chosen with equal probability, A_0 is uniformly split among the M output fibers. Recall that the burst that cannot be forwarded onto the same output wavelength, as this is found busy, attempts to grab an available – if any – wavelength converter and to leave the node onto the same output fiber, but on a different wavelength.

Indicate by A_0^+ the fraction of A_0 traffic that, once rejected, finds a free wavelength converter and can therefore be redirected on any alternative wavelength of the same output fiber. Let A_0' be the traffic load on each *outgoing* wavelength, given by

$$A_0' = A_0 + A_0^+ . \tag{1}$$

Under the hypothesis of Poisson traffic on the outgoing wavelengths, $Pr\{\lambda_j\_busy\}$, the probability of finding the desired output wavelength busy, is expressed by the Erlang B formula referring to the $M/G/1/1$ queueing system loaded by A_0', i.e.,

$$Pr\{\lambda_j\_busy\} = B(1, A_0') = \frac{A_0'}{1 + A_0'} . \tag{2}$$

It follows that the fraction of input traffic A_0 that attempts to resort to wavelength conversion is $A_0 B(1, A_0')$.

Indicate by $Pr\{no\_wc\_free\}$ the probability that a burst is not able to find a free wavelength converter: then,

$$A_0^+ = A_0 B(1, A_0')(1 - Pr\{no\_wc\_free\}) . \tag{3}$$

In order to determine $Pr\{no\_wc\_free\}$, A_{wc}, the traffic loading the pool of converters, is first obtained. To this regard, observe that A_{wc} is given by the overall traffic that cannot be immediately accommodated on the desired output fiber because of the unavailability of the same input wavelength, i.e.,

$$A_{wc} = M \times N \times A_0 B(1, A_0') = M \cdot N \frac{A_0 A_0'}{1 + A_0'}. \tag{4}$$

If the traffic loading the pool of converters is assumed Poisson as well, $Pr\{no\_wc\_free\} = B(wc, A_{wc})$, so that, recalling (2) and (3), (1) becomes

$$A_0' = A_0 + A_0 \cdot \frac{A_0'}{1 + A_0'} \left(1 - B \left(wc, M \cdot N \frac{A_0 A_0'}{1 + A_0'} \right) \right). \tag{5}$$

Expression (5) is an equation in the unknown A_0': for any given value of A_0, it can be numerically solved, adopting, e.g., the iterative procedure briefly described in the Appendix.

Having determined the effective load A_0' onto the generic output wavelength, next step is to evaluate the blocking probability experienced by a burst which arrives the node on the incoming wavelength λ_j of a generic fiber. Given that the same wavelength on the desired output fiber is busy, the burst can be blocked under two circumstances: either no wavelength converter is available, or there is at least one converter free, but all the remaining $(N - 1)$ wavelenghts of the fiber are already in use. Hence:

$$P_b = Pr\{\lambda_j\_busy\} \cdot \{Pr\{no\_wc\} + Pr\{wc\} \cdot$$
$$\cdot Pr\{all\_remaining\_(N - 1)\_wavelengths\_busy\}\}. \tag{6}$$

The circumstance that all the remaining $(N - 1)$ wavelengths are taken occurs with probability:

$$Pr\{all\_(N - 1)\_wavelengths\_busy\} = B(N - 1, (N - 1) \cdot A_0'), \tag{7}$$

so that replacing (7), as well as (2) and $Pr\{no\_wc\_free\}$ expression into the burst blocking probability in (6) leads to:

$$P_b = \frac{A_0'}{1 + A_0'} \cdot \{B(wc, A_{wc}) +$$
$$+ [1 - B(wc, A_{wc})] \cdot B(N - 1, (N - 1) \cdot A_0')\}. \tag{8}$$

In Section V this outcome will be checked against some simulative results, with the following objective: demonstrate that it is feasible and accurate to analytically predict the optical node behavior.

3   Routing and Network Topology

Regarding the network as a whole, information flows are made of bursts routed within the network by means of the Dijkstra algorithm. Bursts are created at the

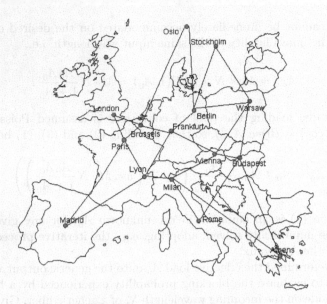

Fig. 2. The reference network.

edge nodes according to three, for simplicity but in general k, classes of service, class 1 carrying time-sensitive data, class 2 and 3 loss-sensitive data. Class 1 and 2 bursts can make use of the limited set of wavelength converters within the core routers. Routing is modified in order to better meet the performance required by each class and a simple QoS routing criterion is introduced: class 2 and 3 bursts are allowed to be deflected and thus re-routed in case of unavailability of a wavelength in the desired output fiber and this may lead to variable edge-to-edge delays. However, this solution is better than increasing the loss of this type of bursts. It is worth noting that variable delays can be managed by the node architecture assumed (Figure 1), employing input FDLs.

In summary, class 1 bursts are given the highest priority through an additional extra-offset and the use of wavelength converters; class 2 bursts have medium priority by using the converters; class 3 bursts have low priority since they have no extra offset and cannot exploit wavelength conversion; on the other hand, class 2 and 3 can use alternative sub-optimal variable delay paths. Also, p_1, p_2 and p_3 represent the occurrence probabilities of bursts of the three classes.

In this paper the network considered covers most European countries (Figure 2). Each node operates as a core router and, in addition, nodes A (London), B (Oslo) and C (Stockolm) are sources of information flows, whereas nodes O (Madrid), P (Rome) and Q (Athens) are the possible destinations; moreover, no flow is supposed to enter or leave the network at intermediate steps.

Section V will show the performance, in terms of overall edge-to-edge burst blocking probability for the three classes of bursts in the OBS network described above, when each node implements the JET reservation mechanism with a limited set of wavelength converters available, under different mixtures of traffic.

4 End to End Performance

The throughput is here studied on end-to-end basis when TCP is assumed as transport layer protocol. In fact, in wide area data networks like Internet congestion control mechanisms have a fundamental role for the global functioning. TCP is a reliable window-based acknowledgment-clocked flow control protocol, thought to avoid to overload the network and to react to a possible congestion at network level. In the system under investigation TCP Reno is assumed to be employed by hosts. TCP Reno is modeled following the approach detailed in [10] where the throughput (bit/s) is approximated by:

$$ Thr_{TCP} = \frac{MSS}{RTT\sqrt{\frac{2bp}{3}} + T_0 min\left(1, 3\sqrt{\frac{3bp}{8}}\right) p(1 + 32p^2)} \tag{9}$$

being MSS the maximum segment size expressed in bits, RTT the round trip time, p the segment loss probability, T_0 the time out and b the number of packets acknowledged by ACKs.

The performance of TCP over OBS networks have been studied in some previous works [11, 12] but it can still be considered an open issue and thus a challenge for the research community. Here we want to add some thoughts to the discussion. Considering the OBS network previously described where the core nodes are bufferless, the edge nodes where the bursts are generated represent the place of greatest impact for the end-to-end performance. Of course, this depends on the burst assembly algorithm adopted but anyway the delay introduced must be taken into account.

Let us assume, as classified in [11], to have only slow TCP sources which emit at most one segment during the interval $(0, T_{max_j})$, where T_{max_j} is the maximum class j assembly time, meaning that a burst is anyway emitted regardless the amount of collected packets once the first arrived packet of the burst has already experienced a delay equal to T_{max_j}. This also means that at most one segment for each connection is contained in a burst generated by edge nodes and injected into the OBS network. Therefore, even if approximated, for this type of sources the segment loss probability p can be assumed equal to the burst blocking probability.

Now, considering the reference network topology (Figure 2) an average link length of 800 Km can be assumed and the number of hops, N_{hops}, is in the range [3..5]. Since the light propagation speed in the fibers is roughly 70% the speed of the light in the vacuum, the propagation delay for each hop, T_{hop}, is 4 ms. Concerning the edge nodes, the general class j assembly time T_{ass_j} is upper bounded by T_{max_j}. Therefore the one-way delay edge-to-edge, from entering the ingress edge to leaving the egress edge, $T_{e2e1way}$ in the OBS network is:

$$ T_{e2e1way} = T_{ass_j} + N_{hops} \times T_{hop} + T_{disass_j} \tag{10}$$

which can be bounded to 30 ms, when T_{max_j} is in the msecs range. If in addition the network has a symmetric behavior RTT is approximately 60 ms. Actually,

RTT has also to consider the delays given by the access networks before entering the OBS network: this means that 60 ms can be considered as a kind of lower bound.

In next Section the behavior of TCP throughput will be shown referring to the above assumptions for p and *RTT*.

5   Numerical Results

The results we first present refer to various symmetric $M \times M$ OBS core nodes featuring a different number of incoming and outgoing fibers. Each of the outgoing fibers is chosen with equal probability $p_F = 1/M$; $N = 8$ wavelengths per fiber are considered. The performance of the optical routers has been investigated by analysis and by simulation, relying upon an *ad-hoc* event-driven C++ object oriented simulator. In order to describe the burst duration, as well as the interarrival time between bursts, the simulator adopts an ON/OFF model with exponentially or Pareto distributed ON (burst duration) and OFF (interarrival time) periods. Figure 3 reports the burst blocking probability P_b, when M is varied from 8 to 16, 32 and 64, for a number of converters equal to $wc = 8$. Note how satisfyingly analysis matches simulation, where simulation results are with ON/OFF sources with exponentially distributed ON and OFF periods. Regarding performance, we outline that, once wc is fixed, the blocking worsens for larger values of M: indeed, an increasing fraction of the overall input traffic, $M \times N \times A_0$, is unable to find one – out of the very few available – free converters. It is also interesting to observe that the ultimate upper bound to this set of curves is given by $P_b = \frac{A_0}{1+A_0}$, i.e., the blocking probability experienced when no converters are available at the optical node. Next, Figure 4 shows the burst

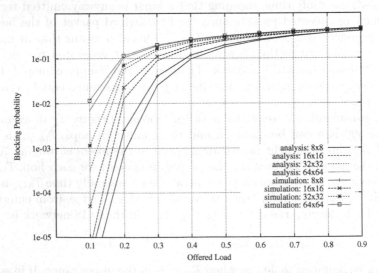

Fig. 3. Burst blocking probability from analysis and simulation as a function of the offered load, when varying the optical node size. $M = 8, 16, 32$ and 64, $wc = 8$.

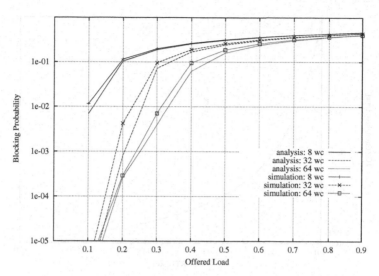

Fig. 4. Burst blocking probability from analysis and simulation varying the number of available wavelength converters for a 64 × 64 node.

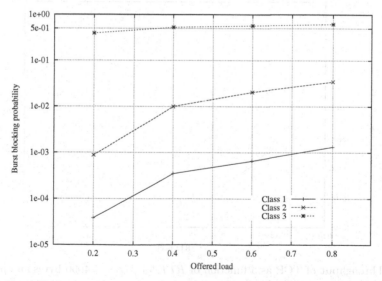

Fig. 5. Total End-to-end burst blocking probability as a function of the offered load for the three burst classes, with reference to the (B,P) pair.

blocking probability values obtained by analysis and simulation for an $M \times M$ node, $M = 64$, when wc is varied: $wc = 8, 32, 64$. Again, analysis closely matches the simulation results. Next, a few words about performance. It is possible to affirm that the optical router provides, by properly limiting its load, reasonable values of P_b: when, e.g., $wc = 64$, P_b is approximately 2×10^{-4} for $A_0 < 0.2$, 10^{-3} for $A_0 < 0.25$ and 10^{-2} for $A_0 < 0.31$.

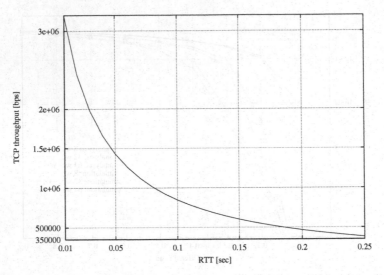

Fig. 6. Throughput of TCP as a function of RTT for $MSS = 1500$ bytes and $p = 1\%$.

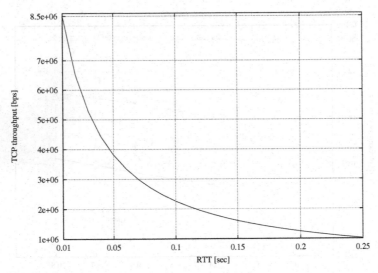

Fig. 7. Throughput of TCP as a function of RTT for $MSS = 4000$ bytes and $p = 1\%$.

Consider now the network reported in Figure 2. The following values for p_1, p_2 and p_3 are considered, $(0.5, 0.2, 0.3)$; moreover, incoming traffic is supposed to have ON/OFF periods with OFF exponentially distributed and ON following the Pareto distribution with the same parameters as above. It is worth reminding that only class 1 and 2 bursts exploit a set of 20 wavelength converters; however, class 2 and 3 can be re-routed in case of unavailability of wavelengths on the outgoing fibre along the least-cost path. Figure 5 shows the total burst blocking probability for the three burst classes having node B as source and node P as

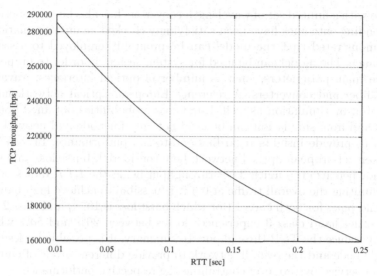

Fig. 8. Throughput of TCP as a function of RTT for $MSS = 4000$ bytes and $p = 10\%$.

destination. In order to have loss values in the range of 10^{-3} for class 1, 1% for class 2 and 50% for class 3, the overall load ρ must be less than 0.4.

Let us now discuss the consequences of the above values of burst blocking probability and network design on end to end performance. Figure 6 shows the throughput of TCP Reno given by (9) as a function of RTT for $T_0 = 1.0$ s, $b = 2$, $MSS = 1500$ bytes and $p = 1\%$. As mentioned in Section IV, for slow TCP sources p can be approximated with the burst blocking probability. Also, the RTT values of interest fall reasonably in the [60..200] ms range, depending on the type of access network. This figure says that even if a very high speed core network is employed the best we can get is a throughput of 1.3 Mbps and it remarkably decreases at 750 kbps as soon as the RTT doubles, or at 500 kbps when the RTT becomes three times, i.e. 180 ms. In order to increase the TCP throughput, some authors [12] have considered a bigger value for MSS. In particular, they suggested $MSS = 4000$ bytes. Of course this impact the TCP implementation and the operating systems but, if widely adopted, can lead to performance improvements. As a matter of fact, Figure 7 reports the TCP throughput with the same previous operating conditions except for $MSS = 4000$. The throughput is roughly three times better than for $MSS = 1500$ for the values within the considered range. Last, Figure 8 shows the effects of a degraded burst blocking probability: if it raises to 10% the TCP performance deteriorates to such values to make the optical backbone much less attractive.

6   Conclusions

This work has evaluated the end-to-end performance of TCP over OBS networks. A model to analytically predict the burst blocking probability of an OBS bufferless node with a limited set of wavelength converters has also been proposed. The corresponding outcomes have been satisfyingly checked against simulation, and have demonstrated that the model can be profitably employed to assess node performance. The model can be used for system design, in order to properly size the node main parameters, such as number of optical interfaces, wavelengths on each fiber and converters. A reference European optical network has been investigated by simulation as well. Burst loss probabilities obtained both from analysis and from simulation can be used not only for node and network design but also to provide insights regarding end-to-end performance. In a WAN like the reference European optical network here considered, burst loss values up to 1% do not penalize too much TCP throughput being the RTT values not critical. By limiting the overall traffic at 0.4 it is possible to achieve high bandwidth end-to-end pipes for both class 1 (for which burst loss is 10^{-3}) and class 2 bursts. On the other hand class 3 experiences losses between 20% and 50% which remarkably reduce the TCP throughput. In summary, it is possible to design both the single node and the overall network to provide different levels of end-to-end quality of service by properly controlling the respective performance.

References

1. B. Bostica, F. Callegati, M. Casoni, C. Raffaelli, "Packet Optical Networks for High Speed TCP-IP Backbones" – *IEEE Communications Magazine*, January 1999, pp. 124-129.
2. J.S. Turner, "Terabit burst switching", *Journal of High Speed Networks*, Vol. 8, No. 1, January 1999, pp. 3-16.
3. C. Qiao, M. Yoo, "Optical Burst Switching (OBS) – a New Paradigm for an Optical Internet," *Journal of High Speed Networks*, No.8, pp.69-84, 1999.
4. M. Yoo, C. Qiao, S. Dixit, "QoS Performance of Optical Burst Switching in IP-Over-WDM Networks," *IEEE Journal on Selected Areas in Communications*, Vol.18, No.10, pp.2062-2071, October 2000.
5. M. Casoni, M.L. Merani, "Resource Management in Optical Burst Switched Networks: Performance Evaluation of a European Network", Proc. of 1st International Workshop on Optical Burst Switching, October 16 2003, Dallas, (Texas, USA).
6. B.Ramamurthy, B. Mukherjee, "Wavelength Conversion in WDM Networking", *IEEE Journal on Selected Areas in Communications*, Vol.16, No.7, September 1998, pp.1061-1073.
7. T.Y. Chai, T. Cheng, C. Lu, G. Shen, S.K. Bose, "On the Analysis of Optical Cross-Connects with Limited Wavelength Conversion Capability", Proc. of IEEE ICC 2001, Helsinki (Finland).
8. Z. Zhang, Y. Yang, "Perfomance Modeling of Bufferless WDM Packet Switching Networks with Wavelength Conversion", Proc. of IEEE Globecom 2003, San Francisco (U.S.A.).
9. Y. Xiong, M. Vandenhoute, H.C. Cankaya, "Control Architecture in Optical Burst-Switched WDM Networks", *IEEE Journal on Selected Areas in Communications*, Vol.18, No.10, October 2000, pp.1838-1851.
10. J. Padhye, V. Firoiu, D. F. Towsley, J. F. Kurose, "Modeling TCP Reno Performance: A Simple Model and its Empirical Validation", *IEEE/ACM Trans. on Networking*, vol.8, no.2, pp.133-145, April 2000.

11. A. Detti, M. Listanti, "Impact of Segment Aggregation on TCP Reno Flows in Optical Burst Switching Networks", Proc. of IEEE Infocom 2002, 23-27 June 2002, New York (U.S.A.).
12. S. Gowda, R. K.Shenai, K. M.Sivalingam, H.C. Cankaya, "Perfomance Evaluation of TCP over Optical Burst-Switched (OBS) WDM Networks", Proc. of ICC 2003, 11-15 May 2003, Anchorage (U.S.A.).

Appendix

The iterative procedure starts by setting $A_0' = A_0$, $A_0^+ = 0$, $eps = 0.001$. Then, a loop begins, which terminates as soon as the required accuracy – dictated by eps – is reached.

do{
$A_{ref} = A_0^+$
$Pr\{\lambda_j\_busy\} = \frac{A_0'}{1+A_0'}$
$A_{wc} = M \times N \times A_0 \times Pr\{\lambda_j\_busy\}$
$A_0^+ = A_0 \times Pr\{\lambda_j\_busy\} \times (1 - B(wc, A_{wc}))$
$A_0' = A_0 + A_0^+$
}while $(A_0^+ - A_{ref} > eps)$

Loss Differentiation Schemes for TCP over Wireless Networks[*]

Fabio Martignon and Luigi Fratta

DEI, Politecnico di Milano, Piazza L. da Vinci 32, 20133 Milan, Italy
{martignon,fratta}@elet.polimi.it

Abstract. The use of loss differentiation schemes within the congestion control mechanism of TCP was proposed recently as a way of improving TCP performance over heterogeneous networks including wireless links affected by random loss. Such algorithms provide TCP with an estimate of the cause of packet losses. In this paper, we propose to use the Vegas loss differentiation algorithm to enhance the TCP NewReno error-recovery scheme, thus avoiding unnecessary rate reduction caused by packet losses induced by bit corruption on the wireless channel.

We evaluate the performance of the so-enhanced TCP NewReno source (TCP NewReno-LP) with both extensive simulation and real test bed measurements, and we compare it with that achieved by existing solutions, namely TIBET [1], TCP Westwood [2] and the standard TCP NewReno. For that purpose, Linux implementations of TCP NewReno-LP, TIBET and TCP Westwood have been developed and compared with an implementation of NewReno.

We show that TCP NewReno-LP achieves higher goodput over wireless networks, while guaranteeing fair share of network resources with classical TCP versions over wired links. Finally, by studying the TCP behavior with an ideal scheme having perfect knowledge of the cause of packet losses, we provide an upper bound to the performance of all possible schemes based on loss differentiation algorithms. The proposed TCP enhanced with Vegas loss differentiation algorithm well approaches this ideal bound.

1 Introduction

The Transmission Control Protocol (TCP) performs well over the traditional network, that is constructed by purely wired links. However, as wireless access networks (like cellular networks and wireless local area networks) are growing rapidly, a heterogeneous environment will get wide deployment in the next-generation wireless networks, thus posing new challenges to the TCP congestion control scheme.

The performance degradation of existing versions of TCP in wireless and wired-wireless hybrid networks is mainly due to their lack of the ability to differentiate the packet losses caused by network congestions from the losses caused by wireless link errors. Therefore, the standard TCP congestion control mechanism reduces, even when not necessary, the transmission rate. To avoid such limitation and degradation, several schemes have been proposed and are classified in [3].

---

[*] This work has been partially supported under the grant of MURST Tango project.

M. Ajmone Marsan et al. (Eds.): QoS-IP 2005, LNCS 3375, pp. 586–599, 2005.

A possible approach to this problem is to modify the TCP congestion control scheme implementing explicit bandwidth estimation [1, 2] and loss differentiation schemes [4–8]. Note that these two approaches are deeply interwined, as we showed in [4] that the most efficient loss differentiation algorithms base their functioning on bandwidth measurements to estimate the cause of packet losses.

We analyzed and discussed in detail the first approach in [1] where we proposed TIBET, a new bandwidth estimation algorithm that allows to obtain accurate and unbiased estimates of the TCP trasmission rate.

In this paper, we propose a new TCP scheme, called TCP NewReno-LP, which is capable of distinguishing the wireless packet losses from the congestion packet losses, and reacting accordingly. TCP NewReno-LP implements an enhanced error-recovery scheme as proposed in [4–8], based on the Vegas Loss Predictor (LP) [9], and it avoids unnecessary rate reduction caused by packet losses induced by bit corruption on the wireless channel. TCP NewReno-LP can be implemented by modifying the sender-side only of a TCP connection, thus allowing immediate deployment in the Internet.

We evaluate the performance of TCP NewReno-LP with both simulation and real test bed measurements. For that purpose, Linux implementations of TCP NewReno-LP, TIBET and TCP Westwood have been developed and compared with an implementation of NewReno. We compare the performance of TCP NewReno-LP with that achieved by TIBET, TCP Westwood and standard TCP NewReno, showing how the proposed enhanced TCP source achieves higher goodput over wireless networks, while guaranteeing fair share of network resources with current TCP versions over wired links.

We also evaluate the behavior of TCP enhanced with ideal loss prediction, assuming perfect knowledge of the cause of packet losses, thus providing an upper bound to the performance of all possible schemes based on different loss differentiation algorithms. The TCP enhanced with Vegas loss predictor well approaches this ideal bound.

The paper is structured as follows: Section 2 presents TCP NewReno-LP. Section 3 presents the simulation network model. Section 4 analyzes the accuracy of TCP NewReno-LP in estimating the cause of packet losses under several realistic network scenarios. Sections 5 and 6 measure the performance of TCP NewReno-LP in terms of achieved goodput, friendliness and fairness, using both simulation and real Test bed scenarios, respectively. The performance of TCP NewReno-LP is compared to existing TCP versions, like TCP NewReno, TIBET and TCP Westwood [2, 10], over heterogeneous networks with both wired and wireless links affected by independent and correlated packet losses. Finally, Section 7 concludes this paper.

2   TCP NewReno Enhanced with Vegas Loss Predictor

The Vegas loss predictor [9] decides whether the network is congested or uncongested based on rate estimations. This predictor estimates the cause of packet losses based on the parameter V_P, calculated as:

$$V_P = (\frac{cwnd}{RTT_{min}} - \frac{cwnd}{RTT}) \cdot RTT_{min} \tag{1}$$

where $cwnd/RTT_{min}$ represents the *expected* flow rate and $cwnd/RTT$ the *actual* flow rate; $cwnd$ is the congestion window and RTT_{min} is the minimum Round Trip Time measured by the TCP source.

Given the two parameters α and β [segments], when $V_P \geq \beta$, the Vegas loss predictor assumes that the network is congested; when $V_P \leq \alpha$, possible losses will be ascribed to transmission random errors. Finally, when $\alpha < V_P < \beta$, the predictor assumes that the network state is the same as in the previous estimation.

We propose to use this predictor within the congestion control of a TCP source as follows: when the source detects a packet loss, i.e. when 3 duplicate acknowledgements are received or a retransmission timeout expires, the Vegas predictor is asked to estimate the cause of the packet loss.

If the loss is classified as due to congestion, the TCP source reacts exactly as a classical TCP NewReno source [11], setting the slow start threshold (*ssthresh*) to half the current flight size. This allows TCP NewReno-LP to behave as fairly as the standard TCP protocol in congested network environments.

On the contrary, if the loss is classified as due to random bit corruption on the wireless channel, the *ssthresh* is first updated to the current flight size value.

Then, if the packet loss has been detected by the TCP source after the receipt of 3 duplicate ACKs, the TCP sender updates the *cwnd* to *ssthresh* + 3 Maximum Segment Sizes (MSS) and enters the fast retransmit phase as the standard TCP NewReno. This allows the source to achieve higher transmission rates upon the occurrence of wireless losses, if compared to the blind halving of the transmission rate performed by current TCP implementations.

If the packet loss has been detected by the TCP source after a retransmission timeout expiration, the congestion window is reset to 1 segment, thus enforcing a friendly behavior of the TCP source toward current TCP implementations.

3   Simulation Network Model

The TCP NewReno-LP scheme described in the previous Section was simulated using the Network Simulator package (ns v.2 [12]), evaluating its performance in several scenarios as proposed in [13].

We assume, as in the rest of the paper, that the Maximum Segment Size (MSS) of the TCP source is equal to 1500 bytes, and that all the queues can store a number of packets equal to the bandwidth-delay product. The TCP receiver always implements the Delayed ACKs algorithm, as recommended in [14].

The network topology considered in this work is shown in Fig. 1. A single TCP NewReno-LP source performs a file transfer. The wired link $S \longleftrightarrow N$ has capacity C_{SN} and propagation delay τ_{SN}. The wireless link $N \longleftrightarrow D$ has capacity C_{ND} and propagation delay τ_{ND}.

We considered two different statistical models of packet losses on the wireless link: independent and correlated losses. To model independent packet losses, the

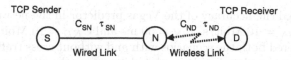

Fig. 1. Network topology in simulations for TCP performance evaluation.

link drops packets according to a Poisson process, causing a packet error rate (PER) in the 10^{-5} to 10^{-1} range.

To account for the effects of multi-path fading typical of wireless environments, we also considered links affected by correlated errors. From the existing literature [15], we modeled the wireless link state (*Good* or *Bad*) with a two-state Markov chain. The average durations of the *Good* and *Bad* states are equal to 1 and 0.05 seconds, respectively. In the *Good* state no packet loss occurs, while we varied the packet error rate in the *Bad* state from 0% to 100%, to take into account different levels of fading.

Finally, we considered two different traffic scenarios: in the first one, no cross traffic is transmitted over the wired link $S \longleftrightarrow N$; in the second scenario, the TCP source shares the wired link with 30 UDP sources having the same priority as the TCP source. Each UDP source switches between ON and OFF states, with Pareto-distributed periods having shape parameter equal to 1.5 and mean durations equal to 100 ms and 200 ms, respectively. During the ON state, each source transmits packets with 1500 byte size at constant bit rate equal to R_{UDP} Mbit/s, while in OFF period the UDP sources do not transmit any packet. In every network scenario with cross traffic on the wired link, the value of R_{UDP} is chosen to leave to the TCP source an available bandwidth that varies randomly during the simulation, with an average equal to half the bottleneck capacity.

4   Accuracy Evaluation

The key feature of Loss Predictor schemes (LP) is to be accurate in estimating the cause of packet losses, as the TCP error-recovery algorithm we introduced in Section 2, based on the Vegas Predictor, reacts more gently or more aggressively than existing TCP sources depending on the LP estimate. Evidently, when the packet error rate is low and most of packet losses are due to congestion, LP accuracy in ascribing losses is necessary to achieve fairness and friendliness with concurrent TCP flows. On the other hand, when the packet error rate is high such as on wireless links, LP accuracy is necessary to achieve higher goodput, defined as the bandwidth actually used for successful transmission of data segments (payload).

TCP sources detect *loss events* based on the reception of triple duplicate acknowledgements or retransmission timeout expirations. We define *wireless loss* a packet loss caused by the wireless noisy channel; a *congestion loss* is defined as a packet loss caused by network congestion.

The overall *accuracy* of packet loss classification achieved by a loss predictor is thus defined as the ratio between the number of correct packet loss classifications and the total number of loss events.

We measured the accuracy of the Vegas predictor in the network topology of Fig. 1, with $C_{SN} = 10$ Mbit/s, $\tau_{SN} = 50$ ms and $C_{ND} = 10$ Mbit/s, $\tau_{ND} = 0.01$ ms. We considered both the scenarios with and without cross traffic on the wired link and both uncorrelated and correlated errors on the wireless link.

As explained in Section 2, the Vegas predictor detects congestion and wireless losses based on two thresholds, α and β. We tested several values for the parameters α and β and we found the best performance for the accuracy of the Vegas predictor for $\alpha = 1$ and $\beta = 3$. We presented a detailed analysis of the accuracy of the Vegas predictor and other loss differentiation algorithms in [4]. In this paper, we summarize only some of the most significant results.

Fig. 2(a) shows the accuracy of packet loss classifications of the Vegas predictor with these parameters as a function of the packet error rate in the scenario with no cross traffic and independent packet losses. Each accuracy value has been calculated over multiple file transfers, with very narrow 97.5% confidence intervals [16]. The vertical lines reported in all Figures represent such confidence intervals for each accuracy value.

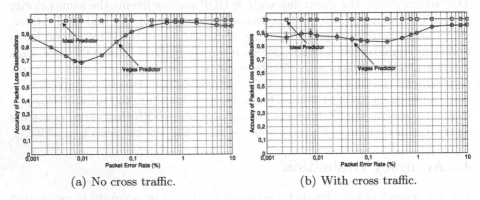

(a) No cross traffic.　　　　　　　　(b) With cross traffic.

Fig. 2. Accuracy of classification of packet losses for the Vegas loss predictor as a function of PER in two scenarios: (a) no cross traffic on the wired link (b) with cross traffic on the wired link.

Fig. 2(b) shows the accuracy for the Vegas predictor in the scenario with cross traffic on the wired link ($R_{UDP} = 0.5$ Mbit/s). We observed that the Vegas predictor is very accurate in discriminating the cause of packet losses for the whole range of packet error rates we considered.

Finally, Fig. 3 shows the accuracy of the Vegas predictor when transmission errors are correlated and modeled as described in Section 3. The Vegas predictor provides high accuracy and approaches an ideal estimator for the whole range of packet error rates.

We have also extended our analysis to more complex network scenarios, with a varying number of TCP connections and multiple hops. For the sake of brevity we do not report these results. In all the scenarios we examined, the accuracy of the Vegas predictor has always been higher than 70%.

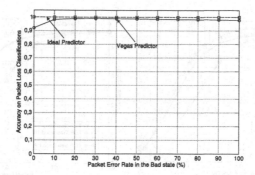

Fig. 3. Accuracy of classification of packet losses for the Vegas loss predictor as a function of PER in the *Bad* state in the scenario with no cross traffic on the wired link.

5   TCP Performance over Wireless Links

So far, this paper has shown that TCP NewReno-LP performs an accurate estimation of the cause of packet losses in various network scenarios. However, as this algorithm is mainly designed to achieve high goodput in the presence of links affected by random errors, a study was made of the performance of this algorithm over wireless links.

To measure TCP NewReno-LP performance, and compare it with other TCP versions, we first considered several simulated network scenarios with long-lived TCP connections, typical of FTP file transfers. In the following we discuss the results obtained by simulation.

5.1   Uncorrelated Losses

Following the guidelines proposed in [13], we considered the topology shown in Fig. 1. We analyzed three network scenarios with different capacity of the wired and wireless link: $C_{SN} = 2$, 5 or 10 Mbit/s and $C_{ND} = 10$Mbit/s. The Round Trip Time (RTT) is always equal to 100 ms and the queue can contain a number of packets equal to the bandwidth-delay product. We considered independent packet losses, modeled as described in Section 3. For each scenario we measured the steady state goodput obtained by TCP NewReno-LP (the bold line), TCP Westwood with NewReno extensions [17] and TCP NewReno. All goodput values presented in this Section were calculated over multiple file transfers with a 97.5% confidence level [16]. The results are shown in Figures 4(a), 4(b) and 5, where the vertical lines represent, as in all the other Figures, the confidence interval for each goodput value.

It can be seen that for all packet error rates and at all link speeds TCP NewReno-LP achieves higher goodput than TCP NewReno. This is due to the Vegas loss predictor that prevents, most of the time, confusion between real network congestion signals, due to queue overflow, and signals due to link errors.

Note that for packet error rates close to zero, when congestion is the main cause of packet losses, TCP NewReno-LP achieves practically the same good-

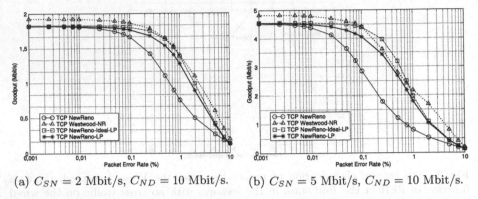

(a) $C_{SN} = 2$ Mbit/s, $C_{ND} = 10$ Mbit/s.   (b) $C_{SN} = 5$ Mbit/s, $C_{ND} = 10$ Mbit/s.

Fig. 4. Goodput achieved by various TCP versions in the topology of Fig. 1 as a function of PER.

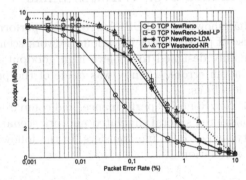

Fig. 5. Goodput achieved by various TCP versions in the topology of Fig. 1 with $C_{SN} = 2$ Mbit/s and $C_{ND} = 10$ Mbit/s as a function of PER.

put as TCP NewReno. This allows TCP NewReno-LP sources to share friendly network resources in mixed scenarios with standard TCP implementations, as it will be shown in Section 5.5.

In all the considered scenarios, we also measured the goodput achieved by a TCP Westwood source with NewReno extensions (TCP Westwood-NR), using the *ns* modules available at [17]. In all simulations this source achieved higher goodput than the other TCP versions, especially when the packet error rate was high. However, we believe that there is a trade-off between achieving goodput gain in wireless scenarios and being friendly toward existing TCP versions in mixed scenarios where the sources use different TCPs. In fact, if a TCP source is too aggressive and achieves a goodput higher than its fair share over a wired, congested link, its behavior is not friendly toward the other competing connections. This behavior will be analyzed, again, in Section 5.5.

To provide a comparison, Figures 4(a), 4(b) and 5 also report the performance achieved by a TCP NewReno based on an ideal estimator that always knows the exact cause of packet losses (TCP NewReno-Ideal-LP). This scheme provides

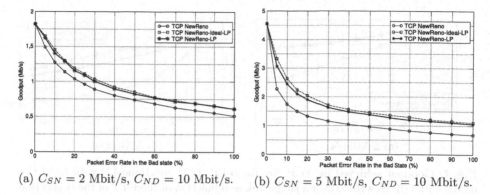

(a) $C_{SN} = 2$ Mbit/s, $C_{ND} = 10$ Mbit/s.   (b) $C_{SN} = 5$ Mbit/s, $C_{ND} = 10$ Mbit/s.

Fig. 6. Goodput achieved by various TCP versions in the topology of Fig. 1 as a function of PER in the *Bad* state.

an upper bound on the performance achievable by every scheme based on loss predictors. Note that our scheme approaches this bound for all the considered scenarios.

5.2 Correlated Losses

To account for the effects of multi-path fading typical of wireless environments, we also investigated the behavior of TCP NewReno-LP in the presence of links affected by correlated errors, modeled as described in Section 3. We considered two different scenarios with wireless link capacities equal to 2 and 5 Mbit/s, and a Round Trip Time equal to 100 ms. Fig. 6(a) shows the steady-state goodput achieved by the TCP versions analyzed in this paper as a function of the packet error rate in the *Bad* state. TCP NewReno-LP achieves higher goodput than TCP NewReno and practically overlaps to the goodput upper bound achieved by the ideal scheme TCP NewReno-Ideal-LP.

A similar behavior was observed in Fig. 6(b) where we reported the goodput achieved by the analyzed TCP versions in the topology shown in Fig. 1 with a 5 Mbit/s link capacity as a function of the packet error rate in the *Bad* state. Note that in this scenario the performance improvement of TCP NewReno-LP over TCP NewReno is higher than in the 2 Mbit/s scenario, as wireless losses affect more heavily TCP NewReno goodput when the bandwidth-delay product of the connection is higher [18].

5.3 Impact of Round Trip Time

Packet losses are not the only cause of TCP throughput degradation. Many studies [19] have pointed out that TCP performance also degrades when the Round Trip Time (RTT) of the connection increases. TCP NewReno-LP allows to alleviate this degradation to improve performance. Fig. 7(a) and 7(b) report the goodput achieved by TCP NewReno, TCP NewReno-LP and TCP NewReno-Ideal-LP sources transmitting over a single link with capacity equal to 2 Mbit/s

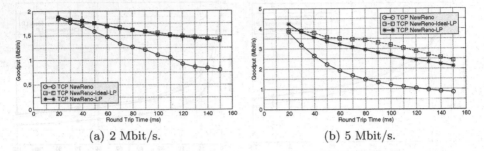

(a) 2 Mbit/s.   (b) 5 Mbit/s.

Fig. 7. Goodput achieved by TCP NewReno-LP, TCP NewReno-LP with Ideal Predictos and TCP NewReno over a single link as a function of the RTT of the connection.

and a 5 Mbit/s, respectively, as a function of the Round Trip Time of the connection. The link drops packets independently with a loss probability constantly equal to 0.5%.

We point out the high goodput gain of TCP NewReno-LP over TCP New-Reno. This behavior is more evident when the Round Trip Time of the connection increases. Note that, even in this scenario, TCP NewReno-LP practically overlaps to the goodput upper bound achieved by the ideal scheme TCP NewReno-Ideal-LP.

5.4   Friendliness and Fairness

So far we have shown that the TCP NewReno-LP scheme estimates accurately the cause of packet losses and that achieves higher goodput than existing TCP versions over wireless links with both uncorrelated and correlated losses.

Following the methodology proposed in [10], we evaluated friendliness and fairness of TCP NewReno-LP in a variety of network scenarios and we compared them by those achieved by TCP Westwood-NR. The term *friendliness* relates to the performance of a set of connections using different TCP flavors, while the term *fairness* relates to the performance of a set of TCP connections implementing the same algorithms.

This section shows how the proposed scheme is able to share friendly and fairly network resources in mixed scenarios where the sources use different TCPs.

To this purpose, we first evaluated TCP NewReno-LP *friendliness* by considering two mixed scenarios: in the first one 5 TCP connections using either TCP NewReno-LP or TCP NewReno share an error-free link with capacity equal to 10 Mbit/s and RTT equal to 100 ms; in the second one the TCP NewReno-LP sources were replaced by TCP Westwood-NR sources.

By simulation we measured the goodput, for each connection, and for all cases. The average goodput of n TCP NewReno-LP and of m TCP NewReno connections, with $n + m = 5$, is shown in Fig. 8(a).

The goodput achieved by both algorithms is very close to the fair share for the full range of sources.

The same experiment was performed with TCP connections using either TCP Westwood-NR or TCP NewReno, and the results are shown in Fig. 8(b).

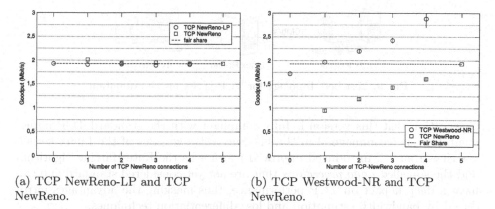

(a) TCP NewReno-LP and TCP
NewReno.

(b) TCP Westwood-NR and TCP
NewReno.

Fig. 8. Average goodput of (a) n TCP NewReno-LP and m TCP NewReno connections and (b) n TCP Westwood-NR and m TCP NewReno connections, with $n + m = 5$, over a 10 Mbit/s link with RTT equal to 100 ms.

In this scenario TCP Westwood-NR sources proved more aggressive toward TCP NewReno sources than TCP NewReno-LP, and achieved a goodput higher than the fair share practically in every case. This behavior evidences the trade off that exists between achieving high goodput gain in wireless scenarios and being friendly in mixed network scenarios.

To measure the level of *fairness* achieved by TCP NewReno-LP we considered the same scenario described above first with 5 TCP NewReno-LP connections and then with 5 TCP NewReno sources sharing a 10 Mbit/s link with RTT equal to 100 ms. In this scenarios congestion is the only cause of packet losses. The Jain's fairness index [21] of 5 TCP NewReno-LP connections was equal to 0.9987, and that achieved by 5 TCP NewReno sources was equal to 0.9995. These results confirm that TCP NewReno-LP achieves the same level of fairness of TCP NewReno.

We also extended our simulation campaign to more complex scenarios with a varying number of competing connections. The results obtained confirm that TCP NewReno-LP achieves an high level of friendliness toward TCP NewReno, thus allowing its smooth introduction into the Internet.

6   Implementation and Test Bed

To get more details on the TCP NewReno-LP implementation we have built a test bed, shown in Fig. 9 that consists of a PC server, a client and a PC router, all connected by 10 Mb/s LAN cables. The PC router emulates a wireless link with the desired delay and packet loss rate using the NIST Net software [22], thus allowing to control and tune the features of the wireless link.

In the PC server, besides the TCP NewReno that is the current TCP implementation in the Linux kernel version 2.2-20, we have implemented TCP NewReno-LP, TIBET and TCP Westwood. The choice to implement the TCP variants detailed above in the Linux kernel version 2.2-20 was motivated by the

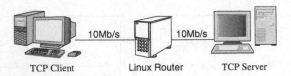

TCP Client   Linux Router   TCP Server

Fig. 9. Test bed Topology for TCP performance evaluation.

observation that this version is fully compliant with the standard TCP implementation as recommended in [14, 11]. Successive versions of the Linux kernel,
starting from 2.4, introduced improved features as the Rate-Halving algorithm
and the so-called *undo procedures* that are not yet considered standard and can
have a deep impact on TCP performance, thus masking the advantages introduced by bandwidth estimation and loss differentiation techniques.

6.1   Uncorrelated Losses

Running the test bed we measured the goodputs achieved by the four TCP
versions. Fig. 10 compares the steady-state goodput achieved by TCP NewReno-
LP, TIBET, TCP Westwood and TCP NewReno connections transmitting data
between the server and the client, with an emulated round trip time equal to
100 ms versus packet loss rates.

The measures on this real scenario validate the results obtained by simulation
(see Fig. 4(a) and 4(b)) and provide a further support on the advantages of TCP
NewReno-LP over TCP NewReno. Fig. 10 also shows the improvement achieved
by TCP NewReno-LP over TIBET, more evident for PER values in the 1% to
4% range.

Note that in this scenario, as well as in all the simulated scenarios presented
in this Section, TCP Westwood obtained a higher goodput than any other TCP

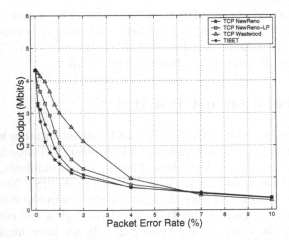

Fig. 10. Goodput achieved by TCP NewReno-LP; TIBET, TCP Westwood and TCP
NewReno in the Test Bed.

version. This behavior is due to its overestimate of the available bandwidth, that leads to aggressive behavior and unfair sharing of network resources, as we showed in the previous Section and as we discussed in detail in [1].

6.2 Correlated Losses

We then considered the same two-state Markov model described in Section 3 to model correlated losses, and we measured the goodput achieved by TCP sources as a function of the packet error rate in the Bad state, to take into account various levels of fading. The results are reported in Figure 11.

These results confirm the improved performance achieved by TCP NewReno-LP even in this network scenario that models very closely real wireless link conditions.

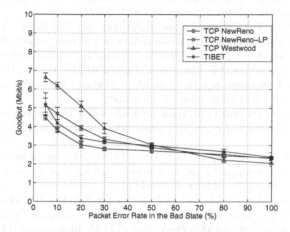

Fig. 11. Goodput Achieved by various TCP versions in the presence of correlated losses.

7   Conclusions

In this work we have discussed and analyzed issues related to the use of Loss Differentiation Algorithms for TCP congestion control. We proposed to use the Vegas loss predictor to enhance the TCP NewReno error-recovery scheme, thus avoiding unnecessary rate reductions caused by packet losses induced by bit corruption on the wireless channel. The performance of this enhanced TCP (TCP NewReno-LP) was evaluated by extensive simulations and real testbeds, examining various network scenarios. Two types of TCP connections were considered, namely long-lived connections, typical of file transfers, and short-lived connections, typical of HTTP traffic. Moreover, we considered two different statistical models of packet losses on the wireless link: independent and correlated losses. We found that TCP NewReno-LP achieves higher goodput over wireless networks, while guaranteeing good friendliness with classical TCP versions over wired links. Moreover, we found that the Vegas loss predictor, embedded in TCP NewReno-LP, proved very accurate in classifying packet losses. Finally, we also defined an ideal scheme that assumes the exact knowledge of packet losses and provides an upper bound to the performance of all possible schemes based on loss differentiation algorithms. The TCP enhanced with Vegas loss predictor well approaches this ideal bound.

References

1. A. Capone, L. Fratta, and F. Martignon. Bandwidth Estimation Schemes for TCP over Wireless Networks. *IEEE Transactions on Mobile Computing*, 3(2):129–143, 2004.
2. C. Casetti, M. Gerla, S. Mascolo, M. Y. Sanadidi, and R. Wang. TCP Westwood: End-to-End Congestion Control for Wired/Wireless Networks. In *Wireless Networks Journal*, volume 8, pages 467–479, 2002.
3. H.Balakrishnan, V.N.Padmanabhan, S.Seshan, and R.H.Katz. A Comparison of Mechanisms for Improving TCP Performance over Wireless Links. In *CIEEE/ACM Transactions on Networking*, volume 5(6), pages 759–769, December 1997.
4. S. Bregni, D. Caratti, and F. Martignon. Enhanced Loss Differentiation Algorithms for Use in TCP Sources over Heterogeneous Wireless Networks. In *Proceedings of GLOBECOM'03*, San Francisco, 1-5 Dec. 2003.
5. S.Cen, P.C.Cosman, and G.M.Voelker. End-to-end Differentiation of Congestion and Wireless Losses. *IEEE/ACM Transactions on Networking*, 11(5):703–717, Oct. 2003.
6. Cheng Peng Fu and S. C. Liew. TCP Veno: TCP enhancement for transmission over wireless access networks. In *IEEE Journal on Selected Areas in Communications*, volume 21(2), pages 216–228, Feb. 2003.
7. K. Xu, Y. Tian, and N. Ansari. TCP-Jersey for Wireless IP Communications. In *IEEE Journal on Selected Areas in Communication*, volume 22(4), May 2004.
8. Eric Hsiao-Kuang Wu and Mei-Zhen Chen. JTCP: Jitter-Based TCP for Heterogeneous Wireless Networks. In *IEEE Journal on Selected Areas in Communication*, volume 22(4), May 2004.
9. L.S. Brakmo and L.L. Peterson. TCP Vegas: End-to-End Congestion Avoidance on a Global Internet. *IEEE Journal on Selected Areas in Communications*, 13(8):1465–1480, October 1995.
10. R. Wang, M. Valla, M.Y. Sanadidi, and M. Gerla. Adaptive Bandwidth Share Estimation in TCP Westwood. In *Proceedings of Globecom'02*, 2002.
11. S. Floyd and T. R. Henderson. The NewReno Modifications to TCP's Fast Recovery Algorithm. *IETF RFC 2582*, 26(4), April 1999.
12. ns-2 network simulator (ver.2).LBL. URL: http://www.isi.edu/nsnam.
13. S. Floyd and V. Paxson. Difficulties in Simulating the Internet. In *IEEE/ACM Transactions on Networking*, volume 9, pages 392–403, August, 2001.
14. M.Allman, V.Paxson, and W.Stevens. TCP Congestion Control. *RFC 2581*, April 1999.
15. A. A. Abouzeid, S. Roy, and M. Azizoglu. Stochastic Modeling of TCP over Lossy Links. In *Proceedings of INFOCOM 2000*, Tel Aviv, Israel, March 2000.
16. K. Pawlikowski, H.-D.Joshua Jeong, and J.-S.Ruth Lee. On Credibility of Simulation Studies of Telecommunication Networks. *IEEE Communications Magazine*, pages 132–139, Jan. 2002.

17. UCLA High Performance Internet Research Group. Tcp westwood home page, URL: http://www.cs.ucla.edu/nrl/hpi/tcpw.
18. T.V.Lakshman and U.Madhow. The performance of TCP/IP for networks with high bandwidth-delay products and random loss. *IEEE/ACM Transactions on Networking*, 5(3):336–350, 1997.
19. J. Padhye, V. Firoiu, D. Towsley, and J. Kurose. Modeling TCP Throughput: A Simple Model and its Empirical Validation. In *Proceedings of ACM SIGCOMM '98*, 1998.
20. N. Cardwell, S. Savage, and T. Anderson. Modeling TCP Latency. In *Proceedings of INFOCOM 2000*, pages 1742–1751, 2000.
21. R. Jain. *The Art of Computer Systems Performance Analysis: Techniques for Experimental Design, Measurement, Simulation and Modeling.* Wiley, New York, 1991.
22. *NIST Net Home Page.* Available at http://snad.ncsl.nist.gov/itg/nistnet/.

Revenue-Based Adaptive Deficit Round Robin

Alexander Sayenko[1], Timo Hämäläinen[1],
Jyrki Joutsensalo[1], and Pertti Raatikainen[2]

[1] University of Jyväskylä, MIT department.
P.O.Box 35, Mattilaniemi 2 (Agora), Jyväskylä, Finland
{sayenko,timoh,jyrkij}@cc.jyu.fi
[2] VTT Information Technology (Telecommunications),
P.O.Box 1202, Otakaari 7B, VTT, Espoo, Finland
pertti.raatikainen@vtt.fi

Abstract. This paper presents an adaptive resource allocation model that is based on the DRR queuing policy. The model ensures QoS requirements and tries to maximize a service provider's revenue by manipulating quantum values of the DRR scheduler. To calculate quantum values, it is proposed to use the revenue criterion that controls the allocation of free resources. The simulation considers a single node with the implemented model that serves several service classes with different QoS requirements and traffic characteristics. It is shown that the total revenue can be increased due to the allocation of unused resources to more expensive service classes. At the same time, bandwidth and delay guarantees are provided. Furthermore, the adaptive model eliminates the need to find the optimal static quantum values because they are calculated dynamically.

1 Introduction

The current development of communication networks can be characterized by several important factors: a) the growth of the number of mobile users and b) the tremendous growth of new services in wired networks. While at the moment most mobile users access services provided by mobile operators, it is possible to predict that they will be eager to use services offered by wired networks. Furthermore, as the throughput of wireless channels grows, more users will access them. In this framework, it is important that users obtain the required end-to-end guarantees, which are referred to collectively as the Quality-of-Service (QoS) requirements. The provision of QoS implies that appropriate arrangements are taken along the packet path, that comprises of network access points, one or more core networks, and interconnections between them. As a packet moves from a source to a destination point, it spends most time in the core networks. As a result, the efficient provision of resources in the core networks is the essential part of QoS.

By now, IETF has proposed several architectures to realize QoS in packet networks. While Integrated Services (IntServ) [1] relies upon the per-flow approach, Differentiated Services (DiffServ) [2] perform the allocation of resources on the per-class basis, thus providing more scalable solutions. However, the presence

M. Ajmone Marsan et al. (Eds.): QoS-IP 2005, LNCS 3375, pp. 600–612, 2005.

of services, such as VoIP and video conferencing, imposes additional constraints on the DiffServ framework. As the output bandwidth is allotted for each service class by the queuing policy in routers along a path, the provided QoS guarantees depend on a scheduler and its parameters. In most cases, static configuration is used, which makes the scheduler irrespective of the number of flows within each traffic class. As a result, a service provider has to *overprovide* its network with resources to meet all the QoS requirements, regardless of the current number of active flows within each service class. Such an approach results in inefficient allocation of resources. Though, it may not be a significant issue for wired providers, it is very critical for the core wireless network, in which resources should be allocated in an optimal way.

The problem of effective allocation of the network resources can be solved if a router exploits adaptive service weight assignment. It means that the minimum departure rate should be adjusted dynamically to reflect changes in the bandwidth requirements of service classes. Obviously, it requires the DiffServ framework to track the number of active data flows within each service class. One possibility is to use the Resource Reservation Protocol (RSVP) from the IntServ framework [3, 4]. Also, other proprietary solutions can be used.

This paper presents a resource sharing model that ensures the QoS requirements and maximizes a service provider's revenue. The objectives of the model are somewhat similar to those considered in [5]: a) to share bandwidth between various service classes with the required level of QoS and b) to distribute free resources in a predictable and controlled way. However, we propose more rigorous bandwidth allocation. While the QoS requirements determine the minimal required amount of resources, prices of network services can control the allocation of free bandwidth. It is intuitively understandable that it is worth of providing more bandwidth for those classes for which end-users are willing to pay more. Furthermore, such an approach may interest wireless providers as they use to charge for all kind of services that are accessed from mobile devices. Thus, the goal of the proposed model is to increase the total revenue by allocating free resources to certain service classes and reducing bandwidth, previously assigned to the other ones.

The proposed adaptive model is based on the Deficit Round Robin (DRR) queuing discipline and functions as a superstructure over that scheduling policy. In fact, the adaptive model is an integer linear programming (ILP) task that calculates the optimal quantum values, based on the QoS requirements of data flows and the pricing information. This approach has been used successfully in our adaptive models for the Weighted Fair Queuing (WFQ) and Weighted Round Robin (WRR) scheduler [6, 7], and in related network optimization problems [8, 9]. In this work, we refine the adaptive model for the DRR scheduler.

The rest of this paper is organized as follows. Section II surveys the basic queuing disciplines, section III describes the adaptive model, and section IV presents the carried out simulation. Finally, the conclusions chapter summarizes the results and discusses about the future research plans.

2 Queuing Disciplines

The choice of an appropriate service discipline is the key in providing QoS because the queuing policy is the basis for allocating resources [10]. The most popular and fundamental queuing policies are First-Come-First-Served (FCFS), Priority Queue (PQ) [11], WRR [12] and WFQ [13]. FCFS determines service order of packets strictly based on their arrival order. Therefore, this policy cannot perform necessary bandwidth allocation and provide the required QoS. The PQ policy absolutely prefers classes with higher priority and, therefore, packets of a higher priority queue are always served first. Thus, if a higher priority queue is always full then the lower priority queues are never served. This problem can be eliminated by using WRR, in which queues of all service classes are served in the round-robin manner. However, if some queue has a longer average packet size than the other queues, it receives more bandwidth implicitly. This disadvantage was overcome with the WFQ technique, which schedules packets according to their arrival time, size, and the associated weight.

Though WFQ provides a way to specify precisely the amount of output bandwidth for each traffic class, it is complicate in implementation. From this viewpoint, WRR does not introduce any computational complexity, but it fails to take the packet size into account. The DRR policy [14] came as the tradeoff between the implementation complexity and precise allocation of resources.

3 Adaptive Model

3.1 Deficit Round Robin

The DRR scheduler works in a cyclic manner serving consequently input queues. During a round, a certain number of packets, determined by the value of the deficit counter, are sent from each queue. As all queues are served, the DRR scheduler updates the deficit counter using the *quantum value* and begins the next cycle.

Suppose, each physical queue of the DRR scheduler has the associated quantum value Q_i. If there are m input queues, then the following expression estimates the average amount of data transmitted from all queues during one round:

$$\sum_{i=1}^{m} Q_i. \tag{1}$$

If B is the bandwidth of the output interface, then it is possible to approximate the bandwidth allocated for the given kth queue:

$$\frac{Q_k}{\sum_i Q_i} B \tag{2}$$

3.2 QoS Requirements

Bandwidth. Assume that each service class is associated with a queue of the DRR scheduler. Then, (2) represents the bandwidth allotted for the whole service class. If class k has N_k active data flows with identical bandwidth requirements, then each data flow obtains the following bandwidth[1]:

$$B_k^f = \frac{Q_k}{N_k \sum\limits_i Q_i} B \tag{3}$$

Here B_k^f can be treated as one of the QoS parameters that specifies the bandwidth to be provided for each data flow. Thus, if B_k^f is given, then a router must allocate a certain amount of resources to satisfy requirements of all flows within each service class.

$$\frac{Q_k}{N_k \sum\limits_i Q_i} B \geq B_k^f, \forall k = \overline{1, m} \tag{4}$$

Since quantum values control the allocation of the output bandwidth between service classes, the task is to find such values of Q_k that all the QoS requirements are satisfied.

$$Q_k \geq \frac{B_k^f N_k}{B - B_k^f N_k} \sum_{\substack{i=1 \\ i \neq k}}^{m} Q_i, \forall k = \overline{1, m} \tag{5}$$

Delay. Along with bandwidth requirements, certain service classes must be provided with delay guarantees. Since the DRR scheduler serves input queues in a cyclic manner, processing of a packet can be delayed by $\sum Q_i/B$ seconds. To decrease this delay, it is possible to introduce the Low Latency Queue (LLQ) that can work in two modes [16]: *strict priority mode* and *alternate priority mode*. In the strict priority mode, the DRR scheduler always outputs packets from LLQ first. However, it is difficult to predict the allocation of bandwidth for other queues in this case. Thus, the alternate priority mode will be considered, in which LLQ is served in between queues of the other service classes. For instance, if there are 3 input queues, numbered from 0 to 2, and queue 0 is LLQ, then queues are served in the following order: 0–1–0–2–.... In this case, processing of a packet can be delayed by

$$\max_i\{Q_i\}/B$$

seconds. As in the DRR scheduler, each queue is allowed to transmit no more than Q_i bytes during a round.

If a router implements LLQ, it is necessary to reformulate (5). Suppose, LLQ is identified by index l, where $1 \leq l \leq m$. Then, it is possible to approximate the amount of data that the DRR+ scheduler[2] outputs in a round.

---

[1] A router has to deploy appropriate mechanisms to provide fairness between data streams within each service class. One of the possible solutions is to combine Stochastic Fair Queuing (SFQ) [15] with the DRR scheduler.

[2] Introduced in [14], the term DRR+ corresponded to the DRR scheduler with LLQ that works in the absolute priority mode. For the sake of simplicity, we use DRR+ to denote the DRR scheduler with LLQ in the alternate priority mode

$$(m-1)Q_l + \sum_{i=1,i\neq l}^{m} Q_i \tag{6}$$

Taking account of the presented above considerations, it is possible to derive an expression for the minimum quantum values that satisfy all the bandwidth requirements of each service class.

$$Q_k \geq \frac{B_k^f N_k}{B - B_k^f N_k} \left((m-1)Q_l + \sum_{\substack{i=1 \\ i\neq k, i\neq l}}^{m} Q_i \right), k \neq l \tag{7}$$

$$Q_l \geq \frac{B_l^f N_l}{(m-1)(B - B_l^f N_l)} \sum_{i=1,i\neq l}^{m} Q_i \tag{8}$$

These constraints only reserve bandwidth for normal queues and LLQ, but they do not provide any delay guarantees. Suppose, that each data flow, which belongs to the class that has the delay requirements, is regulated by the Token Bucket policer [17] with the mean rate ρ and the burst size σ. Thus, it takes the DRR scheduler σ/B seconds to transmit the received burst under ideal conditions. However, if σ is bigger than Q_l, then more time is needed to output the data burst because LLQ will be interrupted by another queue. While the scheduler serves that queue, packets in LLQ can de delayed by

$$\max_{i,i\neq l}\{Q_i\}/B$$

seconds at most. Thus, the queuing delay of packets in LLQ can be estimated by:

$$D_l \leq \frac{\sigma}{B} + \max\left\{\frac{\sigma}{Q_l} - 1, 0\right\} \frac{\max_{i,i\neq l}\{Q_i\}}{B} \tag{9}$$

Here D_l stands for the worst-case delay, experienced by packets in LLQ. The previous inequality does not consider the fact that the initial processing of LLQ can be delayed by the other queue being processed when a LLQ packet arrives to an empty queue. Thus, it is possible to introduce a corrected estimation.

$$D_l \leq \frac{\sigma}{B} + \max\left\{\frac{\sigma}{Q_l} - 1, 0\right\} \frac{\max_{i,i\neq l}\{Q_i\}}{B} + \frac{\max_{i,i\neq l}\{Q_i\}}{B}$$

$$= \frac{\sigma}{B} + \max\left\{\frac{\sigma}{Q_l}, 1\right\} \frac{\max_{i,i\neq l}\{Q_i\}}{B} \tag{10}$$

Based on the value of σ and Q_l, it is possible to consider two distinctive cases.

$$D_l \leq \begin{cases} \dfrac{1}{B}\left(\sigma + \max_{i,i\neq l}\{Q_i\}\right), \sigma \leq Q_l & (11) \\[4ex] \dfrac{\sigma}{B}\left(1 + \dfrac{\max_{i,i\neq l}\{Q_i\}}{Q_l}\right), \sigma > Q_l & (12) \end{cases}$$

The first inequality corresponds to the case when a burst is output completely in one round. However, as a service class aggregates multiple data flows, one could expect that the resulting burst size of the whole service class is bigger than σ and is not bigger than $N_l \sigma$. Since the latter value is usually larger than Q_l, we will consider (12). It is possible to rewrite it in the following form:

$$\left(\frac{B}{N_l \sigma} D_l - 1 \right) Q_l - \max_{i, i \neq l} \{Q_i\} \geq 0 \tag{13}$$

Constraints (7), (8), and (13) form the set of solutions [7] that satisfy the given QoS requirements. However, a criterion is necessary to choose the best solution from a certain viewpoint.

3.3   Charging

As in our previous research [6, 7], we propose to use the revenue criterion to control the allocation of free resources between service classes. To charge customers, a service provider uses a *pricing function*. As presented earlier, (1) approximates the amount of data the DRR scheduler outputs during a round. Suppose, each class has the associated price $C_i()$ measured in monetary units per one unit of data. The price can remain fixed or change over the course of time. It can depend on parameters, such as the time of day, congestion level, and provided bandwidth. Furthermore, in telecommunication networks, it is often the case that price depends on the amount of data transferred. In this study, we assume that price is almost constant, i.e. it changes vary rarely compared to the duration of a round. Thus, the mean revenue, which a service provider obtains during a DRR round, can be given by:

$$r(Q_1 \ldots Q_m) = \sum_{i=1}^{m} C_i() Q_i \text{ [monetary units/round]} \tag{14}$$

The value of function r depends only on the quantum values Q_i because C_i is constant during a round. Thus, by manipulating Q_i, different instantaneous revenue is obtained.

If the DRR scheduler implements LLQ, then (14) can be modified to the form:

$$r(Q_1 \ldots Q_m) = (m-1) C_l() Q_l + \sum_{i=1, i \neq l}^{m} C_i() Q_i \tag{15}$$

3.4   General Model

The general model consists of the pricing function (15) and constraints (7),(8) and (13). It should be noted that the constraints are written in a different form to be suitable for calculating the optimal quantum values.

$$\max \left\{ (m-1) \gamma_l C_l() Q_l + \sum_{i=1, i \neq l}^{m} \gamma_i C_i() Q_i \right\} \tag{16}$$

subject to:

$$(m-1)B_k^f N_k Q_l + B_k^f N_k \sum_{\substack{i=1 \\ i \neq k, i \neq l}}^{m} Q_i + (B_k^f N_k - B)Q_k \leq 0, \ \forall k, k \neq l$$

$$(m-1)(B_l^f N_l - B)Q_l + B_l^f N_l \sum_{i=1, i \neq l}^{m} Q_i \leq 0$$

$$\left(\frac{B}{N_l \sigma_l} D_l - 1\right) Q_l - Q_k \geq 0, \ \forall k = \overline{1,m}, k \neq l$$

Alternatively, if a provider has to provide only bandwidth guarantees, then the given above model can be simplified, i.e. (14) and constraint (5) should be used.

$$\max \left\{ \sum_{i=1}^{m} \gamma_i C_i() Q_i \right\} \tag{17}$$

subject to:

$$B_k^f N_k \sum_{i=1, i \neq k}^{m} Q_i + (B_k^f N_k - B)Q_k \leq 0, \forall k = \overline{1,m} \tag{18}$$

The purpose of a new parameter $\gamma_i \in \{0, 1\}$ in the target function is to disable or enable the allocation of excess resources for the ith service class. Suppose, there is a service class consisting of applications that generate constant rate data streams. Although it may be the most expensive class, all the excess resources, allotted to it, will be shared among the other classes, because the constant rate sources will not increase their transmission rates. Therefore, the allocation of the excess resources can be disabled by setting $\gamma_i = 0$. If more bandwidth is allocated for a service class that consists predominantly of TCP flows, then the applications will increase their window sizes and, as a result, their transmission rates. Thus, it makes sense to set $\gamma_i = 1$.

The presented adaptive models are the ILP. One of the methods that can be used to calculate the optimal values of Q_i is the *branch & bound* algorithm [18]. These problems are not computationally expensive because the number of service classes and, as a result, the number of constraints is not usually large. In the case of (16), the number of constraints is proportional to the number of the service classes and equals to $2m - 1$ (m bandwidth constraints and $m - 1$ delay constraints). In turn, the number of constraints is always equal to the number of the service classes in (17). According to the current DiffServ specification [2], there are at most six aggregates including the best-effort aggregate, and a provider is not obliged to implement all of them. In turn, the IntServ architecture [1] defines only three major classes.

It is worthy of noticing that the adaptive model possesses the characteristics of the admission control module. As the optimization problem is solved, new optimal quantum values are obtained that can be passed to the scheduler. If no feasible solution exists, then a router does not have enough resources and a newly arrived flow can be rejected.

4   Simulation Results

In this section, we study the proposed adaptive resource sharing model and compare it to the ordinary DRR discipline in terms of parameters, such as the provided bandwidth, queuing delay, and obtained revenue. The simulation is carried out by the NS-2 simulator [19]. For these purposes, the adaptive model is implemented in C++ and the appropriate NS-2 interface is created so that the model can be accessed from a simulation script. Since the implementation of DRR in NS-2 lacks the class-based resource sharing, we have created our DRR scheduler that supports the LLQ mode.

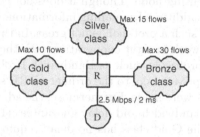

Fig. 1. Simulation environment.

The simulation environment is illustrated in Fig. 1. It consists of a router with the adaptive model, a destination point, and a set of client nodes with applications. To simplify the simulation and avoid mutual interference of applications, each node hosts exactly one application that generates exactly one stream of data, addressed to a destination node. Every node is connected to the router with a link, whose bandwidth and delay are set to 1 Mbps and 2 ms, respectively. It should be noted that the router classifies packets only when they move to the destination node. All responses are sent back to the source applications unclassified.

All client applications are divided into service classes that are referred to as the *Gold*, *Silver*, and *Bronze* class. The details of each service class are presented in Table 1. The Gold class corresponds to the real-time audio and video services. It is simulated by the constant-rate traffic that is transferred over the UDP protocol. The Silver and Bronze classes represent the general purpose services.

Table 1. Parameters of service classes.

Class	Price for 1Mb	Max flows	Flow parameters Band. (Kbps)	Delay (ms)	ON/OFF time (s)
Gold	2	10	100	20	5–10/8
Silver	1	15	50	–	10–20/5
Bronze	0.5	30	10	–	15–25/10

These classes are simulated by FTP-like applications that generate bulk data transferred over the TCP protocol. The rate in Table 1 specifies the minimum bandwidth that a flow must obtain within its service class. To produce a random number of active flows, the simulation uses the on-off model, parameters of which are given in Table 1. The ON-time represents a uniformly distributed random number, taken from an appropriate interval, and the OFF-time follows the exponential distribution with an appropriate mean value.

It should be noted that the simulation scenario does not consider any specific signalling protocol that the client nodes, the router, and the destination node can use to exchange information about the required resources. Instead, inner possibilities of the simulation environment are used to keep track of the number of active flows at the routing node. Though it does not correspond to a real-life scenario, the amount of additional signalling information, which the nodes would exchange in presence of such a protocol, is not great. In the real-life scenario, the adaptive model should work in tight cooperation with the network management entities and signalling protocols, such as Bandwidth Broker and RSVP.

To compare the adaptive model to the ordinary DRR scheduling policy, independent simulation runs were made. Furthermore, the adaptive model was tested in two modes: a) with bandwidth and delay guarantees (16), b) with bandwidth guarantees only, as if the Gold class has no delay requirements at all (17). For the sake of simplicity, we will refer to these modes as A-DRR+ and A-DRR respectively. In order to make this comparison a fair one, the same behaviour patterns of data flows were submitted in each case (see Fig. 2).

When DRR is applied, static quantum values are used. They are chosen so that each service class always has enough bandwidth, regardless of the number of active data flows. It corresponds to the case when the network is overprovisioned. As opposite to this, when the number of flows changes within each service class, the adaptive model recalculates the quantum values for the DRR scheduler. Fig. 3 shows the dynamics of the calculated quantum values. As can be seen, regardless of the type of the adaptive model, the latter always tries to allocate as much resources as possible to the Silver class. Though the Gold class is the most

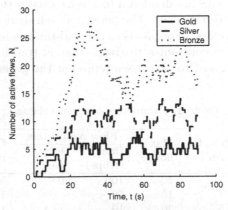

Fig. 2. Dynamics of the number of flows.

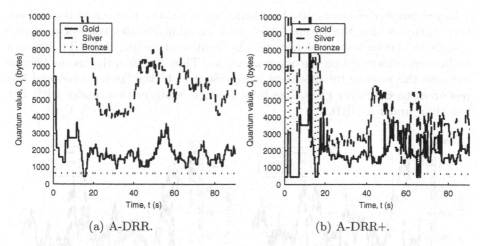

Fig. 3. Dynamics of quantum values.

expensive one, it consists of UDP flows, for which there is no sense to provide excess resources (the value of γ_i for this class is 0). The difference between Fig. 3(a) and Fig. 3(b) is that A-DRR+ imposes additional constraints on the quantum values of the Silver and Bronze classes. That is why the quantum value of the Silver class is not as high as in the case of A-DRR.

Table 2 shows statistical data collected during the simulation. As follows from these results, the same number of Gold packets are transmitted under DRR, A-DRR, and A-DRR+. Thus, the Gold class is provided with the required rate (compare it to the rate in Table 1). The number of the transmitted packets of the Silver and Bronze classes are different. Since DRR relies upon the static quantum values, resources are not allocated optimally. Though all the QoS requirements are guaranteed, the Bronze class has a higher mean per-flow rate than required. As a result, the total revenue is lower than in the A-DRR case. Since the adaptive model calculates the quantum values based on the number of flows and their QoS requirements, better resource allocation is achieved. The Silver class has

Table 2. Simulation results.

Quantity	Discipline	Classes		
		Gold	Silver	Bronze
	DRR	12418	21540	9196
Packets departed	A-DRR	12418	25740	3666
	A-DRR+	12418	23065	7176
	DRR	99.04	149.58	26.73
Mean per-flow rate	A-DRR	99.06	177.48	11.17
	A-DRR+	99.08	162.28	19.49
	DRR		247.6	
Total revenue	A-DRR		261.7	
	A-DRR+		252.7	

a larger number of transmitted packets, and a smaller number of packets are transmitted within the Bronze class. Such an adaptive allocation of resources results in the highest total revenue. As mentioned earlier, A-DRR+ imposes additional constraints on the quantum values. Thus, the adaptive model tries to increase the revenue by allocating more resources to the Bronze class and less resources to the Silver class. The total revenue is bigger than under DRR but less than under A-DRR.

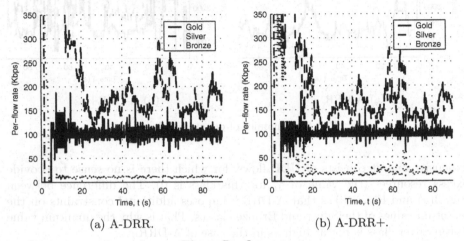

(a) A-DRR.   (b) A-DRR+.

Fig. 4. Per-flow rate.

Fig. 4 illustrates the dynamics of the per-flow rate within each service class. The per-flow rate is calculated by dividing the amount of transmitted data during a sufficiently small time interval by the number of active data flows within a service class. As can be seen, there is a warm-up period that lasts approximately 10-20 seconds. During that period new TCP flows of the Silver and Bronze classes are injected into the network. Under A-DRR and A-DRR+, the Gold class has a per-flow rate that fluctuates near the value of 100 Kbps. The rate fluctuations are explained by the nature of the transmitted data packets and by the fact that flows appear and disappear. The way the DRR scheduler works also has an impact on the rate fluctuation. However, it is noticeable that there are less fluctuations under A-DRR+. Since LLQ is served in between other queues, packets are less likely to be delayed. Depending on the used discipline, different per-flow rates are provided for the Silver and Bronze classes. As shown in Fig. 4(a), A-DRR provides the Silver class with the highest rate – it reaches the value of 300 Kbps. At the same time, less bandwidth is allocated for the Bronze class. Nonetheless, it is noticeable that the Bronze per-flow rate never goes below the required value of 10 Kbps (see Table 1). As A-DRR+ tries to provide all the delay guarantees, the per-flow rate of the Bronze class is slightly larger than the required one.

Fig. 5 illustrates the queuing delay of packets in LLQ when A-DRR+ is in effect. A-DRR+ guarantees the required queuing delay of 20 ms. As in the case with bandwidth, packets experience bigger queuing delay at the beginning

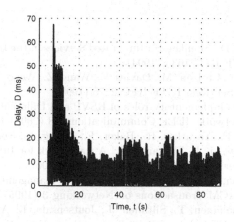

Fig. 5. Queuing delay under A-DRR+.

of the simulation. But as the network stabilizes after the warm-up period, the Gold class is provided with the required delay guarantees.

5   Conclusions

This paper has presented an adaptive resource sharing model that works as the superstructure over DRR and uses the revenue criterion to calculate the optimal quantum values for the scheduler. The model is capable of allocating resources in such a way that the total revenue is maximized and the QoS requirements are ensured including bandwidth and delay. The carried out simulations have demonstrated that the model is capable of working in two different modes. In both cases, all the QoS parameters for service classes are guaranteed regardless of traffic characteristics. The proposed model can be applied directly to the DiffServ and IntServ QoS frameworks.

The adaptive model is capable of working with various linear and non-linear pricing schemes. Though the simulation has presented the simple case with the constant price, the presented model increases the total revenue regardless of the pricing function.

The existent limitation of the model is that only one service class is provided with delay guarantees. It is not a problem for QoS frameworks, such as DiffServ or IntServ, in which only one class has the delay requirements. However, it can be an obstacle for other QoS approaches. Another limitation is that the model is incapable of providing the required jitter.

Our future plan is to provide a comprehensive analysis of the revenue-based adaptive models that work on the top of different schedulers. It will enable a service provider to choose the best solution depending on the requirements of data flows and network characteristics.

References

1. Braden, R., Clark, D., Shenker, S.: Integrated Services in the Internet Architecture: an Overview. IETF RFC 1633 (1994)
2. Blake, S., Black, D., Carlson, M., Davies, E., Wang, Z., Weiss, W.: An Architecture for Differentiated Services. IETF RFC 2475 (1998)
3. Bernett, Y.: The complementary roles of RSVP and Differentiated Services in the full-service QoS network. IEEE Communications **38** (2000) 154–162
4. Bernet, Y., Ford, P., Yavatkar, R., Baker, F., Zhang, L., Speer, M., Braden, R., Davie, B., Wroclawski, J., Felstaine, E.: A framework for Integrated Services operation over DiffServ networks. IETF RFC 2998 (2000)
5. Floyd, S., Jacobson, V.: Link-sharing and resource management models for packet networks. IEEE/ACM Transactions on Networking **3** (1995) 365–386
6. Sayenko, A., Hämäläinen, T., Siltanen, J., Joutsensalo, J.: An adaptive approach for Weighted Fair Queueing with revenue as the optimization criterion. In: The 8th IEEE Symposium on Computers and Communications (ISCC 2003). (2003) 181–186
7. Sayenko, A., Hämäläinen, T., Joutsensalo, J., Raatikainen, P.: Adaptive scheduling using the revenue-based Weighted Round Robin. (In: The 12 IEEE International Conference On Networks (ICON 2004)) in press.
8. Joutsensalo, J., Hämäläinen, T., Pääkkönen, M., Sayenko, A.: QoS- and revenue aware adaptive scheduling algorithm. Journal of Communications and Networks **6** (2004) 68–77
9. Magaña, E., Morató, D., Varaiya, P.: Router scheduling configuration based on the maximization of benefit and carried best effort traffic. Telecommunication systems **24** (2003) 275–292
10. Zhang, H.: Service disciplines for guaranteed performance service in packet-switching networks. Proceeding of IEEE **83** (1995) 1374–1396
11. Kleinrock, L.: Queueing systems. New York: John Wiley & Sons (1975)
12. Hahne, E.: Round Robin scheduling for fair flow control. PhD thesis, MIT, Cambridge (1986)
13. Parekh, A., Gallager, R.: A generalized processor sharing approach to flow control in integrated services networks: The single node case. IEEE/ACM Transactions on Networking **1** (1993) 344–357
14. Shreedhar, M., Varghese, G.: Efficient fair queuing using deficit round-robin. IEEE/ACM Transactions on Networking **4** (1996) 375–385
15. McKenny, P.: Stochastic fairness queueing. In: 9th Annual Joint Conference of the IEEE Computer and Communication Societies. Volume 2. (1990) 733–740
16. Vegesna, S.: IP Quality of Service. 1st edn. Cisco Press (2001)
17. Cruz, R.: A calculus for network delay, Part I: Network elements in isolation. IEEE Transaction on Information Theory **37** (1991) 114–131
18. Thie, P.: An Introduction to Linear Programming and Game Theory. Second edn. John Wiley & Sons, New York (1988)
19. UCB/LBNL/VINT: Network simulator ns-2 (1997) http://www.isi.edu/nsnam/ns.

A Shift Varying Filtering Theory for Dynamic Service Guarantees

Ashok Karumanchi[1], Srinivas Talabattula[1], Kalyan Rao[2], and Sridhar Varadarajan[2]

[1] Department of Electrical Communication Engineering, Indian Institute of Science, Bangalore, Karnataka, India
Ashok_Karumanchi@hotmail.com, srinu@ece.iisc.ernet.in
[2] Satyam Computer Services Limited, #14 Langford Avenue, Lalbagh Road, Bangalore, Karnataka, India
{Kalyan,Sridhar_Dr}@Satyam.com

Abstract. We extend the recently developed filtering theory under (max,+)-algebra to shift varying setting and develop a calculus for dynamic service guarantees for variable length packet networks. By modifying the Chang's traffic characterization to shift varying setting, we introduce two basic network elements for time varying input traffic models: (i) traffic regulators that generate G-regular marked point processes and (ii) dynamic G-server that provide service guarantees for marked point processes. Similar to shift invariant setting under (max,+)-algebra, network elements can be joined by concatenation, "filter bank summation," and feedback to form a composite network element. We illustrate the use of framework by various examples like time varying $G/G/1$ queue, segmentation and reassembly, jitter control, damper and window flow control.

1 Introduction

Telecommunication networks are evolving towards transporting multiple media types like voice, video, fax, and data. Networks need to provide Quality-of-Service (QoS) guarantees for supporting these various media types. There are various analytical frameworks appeared in recent literature that address the problem of providing deterministic QoS in these networks. Cruz [1]-[3] reported that if the input traffic stream described by $A(t)$: the cumulative number of arrivals until time t, is regulated according to the constraint $A(t) - A(s) \leq f(t - s), \forall t \leq s$, bounds can be established on buffer length and delay. An example for the function f is the (σ, ρ)-regulation. Here ρ determines upper bound on long-term average rate of traffic flow. For a given ρ, σ determines upper bound on maximum burstiness. Input traffic is modeled by the arrival curve and the service from a network element is modeled by service curve. Network service curve and tandem queues are described to model service from a sequence of network elements like concatenation of leaky buckets.

Cruz and Sariowan [4] describe service curve based earliest deadline first (SCED) scheduling for guaranteeing QoS. Target output curve concept is used for computing deadlines for individual packets. Also a generic rate based admission control criteria is defined for admitting connections into the network. LeBoudec and Thiran [5] simplify Cruz's traffic regulation in Network Calculus. A new service curve called extended service curve is defined which is same for network of queues and isolated

M. Ajmone Marsan et al. (Eds.): QoS-IP 2005, LNCS 3375, pp. 613–625, 2005.

queues. New operators on arrival and service curves are defined which simplify the derivation of fundamental results like characterization of output flow from the shaper. Chang C.S. [6]-[8] developed filtering theory under (min,+) algebra for similar purpose. In this theory traffic stream is treated as *signal* and the arrival and service curves are treated as *transfer function* of the network. Two basic operators are defined: *min* and *convolution*. Two basic network elements are defined: maximal *f*-regulator and *f*-server using which the arrival curve and service curve concepts can be modeled. A sub-additive closure operator is defined which optimizes the *f*-regulator.

These theories are mainly useful for fixed length packet networks and require special use of *packetizers* for extending to variable length packet networks. With the advent of internet protocol (IP) over wavelength division multiplexing (WDM), there is a requirement for providing QoS guarantees directly for traffic streams comprised of variable length packets. This helps in reducing the protocol redundancy in models like IP over asynchronous transfer mode (ATM) over synchronous optical networking (SONET) over WDM or IP over SONET over WDM. Chang [9] extended filtering theory to the (max,+) algebra setting for modeling performance guarantees in variable length packet networks. A new traffic regulation is introduced for variable length packet traffic streams using marked point processes. In this theory, traffic stream comprised of variable length packets is characterized by a marked point process $\psi = (\tau, l)$ with $\tau = \{\tau(n), n=0,1,2...\}$ and $l = \{l(n), n=0,1,2...\}$ where $\tau(n)$ is the arrival time of $(n+1)$th packet and $l(n)$ is the service requirement (packet length) of $(n+1)$th packet. Define $L(n)$ as the cumulative service requirement of first n packets and $L(0) = 0$. A marked point process $\psi = (\tau, l)$ is said to be g-regular where g is a non-decreasing sequence function if for all $m \leq n$ there holds the inequality

$$\tau(n) - \tau(m) \geq g(L(n) - L(m)), \forall\, m \leq n \tag{1}$$

Using this traffic characterization, two basic network elements are defined (i) g-regulators for regulating the input marked point processes and,(ii) g-servers for providing service guarantees to marked point processes. These network elements can be combined in standard filtering theory operations like concatenation, "filter bank summation" and feedback. This filtering theory is shift invariant and hence not useful for representing time dependent traffic models.

In [11], we have introduced shift varying extensions to filtering theory under (max,+) algebra for modeling the distributed traffic regulation for time varying traffic streams comprised of variable length packets. In this paper, we elaborate this shift varying filtering theory to include dynamic service guarantees and standard filtering theory operations like concatenation, "filter bank summation" and feedback. We modify the traffic characterization in (1) by using a bivariate function $G(m, n)$ to the following

$$\tau(n) - \tau(m) \geq G(L(m), L(n)), \forall\, m \leq n \tag{2}$$

Using this shift varying traffic characterization, we define two basic network elements similar to shift invariant filtering theory: (i) G-regulators for regulating the time varying input marked point processes and,(ii) dynamic G-servers for providing service guarantees to time varying marked point processes. These network elements can be combined in standard filtering theory operations like concatenation, "filter bank summation" and feedback. Our extension of filtering theory under (max,+)

algebra to shift varying setting is in similar lines to the extension of filtering theory under (min,+) algebra to time varying setting [8].

This paper is organized as follows. In Section 2 we extend the (max,+) framework to bivariate setting. Necessary mathematical operations like *max* and *convolution* are defined for bivariate functions and various properties of these operators are explained. In Section 3, we describe the traffic regulation mechanism. Section 4 describes the dynamic G-server for providing dynamic service guarantees. Examples for G-server like maximum delay server are provided. In Section 5, we present the various operations on these network elements like concatenation, "filter bank summation" and feedback. Section 6 concludes and lists future directions of this research.

2 (max,+) Framework for Bivariate Functions

In this section we extend the (max,+)-framework to bivariate setting. Consider the family of bivariate functions

$$\mathcal{S}^l = \{F(.,.) :\ F(m, n) \geq 0, F(m, n) \leq F(m, n + 1), \forall\ 0 \leq m \leq n,$$
$$F(n, n) = 0\ \forall\ n \geq 0,\ F(m + 1, n) \leq F(m, n)\ \forall\ 0 \leq m \leq n\ \}$$

Thus, for any $F \in \mathcal{S}^l$, $F(m, n)$ is non-negative and non-decreasing in n. For any two bivariate functions F and G in $\mathcal{S}^l$, we say $F = G$ (resp.$F \leq G$) if $F(m, n) = G(m, n)$ (resp. $F(m, n) \leq G(m, n)$ for all $0 \leq m \leq n$. Let $\mathcal{S}^0_{inc}$ be the set of increasing single variable functions as defined in [9].

Now consider the following two operations for functions under (max,+) algebra.

(i) (*max*) the point wise maximum of two functions F and G:

   $(F \oplus G) (m, n) = \max [F(m, n), G(m, n)]$.

(ii)(*convolution*) the convolution of two functions F and G with respect to an integer-valued sequence $L \in \mathcal{S}^0_{inc}$:

   $(F *_L G) (m, n) = \max_{m \leq k \leq n} [F(m, k) + G(L(k), L(n))]$.

If $L(n) = n$, we simply write $*_L$ as $*$. Define the zero function $E(m, n) = -\infty\ \forall m \leq n$ and the identity function $I(m, n) = 0$ when $m = n$ and $-\infty$ otherwise.

We have the following properties:

1. (Associativity) $\forall F, G, H \in \mathcal{S}^l$

$$(F \oplus G)\ \oplus H = F \oplus (G \oplus H)$$

2. (Commutativity) $\forall F, G \in \mathcal{S}^l$

$$F \oplus G = G \oplus F$$

3. (Idempotency of addition) $\forall F \in \mathcal{S}^l$

$$F \oplus F = F$$

4. (Distributivity) $\forall F, G, H \in \mathcal{S}^l$

$$(F \oplus G) *_L H = (F *_L H) \oplus (G *_L H)$$
$$H *_L (F \oplus G) = (H *_L F) \oplus (H *_L G)$$

5. (Monotonicity) For $F \leq F^1$ and $G \leq G^1$

$$F \leq F \oplus G \leq F^1 \oplus G^1$$
$$F *_L G \leq F^1 *_L G^1.$$

6. (Zero element) $\forall F \in \mathcal{S}^l$

$$F \oplus E = F$$

We define the convolution of a single variable function $f \in \mathcal{S}^0_{inc}$ and a bivariate function $G \in \mathcal{S}^l$ as follows

$$(f *_L G)(n) = \max_{0 \le m \le n} [f(m) + G(L(m), L(n))]$$

Under such definition, $f *_L G$ is in $\mathcal{S}^0_{inc}$.

Lemma 1. For $f \in \mathcal{S}^0_{inc}$ and $G, H \in \mathcal{S}^l$,

$$(f *_L G) *_L H \le f *_L (G * H)$$

with equality holding if $L(n) = n$ for all n.

Proof.

$$((f *_L G) *_L H)(n)$$
$$= \max_{0 \le m \le n} [\max_{0 \le u \le m} [f(u) + G(L(u), L(m))] + H(L(m), L(n))]$$
$$= \max_{0 \le u \le n} [f(u) + \max_{u \le m \le n} [G(L(u), L(m)) + H(L(m), L(n))]]$$
$$\le \max_{0 \le u \le n} [f(u) + \max_{L(u) \le k \le L(n)} [G(L(u), k) + H(k, L(n))]]$$
$$= f *_L (G * H)(n)$$

The third step in above derivation follows from the properties that $L(n) \ge n$, $\forall\ n$ and the non-decreasing nature of the functions G and H. Cleary equality holds when $L(n) = n$. □

The key difference to the shift invariant filtering theory is that the operator $*$ is not commutative. For the operator $*$, the identity and absorbing functions are defined as follows.

$$F * E = E * F = E$$
$$F * I = I * F = F$$

Now define the shift operator function $I_w(m, n) = 0$ when $m = n - w$ and $-\infty$ otherwise. Also note that as in [9],

$$I_w * I_u = I_{w+u} \tag{3}$$

Now for a single variable function $f \in \mathcal{S}^0_{inc}$, the convolution with shift operator result in the following

$$(f * I_w)(n) = f(n-w) \text{ if } n \ge w$$
$$= -\infty \text{ otherwise} \tag{4}$$

For a bivariate function $F \in \mathcal{S}^l$, the right convolution with shift operator is defined as follows

$$(F * I_w)(m, n) = F(m, n-w) \text{ if } m \le n - w$$
$$= -\infty \text{ otherwise} \tag{5}$$

Similarly for the bivariate function $F \in \mathcal{S}^l$, the left convolution with shift operator is defined as follows

$$(I_w * F)(m, n) = F(m+w, n) \text{ if } m \le n - w$$
$$= -\infty \text{ otherwise} \tag{6}$$

From the above definitions it can be observed that the convolution with shift operator is not commutative as in the case of shift invariant setting.

Remark 2. A bivariate function $F \in \mathfrak{I}^1$ is super-additive if $F(m, l) + F(l, n) \leq F(m, n)$ for all $m \leq l \leq n$. Hence for an super-additive bivariate function, $F * F = F$. In the special case when the inequality is equality in the above equation, the bivariate function is called additive function. □

For any bivariate function $G \in \mathfrak{I}^1$, define the mth self-convolution of G by the following recursive equations:

$$G^{(1)} = G,$$
$$G^{(m)} = G^{(m-1)} * G, m \geq 2$$

Similar to [10], we define the closure of G as follows:

$$G^* = \text{Lim}_{n \to \infty} G^{(n)} \tag{7}$$

Remark 3. There is a difference between the closure defined in (7) and that in [9]. In [9] closure for the single variable function $g \in \mathfrak{I}_{\text{inc}}$ is defined as

$$g^* = \text{Lim}_{n \to \infty} (g \oplus I_0)^{(n)} \tag{8}$$

where I_0 is the identity function in $\mathfrak{I}_{\text{inc}}$. In bivariate case we do not use the corresponding identity function as we assume for all $G \in \mathfrak{I}^1$, $G(m, m) = 0$. □

Since $G(m, m) = 0$, expanding (7) recursively yields,

$$G^*(m, n) = \max\left\{\sum_{i = m, m+1, \ldots n} G(k_i, k_{i+1}) : k_m = m \leq k_{m+1} \leq \ldots \leq k_n \leq k_{n+1} = n\right\} \tag{9}$$

Thus, for example $G^*(m, m+1) = G(m, m+1)$, $G^*(m, m+2) = G * G (m, m+2)$ and $G^*(m, m+3) = G * G * G (m, m+3)$. Also $G^* * G(m, n) = G^*(m, n)$ for all $m \leq n$. Also similar to [11], the closure G^* of the bivariate function $G \in \mathfrak{I}^1$ can be computed using the following recursive formulation.

$$G^*(m, m) = 0, G^*(m, n) = \max_{m \leq k < n}\left[G^*(m, k) + G(k, n)\right] \tag{10}$$

Lemma 4. For any $w > 0$, there exists $k \geq 0$ such that for $F \in \mathfrak{I}^1$ the self convolution of $I_w * F$, k times i.e. $(I_w * F)^{(k)} (m, n) = -\infty$

Proof. Using the formula in (9),

$(I_w * F)^{(k)} (m, n)$

$= \max_{m \leq t1 \leq t2 \ldots \leq tk \leq n} \{ I_w * F (m, t_1) + I_w * F (t_1, t_2) + \ldots I_w * F (t_k, n)\}$

$= \max_{m \leq t1 \leq t2 \ldots \leq tk \leq n} \{ F (m+w, t_1) + F (t_1+w, t_2) + \ldots F (t_k +w, n)\}$

Clearly from (5) it can be seen that all the terms in the above *max* summation becomes $-\infty$ when $k > (n - m) / w$. □

3 Dynamic Traffic Regulation

We follow the similar traffic characterization in [9]. An arrival process to telecommunication network is characterized by two sequences of variables: the arrival times and the service requirements (packet lengths). This can be represented by a marked point process $\psi = (\tau, l)$ with $\tau = \{ \tau(n), n=0,1,2\ldots\}$ and $l = \{l(n), n=0,1,2\ldots \}$ where $\tau(n)$ is the arrival time of $(n+1)$th packet and $l(n)$ is the service requirement (packet length) of $(n+1)$th packet. Define $L(n)$ as the sum of the service requirements of the first n packets. Let $L(0) = 0$. Similar to [9], we assume for a marked point process $\psi = (\tau, l)$, the sequence τ is an increasing sequence and l is a non-negative integer valued sequence. Under such assumption for l, the sequence $L(n)$ is an increasing integer valued sequence with $L(0) = 0$.

Definition 5 (Traffic characterization). A marked point process $\psi = (\tau, l)$ is said to be G-regular for some $G \in \mathfrak{S}^1$ if for all $m \leq n$ there holds $\tau(n) - \tau(m) \geq G(L(m), L(n))$. As $G(n, n) = 0$, this characterization can be rewritten as

$$\tau(n) = \max_{0 \leq m \leq n} [\tau(m) + G(L(m), L(n))] \tag{11}$$

or simply,

$$\tau = \tau *_L G \tag{12}$$

Note that the traffic characterization in Definition 5 provides a lower bound on the distance between two arrivals. □

In the following theorem, we show that one can construct a traffic regulator such that its output G-regular, provided that the bivariate function G is super-additive as defined in Remark 2.

Theorem 6 (Minimal G-regulator). Suppose $G \in \mathfrak{S}^1$ and G is super-additive. Consider a marked point process $\psi = (\tau, l)$. Let $\tau_1 = \tau *_L G$ .i.e.,

$$\tau_1(n) = \max_{0 \leq m \leq n} [\tau(m) + G(L(m), L(n))],$$

Construct the marked point process $\psi_1 = (\tau_1, l)$.

(i) *(Traffic regulation)* ψ_1 is G-regular.
(ii) *(Optimality)* For any G-regular marked point process $\psi_2 = (\tau_2, l)$ with $\tau_2 \geq \tau$, we have $\tau_2 \geq \tau_1$.
(iii) *(Conformity)* ψ is G-regular if and only if $\psi_1 = \psi$. The construction ψ_1 is called minimal G-regulator.

Proof. (i) In view of (12) it suffices to show that $\tau_1 *_L G = \tau_1$. Note that

$$(\tau_1 *_L G)(n) = ((\tau *_L G) *_L G)(n)$$

$$= \max_{0 \leq m \leq n} [\max_{0 \leq u \leq m} [\tau(u) + G(L(u), L(m))] + G(L(m), L(n))]$$
$$= \max_{0 \leq u \leq n} [\tau(u) + \max_{u \leq m \leq n} [G(L(u), L(m)) + G(L(m), L(n))]]$$

As G is super-additive,

$$\max_{u \leq m \leq n} [G(L(u), L(m)) + G(L(m), L(n))] = G(L(u), L(n))$$

Thus,

$$(\tau_1 *_L G)(n) = \max_{0 \leq u \leq n} [\tau(u) + G(L(u), L(n))] = (\tau *_L G)(n) = \tau_1(n).$$

(ii) As ψ_2 is assumed to be G-regular, we have from (12) that $\tau_2 = \tau_2 *_L G$. From the assumption $\tau_2 \geq \tau$ and the monotonicity of $*_L$ it follows that $\tau_2 \geq \tau *_L G = \tau_1$.

(iii) if $\psi_1 = \psi$, it then follows from (i) that ψ is G-regular. To prove the converse, suppose that ψ is G-regular. Then $\tau = \tau *_L G = \tau_1$. Thus, $\psi = \psi_1$. □

Example 7 (Time varying G/G/1 queue). Consider a particular case of G/G/1 queue with time varying capacity governed by service rate at time t: $r(t)$. A G/G/1 queue is a single-server queue with deterministic inter-arrival times and deterministic service requirements. Consider feeding the marked point process $\psi = (\tau, l)$ to the time varying G/G/1 queue. Let $\psi' = (\tau', l)$ be the output marked point process when the customer starts service. Let $R(t)$ be the cumulative service time up to time t, i.e.

$$R(t) = \int_{-\infty}^{t} r(s)ds$$

The time varying G/G/1 queue is governed by Lindley's recursive equation as follows

$$\tau'(0) = \tau(0),$$
$$\tau'(n) = \max[\tau(n), \tau'(n-1) + l^1(n-1)], n \geq 1$$

where $l^1(n-1)$ is the service time of $(n-1)^{\text{th}}$ packet which is given by the following equation

$$l^1(n-1) = \inf\{t \geq 0: R(\tau'(n-1) + t) - R(\tau'(n-1)) \geq l(n-1)\} - \tau'(n-1).$$

Expanding recursively, it can be easily seen that,

$$\tau'(n) = \max_{0 \leq m \leq n}\{\tau(m) + F(L(m), L(n))\} \text{ where } F(L(m),L(n)) \text{ is given by}$$
$$F(L(m),L(n)) = \inf\{t \geq 0: R(\tau(m) + t) - R(\tau(m)) \geq (L(n) - L(m))\}$$

Hence the time varying *G/G/1* is minimal *F*-regulator where $F(L(m),L(n))$ is given by the above equation.

To provide an example of *G*-regulation, we extend the segmentation and reassembly in [9] to shift varying setting. In telecommunication network, some times it is necessary to transmit variable length packets by smaller packets with same size. For example segmentation is required for transporting IP packets over an ATM backbone network. Consider a marked point process $\psi = (\tau, l)$. Assume that $l(n) \geq 1$ for all n. Let $L(n)$ be the cumulative service requirement of first n packets and $L(0) = 0$. Its output after the segmentation is the marked point process, $\psi^s = (\tau^s, l^s)$, where

$$l^s(n) = 1, \text{ for all } n, \tag{13}$$
$$\tau^s(n) = \tau(m), L(m) \leq n < L(m + 1) \tag{14}$$

i.e. cells from the same packet have same arrival times. On the other hand for a segmented marked point process $\psi^s = (\tau^s, l^s)$, its output after reassembly is the marked point process $\psi^r = (\tau^r, l^r)$, where

$$l^r(n) = l(n), \text{ for all } n, \tag{15}$$
$$\tau^r(n) = \tau^s(L(n + 1) - 1), \text{ for all } n \tag{16}$$

i.e. the time to reassemble a packet is the arrival time of the last cell in that packet.

Lemma 8 (Segmentation and Reassembly). Assume that $l(n) > 0$ for all n. Let $l_{\max} = \sup_{n \geq 0} l(n)$.

(i) If ψ is *G*-regular, then ψ^s is G^s-regular, where $G^s(m,n) = G(m, n - (l_{\max} - 1))$.

(ii) If ψ^s is *G*-regular, then ψ^r is G^r-regular, where $G^r(m,n) = G(m + (l_{\max} - 1), n)$.

Proof. (i) For $k_2 < k_1$, assume that $L(n) \leq k_1 < L(n + 1)$ and $L(m) \leq k_2 < L(m + 1)$ for some $m \leq n$. Thus $k_1 \leq L(n) + l_{max} - 1$. It then follows from (14) and G-regularity for ψ that

$$\tau^s(k_1) - \tau^s(k_2) = \tau(n) - \tau(m) \geq G(L(m), L(n))$$
$$\geq G(k_2, L(n)) \geq G(k_2, k_1 - (l_{max} - 1))$$

This shows that ψ^s is G^s-regular.

(ii) Note from (16) and G-regularity for ψ^s that

$$\tau^s(n) - \tau^s(m) = \tau^s (L(n + 1) - 1) - \tau^s (L(m + 1) - 1)$$
$$\geq G(L(m+1) - 1, L(n+1) - 1)$$
$$\geq G(L(m) + (l_{max} - 1), L(n+1) - 1)$$
$$\geq G(L(m) + (l_{max} - 1), L(n))$$

The third and fourth steps in the above inequality follows from the facts that $L(m+1) - 1 \leq L(m) + (l_{max} - 1)$ and $L(n+1) - 1 \geq L(n)$. □

4   Dynamic Service Guarantees

In this section we extend the g-server concept in [9] to shift varying setting for modeling service guarantees in shift varying setting.

Definition 9 (Dynamic G-server): A server is called G-server ($G \in \mathfrak{I}^1$) for an input marked point process $\psi = (\tau, l)$ if its output marked point process $\psi_1 = (\tau_1, l)$ satisfies $\tau_1 \leq \tau *_L G$, i.e.

$$\tau_1(n) \leq \max_{0 \leq m \leq n} [\tau(m) + G(L(m), L(n))] \tag{17}$$

for all n. If the inequality in (17) is satisfied for all input marked point processes, then we say that G-server is universal. If the inequality is equality, the G-server is exact. Clearly, the minimal G-regulator is universal and exact G-server. Also we note that if a server is a G_1-server for a marked point process $\psi = (\tau, l)$ and $G_1 \leq G_2$ for some $G_2 \in \mathfrak{I}^1$, then the server is also a G_2-server for $\psi = (\tau, l)$.

Example 10 (Segmentation and Reassembly). Consider transmitting a marked point process $\psi = (\tau, l)$ via segmentation and reassembly as in Lemma 8. Assume that the segmented marked point process $\psi^s = (\tau^s, l^s)$ is fed into a G-server. Let $\psi^{s,1} = (\tau^{s,1}, l^s)$ be the output from the G-server. From the Definition 9 and $l^s(n) = 1$ from (13), it follows that

$$\tau^{s,1}(n) \leq \max_{0 \leq m \leq n} [\tau^s(m) + G(m, n)] \tag{18}$$

The marked point process $\psi^{s,1}$ is then reassembled via (15) and (16). Let $\psi^1 = (\tau^1, l)$ be the output after the reassembly. From (16) and (18), we have

$$\tau^1(n) = \tau^{s,1} (L(n + 1) - 1)$$
$$\leq \tau^s(m^*) + G(m^*, L(n + 1) - 1)$$

for some $0 \leq m^* \leq L(n + 1)$. Now suppose that $L(m) \leq m^* < L(m + 1)$ for some $0 \leq m \leq n$. Hence from (18)

$$\tau^1(n) \leq \tau(m) + G(m^*, (L(n + 1) - 1))$$
$$\leq \tau(m) + G(L(m), (L(n) + l_{max} - 1))$$
$$\leq \max_{0 \leq k \leq n} [\tau(k) + G(L(k), (L(n) + l_{max} - 1))]$$

Hence ψ^1 is output from G_1-server where $G_1(m, n) = G(m, n + l_{max} - 1)$. $\square$

In the following theorem we prove performance bounds for G-servers.

Theorem 11 (Performance bounds). Consider a G_2-server for marked point process $\psi = (\tau, l)$. Let $\psi^1 = (\tau^1, l)$ be the output. Suppose that ψ is G_1-regular. Let G_1 be super-additive function. Let d be the maximum delay at the G_2-server.

(i) (*Maximum delay*) $d \leq \sup_m \max_{n \geq m} [G_2(m, n) - G_1(m, n)]$

(ii) (*Output characterization*) ψ^1 is G_3-regular where $G_3 = G_1 + \inf_k \min_{l \geq k} [G_1(k, l) - G_2(k, l)]$

Proof. (i) As we assume that the server is a G_2-server for ψ and ψ is G_1-regular

$$\tau^1(n) - \tau(n) \leq \max_{0 \leq m \leq n} [\tau(m) + G_2(L(m), L(n))] - \tau(n)$$
$$= \max_{0 \leq m \leq n} [G_2(L(m), L(n)) - (\tau(n) - \tau(m))]$$
$$\leq \max_{0 \leq m \leq n} [G_2(L(m), L(n)) - G_1(L(m), L(n))]$$
$$\leq \sup_m \max_{n \geq m} [G_2(m, n) - G_1(m, n)]$$

(ii) Now

$$\tau^1(n) - \tau^1(m) \geq \tau(n) - \max_{0 \leq k \leq m} [\tau(k) + G_2(L(k), L(m))]$$
$$= \min_{0 \leq k \leq m} [\tau(n) - \tau(k) - G_2(L(k), L(m))]$$
$$\geq \min_{0 \leq k \leq m} [G_1(L(k), L(n)) - G_2(L(k), L(m))]$$

Since G_1 is assumed to be a super-additive function,

$$\tau^1(n) - \tau^1(m)$$
$$\geq \min_{0 \leq k \leq m} [G_1(L(k), L(m)) + G_1(L(m), L(n)) - G_2(L(k), L(m))]$$
$$= \min_{0 \leq k \leq m} [G_1(L(k), L(m)) - G_2(L(k), L(m)) + G_1(L(m), L(n))]$$
$$= G_1(L(m), L(n)) + \min_{0 \leq k \leq m} [G_1(L(k), L(m)) - G_2(L(k), L(m))]$$
$$\geq G_1(L(m), L(n)) + \inf_k \min_{l \geq k} [G_1(k, l) - G_2(k, l)] .$$

In the following theorem, we provide another example of dynamic G-server called maximum delay server. Similar to shift invariant filtering theory, the dynamic maximum delay server provides an upper bound on delay at the server.

Theorem 12 (Maximum delay server): Let O_D be the bivariate function with $O_D(m, n) = D$ for all m, n. A server guarantees maximum delay D for a marked point process $\psi = (\tau, l)$ if and only if it is an O_D-server for ψ.

Proof. Note that for all n

$$\tau(n) + D = \max_{0 \leq m \leq n} [\tau(m) + D] = \max_{0 \leq m \leq n} [\tau(m) + O_D(L(m), L(n))] = (\tau *_L O_D)(n).$$

Let $\psi_1 = (\tau_1, l)$ be the output from the server. It is then clear that $\tau_1(n) \leq \tau(n) + D$ if and only if $\tau_1(n) \leq (\tau *_L O_D)(n)$. $\square$

5 Shift Varying Filtering Operations

In this section we present the basic filtering theory operations like concatenation, filter-bank summation and feedback for network elements in shift varying setting.

Theorem 13 (Concatenation). A concatenation of a G_1-server for a marked point process $\psi = (\tau, l)$ and a G_2-server for the output from G_1-server is a G-server for ψ where $G = G_1 * G_2$.

Proof. Let $\psi_1 = (\tau_1, l)$ be the output from G_1-server and $\psi_2 = (\tau_2, l)$ be the output from G_2-server. It then follows that

$$\tau_1 \le \tau *_L G_1 \text{ and } \tau_2 \le \tau_1 *_L G_2$$

From the monotonicity of $*_L$ and Lemma 1, we have

$$\tau_2 \le (\tau *_L G_1) *_L G_2 \le \tau *_L (G_1 * G_2)$$

Thus $\psi2$ is the output from a G1 * G2 –server for ψ. $\square$

Theorem 14 (Filter bank summation). Consider a marked point process $\psi = (\tau, l)$. Let $\psi^1 = (\tau^1, l)$ (resp. $\psi^2 = (\tau^2, l)$) be the output from a G_1-server (resp. G_2-server) for ψ. The output from the "filter bank summation", denoted by $\psi^3 = (\tau^3, l)$, is constructed from $\tau^3 = \tau^1 \oplus \tau^2$. Then the "filter bank summation" of a G_1-server for ψ and a G_2-server for ψ, where $G = G_1 \oplus G_2$.

Proof. As we assume that ψ^1 is the output from the G_1-server for ψ,

$$\tau^1 \le \tau *_L G_1$$

Similarly,

$$\tau^2 \le \tau *_L G_2$$

It then follows from distributive property that,

$$\tau^3 = \tau^1 \oplus \tau^2 \le \tau *_L G_1 \oplus \tau *_L G_2 \le \tau *_L (G_1 \oplus G_2).$$

Example 15 (Jitter control). Consider feeding a marked point process $\psi = (\tau, l)$ to an O_D-server. Let $\psi^1 = (\tau^1, l)$ be the output from the O_D-server. The maximum jitter between the ψ^1 and ψ is defined by the maximum difference between the interarrival times of τ^1 and τ, i.e.

$$j_{max} = \sup\nolimits_{n \ge 1}[\tau^1(n) - \tau^1(n-1) - (\tau(n) - \tau(n-1))]$$

Under the causal condition $\tau^1(n) \ge \tau(n)$ for all n, it is easy to see that the maximum jitter j_{max} is bounded above by the maximum delay D. One way to achieve zero jitter is to delay every customer to maximum delay D. To be precise let $\psi^2 = (\tau^2, l)$ with $\tau^2(n) = \max[\tau^1(n), \tau(n) + D]$. As ψ^1 is the output from the O_D-server, $\tau^1(n) \le \tau(n) + D$ and $\tau^2(n) = \tau(n) + D$. Thus there is no jitter between ψ^2 and ψ. In view of this one can have partial jitter control by setting $\psi^3 = (\tau^3, l)$ with $\tau^3(n) = \max[\tau^1(n), \tau(n) + D_1]$ for some $0 \le D_1 \le D$. Equivalently,

$$\tau^3 = \tau^1 \oplus (\tau *_L O_{D1})$$

Thus ψ^3 is the output from the "filter bank summation" of an O_D-server for ψ and an exact O_{D1}-server for ψ. It then follows from Theorem 14 that ψ^3 is the output from an O_D-server as $O_D \oplus O_{D1} = O_D$. On the other hand, one has from the monotonicity of $\oplus$ that $\tau^3 \ge \tau *_L O_{D1}$. Thus we have for all n,

$$\tau(n) + D_1 \le \tau^3(n) \le \tau(n) + D \qquad (19)$$

This implies that

$$\tau^3(n) - \tau(n) + \tau(n - 1) - \tau^3(n - 1) \le D - D_1$$

from which it can be observed that the maximum jitter between ψ^3 and ψ is then bounded above by $D - D_1$, while the maximum delay is bounded above by D. If ψ is G-regular, then it follows from (19) that for $m < n$

$$\tau^3(n) - \tau^3(m) \ge \tau(n) + D_1 - (\tau(m) + D)$$
$$= \tau(n) - \tau(m) - (D - D_1)$$
$$\ge G(L(m), L(n)) - (D - D_1)$$

Thus ψ^3 is G_3-regular, where $G_3(m, n) = (G(m, n) - (D - D_1))^+$.

Example 16 (Damper). Similar to [9], we describe a more sophisticated jitter control called Damper in shift varying setting. A damper for a marked point process $\psi = (\tau, l)$ is a "filter bank summation" of an G_1-server and an exact G_2-server with $G_1 \ge G_2$. Assume that G_2 is super-additive. Since $G_1 \oplus G_2 = G_1$, the damper is also an G_1-server. Let $\psi^1 = (\tau^1, l)$, $\psi^2 = (\tau^2, l)$ and $\psi^3 = (\tau^3, l)$ be the outputs from the G_1-server, the G_2-server and the damper. It then follows from the monotonicity of $\oplus$ that

$$\tau^3 = \tau^1 \oplus \tau^2 \ge \tau^2 = \tau *_L G_2$$

Since the damper is also a G_1-server, we have

$$\tau *_L G_2 \le \tau^3 \le \tau *_L G_1$$

Thus for $m < n$,

$$\tau^3(n) - \tau^3(m) \ge \max_{0 \le u \le n} [\tau(u) + G_2(L(u), L(n))] - \max_{0 \le u \le m} [\tau(u) + G_1(L(u), L(m))]$$

Let m^* be the argument that achieves the maximum in $\max_{0 \le u \le m} [\tau(u) + G_1(L(u), L(m))]$, i.e.

$$\max_{0 \le u \le m} [\tau(u) + G_1(L(u), L(m))] = \tau(m^*) + G_1(L(m^*), L(m)).$$

Since $0 \le m^* \le m$, we have for $m < n$,

$$\tau^3(n) - \tau^3(m) \ge \tau(m^*) + G_2(L(m^*), L(n)) - \tau(m^*) - G_1(L(m^*), L(m))$$
$$= G_2(L(m^*), L(n)) - G_1(L(m^*), L(m))$$
$$\ge G_2(L(m^*), L(m)) + G_2(L(m), L(n)) - G_1(L(m^*), L(m))$$
$$\ge G_2(L(m), L(n)) + \inf_k \min_{l \ge k} [G_2(k, l) - G_1(k, l)]$$

The last step in the above derivation follows a similar argument as in Theorem 11 (ii). Hence the output ψ_3 from the damper is G_3-regular where $G_3(m, n) = G_2(m, n) + \inf_k \min_{l \ge k} [G_2(k, l) - G_1(k, l)]$. □

In the following theorem, we extend the g-server with feedback in [9] to shift varying setting.

Theorem 17 (Feedback). Consider a feedback system with the input $\psi = (\tau, l)$ and the output $\psi^1 = (\tau^1, l)$. In the feedback system, $\psi^2 = (\tau^2, l)$ is the output from $I_w * G$ server for ψ^1, and the feedback system is joined by the following equation

$$\tau^1 = \tau \oplus \tau^2 \tag{20}$$

If $w > 0$ and $L(n) < \infty$ for all n, then ψ^1 is the output from a $(I_w * G)^*$-server for ψ.

Proof. As ψ^2 is the output from a $I_w * G$ –server for ψ^1

$$\tau^2 \leq \tau^1 *_L (I_w * G)$$

From the monotonicity of $\oplus$, we have from (20) that

$$\tau^1 \leq \tau \oplus (\tau^1 *_L (I_w * G)) \tag{21}$$

Iterating (21) yields

$$\tau^1 \leq \tau \oplus ((\tau \oplus (\tau^1 *_L (I_w * G))) *_L (I_w * G))$$
$$\leq \tau \oplus (\tau *_L (I_w * G)) \oplus (\tau^1 *_L (I_w * G)^{(2)}$$

where we use the distributivity and Lemma 1. Thus iterating (21) m times yields

$$\tau^1 \leq \tau \oplus (\tau *_L (I_w * G)) \oplus \ldots\ldots \oplus (\tau *_L (I_w * G)^{(m)}) \oplus (\tau^1 *_L (I_w * G)^{(m+1)}) \tag{22}$$

As $w > 0$ and $L(n) < \infty$, from Lemma 4, (22) can be expanded until the last term becomes $-\infty$. Now (22) is reduced to

$$\tau^1 \leq \tau \oplus (\tau *_L (I_w * G)) \oplus \ldots\ldots \oplus (\tau *_L (I_w * G)^{(m)}) \leq \tau *_L (I_w * G)^*$$

The last derivation follows from the closure property of $*$ and the monotonicity of $*_L$. This shows that feedback system is indeed a $(I_w * G)^*$ server. □

Example 18 (Window Flow Control). Let $\psi = (\tau, l)$ and $\psi^1 = (\tau^1, l)$ be the input and output to a network. Assume that the service requirements are all fixed 1, i.e. $l(n) = 1$ for all n. This in turn implies $L(n) = n$ for all n. Let $q(t)$ be the total number of customers at time t. A window flow control mechanism with windows size w works as follows: a customer after it arrives is allowed to enter the network at time t if $q(t) \leq w$ immediately after its entrance. Thus in a window flow control network with window size w, the total number of customers in the network never exceeds w. To avoid triviality, we assume that $w > 0$. To model such mechanism, $\psi^2 = (\tau^2, l)$ with

$$\tau^2 = \tau^1 * I_w \tag{23}$$

The effective input to the network $\psi^3 = (\tau^3, l)$ is then controlled by

$$\tau^3 = \tau \oplus \tau^2 \tag{24}$$

Note from (23) that $\tau^2(n) = -\infty$ if $n < w$, and $\tau^2(n) = \tau^1(n - w)$ if $n \geq w$. Thus $(n+1)^{\text{th}}$ customer is allowed to enter the network at

$$\tau^3(n) = \tau(n) \text{ if } n < w$$
$$= \max[\tau(n), \tau^1(n - w)] \text{ if } n \geq w$$

Suppose that the network is a G-server for the effective input ψ^3, i.e.

$$\tau^1 \leq \tau^3 * G \tag{25}$$

In view of (23)-(25), we have

$$\tau^1 \leq \tau^3 * G \leq (\tau \oplus \tau^2) * G$$
$$\leq (\tau \oplus \tau^1 * I_w) * G = (\tau * G) \oplus (\tau^1 * I_w * G)$$

Expanding the above similar to in Theorem 17 results in the following result

$$\tau^1 \leq \tau * G * (I_w * G)^* \tag{26}$$

Hence the window flow control network is a $G * (I_w * G)^*$ server. □

6 Conclusions and Future Work

We have extended filtering theory under (max,+) algebra to shift varying setting. We have introduced two basic network elements: G-regulator and dynamic G-server. We have shown that these network elements can be joined by filtering theory operations like concatenation, "filter bank summation" and feedback. We have illustrated the use of this framework by examples like segmentation and reassembly, jitter control, damper and window flow control. Our extension of filtering theory is mainly useful for modeling service guarantees for time dependent traffic streams comprising variable length packets. In the future we would like to extend this framework to stochastic traffic models.

References

1. R L. Cruz, "A calculus for network delay, Part I: Network elements in isolation," IEEE Trans. inform. theory, vol. 1, pp. 114–131, January 1991.
2. R L. Cruz, "A calculus for network delay, Part II: Network Analysis," IEEE Trans. inform. theory, vol. 1, pp. 132–141, January 1991.
3. R L. Cruz, "Quality of service guarantees in virtual circuit switched networks," IEEE Trans. inform. theory, IEEE J. selected areas in communications, vol. 13, pp. 1048-1056, August 1995.
4. Hanrijanto Sariowan, Rene L. Cruz, and George C. Polyzos, "SCED: A generalized scheduling policy for guaranteeing quality-of-service," IEEE/ACM Transactions on networking, vol. 7, pp. 669-684, October 1999.
5. Jean-Yves Le Boudec, Patrick Thiran, "Network calculus," Springer-Verlag
6. Cheng-Shang Chang, "On deterministic traffic regulation and service guarantees: A systematic approcah by filtering," IEEE Trans. inform. theory, vol. 5, pp. 1097–1110, May 1998.
7. Cheng-Shang Chang, "On deterministic traffic regulation and service guarantees: A systematic approcah by filtering," IEEE Trans. inform. theory, vol. 5, pp. 1097–1110, May 1998.
8. Cheng-Shang Chang, "Matrix extensions of the filtering theory for deterministic traffic regulation and service guarantees," IEEE J. selected areas in communications, vol. 16, pp. 708-718, June 1998.
9. Cheng-Shang Chang, Yih Haur Lin, ``A general framework for deterministic service guarantees in telecommunication networks with variable length packets," IEEE Trans. on automatic control, vol. 46, pp. 210-221, 2001.
10. Cheng-Shang Chang, Rene L. Cruz, Jean-Yves Le Boudec and Patrick Thiran "A min-plus system theory for constrained traffic regulation and dynamic service guarantees," IEEE/ACM Transactions on Networking, vol. 10, pp. 805-817, October 2002.
11. R.L. Cruz, M. Taneja, "An analysis of traffic clipping," Proceedings of international conference on information sciences and systems, Princeton University (1998).
12. Ashok Karumanchi, Sridhar Varadarajan and Srinivas Talabattula, "Distributed traffic regulation and delay guarantees with a centralized controller in WDM passive optical networks," OSA Journal of optical networking, vol. 2, no. 7, pp. 202-212, July 2003.

Robust Delay Estimation
of an Adaptive Scheduling Algorithm

Johanna Antila and Marko Luoma

Networking Laboratory of Helsinki University of Technology,
Otakaari 5A, Espoo, FIN 02015, Finland
Tel: +358 9 451 6097, Fax: +358 9 451 2474
{johanna.antila,marko.luoma}@hut.fi
http://www.netlab.hut.fi/~jmantti3

Abstract. In this paper we propose three new packet delay estimators for an adaptive, delay-bounded HPD (DBHPD) scheduling algorithm in the DiffServ context: simple Exponential Weighted Moving Average (EWMA) estimator, EWMA estimator with restart (EWMA-r) and EWMA based on proportional error of the estimate (EWMA-pe). We compare these estimators with the original, simple sum estimator with ns2-simulations using several traffic mixes. We show that the simple sum and EWMA estimators often lead to false scheduling decisions. On the other hand, the EWMA-r and especially the EWMA-pe estimator provide good estimates of the packet delay regardless of the traffic mix.

1   Introduction

During the recent years, adaptivity has become a key word in network provisioning and management. Adaptive mechanisms are attractive since they enable a self-configurable network that is able to adjust itself to different conditions, such as changes in traffic trends, load fluctuations due to time-of-day effects or attachment of new customers. The operation of adaptive mechanisms relies on measurements: the state of the network is monitored either off-line or on-line to produce an estimate of a desired quantity, such as link utilization, average packet loss or average queueing delay. These measurements can be used in different time scales of network control, such as in routing and load balancing, admission control and packet scheduling.

In this paper, we focus on the use of measurements in adaptive packet scheduling in the Differentiated Services (DiffServ) [1] context. The basic idea of adaptive scheduling algorithms is to dynamically adjust the class resources either periodically or on a packet per packet basis, so that the policy chosen by the operator will be fulfilled regardless of the traffic conditions. A few adaptive scheduling algorithms have been proposed in the literature: Christin et al. [2] have proposed a Joint Buffer Management and Scheduling (JoBS) mechanism that provides both absolute and proportional differentiation of loss, service rates and packet delay. In JoBS the buffer management and scheduling decisions are interdependent and are based on a heuristically solved optimization problem. Liao

M. Ajmone Marsan et al. (Eds.): QoS-IP 2005, LNCS 3375, pp. 626–642, 2005.

et al. [3] have defined a dynamic core provisioning method that aims to provide a delay guarantee and differentiated loss assurance for the traffic classes. The algorithm utilizes measurement based closed-loop control and analytic models of an M/M/1/K queue for the resource allocation decisions. Moore et al. [4] have developed an adaptive scheduling algorithm that also aims to provide a delay guarantee for one class and loss assurance for the other classes. The idea of their algorithm is to allocate the resources with a rate based scheduler based on the calculated theoretical equivalent bandwidth that would be required to support the delay-bound or loss rate. All these algorithms [2–4] use rate-based scheduling, such as Worst Case Weighted Fair Queueing (WF^{2}Q) [5] or Deficit Round Robin (DRR) [6] that adapts the capacity allocation of the classes.

Our approach is to base the resource adaptation on packet delay: In [8], we have proposed a delay-bounded HPD (DBHPD) scheduling algorithm for combined absolute and proportional delay differentiation. In this algorithm, the most delay sensitive class is assigned an absolute delay bound. If this bound is about to be violated, a packet is directly dispatched from this class, otherwise the operation is based on the delay ratios between classes according to [7]. The main reason for using delay-based provisioning is that nowadays most of the traffic is time-critical to some extent. Another important advantage of delay-based algorithms is that over provisioning of resources is not required in order to guarantee small delays, which is the case with rate-based algorithms. We have evaluated the DBHPD algorithm with extensive simulations in [8], [9] and [10] and showed that it performs better than static scheduling algorithms. However, the largest problem with the DBHPD algorithm seems to be the robust estimation of queueing delays. The current simple sum estimation approach used in the algorithm [7] easily leads to overflows in counters and to scheduling decisions that do not support the policy chosen by the operator.

In this paper, we solve this problem by developing more robust delay estimators for the algorithm. We compare four possible estimation approaches: simple sum, simple EWMA, EWMA with restart (EWMA-r) and EWMA based on proportional error of the estimate (EWMA-pe), and show that the new estimators perform better than the simple sum estimator. We propose that the EWMA-r or EWMA-pe estimator should be used for delay estimation in the DBHPD algorithm since they give the best results regardless of the traffic mix.

The rest of this paper is structured as follows: Section 2 describes in detail the DBHPD algorithm and Section 3 presents the possible estimation approaches for the algorithm. Section 4 describes the simulation setups and Section 5 summarizes the results of the tested estimators. Finally, Section 6 concludes the paper.

2   The Algorithm

The delay-bounded HPD (DBHPD) algorithm first checks if the packet in the highest class (class 0) is about to violate its deadline. Denote by d_{max} the delay bound in the highest class, by t_{safe} a safety margin for the delay bound, by

t_{in} the arrival time of the packet in the highest class queue and by t_{curr} the current time. The packet in the highest class queue is considered to be violating its deadline if

$$t_{in} + d_{max} < t_{curr} + t_{safe}. \tag{1}$$

If delay violation is not occurring, the algorithm takes into account the delay ratios between the other classes. Denote by $\bar{d}_i$ the average queueing delay of class i packets and by δ_i the Delay Differentiation Parameter (DDP) of class i. The ratio of average delays in two classes i and j should equal the ratio of DDPs in these classes

$$\frac{\bar{d}_i}{\bar{d}_j} = \frac{\delta_i}{\delta_j}, \quad 1 \le i, j \le N. \tag{2}$$

In [7] this is interpreted so that the normalized average delays of traffic classes must be equal, i.e.,

$$\tilde{d}_i = \frac{\bar{d}_i}{\delta_i} = \frac{\bar{d}_j}{\delta_j} = \tilde{d}_j, \quad 1 \le i, j \le N. \tag{3}$$

The DBHPD algorithm selects for transmission at time t, when the server becomes free, a packet from a backlogged class j with the maximum normalized hybrid delay [7]:

$$j = \arg\max\left(g\tilde{d}_i(m) + (1 - g)\tilde{w}_i(m)\right), \tag{4}$$

where $\tilde{d}_i(m)$ and $\tilde{w}_i(m)$ denote the normalized average queueing delay and the normalized head waiting time (i.e. the waiting time of the first packet) of class i when m packets have departed and $0 \le g \le 1$ is a weighting coefficient. Thus, the algorithm utilizes measurements of both short and long term queuing delays ($\tilde{d}_i(m)$ and $\tilde{w}_i(m)$) in the scheduling decisions. The operation of the algorithm depends largely on how the average delay $\bar{d}_i(m)$ is calculated, because it determines the amount of history that is incorporated into the scheduling decisions. In the next subsections we will present possible estimators for $\bar{d}_i(m)$ and asses their strengths and weaknesses.

3   The Estimators

An ideal estimator should be stable, agile and simple to implement. Stability means that the estimator should provide a smoothed, long term estimate of the desired quantity. Agility on the other hand refers to the ability of the estimator to follow the base level changes accurately enough. The most well known estimators are the Token Bucket (TB) estimator and the exponential weighted moving average (EWMA) estimator [11]. The main benefit of these estimators is their simplicity. However, these estimators can not be tuned to be both stable and agile, since they rely on a static parameter (a weighting factor) that determines how much history the estimate will incorporate. Some estimators based on neural networks have been developed to overcome this problem. However, these estimators are often too complex to implement in router software/hardware. Thus, our basic idea is to maintain the simplicity of the EWMA estimator but modify it in such a way that the estimator operates properly in different regions.

3.1   Simple Sum Estimator

In the original form [7] the average delay of class i after m packet departures, $\bar{d}_i(m)$, is calculated by a simple sum estimator as follows: Denote by $D_i(m)$ the sequence of class i packets that have been served and by $d_i(m)$ the delay of the m'th packet in $D_i(m)$. Then, assuming that at least one packet has departed from class i before the m'th packet

$$\bar{d}_i(m) = \frac{\sum_{m=1}^{|D_i(m)|} d_i(m)}{|D_i(m)|}. \tag{5}$$

However, this kind of calculation to infinity is not feasible in practice, since the counter for the sum of delay values easily overflows as enough packets have been departed from a certain class. Also, we do not want to incorporate infinite history into the estimator and thus into the scheduling decisions.

3.2   Simple EWMA Estimator

A simple approach to eliminate the overflow problem in the sum estimator is to update $\bar{d}_i(m)$ in each packet departure with exponential smoothing as follows

$$\bar{d}_i(m) = (\gamma_i d_i(m) + (1 - \gamma_i)\bar{d}_i(m-1)), \tag{6}$$

where $0 \leq \gamma_i \leq 1$. Now calculation to infinity is not required and the amount of history can be determined by the selection of the γ_i parameters. We believe that separate γ_i should be used for each traffic class, since the traffic characteristics of the classes can be totally different. In principle, the value of γ_i should be related to the regeneration period of the class queue, since it reflects the timescales of the arriving traffic. However, since determining the regeneration period would require additional measurements, we propose to use a fixed system parameter, namely the queue size, to determine γ_i. Denote by q_i the physical queue size of class i. Then, γ_i can be determined by an approximative function

$$\gamma_i(q_i) = \frac{1}{N * \sqrt{q_i} * ln(q_i)}. \tag{7}$$

If the queue size is small, it can be assumed that the scheduling decision should not depend too much on history, and the value for γ_i will be higher. The square root and logarithm function and the number of classes N are used in Eq. (7) for scaling the γ_i values to a reasonable range, assuming that the queue lengths in a router can range approximately from 10 packets to 10000 packets. It should be noted that Eq. (7) is not an exact, analytically derived expression. However, it provides a good guideline for setting the γ_i values. The value of γ is depicted as a function of the queue length in Figure 1, assuming 4 service classes.

We also argue that the g parameter in Eq. (4) should be separate for each traffic class, since the g and γ parameters together determine how much the scheduling decision depends on history or the current situation. We propose

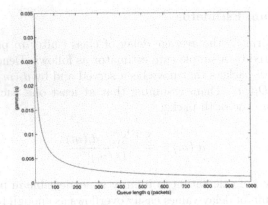

Fig. 1. The gamma function.

that the g parameters should be related to the δ parameters defined in Eq. (2) that reflect the policy of the operator and the real urgency of the packets. The parameters could be set as follows:

$$g_1 = 0.75, \tag{8}$$

$$g_i = g_{i-1} * \frac{\delta_{i-1}}{\delta_i} \tag{9}$$

If according to the policy the packets in some class are urgent, more weight will be given to the current measurement values than to history. In practice the delay-bounded class (class 0) will be used for extremely time-critical traffic, such as VoIP and network control traffic. Class 1 on the other hand will be used for less urgent real-time traffic, such as video. For video traffic it would be reasonable that the scheduling decision is based on the current waiting time with 75% and on history with 25%.

3.3   EWMA Estimator with Restart (EWMA-r)

One problem with both the simple sum and EWMA estimator is that they do not take into account the times when the queue becomes idle. If the queue of a traffic class is idle for a long time, the delay history of that class should not be taken into account. This is because otherwise a class that recently became active and that experiences only little congestion would be served and steal capacity from other classes just because it had large delays in the history. Thus, when a queue becomes active after an idle period, the EWMA estimator should be resetted. The idea is that after resetting, an average of the queuing delay is calculated fast until a certain threshold of packets, p_tresh, has been departed. After the threshold has been reached the delay is smoothed again with the low pass filter, where the γ_i parameters are determined by Eq. (7). The reason for not performing smoothing below the threshold is that if smoothing is started too early, the smoothed value will lag too much behind the real delay experienced. We set p_tresh to $0.25 * q_i$.

In order to know when to restart the estimator, it must be determined when the queue has been empty long enough. Simply restarting each time the queue becomes empty would result in unstable behavior, especially if the incoming traffic is bursty. Thus, we define for each class a variable called $cycle_i$ that indicates when the estimator will be restarted:

$$cycle_i = \frac{abs\_factor_i * q_i * s_i}{C}, \tag{10}$$

where s_i is the mean packet size of the class, C is the link capacity and abs_factor_i is an absorption factor. When absorption factor is 1, $cycle_i$ tells how long it would take to serve the queue of this class if it was full with the capacity of the link.

Denote by $qlen_i$ the queue length of class i and by q_idle_i the time when class i goes idle. Then, EWMA-r operates as follows in traffic class i:

1: **Initialization:**
2: $lowpass\_delay_i = 0.0, p\_samples_i = 0.0, sample\_sum_i = 0.0, qlen_i = 0,$
   $q\_idle_i = 0.0$
3: **Upon each packet departure:**
4: **if** $p\_samples_i < p\_tresh_i$ **then**
5:    $p\_samples_i + = 1$
6:    $sample\_sum_i + = d_i$
7:    $lowpass\_delay_i = sample\_sum_i/p\_samples_i$
8: **else**
9:    $lowpass\_delay_i = \gamma_i * d_i + (1 - \gamma_i) * lowpass\_delay_i$
10: **end if**
11: **if** $qlen_i == 0$ **then**
12:    $q\_idle_i = now$
13: **end if**
14: **Upon each packet arrival:**
15: **if** $q_i == 0$ **then**
16:    $idle\_period = now - q\_idle_i$
17: **end if**
18: **if** $idle\_period \geq cycle_i$ **then**
19:    $lowpass\_delay_i = 0.0, p\_samples_i = 0.0, sample\_sum_i = 0.0$
20: **end if**

3.4 EWMA Estimator Based on Proportional Error of the Estimate (EWMA-pe)

The EWMA estimator with restart provides an updated estimate for the delay when the queue becomes active after an idle period. However, the γ_i and g_i parameters that determine the timescale of the algorithm are still both static. This means that the estimator of each class can be tuned to be either agile or stable but not both. At some point of time the traffic might be bursty and at some point smooth, even if it is assumed that the traffic in each class consists only of one traffic type. Thus, a single filter may not be suitable for the estimation, even if the parameter selection is well argued.

One alternative is to change also the memory of the estimator, γ_i (determined by Eq. (7)) packet per packet based on how much the predicted queuing delay deviates from the real queuing delay value. The memory of the estimator is adapted as follows:

$$\bar{d}_i(m) = (n * \gamma_i d_i(m) + (1 - n * \gamma_i)\bar{d}_i(m-1)), \tag{11}$$

where $0 \leq \gamma_i \leq 1$ is the base weight of class i and n is a multiplier for the base weight. The idea is that the base weight ($n = 1$) determines the longest possible memory for the estimator. The base weight estimator is suitable when there are only little changes in the traffic process and the system is stable. However, if the traffic process is more variable, the value of n is increased and thus the estimator will be more aggressive. The value of n is determined by observing the proportional error of the estimated average value $\bar{d}_i(m)$ to the actual measured delay value $d_i(m)$.

$$n = \begin{cases} 7, \text{ if } 0.4\bar{d}_i(m) > d_i(m) > 1.6\bar{d}_i(m); \\ 6, \text{ if } 0.4\bar{d}_i(m) \leq d_i(m) \leq 1.6\bar{d}_i(m); \\ 5, \text{ if } 0.5\bar{d}_i(m) \leq d_i(m) \leq 1.5\bar{d}_i(m); \\ 4, \text{ if } 0.6\bar{d}_i(m) \leq d_i(m) \leq 1.4\bar{d}_i(m); \\ 3, \text{ if } 0.7\bar{d}_i(m) \leq d_i(m) \leq 1.3\bar{d}_i(m); \\ 2, \text{ if } 0.8\bar{d}_i(m) \leq d_i(m) \leq 1.2\bar{d}_i(m); \\ 1, \text{ if } 0.9\bar{d}_i(m) \leq d_i(m) \leq 1.1\bar{d}_i(m). \end{cases} \tag{12}$$

The selection of the regions for different values of n in Eq. (12) determine how much the estimator reacts to small or to large, sudden changes.

4   Simulations

We have implemented all four delay estimators along with the DBHPD algorithm in the ns2-simulator. We test each estimator with three traffic mixes: pure CBR-traffic, pure Pareto-ON-OFF traffic and mixed traffic from several real applications. Different traffic mixes are used in the evaluation since the performance of the estimators depends largely on the characteristics of incoming traffic. We want to ensure that our results are applicable to more than one particular traffic type.

4.1   General Simulation Parameters

The topology used in the simulations for traffic mix 1 and 2 is depicted in Figure 2. The topology used for traffic mix 3 is the same except that each client and server has a separate access link. We have kept the topology simple, since the aim is not to collect end-to-end performance results but to investigate the queueing delay and delay estimate time-series in the bottleneck link. For this purpose, a complicated topology would add only of little value. Table 1 shows the parameters assigned for the scheduler in the bottleneck link. The delay bound

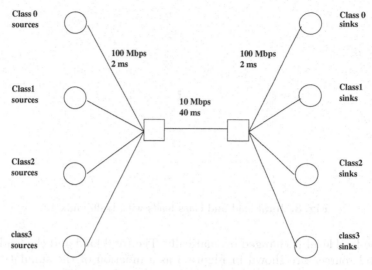

Fig. 2. The topology with traffic mix 1 and 2.

for the class 0 is 5 ms and the target ratio for delays between consecutive classes is 4. The parameters s and *absorption_factor* are used only in the EWMA estimator with restart. All other parameters of the different estimators can be directly derived from the parameters presented in Table 1. The theoretical total offered load in the bottleneck link is 1.0 in traffic mix 2 and 0.8 in traffic mix 3. The reason for using a smaller total load with traffic mix3 is that the TCP retransmissions will increase the theoretical load. In traffic mix 1 the load shares of the classes change over time. The packet sizes with traffic mix 1 and 2 are following: (class 0: 200 bytes, class 1: 750 bytes, class 2: 1000 bytes, class 3: 1000 bytes). In traffic mix 3, the packet sizes are determined by the applications.

Table 1. Scheduler parameters.

Class	$d_{max}(s)$	δ	q (packets)	s (bytes)	absorptionfactor
0	0.005	0.015625	15	200	1.0
1	–	0.0625	15	750	1.0
2	–	0.25	100	1000	1.0
3	–	1.0	100	1000	1.0

4.2   Traffic Mix 1

Traffic mix 1 represents the most simple setup, where the load in each class consists of the traffic sent by a single CBR-source. Thus, the incoming stream will be deterministic during the active periods and easy to predict. We alternate the load level of each source periodically so that at some point of time the source sends traffic with a high, constant speed while in another moment the source is idle. We want to see how the estimators react during the transition periods,

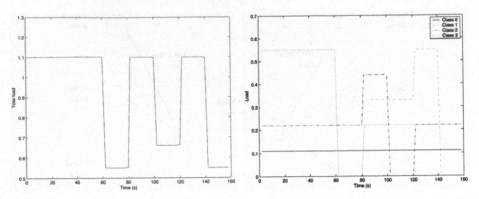

Fig. 3. Total load and class loads with traffic mix 1.

when the load level is changed dramatically. The total load and the loads of the individual sources are shown in Figure 3 as a function of the simulation time. The total simulation time is only 160 s in this scenario, since the traffic process is fully predictable.

4.3 Traffic Mix 2

Traffic mix 2 consists of 15 multiplexed Pareto-ON-OFF sources for each traffic class. The parametrization of the sources is shown in Table 2. It has been shown that multiplexing several Pareto-ON-OFF sources results in self-similarity, which is a fundamental characteristic in Internet traffic. Prediction of self-similar traffic is considerably more difficult than prediction of CBR-traffic, since self-similar traffic is correlated over several time scales. The traffic shares of each class are set as follows: (class 0: 0.1, class 1: 0.2, class 2: 0.4, class 3: 0.3). The simulation time with traffic mix 2 is 2000 s.

Table 2. Parameters for the Pareto sources.

class	on_time (s)	off_time (s)	shape_on	shape_off
0	1.0	1.0	1.2	1.2
1	0.4	0.4	1.5	1.5
2	0.6	0.6	1.4	1.4
3	0.8	0.8	1.3	1.3

4.4 Traffic Mix 3

In traffic mix 3 we use 'real' applications in order to produce a realistic mix of Internet traffic. The mix consists of five different traffic types: FTP, HTTP, Video, VoIP and control traffic (small and large control messages). FTP transfers are also used to represent P2P traffic, which is becoming popular in the Internet. Control and VoIP traffic are mapped to class 0 while Video is mapped to class 1,

HTTP to class 2 and FTP to class 3. The generation of HTTP-traffic is based on the webcache-model implemented in ns2. In this model it is assumed that a HTTP session consists of a number of page requests, each possibly containing several objects. We have also used the webcache-model for FTP-traffic, except that in FTP there is no reading time between page requests, and a page always contains one object. Video traffic generation is based on a real trace of mpeg4 coded BBC news, from which the traffic stream has been regenerated by using a Transform Expand Sample (TES) process. In the simulations the traffic flows to both directions: there is one HTTP client and server (constantly creating new page requests), one FTP client and server, two control traffic sources and sinks and 10 video and VoIP sources and sinks in both sides of the network. The simulation time with traffic mix 3 is 1200 s, and the shares of the different applications are the following: (FTP: 9%, HTTP: 71%, Video: 9%, VoIP: 10%, Control: 1%). This corresponds to the current situation, where the majority of traffic is HTTP. However, in the future, the amount of P2P traffic is assumed to increase dramatically.

5   Results

In this section, we present the simulation results for the different estimators. We show snapshots of both the instantaneous queueing delay time-series and the estimated queueing delay time-series to evaluate the goodness of the estimators. It should be noted that the estimated delays in the figures are calculated according to Eq. (4), i.e. the effect of both the head of line packet delay and the long term delay is taken into account. This is because the scheduling decision is based on the joint effect of these delays, not only on the filtered delay $\bar{d}_i(m)$. Both the instantaneous and estimated queueing delays are real values, i.e. they have not been normalized with δ_i.

5.1   Traffic Mix 1

Figures 4, 5, 6 and 7 depict the instantaneous and estimated queueing delays with different estimators in simulation scenario 1, where the incoming traffic is pure CBR. From Figure 4 (a) and Figure 5, it can be observed that the simple sum estimator leads to false scheduling decisions especially during the moments when a class becomes active after an idle period. Since the sum estimator incorporates an infinite history, the estimate can have a high value even when the real queueing delays are nearly zero. Thus, a packet will be served from a class that has virtually no congestion. In Figure 4 (b) the behavior of the EWMA estimator is shown. For class 1 it leads to a considerably better estimate than the simple sum estimator. However, this is due to the selection of a separate g parameter for each class, defined in Eq. (9). For class 2 and 3 that have small g and thus give only little weight for the head of line packet delay the results resemble closely the simple sum estimator. Thus, the only advantage that the EWMA estimator provides compared with the simple sum estimator is the smaller implementation

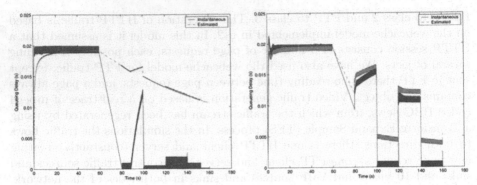

Fig. 4. Instantaneous and estimated delays with simple sum (a) and EWMA (b) estimators for class 1.

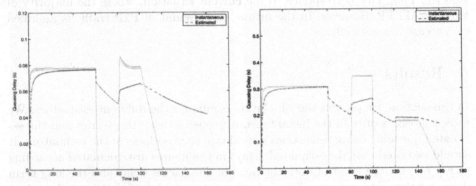

Fig. 5. Instantaneous and estimated delays with simple sum estimator for class 2 and 3.

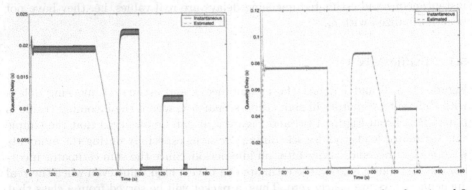

Fig. 6. Instantaneous and estimated delays with EWMA-r estimator for class 1 and 2.

complexity. Figure 6 and 7 depict the instantaneous and estimated delay values with the EWMA-r and EWMA-pe estimator. It is obvious that both of these estimators lead to an accurate prediction and are able to take into account the idle periods. Thus, the scheduling decision will follow the selected differentiation policy.

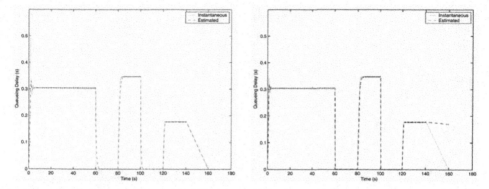

Fig. 7. Instantaneous and estimated delays with EWMA-r (a) and EWMA-pe (b) estimator for class 3.

5.2 Traffic Mix 2

The purpose of simulation scenario 1 was to show how the estimators perform when the classes become active after an idle period. However, since the traffic in scenario 1 was CBR, the queueing delays during the active periods remained almost constant. Thus, a good estimator always matches the actual queueing delays very closely. In scenario 2 the traffic mix consists of Pareto-ON-OFF sources that produce variable rate traffic over several timescales. With this kind of a traffic mix the responsiveness of the estimators to sudden bursts and longer term variations can be explored. Ideally, the estimator should follow the changes in queueing delays but not react too aggressively to sudden peaks. We study the performance of the most promising estimators, the EWMA-r and the EWMA-pe estimator in this scenario.

Figures 8, 9 and 10 present the behavior of the EWMA-r estimator in two timescales: 40 seconds (approximately 67000 packet departures assuming a mean packet size of 750 bytes) and 5 seconds (approximately 8300 packet departures). In the 40 second timescale the estimator seems to follow the instantaneous queue

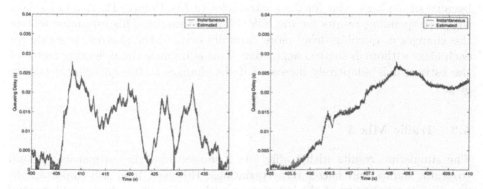

Fig. 8. Instantaneous and estimated delays with EWMA-r estimator in two timescales for class 1.

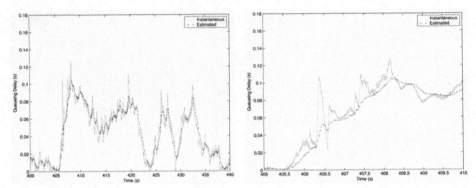

Fig. 9. Instantaneous and estimated delays with EWMA-r estimator in two timescales for class 2.

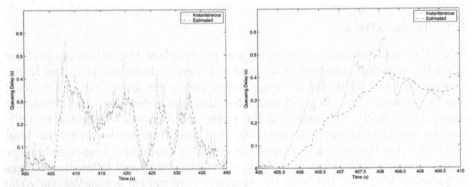

Fig. 10. Instantaneous and estimated delays with EWMA-r estimator in two timescales for class 3.

length quite closely in all traffic classes. However, in the 5 second timescale it can be observed that in class 2 and 3 the estimator smoothes the values quite roughly at some points and ignores the changes in the queueing delays. In class 1 the estimated delay matches the real queueing delay even in the shorter timescale because of the high value for the g parameter (0.75). Figures 11, 12 and 13 show the corresponding results for the EWMA-pe estimator. This estimator follows the changes in queuing delay more carefully even in the shorter timescale for each class without being too aggressive. This is because the weighting factor of the estimator is adaptively increased if the changes in the queueing delay are large.

5.3   Traffic Mix 3

The simulation results with traffic mix 2 showed how the estimators respond to traffic bursts and longer term variations. However, it is still important to examine the behavior of the estimators with real traffic, where a substantial part of the total traffic is carried on top of the TCP-protocol.

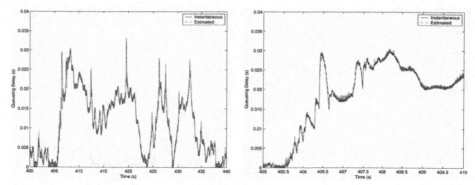

Fig. 11. Instantaneous and estimated delays with EWMA-pe estimator in two timescales for class 1.

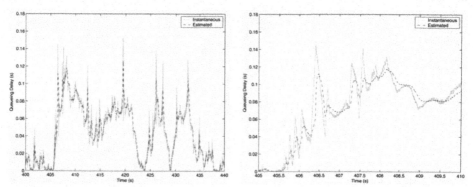

Fig. 12. Instantaneous and estimated delays with EWMA-pe estimator in two timescales for class 2.

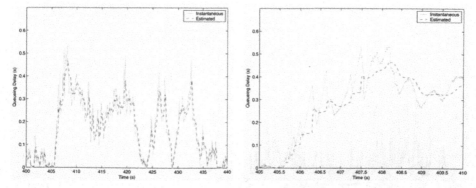

Fig. 13. Instantaneous and estimated delays with EWMA-pe estimator in two timescales for class 3.

Figures 14, 15 and 16 depict the instantaneous and estimated delay time-series with the simple sum, EWMA-r and EWMA-pe estimator in this scenario for class 2 and 3. The figures for class 1 are not shown since in EWMA-r and EWMA-pe estimator the instantaneous and estimated delay time-series are very close to each other due to the high value for the g parameter. The time-scale

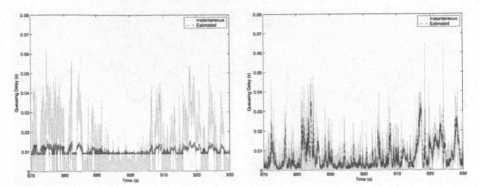

Fig. 14. Instantaneous and estimated delays with sum (a) and EWMA-r (b) estimator for class 2.

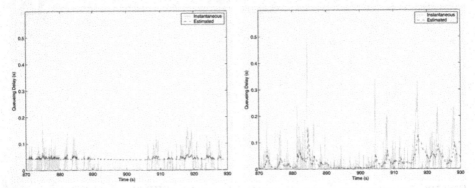

Fig. 15. Instantaneous and estimated delays with sum (a) and EWMA-r (b) estimator for class 3.

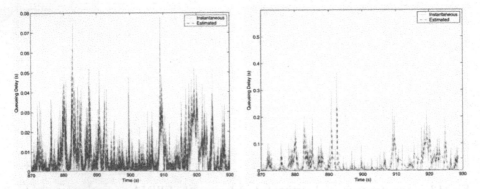

Fig. 16. Instantaneous and estimated delays with EWMA-pe estimator for class 2 and 3.

used is relatively long (60 seconds). Again, the simple sum estimator is either considerably below or above the instantaneous queueing delay value. The effects of the false scheduling decisions can be seen especially for class 3 (FTP traffic): with simple sum estimator the delays of class 3 remain lower than with EWMA-r and EWMA-pe estimator, since the estimated value is much larger than it should be. Thus, the scheduling algorithm decides that a packet must be selected from class 3 because it has high delays. This wrong conclusion also reflects to the other classes, that have correspondingly somewhat larger delays during those periods. However, it should be noted that the absolute delay values with the different estimators can not be directly compared, since the estimators result in different packet loss patterns and thus different queueing delays. To conclude, also with this traffic mix the EWMA-pe estimator gives the best estimation result in the sense that it is agile but still stable enough.

6   Conclusions

In this paper we proposed new delay estimators for the delay-bounded HPD (DBHPD) algorithm: simple Exponential Weighted Moving Average (EWMA) estimator, EWMA estimator with restart (EWMA-r) and EWMA based on proportional error of the estimate (EWMA-pe). We compared these estimators with the original, simple sum estimator with ns2-simulations. We used three traffic mixes in the simulations: pure CBR-traffic, pure Pareto-ON-OFF traffic and mixed traffic from several real applications.

According to the evaluation the simple sum and EWMA estimators often lead to false scheduling decisions. Thus, they are not appropriate solutions for the delay estimation problem. On the other hand, with all traffic mixes the EWMA-r and especially the EWMA-pe estimator proved to be promising alternatives. Furthermore, both the EWMA-r and EWMA-pe require only small changes to the original EWMA estimator, which is extremely simple to implement in practice. However, in order to be able to judge which one of the EWMA-r and the EWMA-pe estimators is better, network level performance evaluations with real traffic should be conducted with both estimators: Even if the estimator approximates the real delays well it might result into undesirable effects such as correlated packet loss patterns and oscillatory behavior especially for TCP-flows. This is because the adaptive scheduling algorithm is a closed-loop system: the estimated delays affect the real queueing delays, and the real queueing delays affect the estimate.

Acknowledgements

This work has been partly supported by the European Union under the E-Next Project FP6-506869 and by the Ironet project funded by TEKES.

References

1. Blake, S., Black, D., Carlson, M., Davies, E., Wang, Z., Weiss, W.: An Architecture for Differentiated Services. IETF RFC 2475, December(1998)
2. Christin, N., Liebeherr, J., Abdelzaher, T.: A Quantitative Assured Forwarding Service. Proceedings of IEEE Infocom, (2002)

3. Liao, R., Campbell, A.: Dynamic Core Provisioning for Quantitative Differentiated Service. Proceedings of IWQoS, (2001)
4. Moore, A.: Measurement-based Management of Network Resources, Ph.D Thesis, University of Cambridge, Computer Laboratory, April (2002)
5. Bennett, J., Zhang, H.: WF^{2}Q: Worst-case Fair Weighted Fair Queueing. Proceedings of IEEE Infocom, 120–127, March(1996)
6. Shreedhar, M., Varghese, G.: Efficient Fair Queuing using Deficit Round Robin. Proceedings of ACM SIGCOMM, (1995)
7. Dovrolis, C., Stiliadis, D., Ramanathan, P.: Proportional Differentiated Services: Delay Differentiation and Packet Scheduling. IEEE/ACM Transactions on Networking 10:2:12–26 (2002)
8. Antila, J., Luoma, M.: Scheduling and quality differentiation in Differentiated Services. Proceedings of Multimedia Interactive Protocols and Systems (MIPS) 2003
9. Antila, J., Luoma, M.: Adaptive Scheduling for Improved Quality Differentiation. To appear in the Proceedings of Multimedia Interactive Protocols and Systems (MIPS) 2004
10. Antila, J., Luoma, M.: Differentiation of Internet Traffic. Proceedings of ICN 2004
11. Young, P.: Recursive estimation and time-series analysis, Springer-Verlag, 1984.

Input Register Architectures of High Speed Router for Supporting the PHB of Differentiated Services*

Sungkwan Youm[1], Seung-Joon Seok[2], Seung-Jin Lee[3], and Chul-Hee Kang[1]

[1] Department of Electronics Engineering, Korea University
1, 5-ga, Anam-dong, Sungbuk-gu, 136-701, Seoul, Korea
{skyoum,chkang}@widecomm.korea.ac.kr
[2] Department of Computer Engineering, Kyungnam University
449 wolyong-dong, masan, 631-701, kyungnam, Korea
sjseok@kyungnam.ac.kr
[3] LG Electrics, Hogye-ldong, Dongan-gu, Anyang-shi, 431-749, Korea
linuz@lge.com

Abstract. In this paper, we presents a new router architecture for supporting the Differentiated Services(DiffServ) in the input-registered router, which has multiple Virtual Output Registers(VORs) for buffering packets ahead of the non-blocking switching fabric. A queuing discipline function block needed on the DiffServ router is added in the input-registered router in order to support Per-Hop Behaviors(PHBs) of DiffServ. Also a new matching algorithm, First Scheduled First Matching (FSFM), is considered to match input packets and output ports in the proposed VOR architecture. The simulation results by using ARENA show that the proposed architecture offer packet loss rate and delay close to the results of output-queued router with N time speed-up switch fabric in various PHBs on DiffServ.

1 Introduction

With the rise of the Internet's popularity, there are many studies ongoing on two issues: One is a need for a faster switching/routing infrastructure. The other is a great demand to provide Qualities-of-Service (QoS). As for the former studies, the input-queued switching architecture is becoming an attractive alternative to design very high-speed routers for the reason that the output-queued switching architecture requires the switch fabric with high speed and performances. Unfortunately, input-queued switches suffer from a blocking phenomenon, called Head of Line blocking (HOL), which limits the throughput of the outgoing links to less than 58.6% on uniform traffics [8]. Some approaches have been proposed to address this problem and proven that the HOL blocking can be completely eliminated by adopting virtual output-queuing, in which multiple Virtual Output Queue (VOQ) directed to different outputs are maintained at each input [4]-[8]. In the latter studies, the Differentiated Services (DiffServ) have been introduced as a scarable solution for QoS provisioning in the Internet [1]-[3]. The basic principle of DiffServ is that routers in DiffServ Domains handle packets of different traffic aggregates by applying different Per-Hop

---

* This research was supported by University IT Research Center Project.

M. Ajmone Marsan et al. (Eds.): QoS-IP 2005, LNCS 3375, pp. 643–653, 2005.

Behaviors (PHB) to the different forwarding mechanism. There are many literatures related to guaranteeing QoS in DiffServ architecture.

Both researches specifically focus on scalability of performance on heavy traffic as one of its fundamental goal but they have been discussed separately; that is, all most of researches on DiffServ have been discussed on output-queued routers. Indeed, recent architecture of DiffServ can not be implemented directly to routers based on input-queued switching architecture. A router with input-queued switching architecture could be designed to provision QoS guarantees with DiffServ. In this paper, we propose a new router architecture with multiple Virtual Output Register(VOR) for provisioning QoS guarantees. We also consider a new matching algorithm, called First Scheduled First Matching (FSFM), which is based on stable matching algorithm, which is first proposed by Gale and Shapley to match input packets and output ports[8].

The rest of the paper is organized as follows. Firstly in section 2, we give the overview of researches on DiffServ and input-queued router architecturees. Section 3 describes the proposed input-registered router architecture for supporting QoS guarantees with DiffServ. In section 4, we describe FSFM to find matching input packets and output ports of switching fabric in the proposed architecture. And we evaluate the performance of the proposed architecture by using ARENA in section 5. Finally we make conclusions and discuss about future works in section 6.

2 Related Works

DiffServ introduced in IETF has developed a simple model to differentiate packets marked with different priorities [1]. In this standard, an individual flow is combined with similar service requirements into a single aggregation. And routers in DiffServ domains handle packets of different mark for applying different PHBs to packet forwarding services.

At each boundary nodes in DiffServ domain, the traffic are passed to packets classifier, which may classifies the packet based on the Differentiated Services Code Point (DSCP) of IP header. After the class of flow is identified, each packet belonging to the same class will be conditioned at a traffic conditioning block according to the traffic specification and profile. Thereafter, each packet is classified according to DSCP and put into the appropriate queue for each PHB at the corresponding output interface [1]-[3].

A behavior aggregate classifier in the interior node of DiffServ domain will classify packets based on the DSCP. At each egress interfaces of this DiffServ node, a scheduling discipline that perform the PHB for each behavior aggregate dispatches the packets among a set of inter-dependent queues. And, a queuing mechanism such as Weighted Fair Queuing or Strict Priority Queuing must collect packets from the different queues and schedule them for transmission over output links.

The backbone needs routers capable of transmission with high speed. Generic backbone router of today has an internal switching fabric, which interconnects input interfaces with output interfaces. In output-queued routers, contentions were occurred among packets to be aggregated on the output interfaces. Hence, a queuing discipline is placed at the output side on the router. Most of previous studies on DiffServ focused on the output scheduling owing to its conceptual simplicity.

The alternative to an output queuing mechanism is an input queuing mechanism, where packets are buffered in queues at the input ports until they are transmitted to output ports. Recently, there are some researches on the provisioning of QoS guarantees in input-queued routers or switches.

In those researches, the Weighted Probabilistic Iterative Matching (WPIM) enables flexible allocations of bandwidth among different links [4]. The WPIM is able to guarantee that traffic from each input port receive its defined shared of the bandwidth of the output link. However, WPIM has a time complexity of $O(N^2)$ for running once in each time slot. The Lowest Output Occupancy First Algorithm (LOOFA) is based on Parallel Iterative Matching (PIM)-like maximal matching algorithm. LOOFA maintains a global input preference list to resolve contentions occurring at each input port. However, LOOFA requires an internal 4 or 6 times speed-up to bound packet delay with or without dependence on switch size respectively.

Instead of using PIM-like algorithms to find matching of input packets and output ports, the Cale-shapley algorithm (GSA) was introduced to address the stable matching problem. An input i port and output j port are said to be a blocking pair for a matching M, if i and j does not match in M, but both i and j prefer each other to their current match in a matching M. A matching is said to be stable if there is no blocking pair of input and output; otherwise it is called unstable[8]. The GSA is able to find stable matching of input and outputs based on the input or output preference lists, which ranks input or output packets in order of preference. This property of stable matching forms the basis of designing schemes for provisioning QoS guarantees.

These recent researches concentrate on the scheduling algorithms for QoS guarantees including throughput, packet delay, and jitter and the router performance. Indeed, very little has been done to implement these QoS scheduling polices on scalable high-speed routers or switches. Moreover, most of above algorithms can't be used directly to implement DiffServ architecture. In this paper, we merge these two mechanisms on a new router architecture with VOR for supporting QoS guarantees specified in Diff-Serv and propose a new matching algorithm, FSFM.

3 The Input-Registered Router

We now explain a new router architecture for provisioning QoS defined in DiffServ. We proposes a high speed DiffServ router with multiple VORs, referred as the input-registered router, for performing PHBs of DiffServ on backbone routers. VORs have a similar architecture with Virtual Output Queues (VOQs) in the input-queued router except that each VOR is various-size temporary registers for buffering packets to be switched through the switching fabric. Fig. 1 shows the conceptual VORs block at the input interfaces. We will describe details of VORs in Fig. 2.

Fig. 1 shows the interior router architecture we proposed. Compared with traditional input-queued routers, a queuing discipline function block, PHB scheduler, is added ahead of VORs at the input interface. This scheduler performs the queuing discipline required for a specific service of each packet in each router. It is to provide QoS guarantees with DiffServ in the input registered router. The queuing discipline function block is similar with that of output-queued routers. In case of an ingress router, an additional function blocks are required as Fig. 1; that is, traffic conditioner function block has to be added ahead of queuing discipline. Similar with the tradi-

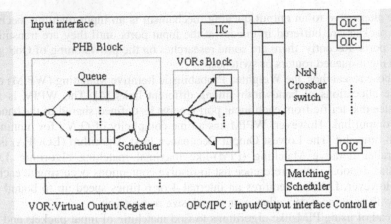

Fig. 1. The input-registered router architecture

tional router architecture, when a packet arrives at an input interface in the ingress router, it is classified, marked with an appropriate DSCP, shaped and policed according to DSCP in the traffic conditioner function blocks. And packets are put to the corresponding buffer in the queuing discipline function block. These components can be implemented by using general DiffServ function blocks.

Provisioning QoS with DiffServ in the input-registered router is based on the concept of PHB, which is defined by a set of forwarding behaviors for the aggregation of traffic with similar or same characteristics in QoS requirements at each router along the path. One of aspect of the proposed architecture is the concept of the "Queuing discipline in the input side." In the case of input-registered routers, a conflict occur among packets at the input side to be switched to the corresponding output interface whereas packets in the output-queued router contend to be transmitted to the output link at the output interface. Therefore, the queuing discipline is placed at the input side in the input-registered router.

Any queuing discipline algorithm is to be applied in the proposed architecture; that is, the Expedited Forwarding (EF) queuing discipline can be implemented using a simple FIFO strategy because EF requires assured bandwidth with low loss, low latency and low jitter in an end-to-end connection. And the Assured Forwarding (AF) queuing discipline can be served according to a complex strategy such as RIO.

The proposed architecture has two phases in each time slot: scheduling phase, matching phase. In scheduling phase, packets are scheduled according to the PHB and scheduler chooses one packet from queues to be put to the VORs function block. Thereafter, the packet is inserted to the specific register in VORs function block. In matching phase, packets in VORs are matched by performing the proposed matching algorithm and switched to the corresponding output ports. The VORs function block caches packets in registers until those packets is switched to the output through the switching fabric.

Fig. 2 depicts various types of VORs in a general model of an NxN router. All VORs is based on the architecture shown in Fig. 1; that is, when more than one packet simultaneously arrives at same input port and are destined for the same output port, they are buffered at some registers in VOR. Therefore, VORs can be free from contention at the input interface. And we consider adapting various resister sizes for scaling traffics on input ports as shown in Fig 2.

AR (Auxiliary Register) buffers the packets that are came from PHB schedulers. In scheduling phase, PHB scheduler inserts the packet to AR if and only if AR is empty in the end of the matching phase. Therefore, AR always contains a packet should be put to the specific VOR in the beginning of each matching phase. Firstly a packet in AR is transferred to its destining VOR if the VOR is empty in that time slot. And next, matching algorithm match input packets and output ports in some manner and packets in VORs could be switched to the corresponding output ports.

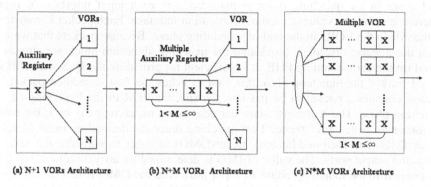

(a) N+1 VORs Architecture   (b) N+M VORs Architecture   (c) N*M VORs Architecture

Fig. 2. Virtual Output Registers

In the previous works on the input-queued router architecture, the basic objective of scheduling at the input side of routers is to find a contention-free match based on the connection requests. This kind of scheduler, so-called matching controller, selects a match between input packets and output ports with the constraints of unique pairing (i.e., at most one input packet can be matched to each output port and vice versa). Similarly, a new matching algorithm for the input-registered router is discussed in Section 4.

Also, any types matching algorithm based on stable matching can be used with the proposed architecture. The proposed matching algorithm to match inputs and outputs in the input-registered router is presented in the next section.

4 First Scheduled First Matching (FSFM)

In this section we propose a new matching algorithm to match input packets and output ports at the switching fabric in the input-registered router architecture. The switching fabric under this investigation is assumed to be non-blocking NxN switches; that is, only external conflicts can be occurred in the input or output inter-face of the switch. It is also assumed that time slots are divided in slots and packets arrive to the switch at the beginning of time slot. Each time slot has two phases: scheduling phase, matching phase. We now consider a matching algorithm, referred as First Scheduled First Matching (FSFM), to match input packets and output ports in input-registered router with VORs.

There are two important concepts in FSFM. Firstly, we define the matching order of packets, which are used to match input packets and output ports. The matching order is a switching preference value of each packet in VOR at the input interface. Each register in VORs has at least one register for matching order.

Definition 1. The matching order of packet at each input interface represents a priority of packet for being switched through the switching fabric in the input-registered router. It is ordered by the arrival times at corresponding VOR at that input interface.

The matching order is rearranged in each time slot. The similar value is ordered by the packet arrival times at the input interface. We however define matching order using the packet arrival sequences at a corresponding VORs at each input interface. And a set of the matching order is maintained in each input interface of routers whereas a set of the value is used at each output interface. Each packet is transferred to the AR in the VORs in the end of scheduling phase. Because packets that were put from the queuing discipline block into the matching algorithm block are already ordered on the corresponding PHB, it is sufficient for providing the QoS with DiffServ if and only if the matching algorithm first switches the first scheduled packet. AR always contains a packet to be put into the specific VOR in the beginning of each matching phase. The matching controller determines matching pairs by using matching order as switch preference. Two matching order are defined as Input Matching Order(IMO) and Output Matching Order(OMO) in order to consider the status of input and output ports. The value of IMO is determined by arrival sequence of packets entered to the matching phase of input port. And the OMO is determined by the arrival sequence of packets destined to the same output ports at all input ports.

At the beginning of each matching phase, the matching controller assign IMO on the packet which was transferred from scheduling phase at each port and assign the value of OMO on a packet based on arrival sequence at the corresponding output port number. In the end of each matching phase, the matching controller reallocated packets to each VORs after the packets in VORs switch to the each output port. The following procedure repeat to find matching input packets and output ports as shown in Fig. 3.

(1) First, each output port search inputs to find its corresponding VOR
  1) Each outputs sends the requests to input ports with the packet that has a packet with smallest OMO value in the corresponding VORs at all input.
  2) If more than one input has the packet destined the same output with the same OMO then output chooses the one that has a packet with more precedent PHB.
  3) If there is a tie between them, the input with the smallest interface number wins.
(2) If more than one output simultaneously requests the same input port then input that has a packet with the smallest IMO value is granted.
(3) Next, several iterations are performed. (Go Step 2.) When there is no more available matching of inputs and outputs, packets are switched.

From above statements, packets registered in the VOR at input interfaces are scheduled according to its matching order. The packet with lowest matching order value has a higher priority than others.

Fig. 4 shows the example for a 3x3-nonblocking switch with N+1 VORs. Lettered box at each input interface denotes packet to be located in that register. The letter in each packet indicates the output interface destined by that packet. The right and downside number denotes the OMO value of the packet at each input interface. The

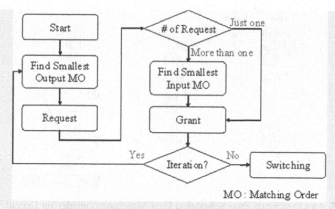

MO : Matching Order

Fig. 3. The flow chart of FSFM algorithm

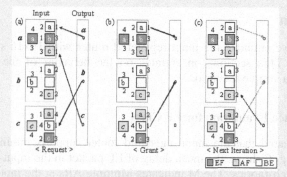

Fig. 4. An example of First Scheduled First Matching

left and upside number indicate the IMO value. The color of packet denotes the PHB of packet. The darker has the more precedent PHB; that is, dark gray color indicates EF packets and light gray color indicates AF packets and the other is BE packets.

Fig. 4-(a) shows that every output ports send requests to the input VORs which has the smallest OMO. Two requests in input *a* ports are arrived as the same time. Fig. 4-(b) depicts that input *a* port sends a grant to the output a port whose IMO assigned lower than output c port according to FSFM step (2). Fig. 4-(c) shows that the rest output c port is matched during the next iteration.

If a packet in VOR destined by the packet in AR is not empty, an existing packet in that VOR should be firstly matched because router must provide the bandwidth specified in traffic profile for a corresponding PHB. In this case, matching controller sets the OMO value of that packet with zero referred as Urgent Value (UV). And then packets in VORs are matched by FSFM algorithm with OMO value. Finally, non-zero matching order is rearranged in order of current value in end of a time slot

Fig. 3-(a) shows that two packets in input *a* and *c* ports are reordered as UV in the beginning of time slot. Fig. 3-(b) depicts that output *a*, *b* and *c* port also sends a requests to the input *a c* and *c* port respectively according to FSFM step (1)-1) and (1)-2). Fig. 3-(c) shows that the input *c* port selects the output port *c* based on UV. The rest input *b* port is matched during the next iteration.

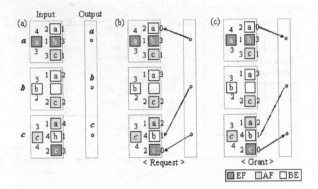

Fig. 5. An example of First Scheduled First Matching considering Urgent Value

5 Simulation

In this section we simulate the input-registered router with 16x16 switch fabrics by using ARENA. This section concentrates on the behavior of the input-registered router with just a simple scenario.

5.1 Performance Evaluation for EF Class

Firstly we evaluated that the input-registered router supports minimum mean delay for EF class. Fig. 6 shows the mean delay of EF packet in the input-registered router with 16x16 switch fabrics. The M indicates Multiple ARs at the input register router. We assume that the bandwidth of input link is 10Gbps at our simulation. As to maintain the average load of link at 98%, the traffic load of EF class increase from 10% to 90% where the traffic load of background decreases from 90% to 10%. The average delay of EF packet increase as the traffic load of EF class increases in the case of the output queued router. On the contrary, the average delay of EF packet at the input-registered router is nearly uniformly distributed regardless of the traffic load of EF class. It proves that the suggested structures are more reliable for the profile of EF traffic than the output-queue router.

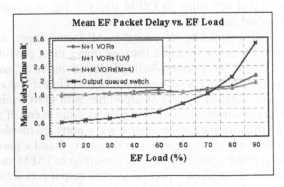

Fig. 6. Mean EF packet delay

5.2   Performance Evaluation for AF Class

Mean AF Packet Loss

We evaluated that the input-registered router support the proper packet loss of AF class. At this time, we assume that the bandwidth of each input links, the average load of link and the percentage of EF traffic with total traffic loads are 10Gbps, 98% and 30% respectively. The load of background traffic decreases from 70% to 10% as the AF class traffic increases.

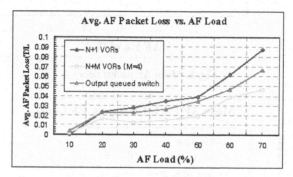

Fig. 7. Average AF packet loss

Fig.7 shows that all of the average AF packet loss increases as the traffic load of AF class increases. It was caused by the limited RED queue size as 30 Kbytes. Because the average length of AF queue exceeds the RED threshold packet losses occur. In case of N+1 VORs structure, packet loss rate is larger than the output queued router, but the output queued switch cannot afford to support a switching fabric with N times higher speed. Indeed, Fig. 7 stands for ideal situation. From this point of view, N+1 VORs structure also supports the suitable service for AF class.

Mean AF Packet Delay

Fig. 8 illustrates the mean delay of AF packets when the traffic load of AF class varies from 10% to 70%. We can see that the mean AF packet delay decreases at all case because of the property of PHB buffer using a WRR algorithm. In this algorithm, a weight for AF traffic increases as the AF class traffic increases. On the other hand, In a RED queue in the AF class traffic management packet loss increases as AF class traffics increases. So, the mean delay of AF packet decreases.

The output queued router indicates the best performance in Fig.8. But it is the ideal case only if the switch fabric has N times faster than the input registered router. So this figure tells us that the supposed architecture could serve the AF class service with supporting the same delay of output queued router.

Mean AF Packet Delay

In this case, a PHB buffer scheduler works as Strict Priority Queuing and Weighted Round-Robin. Accordingly, if a packet exists in the EF queue, it is served first. Therefore, the characteristics of AF class traffic service are sensitive of amount of EF class traffic.

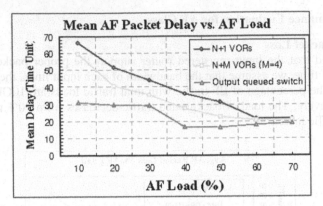

Fig. 8. Mean AF packet delay

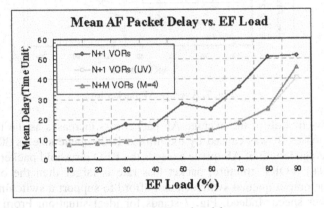

Fig. 9. Mean AF packet delay

Fig. 9 shows the mean delay of AF packet at the input-registered router with 16x16 switch fabric. We assumed that there are no BE traffics took service and the bandwidth and average load of link are 10Gbps and 98% respectively. To remove the influence of BE traffic the traffic load of EF class increases from 10% to 90% where the load traffic of AF class decreases from 90% to 10%.

We could see the mean delay of AF packet increases as the EF class traffic increases. The router requires more times to serve the EF class traffic because the amount of packet increases in the EF queue. But the increase rate is small when the load traffic of EF class is smaller than 80% of the entire bandwidth. The proposed architecture could serve the AF class traffic with the proper mean delay.

6   Conclusion

In this paper, a study on the input-registered router architecture to provide QoS guarantees with DiffServ is described. The input-registered router has a similar architecture with input-queuing router except that the former use registers instead of queues at the input side. We refer these registers as VORs, which is similar with VOQs of the input-queued router. However VOR is temporary registers for buffering packets to be switched by the switching fabric. Another aspect of the proposed architecture is the concept of the "queuing discipline in the input side." It is located at input interfaces on the router and schedules packets to be switched through the switching fabric.

Also we propose a new matching algorithm, referred as FSFM, for the input-registered router architecture. The FSFM is designed based on GSA, using the matching order of individual packet in virtual output queues. The main idea of FSFM is that it is sufficient for providing the QoS with DiffServ in the proposed architecture if and only if the matching algorithm first switches the first scheduled packet. The simulation results performed by ARENA show that the proposed architecture offer packet loss rate and delay close to the results of output-queued router with N time speed-up switch fabric in various PHBs on the differentiated services.

References

1. Blake, S., Black, D., Carlson, M.: An Architecture for Differentiated Service, RFC 2475, Internet Engineering Task Force (1998)
2. Kumar, V. P., Lakshman, T. V.: Beyond Best Effort: Architectures for the Differentiated Services of Tomorrow's Internet, IEEE Communications, Vol. 36, No. 5 (1998) 151-164
3. Weiss, W.: QoS with Differentiated Services, Bell Labs Technical Journal, Vol. 3, No. 4 (1998) 48-62
4. Stiliadis, D., Varma, A.: Providing Bandwidth Quarantees in an Input Buffered Crossbar Switch, Proc. IEEE INFOCOM'95, Vol. 3 Boston (1995) 960-968
5. Prabhakar, B., McKeown, N.: On the Speedup Required for Combined Input and Output Queued Switching, Technical report, Computer Science Laboratory, Stanford University (1997)
6. Li, S., Ansari, N.: Scheduling Input-Queued ATM Switches with QoS Features, Proc. ICCCN'98 (1998) 107-112
7. Stoica, I., Zhang, H.: Exact Emulation of an Output Queueing Switch by a Combined Input Output Queueing Switch, IWQoS'98, USA (1998) 218-224
8. Nong, G., Hamdi, M.: On the Provision of Quality-of-Service Guarantees for Input Queued Switches, IEEE Communications Magazine, Vol. 38, Issue. 12 (2000) 62-69

Author Index

Lecture Notes in Computer Science

For information about Vols. 1–3281

please contact your bookseller or Springer